Principles of Plant-Microbe Interactions

Ben Lugtenberg
Editor

Principles of Plant-Microbe Interactions

Microbes for Sustainable Agriculture

 Springer

Editor
Ben Lugtenberg
Molecular Microbiology and Biotechnology
Leiden University, Sylvius Laboratory
Leiden
The Netherlands

ISBN 978-3-319-08574-6   ISBN 978-3-319-08575-3 (eBook)
DOI 10.1007/978-3-319-08575-3
Springer Cham Heidelberg New York Dordrecht London

Library of Congress Control Number: 2014956088

Printed on acid-free paper

Springer is part of Springer Science+Business Media (www.springer.com)

I dedicate this book to my wife Faina and my children Annelieke, Martijn and Marjolein

Preface

The field of Plant Microbe Interactions is very broad. It covers all topics in which microbes influence or even determine plant activities. Plant enemies can be pathogenic viruses, microbes or insects which cause pests. Fortunately, these enemies in turn have natural enemies in the form of beneficial microbes, which can protect plants against pathogens and pests. As is rather common in this field, we included nematodes and insects in the book. Although they are not microbes, they have in common with microbes that some can cause harm to, and others help protect, the plant. Another group of microbes is beneficial for plant growth. Some microbes promote plant growth, for example by producing "plant" hormones or by making nutrients available to the plant. Other beneficial microbes can alleviate plant stress or can inactivate environmental pollutants, thereby cleaning the environment and allowing plants to grow without toxic residues. The present market share of biologicals is estimated at 1.6 billion USDs and is growing fast. In the past years the trend is that major chemical companies buy smaller biotech companies.

For this book I have invited the world's top scientists to summarize the basic principles of all these topics in brief chapters which give a helicopter view on the subjects. The book also contains important techniques, success stories and future prospects. The topics include basic as well as applied aspects. Hereby we make an attempt to close the gap that still exists between fundamental and applied research. In my opinion the two fields need each other and cooperation will create a win-win situation for both parties. Since space is limited, the authors have often referred to reviews. For more detailed information, the reader can consult primary articles listed as references in these reviews.

This book is meant for everybody who is interested in plant-microbe interactions and in the roles microbes can play in making agriculture and horticulture more sustainable. These include academic scientists, industrial professionals working in agriculture, horticulture, biotech and food industry, students, teachers, as well as government officials and decision makers who quickly want to make themselves familiar with particular aspects of this broad field. Using this information as a basis, also a non-specialist reader should be able to understand more complicated articles and to discuss selected topics with colleagues. To read the book, basic knowledge of plant science, microbiology, biochemistry, and molecular biology is helpful.

Ben Lugtenberg, editor

Acknowledgement

I am very much indebted to all authors for their contributions. I am particularly thankful to the following people who have contributed by useful advice and discussions: Gabriele Berg, Rainer Borriss, Frans de Bruijn, Faina Kamilova, Christoph Keel, Corné Pieterse, and Clara Pliego. I am greatly obliged to Izabela Witkowska and Melanie van Overbeek of Springer Dordrecht for their help and patience during the preparation of the manuscript.

The following sponsors made the editing of the book more pleasant. Their contributions will go to a foundation which supports the promotion of knowledge about plant-microbe interactions and their applications.

DIAMOND SPONSORS

GOLD SPONSORS

INSTITUTE of BIOLOGY LEIDEN

BISOLBI-INTER LLC

Contents

Contributors

Tom Adams Monsanto Company, St. Louis, MO, USA

Carmen Büttner Division Phytomedicine, Faculty of Life Sciences, Humboldt-Universität zu Berlin, Berlin, Germany

Esther Badosa Laboratory of Plant Pathology, Institute of Food and Agricultural Technology, University of Girona, Girona, Spain

Jaap Bakker Laboratory of Nematology, Wageningen University, Wageningen, The Netherlands

Peter A. H. M. Bakker Plant-Microbe Interactions, Institute of Environmental Biology, Utrecht University, Utrecht, The Netherlands

Martina Bandte Division Phytomedicine, Faculty of Life Sciences, Humboldt-Universität zu Berlin, Berlin, Germany

José-Miguel Barea Soil Microbiology and Symbiotic Systems Department, Estación Experimental del Zaidín, CSIC, Granada, Spain

Gabriele Berg Institute of Environmental Biotechnology, Graz University of Technology, Graz, Austria

Anna Bonaterra Laboratory of Plant Pathology, Institute of Food and Agricultural Technology, University of Girona, Girona, Spain

Paola Bonfante Department of Life Sciences and Systems Biology, University of Torino, Torino, Italy

Rainer Borriss ABiTEP GmbH, Berlin, Germany

Inge Broer Agricultural and Environmental Faculty, University of Rostock, Rostock, Germany

Massimiliano Cardinale Institute of Applied Microbiology, Justus-Liebig-University Giessen, Giessen, Germany

Aurelien Carlier Institute of Plant Biology, University of Zurich, Zurich, Switzerland

Woo-Suk Chang Department of Biology, University of Texas-Arlington, Arlington, Texas, USA

Daniele Daffonchio DeFENS, Department of Food, Environmental and Nutritional Sciences, University of Milan, Milan, Italy

BESE Division, King Abdullah University of Science and Technology, Thuwal, Kingdom of Saudi Arabia

Solke H. De Boer Emeritus Scientist, Charlottetown, PE, Canada

Frans J. de Bruijn INRA/CNRS Laboratory of Plant-Microbe Interactions, Castanet-Tolosan Cedex, France

Ruud A. de Maagd Plant Research International, Wageningen UR, Wageningen, The Netherlands

Dulce Eleonora de Oliveira VIB–Institute of Plant Biotechnology Outreach, Department of Biotechnology and Bioinformatics, Ghent University, Gent-Zwijnaarde, Belgium

Sandra de Weert Koppert Biological Systems, Berkel en Rodenrijs, The Netherlands

Pierre J. G. M. de Wit Laboratory of Phytopathology, Wageningen University, Wageningen, The Netherlands

Alessandro Desirò Department of Life Sciences and Systems Biology, University of Torino, Torino, Italy

Leo Eberl Institute of Plant Biology, University of Zurich, Zurich, Switzerland

Namis Eltlbany Julius Kühn-Institut, Federal Research Centre for Cultivated Plants (JKI), Braunschweig, Germany

Armin Erlacher Institute of Environmental Biotechnology, Graz University of Technology, Graz, Austria

Institute of Plant Sciences, University of Graz, Graz, Austria

Jesús Francés Laboratory of Plant Pathology, Institute of Food and Agricultural Technology, University of Girona, Girona, Spain

Bernard R. Glick Department of Biology, University of Waterloo, Waterloo, ON, Canada

Francine Govers Laboratory of Phytopathology, Wageningen University, Wageningen, The Netherlands

Martin Grube Institute of Plant Sciences, University of Graz, Graz, Austria

Johannes Helder Laboratory of Nematology, Wageningen University, Wageningen, The Netherlands

Heribert Hirt BESE Division, King Abdullah University of Science and Technology, Thuwal, Kingdom of Saudi Arabia

Paul J. J. Hooykaas Institute of Biology, Sylvius Laboratory, Leiden University, Leiden, The Netherlands

Katja Hora Koppert Biological Systems, Berkel en Rodenrijs, The Netherlands

Xiaowei Huang Yunnan University, Kunming, People's Republic of China

Mariangela Hungria Embrapa Soja, Londrina, Paraná, Brazil

Mike S. M. Jetten Department of Microbiology, Institute of Water and Wetland Research, Faculty of Science, Radboud University Nijmegen, Nijmegen, The Netherlands

Faina Kamilova Koppert Biological Systems, Berkel en Rodenrijs, The Netherlands

Eva Kondorosi Institute of Biochemistry, Biological Research Centre of the Hungarian Academy of Sciences, Szeged, Hungary

Martine A. R. Kox Department of Microbiology, Institute of Water and Wetland Research, Faculty of Science, Radboud University Nijmegen, Nijmegen, The Netherlands

Hae-In Lee Department of Biology, University of Texas-Arlington, Arlington, Texas, USA

Johan H. J. Leveau Department of Plant Pathology, University of California, Davis, CA, USA

Guohong Li Yunnan University, Kunming, People's Republic of China

Matteo Lorito Department of Agriculture, University of Naples Federico II, and Institute of Plant Protection IPP–CNR, Portici (NA), Italy

Ben Lugtenberg Institute of Biology, Sylvius Laboratory, Leiden University, Leiden, The Netherlands

Jesús Mercado-Blanco Department of Crop Protection, Institute for Sustainable Agriculture, Agencia Estatal Consejo Superior de Investigaciones Científicas (CSIC), Córdoba, Spain

Emilio Montesinos Laboratory of Plant Pathology, Institute of Food and Agricultural Technology, University of Girona, Girona, Spain

Yaacov Okon Department of Plant Pathology and Microbiology, Faculty of Agriculture, Food and Environment, The Hebrew University of Jerusalem, Rehovot, Israel

Vladimír Půža Institute of Entomology, Biology Centre of the AS CR, České Budějovice, Czech Republic

Gabriella Pessi Institute of Plant Biology, University of Zurich, Zurich, Switzerland

Corné M. J. Pieterse Plant-Microbe Interactions, Department of Biology, Faculty of Science, Utrecht University, Utrecht, The Netherlands

Casper Quist Laboratory of Nematology, Wageningen University, Wageningen, The Netherlands

Jos M. Raaijmakers Department of Microbial Ecology, Netherlands Institute of Ecology, Wageningen, The Netherlands

Willem J. Ravensberg International Biocontrol Manufacturers Association (IBMA), Brussels, Belgium

Koppert Biological Systems, Berkel en Rodenrijs, The Netherlands

Alan E. Richardson CSIRO Plant Industry, Canberra, Australia

Katarzyna Rybarczyk-Mydłowska Museum and Institute of Zoology PAS, Warsaw, Poland

Choong-Min Ryu Molecular Phytobacteriology Laboratory, KRIBB, Daejeon, Republic of Korea

Thomas Schäfer Novozymes A/S, Bagsvaerd, Denmark

Charikleia Schoina Laboratory of Phytopathology, Wageningen University, Wageningen, The Netherlands

Susanne Schreiter Julius Kühn-Institut, Federal Research Centre for Cultivated Plants (JKI), Braunschweig, Germany

Kornelia Smalla Julius Kühn-Institut, Federal Research Centre for Cultivated Plants (JKI), Braunschweig, Germany

Geert Smant Laboratory of Nematology, Wageningen University, Wageningen, The Netherlands

Stijn Spaepen Centre of Microbial and Plant Genetics, KU Leuven, Heverlee, Belgium

Department of Plant Microbe Interactions, Max Planck Institute for Plant Breeding Research, Köln, Germany

Sofie Thijs Centre for Environmental Sciences, Hasselt University, Diepenbeek, Belgium

Linda Thomashow USDA-ARS, Root Disease and Biological Control Research Unit, Washington State University, Pullman, WA, USA

Jan Tommassen Section Molecular Microbiology, Department of Biology, Utrecht University, Utrecht, The Netherlands

Eddy van der Meijden Institute of Biology, Leiden University, Leiden, The Netherlands

Jan van der Wolf Plant Research International, Wageningen, The Netherlands

Hanny van Megen Laboratory of Nematology, Wageningen University, Wageningen, The Netherlands

Marc Van Montagu VIB–Institute of Plant Biotechnology Outreach, Department of Biotechnology and Bioinformatics, Ghent University, Gent-Zwijnaarde, Belgium

Saskia C. M. Van Wees Plant-Microbe Interactions, Department of Biology, Faculty of Science, Utrecht University, Utrecht, The Netherlands

Jaco Vangronsveld Centre for Environmental Sciences, Hasselt University, Diepenbeek, Belgium

Muthusubramanian Venkateshwaran School of Agriculture, University of Wisconsin-Platteville, Platteville, WI, USA

Mariëtte Vervoort Laboratory of Nematology, Wageningen University, Wageningen, The Netherlands

Susanne von Bargen Division Phytomedicine, Faculty of Life Sciences, Humboldt-Universität zu Berlin, Berlin, Germany

Han A. B. Wösten Section Molecular Microbiology, Department of Biology, Utrecht University, Utrecht, The Netherlands

David M. Weller United States Department of Agriculture-Agricultural Research Service, Root Disease and Biological Control Research Service, Pullman, Washington, USA

Sheridan L. Woo Department of Agriculture, University of Naples Federico II, and Institute of Plant Protection IPP–CNR, Portici (NA), Italy

Zefen Yu Yunnan University, Kunming, People's Republic of China

Keqin Zhang Yunnan University, Kunming, People's Republic of China

Abbreviations

ABA	abscisic acid
ACC	1-aminocyclopropane-1-carboxylate
AFM	atomic force microscopy
AHL	N-acyl homoserine lactone
AMF	arbuscular mycorrhizal fungus
ARISA	Automated rRNA Intergenic Spacer Analysis
BC	biocontrol
BCA	biocontrol agent
BCPC	British Crop Production Council
BNF	Biological nitrogen fixation
BPSG	BioPesticide Steering Group
Bt	*Bacillus thuringiensis*
CFU	colony forming units
CIPAC	Collaborative International Pesticides Analytical Council
CK	cytokinin
c-LP	cyclic lipopeptide
CLSM	confocal laser scanning microscopy
CMV	Cassava mosaic virus
CNN	competition for nutrients and niches
CRAfT	Cre Reporter Assay for Translocation
CSP	Common Symbiotic Pathway
DAMP	damage-associated molecular patterns
DAPG	2,4-diacetylphloroglucinol
DGGE	denaturing gradient gel electrophoresis
DIC	Differential Interference Contrast
DSF	diffusible signal factor
EBIC	European Biostimulant Industry Council
ECM	Ectomycorrhizal fungi
EPA	Environmental Protection Agency
EPN	Entomopathogenic nematode
EPPO	European and Mediterranean Plant Protection Organization
ET	Ethylene

ETI	effector-triggered immunity
ETS	effector-triggered susceptibility
FAME	Fatty Acid Methyl Esters
FAO	Food and Agriculture Organization
FeMoCo	Iron-molybdenum cofactor
GA	gibberellin
GAP	Good agricultural practice
GAs	gibberellins
GFP	Green fluorescent protein
GM	genetically modified
GMO	genetically modified organism
GOGAT	Glutamine oxoglutarate aminotransferase
GS	Glutamine Synthase
HR	Homologous recombination
HR	hypersensitive response
HSL	homoserine lactone
IAA	indole-3-acetic acid
IAM	indole-3-acetamide
IBCA	invertebrate biocontrol agent
IJ	infective juvenile
IncP-1	incompatibility groups of plasmids
InsP5	1D-myo-inositol 1,2,4,5,6 pentakisphosphate
IPM	Integrated pest management
IPyA	indole-3-pyruvate
ISO	international organization for standardization
ISR	induced systemic resistance
ITS fragments	internal transcribed spacer ribosomal DNA (used as the universal barcode for fungi)
JA	Jasmonic Acid
LCO	lipochitooligosaccharide
LPS	lipopolysacchide
LysM	lysine-motif
MALDI MSI	Matrix-Assisted Laser Desorption/Ionization Mass Spectrometry Imaging
MAMP	microbe associated molecular pattern
MGE	mobile genetic element (DNA fragments, which can be transferred from cell to cell)
MHB	mycorrhiza helper bacteria
MRL	maximum residue limit
Myc factor	mycorrhization factor
N_2H_4	hydrazine
N_2O	nitrous oxide
NB-LRR	nucleotide binding-leucine-rich repeat
NH_2OH	hydroxylamine
NHEJ	Non-homologous end-joining

nif gene	nitrogen fixation gene
NO	nitric oxide
Nod factor	nodulation factor
NO_x	nitrogen oxides
OECD	organisation for economic co-operation and development
OMZ	oxygen minimum zone
OTU	operational taxonomic unit (used as species equivalent)
PAHs	polycyclic aromatic hydrocarbons
PAMP	pathogen-associated molecular pattern
PCA	phenazine-1-carboxylic acid
PCB	poly-chlorinated biphenyls
PCN	phenazine-1-carboxamide
PCR	polymerase chain reaction
PCWDE	plant cell wall degrading enzyme
PGP	plant growth-promotion
PGPB	plant growth-promoting bacteria
PGPF	plant growth promoting fungi
PGPM	plant growth promoting microbe
PGPR	plant growth-promoting rhizobacteria
PhyloChip	microarray for the comprehensive identification of microbial organisms
Pi	inorganic phosphate
PIP	plant-incorporated protein
PIR	protein with internal repeat
PIT	pre-infection thread
PKS	polyketide synthase
PMRA	pest management regulatory agency
PPA	pre-penetration apparatus
PPP	plant protection product
PP-transferase	phospho-pantetheinyl-transferase
PQQ	pyrroloquinoline quinone
PR	pathogenesis related
PRR	pattern recognition receptor
PSM	phosphate solubilising microorganism
PTI	pathogen-triggered immunity
PZN	plantazolicin
qPCR	quantitative real-time PCR (provides copy-number of the target organism)
QS	quorum-sensing
rDNA	ribosomal DNA
RLP	receptor-like protein
RNAi	ribonucleic acid interference
ROL	radial oxygen loss
ROS	reactive oxygen species
RP	rock phosphate

SA	salicylic acid
SAM	S-adenosylmethionine
SAR	systemic acquired resistance
SIMS	secondary ion mass spectrometry
SIP	stable-isotope probing
SL	strigolactone
SRP	signal recognition particle
SSCP	single strand conformation polymorphism (method to analyze composition of microbial communities)
SWOT	strengths, weaknesses, opportunities and threats
SYM genes	common symbiosis genes
SYM pathway	symbiosis pathway
T1-6SS	type 1-6 secretion system
T4SS	type 4 secretion system
TAD	take-all decline
TC-DNA	DNA obtained from the total community present in one sample
T-RFLP	terminal restriction fragment length polymorphism (method to analyze the composition of microbial communities)
TTSS	type III secretion system
US EPA OCSPP	United States Environmental Protection Agency Office of Chemical Safety and Pollution Prevention
USDA	US Department of Agriculture
VAMPs	vesicle-associated membrane proteins
VBNC	viable but non-culturable
VOC	volatile organic compound

Chapter 1
Introduction to Plant-Microbe Interactions

Ben Lugtenberg

Abstract Pathogenic microbes and pest organisms as well as unfavorable growth conditions can be a threat for plant growth. Other beneficial microbes and small organisms can be used to protect plants against these attackers or to assist the plant to overcome the unfavorable conditions. These plant-beneficial organisms can be divided into classes which (i) reduce plant diseases, (ii) which regulate plant growth, (iii) help plants to overcome stresses, and (iv) inactivate soil pollutants which inhibit plant growth or make (parts of) the plant unsuitable for consumption.

Plant Pathogens and Pest Organisms are a Threat for Plant Growth In this book we discuss pathogens and pest organisms which are a threat for plant growth. We highlight the roles which microbes can play in making agriculture and horticulture more sustainable. Selected microbes are able to (partly) replace most chemicals which are presently used in agriculture. In addition, microbes can often be used against diseases for which no chemicals are available. In this book, the following activities and applications of microbes will be discussed.

Biological Control of Plant Diseases Approximately 25 % of the world's crop yield is lost every year, mainly due to diseases caused by fungi, by other pathogens, and by pests. Plant protection products are on the market to fight these diseases. Presently these are mainly chemicals. Their use can be threatening the health of people and polluting the environment. Disease control with beneficial microbes is an alternative which allows sustainable crop production. The use of microbial plant protection products is growing and their importance will strongly increase because of political and public pressure.

Regulation of Plant Growth The world population is growing and the amount of food needed by 2050 will be the double of what is being produced now, whereas the area of agricultural land is decreasing. We have to increase crop yield in a sustainable way, i.e. chemical plant growth regulators have to be replaced by microbiological

B. Lugtenberg (✉)
Institute of Biology, Sylvius Laboratory, Leiden University,
Sylviusweg 72, 2333 BE Leiden, The Netherlands
Tel.: +31629021472
e-mail: Ben.Lugtenberg@gmail.com

© Springer International Publishing Switzerland 2015
B. Lugtenberg (ed.), *Principles of Plant-Microbe Interactions,*
DOI 10.1007/978-3-319-08575-3_1

products. Also here, the use of microbial products is growing and their importance will strongly increase.

Control of Plant Stress by Microbes An increasing area of agricultural land is arid and/or salinated. Global warming will increase this area. Plant growth is inhibited, or even made impossible, by drought and salt. It has been proven already that microbes can be used successfully to alleviate such stresses.

Microbial Cleaning of Polluted Land Chemical pollution of land can make plant growth difficult or even impossible. But even when crop plants grow on such lands, their products are often polluted and not suitable for consumption. Selected microbes have been already been used successfully to detoxify chemical pollutants in soil and to remove heavy metals, thereby allowing the growth of healthy plants.

The field of Plant-Microbe Interactions has made important progress thanks to the development of new technologies. Attention to state-of-the-art DNA and visualization techniques is paid in two separate chapters. Moreover, successful examples of progress are presented under Paradigms of Plant-Microbe Interactions. The book ends with the presentation of a number of real innovative research projects of which the future will show whether these are dreams or big steps forwards.

Part I
Introductory Chapters

Chapter 2
The Importance of Microbiology in Sustainable Agriculture

Thomas Schäfer and Tom Adams

Abstract Deriving from various naturally-occurring microorganisms such as bacteria and fungi, microbial technologies can protect crops from pests and diseases and enhance plant productivity and fertility. They enable farmers to increase yield and productivity in a sustainable way and are expected to play a significant role in agriculture.

As the global population's rapid growth is set to continue, the need to significantly increase agricultural output without increasing pressure on the environment also grows. Microbial solutions enable farmers to drive yield and productivity in a sustainable way. Deriving from various naturally-occurring microorganisms such as bacteria and fungi, these solutions can protect crops from pests and diseases and enhance plant productivity and fertility.

Microbial solutions make up approximately two thirds of the agricultural biologicals industry. Representing roughly US$ 2.3 billion in annual sales, agricultural biologicals have posted double-digit sales growth each of the last several years. There are numerous biological products currently on the market that contain microorganisms as active ingredients, including seed treatment and foliar applied products. Microbial technologies can help improve nutrient acquisition, promote growth and yield, control insects and protect against disease. These emerging agricultural biological technologies complement the integrated systems approach that is necessary in modern agriculture, bringing together breeding, biotechnology and agronomic practices to improve and protect crop yields.

There has been significant interest in agricultural biologicals in the past few years from major crop chemical manufacturers, including Bayer's acquisition of Agraquest, BASF's acquisition of Becker Underwood, and Syngenta's acquisitions of Pasteuria and Devgen. Most recently, Novozymes and Monsanto established The

T. Schäfer (✉)
Novozymes A/S, Brudelysvej 32, 2880 Bagsvaerd, Denmark
Tel.: + 45 44460000
e-mail: TSch@novozymes.com

T. Adams
Monsanto Company, 800 N. Lindbergh Blvd., St. Louis, MO 63167, USA
Tel.: + 1 (314) 258-1547
e-mail: tom.h.adams@monsanto.com

© Springer International Publishing Switzerland 2015
B. Lugtenberg (ed.), *Principles of Plant-Microbe Interactions,*
DOI 10.1007/978-3-319-08575-3_2

BioAg Alliance in December 2013, with a goal to discover, develop and sell microbial solutions that enable farmers worldwide to increase crop yields with less chemical input. Novozymes brought an established product portfolio and strengths within microbial discovery, application development and fermentation to this partnership. Combined with Monsanto's highly-developed seeds and traits discovery, field-testing and extensive commercial network, the aim is to deliver a comprehensive research, development and commercial collaboration from which agriculture, consumers, the environment and society at large can benefit.

Microbial solutions provide more choice for farmers and help meet the demand for more sustainable agricultural practices. Such solutions can increase crop yields and develop a more sustainable industry impact profile, ultimately resulting in more food to feed the growing world and new opportunities to protect the planet.

Chapter 3
Life of Microbes in the Rhizosphere

Ben Lugtenberg

Abstract Life of microbes in the rhizosphere is best characterized as starvation for nutrients and attempts to survive. All microbes are hunting for food of which a substantial amount is supplied by the root in the form of exudate. The most successful microbes are attracted to food sources, such as to the root and to each other, by chemotaxis to specific exuded compounds. Subsequently they colonize the target organism. Specific exudate compounds can also initiate communication between organisms as the start of more specialized interactions, such as nodulation, pathogenesis, DNA transfer and the production of antibiotics. Some target organisms have developed defense reactions against such attacks which allows them to survive. All these processes are described in this chapter.

3.1 The Rhizosphere

The rhizosphere was defined by Lorentz Hiltner as "the soil compartment influenced by the roots of growing plants". The rhizosphere is supposed to be no more than a few mm thick. It is 10- to 100-fold richer in microbes than the surrounding "bulk" soil because 6–21 % of the carbon fixed by the plant is secreted by the root. This phenomenon is called the rhizosphere effect. It is good to realize that the concentration of nutrients in the rhizosphere is still 100-fold lower than that in the usual laboratory media. The life style of microbes in the rhizosphere is therefore best characterized as starvation. Recent reviews on the rhizosphere are those of Haas and Défago (2005), Lugtenberg and Bloemberg (2004), Lugtenberg and Kamilova (2009), and Pinton et al. (2007).

Roots of many plants are colonized by mycorrhizal fungi which can function as fine extensions of the root and allow the plant to reach nutrients which cannot be reached by the thicker roots. The combination of root surface and attached mycorrhizal fungi is designated as the mycorrhizosphere (Chap. 25).

B. Lugtenberg (✉)
Institute of Biology, Sylvius Laboratory, Leiden University, Sylviusweg 72,
2333 BE Leiden, The Netherlands
Tel.: + 31629021472
e-mail: Ben.Lugtenberg@gmail.com

© Springer International Publishing Switzerland 2015
B. Lugtenberg (ed.), *Principles of Plant-Microbe Interactions,*
DOI 10.1007/978-3-319-08575-3_3

Plant life is affected by both abiotic and biotic conditions. For the estimation of local conditions, bioreporters have been developed. These are bacterial derivatives harboring a promoter that reacts on the compound or condition of choice, and that is fused to a reporter gene encoding a protein which can easily be detected and quantified (for example *gfp, lux, lacZ* or *inaZ*). Reporter constructs respond for example to the presence of certain sugars, amino acids, or to conditions such as pH or the bioavailability of carbon, phosphate and oxygen (Chap. 10 in ref. Pinton et al. 2007).

Abiotic Conditions Abiotic conditions affecting plant growth include temperature, pH, soil type, water potential, and concentrations of bioavailable essential nutrients and salts. Soils can be rich or poor for plant growth. Rich soils contain sufficient water and nutrients. Drought is a major and increasing problem for plant growth (Chap. 28), and so is salination (Chap. 28). Nitrogen and phosphorous are the major nutrients whereas ions of potassium, iron and micronutrients are also required for plant life. Poor soils can be fertilized chemically, for example by adding N-P-K fertilizer. This can increase plant growth enormously but is not sustainable. Therefore the trend is to replace such chemicals by other means, for example by bacteria which generate nutrients in forms that can be used by the plant (Chap. 23 and 24). A neutral or high pH makes ferric iron ions insoluble and will therefore reduce plant growth. In contrast, at acid pH these ions will be soluble.

Biotic Conditions The rhizosphere contains microbes (bacteria and fungi) as well as small animals, such as amoebae, insect larvae, mites, nematodes and protozoa (Bonkowski et al. 2009). The microbes are collectively called the rhizosphere microbiome (Mendes et al. 2013) (Chap. 30 and 43). The main factors shaping the rhizosphere microbiome are the soil type and the plant genotype. The microbes can have a beneficial, neutral, or negative effect on plant growth. Some of the beneficial microbes have been cultured and formulated and are being sold as commercial products (Chaps. 32–34) and are applied as biopesticides or plant protection products (Chap. 18), biofertilizers (Chaps. 23 and 24), rhizoremediators (Chap. 29), phytostimulators (Chaps. 25 and 26), or stress controllers (Chap. 27). Beneficial microbes include strains of the bacteria *Bacillus* and *Pseudomonas*, of Arbuscular Mycorrhizal Fungi (AMF) (Chap. 25), and of the fungus *Trichoderma* (Chap. 36). Plant pathogens can be viruses (Chap. 13), bacteria (Chap. 9), fungi and oomycetes (Chap. 10), or nematodes (Chap. 11). Some insects cause pests (Chap. 12).

A healthy soil hardly contains pathogens. Soils harboring pathogens are called disease-conducive soils. However, in some cases, soils harbor pathogens but plants growing in this soil remain healthy. These disease-suppressive soils contain beneficial microbes which suppress the action of the pathogens (Chap. 38). Some soils contain human pathogens (Berg et al. 2005) which can be risky for agricultural workers as well as for consumers. It has been suggested that treatment of seeds of crop plants with products containing enhanced colonizing bacteria will reduce the number of pathogens on the root and, therefore, reduce this risk for humans (Egamberdieva et al. 2008).

Nutrition of Rhizosphere Microbes Between 6 and 21 % of the carbon fixed by the plant can be released by the plant root. This material consists of sloughed-off cells, macromolecules and small molecules. The latter are the favorite food sources of the rhizosphere microbes. Based on mutational studies it was concluded that the most important food sources exuded by tomato roots are organic acids, such as citric, malic, lactic, succinic, oxalic, and pyruvic acids. Other root exudate compounds include sugars (such as glucose, xylose, fructose, maltose, sucrose and ribose), amino acids, fatty acids, nucleotides, putrescine, and vitamins. A special group of exudate compounds are the signal molecules which are used for communication between the plant and microbes (see Sect. 3.4).

The qualitative and quantitative composition of root exudate is heavily influenced by the presence of microbes in the rhizosphere in a microbe species-dependant way. For example, the addition of the pathogenic fungus *Fusarium oxysporum* f. sp. *radicis-lycopersici* to a mono-axenic tomato system strongly decreased the amount of citric acid and increased the amount of succinic acid but did not influence the total amount of organic acids in exudate. In contrast, when the *Pseudomonas* biocontrol strain WCS365 was added in amounts sufficient for biocontrol, the total amount of organic acids—especially of citric acid—strongly increased whereas the level of succinic acid decreased dramatically (Kamilova et al. 2006).

3.2 Interactions Between Organisms in the Rhizosphere

Using sophisticated techniques (Chap. 31), interactions occurring between organisms in the rhizosphere can be visualized.

Fungus—Plant Interactions A fungal spore in the soil will germinate when it senses the presence of a plant root through certain chemicals secreted by the root (Lugtenberg and Kamilova 2009). Subsequently, the formed mycelium will grow into the direction of the root, attach to the root, colonize the root surface, and penetrate the root. This process is shown for the fungus *Fusarium oxysporum* f. sp. *radicis-lycopersici* in Fig. 3.1a–3.1d.

Bacterium—Plant Interactions During treatment of seeds with microbes, use is made of the fact that microbes which are present on the seed are in the best position to colonize the roots of the emerging seedling. Therefore, the best way to colonize the root with a beneficial bacterium is to surface-sterilize the seed and coat it subsequently with cells of the beneficial bacterium of choice. After growth, the result is that the top of the seedling root is best colonized (with approx. 10^6 bacteria per cm root). A good colonizing bacterium can even reach the growing root tip. Coming from the seed, a bacterium colonizes the root, initially as individual cells (Fig. 3.1e) which subsequently multiply and grow out to micro-colonies (Fig. 3.1f), presently designated as biofilms (Chap. 7). Only approximately 15 % of the plant root surface can be covered by bacteria (Fig. 3.1f). Most bacteria are found on the junctions between epidermal cells (Fig. 3.1e). Mature biofilms usually consist of multiple layers of cells (Fig. 3.1f) and are covered with a mucous layer (Fig. 3.1g,h).

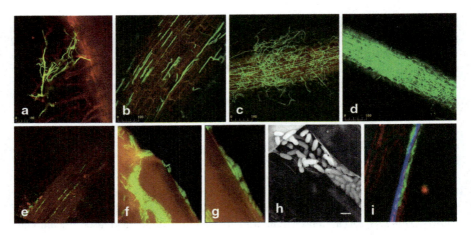

Fig. 3.1 Visualization of plant-microbe and microbe-microbe interactions during biocontrol. Confocal Laser Scanning Microscopy (**a–g** and **i**) and scanning electron microscopy (**h**) were used to visualize control of tomato foot and root rot caused by *Fusarium oxysporum* f. sp. *radicis-lycopersici* by *Pseudomonas* biocontrol bacteria. For explanation, see text. Panels **a**, **c**, and **d** were reproduced from Lagopodi et al. (2002), panel **b** from Bolwerk et al. (2003), and panel **h** from Chin-A-Woeng et al. 1997. Panel **e** is from Bolwerk, Lagopodi and Bloemberg, *unpublished*. Panels **f** and **g** are from Bloemberg et al. 1997; Copyright ©American Society for Microbiology. Reproduced from Lugtenberg and Girard (2013) by permission of the publisher

Bacterium—Fungus Interactions Some bacteria attach to fungal hyphae in order to use them as a food source (Fig. 3.1i).

3.3   Competitive Root Tip Colonization

In order to be successful in their beneficial action, cells of applied microbes have to reach the root in high numbers. Therefore, they have to win the competition with indigenous microbes for nutrients secreted by the root and for niches on the root. In order to identify bacterial traits important for competitive root colonization, Simons et al. (1996) screened individual random mutants of the efficient *P. fluorescens* root colonizing strain WCS365 in a mono-axenic sand system. Surface-sterilized tomato seeds were coated with a 1:1 mixture of one of the mutants and the wild type strain. After the root system had developed, the root tip was isolated and the ratio of wild type to mutant cells was determined. Competitive colonization mutants loose this competition and were found in lower numbers than the wild type. The mutated genes were characterized and the traits lost in the mutants were identified. It turned out that the number of traits involved in competitive root colonization is high. The best understandable traits can be grouped as follows. An excellent root tip colonizer should (i) be able to synthesize amino acids, vitamin B1, uracil, as well as the O-antigenic side chain of lipopolysaccharide; (ii) grow fast in root exudate, show a chemotactic response towards the root, and adhere to the root; (iii) have a number of properties related to secretion (see Chap. 6), such as the *secB* gene involved in a protein secretion pathway, the type three secretion system, and an intact ColR/ColS two-component system. The latter property is supposed to be related to keeping protein pores in the outer membrane open. The most likely explanation for the need of the type three secretion system is that the needle of this system is required for tapping nutrients from the plant cell. For better understanding of these traits, see Chap. 6.

Good tomato root colonizers have also been tested on other crop plants. In contrast to general predictions, pseudomonads colonize roots of many other plants—such as cucumber and wheat—well, so there is hardly any host plant specificity involved in colonization. Moreover, mutants which are not colonizing tomato roots well, are also impaired in the colonization of wheat and cucumber roots, indicating that bacterial colonization traits are common for several plants.

Interestingly, it is possible to increase the colonization ability of bacteria or to select strongly enhanced root tip colonizers. (i) When cells of a transposon mutant library of *P. fluorescens* strain WCS365 were put on a seedling and selected for enhanced root tip colonizers, the best colonizing mutant was characterized as a strain carrying a mutation in the gene encoding MutY, an enzyme which repairs mutations in the DNA. We assume that this enhanced colonizing *mutY* derivative has collected a combination of mutations which have fine-tuned the strain in such a way that it is better adapted to the conditions on the root (De Weert et al. 2004a). It would be interesting to compare the nucleotide sequences of these strains in order to evaluate which traits contribute to enhanced colonization. (ii) Using the same mono-axenic system and a mixture of numerous rhizosphere strains, it appeared possible to select strains which are enhanced colonizers (Fig. 3.2). Half of these strains are even able to control the disease tomato foot and root rot by a mechanism designated as Competition for Nutrients and Niches (CNN) (Kamilova et al. 2005; Fig. 3.2; Chap. 18).

3.4 Communication Between Organisms by Chemical Signaling

Many compounds secreted by plant roots and microbes function as signals that play a role in the communication between plant and microbe or between microbes. A selection of examples is given below.

Chemoattraction Bacteria in the rhizosphere usually live under conditions of starvation for nutrients and, consequently, are hunting for food. They have developed sensing systems which guide them towards compounds secreted by plant roots and by fungal hyphae. *Pseudomonas* cells from soil find the tomato root because they are attracted by root exudate. In order to identify the exudate component which is most active in the tomato rhizosphere, the chemo attractant activity of the individual representatives of the major groups of exudate components of tomato, namely organic acids, sugars, and amino acids, was analyzed. Experiments showed that

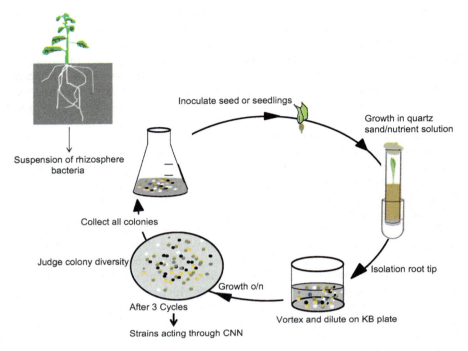

Fig. 3.2 Enrichment of bacteria which compete efficiently for nutrients and niches. Starting from a seed on which a crude mixture of rhizosphere bacteria is applied, enhanced competing bacteria are enriched for by repeatedly selecting for those cells which—after application to a sterile seed— reach the root tip first. Reproduced from Pliego et al. (2011 by permission of the publisher), after modification by Clara Pliego

sugars are inactive, that dicarboxylic acids are active, and that amino acids (especially L-leucine) are the most active chemoattractants. However, when this data is corrected for the levels of the individual components estimated to be present in the tomato rhizosphere, it was concluded that malic acid and citric acid are the major chemoattractants for *Pseudomonas* cells in the rhizosphere (De Weert et al. 2002). After reaching the root surface, the bacteria colonize it (Fig. 3.1e,f).

After reaching the fungal hyphae, the bacteria colonize the surface (Fig. 3.1i). Bacterial cells also try to find fungal hyphae because they can be used as food sources. They have developed a system to detect specific fungal products. This process has been extensively studied in the case of *Pseudomonas* cells trying to find hyphae of the fungus *Fusarium oxysporum* fsp. *radicis-lycopersici*. Using a series of *Fusarium oxysporum* f. sp. *radicis-lycopersici* strains which differ in the levels of secreted fusaric acid, it was found that the more fusaric acid is secreted, the stronger bacteria are attracted to the fungus. Finally, using synthesized chemically pure fusaric acid, the notion that fusaric acid is the major chemo attractant for *Pseudomonas* cells, was confirmed (De Weert et al. 2004b).

Initiation of the Nodulation Process Certain flavonoids secreted by roots of leguminous plants signal to *(Brady)rhizobium* bacteria that a root of their leguminous host plant is close by and that the process of nodule formation can be initiated. These flavonoids induce the nodulation (*nod*) genes in the bacterium which encode the enzymes for the synthesis of Nod-metabolites or lipochitin oligosaccharides (LCOs). These LCOs initiate the formation of nodules in which bacteria, in the form of bacteroides, fix atmospheric nitrogen for the plant.

Induction of *vir* gene Expression in *Agrobacterium* Phenolic compounds released from plant wound sites, such as 3',5' dimethoxy-4'-hydroxy acetophenone (commonly known as acetosyringone) and α-hydroxy acetosyringone, are the key inducers of bacterial virulence genes. A pH of around 5.5 and a temperature below 30 °C are also required. Particular sugars enhance the level of induction (Chap. 37).

N-Acyl Homoserine Lactones (AHLs) and Quorum Sensing (Chap. 7) AHLs are signal molecules secreted by many Gram-negative bacteria. They are used to recognize bacteria of their own kind. These molecules are supposed to diffuse more or less freely over the membranes. If the bacterial cell concentration reaches a certain density, the quorum, the intracellular AHL concentration reaches a level sufficiently high to initiate the synthesis of increased amounts of AHL, which in turn initiates the production of several proteins, including those involved in the synthesis of many antibiotics (Chap. 18) as well as virulence factors. AHLs also play a role in exchanging DNA by inducing F-pilus-mediated conjugation .

AMF-Plant Interaction Flavonoids present in root exudates can initiate interactions of plant roots with AMF such as stimulation of spore germination and hyphal branching. More recently, it was reported that secreted representatives of a group of plant hormones, the strigolactones, can also cause branching of neighboring AMF spores, thereby increasing their chance to encounter a plant root. In addition, the AMF releases signal molecules, identified as lipochito-oligosaccharides or Myc factors which stimulate root growth and branching (Chap. 25).

Volatiles Both plants and microbes can produce a range of volatile organic compounds, briefly designated as volatiles. Whereas volatile plant hormones such as ethylene, methyl jasmonate, and methyl salicylate function as airborne signals in mediating plant communication, several bacterial volatiles play a role in biocontrol (HCN) or induce systemic resistance in plants (such as 2,3-butanediol and its precursor acetoin) (Chap. 8).

War in the Rhizosphere Microbes in the rhizosphere can be attacked by their colleagues but they are not always simple victims. Like bacteria, some fungi have developed resistance mechanisms against antibiotics. The following mechanisms of fungal and bacterial tolerance or resistance have been shown. (i) Detoxification of the antibiotic. Some biocontrol strains produce the antibiotic 2,4-diacetyl phloroglucinol, for example to kill *Fusarium* strains. Some *Fusarium* strains produce an enzyme that deacetylates the antibiotic to mono-acetyl phloroglucinol which is a lot less fungitoxic. (ii) Efflux of the antibiotic. Upon exposure to phenazine antibiotics, some strains of the fungus *Botrytis cinerea* induces an efflux pump for the antibiotic. (iii) Repression of the synthesis of an antibiotic. The biocontrol fungus *Trichoderma atroviride* P1 produces chitinase enzymes to attack the cell wall of fungi (Chap. 6). The *Fusarium* mycotoxin deoxynivalenol inhibits the expression of the chitinase genes *ech42* and *nag1* which contribute to the biocontrol activity. Another example of repression of antibiotic synthesis is the *Fusarium* metabolite fusaric acid which inhibits the syntheses of the antibiotics 2,4-diacetyl phloroglucinol and phenazine-1-carboxamide produced by some *Pseudomonas* biocontrol strains. Finally, AHLs are often required for the synthesis of antibiotics and virulence factors. Some bacteria protect themselves by enzymatic inactivation of AHL.

Gene Transfer in the Rhizosphere A well known form of gene transfer between bacteria, namely F-pilus mediated conjugation, requires AHL. In biofilms, bacteria are very close to each other and covered by a mucous layer (Fig. 3.1f–h). These conditions seem ideal for keeping the intracellular AHL concentration high and therefore for stimulating conjugation. Indeed, it has been reported that gene transfer between bacteria in the rhizosphere is very efficient (van Elsas et al. 1988).

Acknowledgements This research was supported by Leiden University as well as by numerous grants, especially from the European Commission, EET, INTAS as well as from the NWO departments of ALW, CW, and STW.

References

Berg G, Eberl L, Hartmann A (2005) The rhizosphere as a reservoir for opportunistic human pathogenic bacteria. Environ Microbiol 7:1673–1685

Bloemberg GV, O'Toole GA, Lugtenberg BJJ et al (1997) Green fluorescent protein as a marker for *Pseudomonas* spp. Appl Environ Microbiol 63:4543–4551

Bolwerk A, Lagopodi AL, Wijfjes AHM et al (2003) Interactions in the tomato rhizosphere of two *Pseudomonas* biocontrol strains with the phytopathogenic fungus *Fusarium oxysporum* f. sp. *radicis-lycopersici*. Mol Plant Microbe Interact 16:983–993

Bonkowski M, Villenave C, Griffiths B (2009) Rhizosphere fauna: the functional and structural diversity of intimate interactions of soil fauna with plant roots. Plant Soil 321:213–233

Chin-A-Woeng TFC, de Priester W, Van der Bij AJ et al (1997) Description of the colonization of a gnotobiotic tomato rhizosphere by *Pseudomonas fluorescens* biocontrol strain WCS365, using scanning electron microscopy. Mol Plant Microbe Interact 10:79–86

De Weert S, Vermeiren H, Mulders IHM et al (2002) Flagella-driven chemotaxis towards exudate components is an important trait for tomato root colonization by *Pseudomonas fluorescens*. Mol Plant Microbe Interact 15:1173–1180

De Weert S, Dekkers LC, Kuiper I et al (2004a) Generation of enhanced competitive root tip colonizing *Pseudomonas* bacteria through accelerated evolution. J Bacteriol 186:3153–3159

De Weert S, Kuiper I, Lagendijk EL et al (2004b) Role of chemotaxis towards fusaric acid in colonisation of hyphae of *Fusarium oxysporum* f. sp. *radicis lycopersici* by *Pseudomonas fluorescens* WCS365. Mol Plant Microbe Interact 16:1185–1191

Egamberdieva D, Kamilova F, Validov S et al (2008) High incidence of plant growth-stimulating bacteria associated with the rhizosphere of wheat grown in salinated soil in Uzbekistan. Environ Microbiol 10:1–9

Haas D, Défago G (2005) Biological control of soil-borne pathogens by fluorescent pseudomonads. Nat Rev Microbiol 3:307–319

Kamilova F, Validov S, Azarova T et al (2005) Enrichment for enhanced competitive plant root tip colonizers selects for a new class of biocontrol bacteria. Environ Microbiol 7:1809–1817

Kamilova F, Kravchenko LV, Shaposhnikov AI et al (2006) Effects of the tomato pathogen *Fusarium oxysporum* f. sp. *radicis-lycopersici* and of the biocontrol bacterium *Pseudomonas fluorescens* WCS365 on the composition of organic acids and sugars in tomato root exudate. Mol Plant Microbe Interact 19:1121–1126

Lagopodi AL, Ram AFJ, Lamers GE et al (2002) Novel aspects of tomato root colonization and infection by *Fusarium oxysporum* f. sp. *radicis-lycopersici* revealed by confocal laser scanning microscopic analysis using the green fluorescent protein as a marker. Mol Plant Microbe Interact 15:172–179

Lugtenberg BJJ, Bloemberg GV (2004) Life in the rhizosphere. In: Ramos JL (ed) Pseudomonas, vol 1. Kluwer/Plenum, New York, pp 403–430

Lugtenberg B, Girard G (2013) Role of Phenazine-1-carboxamide produced by *Pseudomonas chlororaphis* PCL1391 in the control of tomato foot and root rot. In: Chincholcar S, Thomashow L (eds) Microbial phenazines. Springer, Berlin, pp 163–175

Lugtenberg B, Kamilova F (2009) Plant growth-promoting rhizobacteria. Annu Rev Microbiol 63:541–556

Mendes R, Garbeva P, Raaijmakers JM (2013) The rhizosphere microbiome: significance of plant beneficial, plant pathogenic, and human pathogenic microorganisms. FEMS Microbiol Rev 37:634–663

Pinton R, Varanini Z, Nannipieri P (2007) The rhizosphere. Biochemistry and organic substances at the soil plant interface, 2nd edn. CRC press, Taylor and Francis Group, Boca Raton

Pliego C, Kamilova F, Lugtenberg B (2011) Plant growth-promoting bacteria: fundamentals and exploitation. In: Maheshwari DK (ed) Bacteria in agrobiology: crop ecosystems. Springer, Germany, pp 295–343

Simons M, Van der Bij AJ, Brand I et al (1996) Gnotobiotic system for studying rhizosphere colonization by plant-growth promoting *Pseudomonas* bacteria. Mol Plant Microbe Interact 7:600–607

van Elsas JD, Trevors JT, Starodub ME (1988) Bacterial conjugation between pseudomonads in the rhizosphere of wheat. FEMS Microbiol Lett 53:299–306

Chapter 4
Life of Microbes on Aerial Plant Parts

Johan H. J. Leveau

Abstract The phyllosphere, or leaf environment, is a temporally erratic, spatially heterogeneous, and inherently transient habitat that supports a large and diverse population of microorganisms. This chapter offers an introductory exploration of the leaf surface as a microbial biome and of the genes and gene functions that underlie the unique adaptations for epiphytic survival in this inhospitable milieu. Also reviewed are the various ways in which host plant and environmental conditions affect the assembly, structure, and function of microbial communities on plant foliage. Special emphasis is placed on the challenges of studying microbial life on leaf surfaces, on the impact of leaf-associated microbiota on the ecosystem services provided by plant leaves, and on the interactions of epiphytic microorganisms with their host, each other, and plant and human pathogens in the context of food security and food safety.

4.1 The Phyllosphere

As an ecosystem or 'biome' for microorganisms, leaves and other above-ground plant parts are referred to as the 'phyllosphere', a term coined in the 1950s and inspired by Hiltner's 'rhizosphere' (Chap. 3). The prefix 'phyllo' is derived from the Greek φύλλο for 'leaf' and shares the same base as 'filo', i.e. the dough used for flaky pastries such as baklava and apfelstrudl. Many use the term 'phyllosphere' (sometimes 'phylloplane') exclusively for plant foliage, and refer to other above-ground plant parts with alternative language, for example the anthosphere (flower), carpo- or fructosphere (fruit), caulosphere (bark/stem), calusphere (bud), and spermosphere (seed). This chapter will stick to the original definition of phyllosphere and only cover microbial life on leaves (Leveau 2006; Meyer and Leveau 2012; Vorholt 2012), not

J. H. J. Leveau (✉)
Department of Plant Pathology, University of California, One Shields Avenue,
476 Hutchison Hall, Davis, CA 95616, USA
Tel.: +1 530 752-5046
e-mail: jleveau@ucdavis.edu

© Springer International Publishing Switzerland 2015
B. Lugtenberg (ed.), *Principles of Plant-Microbe Interactions*,
DOI 10.1007/978-3-319-08575-3_4

other above-ground plant parts. Also, its narrative will be limited to epiphytic microorganisms (i.e. surface-associated, in contrast to endophytes, see also Chap. 5) and focus on land-based plants only, not submerged plants or macro-algae.

Based on satellite imagery of Earth's terrestrial foliage, the phyllosphere microbiome is estimated to encompass about half a billion square kilometers. However, the 'coastline paradox' stipulates that this value is many times larger if one considers that the leaf surface represents an intricate three-dimensional topography at the scale at which it is occupied and experienced by its microscopic inhabitants, i.e. the phyllosphere microbiota, which include bacteria, filamentous fungi, yeast, oomycetes, algae, lichens, protists, and protozoa. The extraordinary structural heterogeneity of the leaf surface is a key driver of the nonrandom, micrometer-scale spatial distribution of epiphyllous microorganisms. For example, the cuticle that covers the leaf surface varies laterally in composition, thickness, and permeability, which contributes to differential leaching of plant substances (including photosynthates) to the surface where they become available for exploitation by microorganisms (Leveau and Lindow 2001). Spatial variation in the topography-driven 'waterscape' on the leaf surface also impacts the ability of microorganisms to avoid drought, to laterally disperse, and to sense, respond to and interact with other colonizers on the same leaf surface.

Microbial Colonization Patterns on Leaves Bacteria are by far the most abundant microbial colonizers of leaf surfaces, an estimated 10^{26} cells on the foliage of all terrestrial plants combined. Characteristic of bacterial life in the phyllosphere is that the majority of bacteria occur not as single cells but in large clusters (often referred to as 'aggregates') of up to thousands of cells. The term 'aggregate' is somewhat misleading, as the formation of these clusters is not just a function of cells aggregating (in the literal sense, i.e. bacteria coming together, actively or passively, to form a cluster), but also of cells staying together, i.e. the formation of offspring by a single immigrant to the leaf surface into a colony or cluster. The observation that trichomes (leaf hairs) often support large clusters of bacteria at their base is probably a result of both mechanisms: increased leaching at these sites favors the replication and staying together of cells, while the process of evaporation forces accumulation of cells in areas where water is retained longest, including the base of trichomes. The distribution curve of bacterial cluster sizes on leaves is typically right-hand skewed, with many small clusters and few very large clusters. Underlying this pattern is the ability of individual bacteria to leave larger clusters and start a new one elsewhere on the leaf (Perez-Velazquez et al. 2012; van der Wal et al. 2013). In this, bacterial life in aggregates on leaf surfaces resembles that in biofilms on variably-saturated surfaces (Chap. 7).

Processes of Leaf Inoculation There are several processes by which leaves get colonized. Some microorganisms are already present in buds and therefore among the first to explore the developing leaf. Most microorganisms will arrive after leaf emergence, by various mechanisms and from different sources, including wind, rain, dust, soil splash, and deposition from the air and by insects. As leaves get older, surface topography, cuticle waxiness, availability of water and nutrients, and microbial colonization patterns all change and affect the ability of newly arriving immigrants to the leaf to settle, disperse or start reproducing. An underappreciated factor in the assembly of microbial communities on leaf surfaces is chance. The leaf represents a patchy environment in terms of availability of nutrients, presence or absence of water, and microbial colonization, so an incoming microorganism may just have to be lucky to land in a spot that isn't already taken by other microorganisms and that (still) offers food and shelter. For bacterial immigrants to bean leaves, such luck has been quantified (Remus-Emsermann et al. 2012): there is a 3 % chance that they land in a location that allows five or more doublings, compared to an approximately 50 % chance of landing somewhere where they double once or less. Precolonization significantly reduces immigrant ability to produce offspring, indicating that early arrival by phyllosphere-fit microorganisms offers greatest likelihood to benefit from available nutrients and to dominate the community structure. The stochastic nature of arrival probably underlies the often observed variation in bacterial community composition on individual leaves of plants in the same field or greenhouse (Maignien et al. 2014).

4.2 Structure and Function of Phyllosphere Microbiota

The leaf surface is an extreme habitat. In the course of a single day, leaves may go from wet to dry, dark to light, and cold to hot. On a longer time scale, plant leaves will eventually senesce and shed, so epiphyllous microorganisms also need to avoid, anticipate, or survive life outside of the leaf environment, for example in the soil or air. This picture of the leaf surface as a harsh, quickly changing, and ephemeral environment predicts that phyllosphere microorganisms must be well adapted to deal with phyllosphere stresses. Indeed, many possess genes for the production of pigments that protect against ultraviolet radiation, or DNA repair systems that deal with the damage caused by it. The ability to accumulate solutes to withstand drought is a common property among bacteria and fungi isolated from foliage. Many phyllosphere-related adaptations don't convey mere tolerance to leaf stresses, but contribute to habitat modification in favor of microbial survival and growth: biofilm formation, ice nucleation (see Box 4.1), and surfactant production are all examples of ecosystem engineering by epiphytic microorganisms, designed to provide shelter, release nutrients, and/or facilitate dispersal.

Box 4.1
Bacterial ice nucleation was discovered as a property of plant pathogens of the species *Pseudomonas syringae* which cause frost damage on crop foliage. Pioneering work by Dr. Steve Lindow showed that this activity was due to a single gene and that mutants lacking this gene (so-called ice-minus bacteria) not only lost the ability to cause frost, but in the first deliberate release of a genetically modified organism also protected plants against ice-positive *P. syringae*. Bacterial ice nucleation has been commercially exploited; for example, artificial snow at the 2014 Sochi Olympic Games was generated with the use of devitalized ice-nucleating bacteria.

Phyllosphere Functional Genomics Whole-genome sequencing, transcriptional profiling, and proteomic analysis of phyllosphere-competent microorganisms such as bacteria from the genera *Pseudomonas*, *Erwinia*, *Pantoea*, *Methylobacterium*, and *Arthrobacter* provide valuable insights into leaf surface-specific adaptations of these organisms: they confirm the contribution or involvement of gene functions hypothesized to contribute to phyllosphere fitness ('epiphitness'), or they reveal new ones. For example, *Arthrobacter chlorophenolicus* upregulates genes for the utilization of hydroquinone, a compound not previously detected (looked for) on leaf surfaces but now confirmed to be present in low concentrations (Scheublin et al. 2013). The phyllosphere-induced expression of genes for phenylalanine degradation by *Pseudomonas syringae* is presumed to counter phenylpropanoid-based plant defenses (Yu et al. 2013). Proteomics-identified PhyR is a master regulator for phyllosphere-survival of *Methylobacterium extorquens*, with a pleiotropic role in resistance to desiccation, UV, and heat (Gourion et al. 2008). Metagenomic approaches to understanding life on leaf surfaces are on the rise (Delmotte et al. 2009) and beginning to uncover the functional diversity contained within entire epiphytic microbial communities, not just individual phyllosphere isolates.

Plant and Human Pathogens Much of the interest in studying leaf microorganisms comes from the realization that plant foliage provides many ecosystem services, be it directly, for example as food, shade, or drugs, or indirectly as a harvester of solar energy to produce flowers, fruits, fibers, seeds, roots, and other plant parts. Historically, the field of phyllosphere microbiology has seen overwhelming emphasis on understanding foliar pathogens: their ability to cause chlorosis, wilt, browning, and other symptoms that interfere with proper leaf function represents a great threat to many of the above services. An emerging concern is the foliar contamination of leafy greens with human pathogens (Brandl and Sundin 2013). Research in this area has contributed significantly to the latest phyllosphere-related scientific literature, offering new insights into the sources of disease outbreaks on raw-consumed leafy greens such as spinach and lettuce and the factors that impact the foliar establishment of such pathogens as *E. coli* and *Salmonella*. Little is known still about the interactions that plant and human pathogens have with other members of leaf-surface microbial communities, and how such interactions affect the abundance and activity of said pathogens. Positive as well as negative correlations between pathogen presence and community composition have been reported (Rastogi et al. 2012), and experimental evidence exists for facilitative effects of natural leaf associates on pathogen establishment (Cooley et al. 2006), or antagonistic effects, be it direct, for example through resource competition (Innerebner et al. 2011) or indirect, for example through induction of plant defense genes (Kurkcuoglu et al. 2007).

DNA-Based Community Profiling Early efforts to describe microbial diversity on plant foliage relied on differences in the number and appearance of microbial colonies that formed after pushing a leaf or plating leaf washes onto an agar surface. This culture-dependent approach has dominated the field for a long time, although additional techniques allowed the differentiation and naming of isolates based on properties other than colony appearance, including cell composition (for example, fatty acid methyl esters), phenotype (for example, substrate utilization), or genotype (for example, small-subunit rRNA genes or internal transcribed spacer sequences). Not until the availability and embrace of techniques such as Denaturing Gradient Gel Electrophoresis (DGGE), Terminal Restriction Fragment Length Polymorphism (T-RFLP), Automated rRNA Intergenic Spacer Analysis (ARISA), phylochips, clone library sequencing, and most recently massively parallel amplicon sequencing, transcriptional profiling, and community proteogenomics, did questions about size, structure, and activity of microbial communities on leaves become answerable without having to depend on ability or willingness of microbes to grow in the lab (Rastogi et al. 2013).

Culture-Independent versus Culture-Dependent Data As do other microbiomes, the phyllosphere reveals greater population sizes and a richer biodiversity when analyzed by culture-independent (i.e. DNA-based) approaches than culture-dependent ones (Rastogi et al. 2010). One explanation for this is that some fraction of the bacterial population is truly unculturable, i.e. these bacteria simply cannot form a colony on the media that are used in the lab. A relevant example are bacteria from the genus *Alkanindiges* which have been detected on the leaves of several plant species, but which have been described as obligate alkane degraders, so they cannot be expected to form colonies on standard media. Another explanation for the culture-dependent/independent discrepancy is that the harsh conditions on the leaf surface forces some bacteria into a state of nonculturability. Evidence in support of this notion is that DNA-based community profiles often reveal abundant representation of bacteria from genera such as *Pseudomonas*, *Erwinia*, and *Pantoea*, which one would expect to form colonies if they were alive and well. Another reason for culturable-based underestimation is the incomplete breakup of bacterial aggregates from the leaf surface, as one such aggregate will be counted as one colony-forming unit, although it might consist of hundreds or even thousands of individual bacterial cells. Culture-independent counts may be overestimated because most culture-independent approaches are based on relative abundances of small-subunit rRNA gene amplicons, and most bacteria possess more than one copy of this gene. A DNA-based challenge that is unique to the phyllosphere involves the fact that plant leaf cells contain chloroplasts that have their own genome and small-subunit rRNA genes. A single plant cell may harbor as many as 10^5 copies of the chloroplast 16S rRNA gene, which is a huge source of potential contamination considering that many leaves carry the same number of bacterial 16S rRNA genes per gram of tissue (Rastogi et al. 2010).

4.3 Drivers of Microbial Community Structure on Plant Leaves

Two major questions in phyllosphere microbiology that DNA-based profiling of leaf-associated microbial communities is finding answers to are: (1) to what extent does plant genotype drive community structure, and (2) how is microbial community structure impacted by the environment in which the plant is growing?

Plant Genotype The first question can be addressed in several ways. One is the 'common garden' experiment, where plants from different species are grown or already available under similar environmental conditions. Such studies have indeed demonstrated that different plant species and even plant cultivars vary in the types and abundances of microorganisms that they carry on their foliage. Many labs focus on finding the genetic factors or loci that contribute to these differences. Quantitative trait loci mapping is used to identify genes that contribute to differences in bacterial diversity on leaves. Another approach is the use of mutant plants to evaluate the effect of the mutation on leaf surface microbiology. For example, bacterial communities on the leaves of *Arabidopsis* mutants with an altered cuticular wax biosynthesis are demonstrably distinct from those on wildtype plants. The notion that one or more plant genes can alter the diversity of leaf-associated microbiota feeds the idea that the structure and function of microbial communities have potential as a breeding target. To realize this potential, there is a need not only for knowing the genes that underlie foliar differences in colonization, but also for a definition of what constitutes desirable microbiota. For example, are there combinations of bacterial taxa that consistently protect plant leaves from infection by a plant pathogen? The answers will come from a continued investment in correlative surveying of phyllosphere microbiota on different types of plants, in different states of health, and under different environmental conditions.

Environmental Impacts The fact that demonstration of a plant genotype impact on epiphytic microbiota often requires tight control of environmental conditions suggests that the environment can have a significant impact also. Three such impacts can be distinguished. One is that of the environment as a source of microorganisms (inoculum): different locations harbor different microbial communities in the air, soil and on nearby plants, and represent different sink-source dynamics. The environment may also play a facilitating role in the delivery of new immigrants (inoculation): rain splash, overhead irrigation, wind, and insects all may deposit microorganisms to the leaf surface, and geographical variation in these factors may drive variation in microbiota structure. A third type of environmental impact is that as a driver of microbial activity on the leaf surface (incubation); this includes factors such as temperature and relative humidity. Distance-decay relationships among phyllosphere microbiota (Finkel et al. 2012; Rastogi et al. 2012) are probably best explained by the existence of environmental gradients along different geographical scales in terms of inoculum, inoculation, and incubation, as defined above.

4.4 Future Aspects

The future of phyllosphere microbiology will hopefully bring an integration of many of the approaches and insights described in this chapter. One of the field's future challenges will be to connect knowledge about microbial community structure to what we have come to understand about the heterogeneous nature of the biotic and abiotic leaf environment. In other words, it is time to start asking questions about the spatial distribution of all this recorded microbial diversity on leaves, in order to get basic answers about who is interacting how with whom and where in the phyllosphere, and how these interactions scale up to matters of food security and food safety.

References

Brandl MT, Sundin GW (2013) Focus on food safety: human pathogens on plants. Phytopathology 103:304–305

Cooley MB, Chao D, Mandrell RE (2006) *Escherichia coli* O157: H7 survival and growth on lettuce is altered by the presence of epiphytic bacteria. J Food Prot 69:2329–2335

Delmotte N, Knief C, Chaffron S et al (2009) Community proteogenomics reveals insights into the physiology of phyllosphere bacteria. Proc Natl Acad Sci U S A 106:16428–16433

Finkel OM, Burch AY, Elad T et al (2012) Distance-decay relationships partially determine diversity patterns of phyllosphere bacteria on *Tamarix* trees across the Sonoran desert. Appl Environ Microbiol 78:6187–6193

Gourion B, Francez-Charlot A, Vorholt JA (2008) PhyR is involved in the general stress response of *Methylobacterium extorquens* AM1. J Bacteriol 190:1027–1035

Innerebner G, Knief C, Vorholt JA (2011) Protection of *Arabidopsis thaliana* against leaf-pathogenic *Pseudomonas syringae* by *Sphingomonas* strains in a controlled model system. Appl Environ Microbiol 77:3202–3210

Kurkcuoglu S, Degenhardt J, Lensing J et al (2007) Identification of differentially expressed genes in *Malus domestica* after application of the non-pathogenic bacterium *Pseudomonas fluorescens* Bk3 to the phyllospere. J Exp Bot 58:733–741

Leveau JHJ (2006) Microbial communities in the phyllosphere. In: Riederer M, Mueller C (eds) Biology of the plant cuticle. Blackwell, Oxford, pp 334–367

Leveau JHJ, Lindow SE (2001) Appetite of an epiphyte: quantitative monitoring of bacterial sugar consumption in the phyllosphere. Proc Natl Acad Sci U S A 98:3446–3453

Maignien L, DeForce EA, Chafee ME et al (2014) Ecological succession and stochastic variation in the assembly of *Arabidopsis thaliana* phyllosphere communities. MBio 5:e00682–e00613

Meyer KM, Leveau JHJ (2012) Microbiology of the phyllosphere: a playground for testing ecological concepts. Oecologia 168:621–629

Perez-Velazquez J, Schlicht R, Dulla G et al (2012) Stochastic modeling of *Pseudomonas syringae* growth in the phyllosphere. Math Biosci 239:106–116

Rastogi G, Tech JJ, Coaker GL et al (2010) A PCR-based toolbox for the culture-independent quantification of total bacterial abundances in plant environments. J Microbiol Methods 83: 127–132

Rastogi G, Sbodio A, Tech JJ et al (2012) Leaf microbiota in an agroecosystem: spatiotemporal variation in bacterial community composition on field-grown lettuce. ISME J 6:1812–1822

Rastogi G, Coaker GL, Leveau JHJ (2013) New insights into the structure and function of phyllosphere microbiota through high-throughput molecular approaches. FEMS Microbiol Lett 348:1–10

Remus-Emsermann MNP, Tecon R, Kowalchuk GA et al (2012) Variation in local carrying capacity and the individual fate of bacterial colonizers in the phyllosphere. ISME J 6:756–765

Scheublin TR, Deusch S, Moreno-Forero SK et al (2013) Transcriptional profiling of gram-positive *Arthrobacter* in the phyllosphere: induction of pollutant degradation genes by natural plant phenolic compounds. Environ Microbiol 16:2212–2225

van der Wal A, Tecon R, Kreft J-U et al (2013) Explaining bacterial dispersion on leaf surfaces with an individual-based model (PHYLLOSIM). PLoS One 8:e75633

Vorholt JA (2012) Microbial life in the phyllosphere. Nat Rev Microbiol 10:828–840

Yu XL, Lund SP, Scott RA et al (2013) Transcriptional responses of *Pseudomonas syringae* to growth in epiphytic versus apoplastic leaf sites. Proc Natl Acad Sci U S A 110:E425–E434

Chapter 5
Life of Microbes Inside the Plant

Jesús Mercado-Blanco

Abstract A hidden microbial world is present in the interior of all plants. Myriads of bacteria and fungi live inside them without causing apparent deleterious effects to their hosts. They are designated as endophytes. Endophytic communities are variable and diverse. Their structure and composition are shaped by a number of (a)biotic factors. Endophytes have found evolutionary solutions to cope with defensive responses deployed by host plants to face colonization by microbes. In return, they live within an ecological niche that provides better protection against a number of stresses and a reliable and constant source of nutrients. Endophytes seem to contribute to plant fitness and development, displaying beneficial traits that can be exploited in agricultural biotechnology. However, many questions related to the endophytic lifestyle remain to be answered. This chapter summarizes present knowledge on how endophytes are able to establish and endure within plants. Potential biotechnological applications are also briefly presented.

5.1 Introduction: Beneficial Endophytes Defined

Plants live in close association with a huge diversity of microorganisms. In fact, the composite genome of these accompanying microbial communities is far larger than that of the host plant, and thus is also referred to as the plant's second genome. Most of the components of the plant-associated 'microbiome' (Chap. 30) are only able to colonize and persist on plant tissue surfaces or in the soil rhizosphere (Chap. 3). However, some can also establish themselves as non-deleterious endophytes. It is likely that all plants carry endophytes, which play an important role in plant fitness and development. Plants can thus be considered as super organisms of which both the plant and its endophytic microbiome work coordinately to shape and sustain an extraordinary ecosystem. It is generally recognized that endophytes represent just a minor fraction of the microbiota inhabiting plant surfaces or living in close

J. Mercado-Blanco (✉)
Department of Crop Protection, Institute for Sustainable Agriculture,
Agencia Estatal Consejo Superior de Investigaciones Científicas (CSIC),
Campus 'Alameda del Obispo' s/n, Apartado 4084, 14080 Córdoba, Spain
Tel.: +34957499261
e-mail: jesus.mercado@ias.csic.es

© Springer International Publishing Switzerland 2015   25
B. Lugtenberg (ed.), *Principles of Plant-Microbe Interactions*,
DOI 10.1007/978-3-319-08575-3_5

proximity to them, and comprise mainly microbes originating from the soil region associated with roots (i.e. the rhizosphere, Chap. 3). In fact only small subpopulations of rhizosphere and phyllosphere microorganisms are able to enter and live inside the plant. Relevant reviews suggested are: Bacon and Hinton (2006); Schulz et al. (2006); Rosenblueth and Martínez-Romero (2006); Hardoim et al. (2008); Ryan et al. (2008); Reinhold-Hurek and Hurek (2011); VV.AA. (2013).

The word *endophyte* means 'in the plant' and is derived of the Greek words *endon* (within) and *phyton* (plant). Endophytes have been defined by various authors in somewhat different ways (Bacon and Hinton 2006; Rosenblueth and Martínez-Romero 2006; Schulz et al. 2006; Mercado-Blanco and Lugtenberg 2014). It is generally agreed that they are bacteria and fungi that can be detected at any moment within the tissues of healthy plants, and that do not produce disease symptoms. Microbial phytopathogens, nodule-producing microbes, and mycorrhizal fungi may display endophytic lifestyles during part of their lives but they are not considered here as endophytes. Moreover, this chapter is only focused on bacterial endophytes (for endophytic fungi see, for instance, VV.AA. 2013). Mere isolation from surface-disinfected tissues is not enough to claim a true endophyte. Plant surface sterilization protocols must be sufficiently stringent to eradicate the external microbiota without killing bacteria with the tissue. Moreover, the 'candidate endophyte' must be shown to be a true endophyte by both its ability to re-infect disinfected seedlings, and by microscopic evidence (Reinhold-Hurek and Hurek 1998).

The vast majority of bacterial endophytes are non-culturable or VBNC. This may represent a survival strategy to persist hostile conditions within the plant. By implementing culture-independent and metagenomics approaches our understanding of endophytes is being steadily enriched. These methodologies will undoubtedly continue to reveal a much wider diversity and abundance of endophytic communities than that uncovered by traditional culture-dependent methods.

5.2 How Do Endophytes Get into the Plant and Spread to Distant Tissues?

Endophytes gain entrance into plants predominantly through the roots but also through leaves, flowers, stems or cotyledons. Indeed, the vast majority of endophytes are soil-inhabitants and plant colonization seems to mainly originate from the rhizosphere. Evidence confirming this possibility has been obtained by combining biotechnological and microscopy tools. Some endophytes have thus been shown to spread systemically from the original penetration site(s), and be found in distant plant tissues and organs. Consequently, the population density of endophytes is usually higher in roots than in any other plant organ. It is important to stress that an amazing variety of endophytic bacteria and fungi (hundreds of different taxa) can be found in diverse organs/tissues of any individual herbaceous, woody or moss species (Hallmann and Berg 2006; Schulz et al. 2006; Mercado-Blanco and Lugtenberg 2014).

Overall, our knowledge about the specific sites at which endophytic bacteria attach and penetrate into root tissues is scant. Nevertheless, it is generally assumed that bacteria invade roots passively using cracks or wounds located, for instance, at the emergence points of lateral roots. Such root cracks can also be produced by microbial, nematode or arthropod activities. Preferential sites for rhizosphere bacteria attachment and subsequent entry can also be the thin-walled surface layers located in the apical root region, including the root differentiation, elongation and root hair zones as well as the intercellular spaces of the root epidermis. Specific bacterial components that are known to be involved in endophyte attachment to plant tissue include Type IV pili, lipopolysaccharides, and exopolysaccharides (Hardoim et al. 2008; Reinhold-Hurek and Hurek 2011; Mercado-Blanco and Lugtenberg 2014).

Besides being an important attachment structure, root hairs play also a role in root endophytic colonization. Using fluorescently-tagged bacteria and CLSM allowed demonstration that endophytic *Pseudomonas* spp. strains can internally colonize olive root hairs (Prieto et al. 2011) prior to becoming established within the intercellular spaces of the root cortex (Fig. 5.1). Despite root hairs seeming to play a role in bacterial entrance into the roots, several questions remain to be elucidated: (i) the exact timing of and site(s) used for bacterial penetration of the root hair cell; (ii) how the bacteria move to the intercellular spaces of the root cortex; and (iii) whether these bacteria enter root hairs *via* active or passive mechanisms (Mercado-Blanco and Prieto 2012).

Mechanisms used by endophytic bacteria to enter the plant are largely unknown, although several bacterial traits have been proposed to participate in endophytic colonization of plant roots. Once bacteria overcome the exodermal barrier, they may remain either at the site of entry or move towards the intercellular space of the cortex and even to distant parts (Compant et al. 2005; Hardoim et al. 2008; Reinhold-Hurek and Hurek 2011).

5.3 How Do Endophytes Cope with and Adapt to the Inner Plant Environment?

Living inside plant tissues requires adaptation to an environment that provides food and low exposure to (a)biotic stresses. Therein, competition among endophytic microorganisms can be expected, although nothing is yet known about trophic interactions within such microbial communities. However, compared to the highly competitive/predatory environment found outside the plant, its interior is a "safe heaven" for endophytes since it is a reliable and constant source of nutrients. Endophytes have thus evolved to adapt themselves to nutrients available inside plant tissues (Bacon and Hinton 2006; Mercado-Blanco and Lugtenberg 2014).

Data on the ability of endophytes to utilize nutrients found in the plant interior are not abundant. Comparing the abilities of endophytic and highly-related non-endophytic strains to utilize nutrient sources is a strategy to unravel the feeding capabilities of the endophytes. For instance, utilization of L-arabinose has been

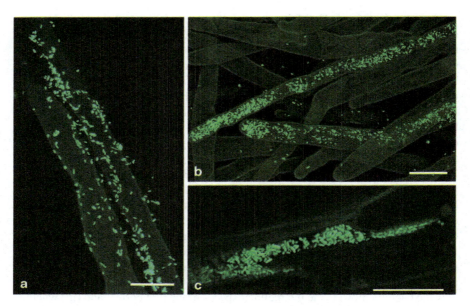

Fig. 5.1 CLSM images of *in vitro* micropropagated olive roots (cv. Manzanilla) colonized by enhanced green fluorescent protein (*EGFP*)-tagged *Pseudomonas fluorescens* PICF7. **a** Surface colonization of root hairs by PICF7 cells; **b** Detection of *EGFP*-tagged PICF7 cells inside two root hairs; **c** Intercellular colonization of the root cortical tissue by PICF7 cells. Scale bar represents 50 µm in panel **a** and 20 µm in panels **b** and **c**. For details on olive roots-PICF7 colonization bioassays, tissue sectioning and CLSM imagery, see Prieto et al (2011). These CLSM microphotographs are reproduced from Prieto et al. (2011), doi:10.1007/s00248-011-9827-6

suggested to be an important trait contributing to the endophytic lifestyle of several *Pseudomonas* spp. strains in cucumber (Malfanova et al. 2013). It is likely that other carbon sources may play similar roles in other plant-endophyte associations, and evidence/suggestions for such a role are available (Mercado-Blanco and Lugtenberg 2014). Because of the ability of endophytes to preferentially metabolize specific carbon sources, competition for these nutrients can be a determining factor in shaping the endophytic community in a given host plant or tissue. Indeed, adaptation to specific nutrients' availability may limit the number of endophytic microbes that can be taken up and survive inside plant tissues either under natural conditions or upon artificial inoculation. Comparative genomics approaches aimed to unravel specific traits linked to endophytic lifestyles can provide relevant information on this issue (Mitter et al. 2013; Brader et al. 2014; Ali et al. 2014).

As a toll to be paid for their *in planta* protection, endophytes must develop strategies to silence or evade the various responses that plants use to confront 'non-hostile' endophyte colonization, or attacks by phytopathogens. The strategies that endophytes have evolved to elude such responses and be recognized as innocuous invaders occur by an as yet poorly-understood modulation of the plant-deployed immune response (Zamioudis and Pieterse 2012). It is conceivable that the prevalence of non-culturable and/or VBNC states encountered among endophytes could be a survival strategy to prevail over stresses operating within plant tissues.

With respect to how endophytes cope with plant defense responses our knowledge is also scant (Reinhold-Hurek and Hurek 2011). Inner colonization and persistence of endophytes within plant tissues likely entail broad changes at the transcriptomic level for both partners. However, little information is available about the genetics changes that an endophyte provokes in the host plant and how colonization modifies the behavior of the new resident. Likewise, the effects that introduced endophytes may provoke on a pre-existing endophytic microbiome and *vice versa* are largely unknown. However, some studies are available and results indicate that colonization of plant tissues by an endophyte triggers, among other changes, a broad range of defensive responses (Conn et al. 2008; Schilirò et al. 2012).

Finally, it is worth mentioning that the composition, abundance, distribution and functionality of the endophytic microbiome found in a given host plant, plant tissue or organ is not static. On the contrary, an endophytic microbiome can be modified over time by the plant growth phase and physiological state, and/or diverse environmental, biological and physical-chemical factors (see, for instance, Van Overbeek and van Elsas 2008; Ardanov et al. 2012).

5.4   The Search for Endophytic-Specific Traits

An important question still to be solved is whether endophytes have specific traits that define their lifestyle. An environment providing specific nutrients, but hostile because of active defense responses, should be a driving force selecting adapted phenotypes (see previous section). Similarly, the presence of microbe-specific machineries for plant tissue penetration could be necessary for establishment as beneficial endophytes, although it is likely that some of the mechanisms could be shared by pathogens. It is also possible that, in contrast to facultative endophytes, obligate endophytes might carry genetic and metabolic determinants more essential to the endophytic lifestyle. Comparative genomics and bioinformatics approaches could definitively help in unraveling specific traits linked to endophytism. Even though studies like these show that some characteristics seem to be shared by different endophytes, the emerging picture is of a broad genetic diversity (Mitter et al. 2013; Ali et al. 2014).

Quorum-sensing (QS) (Chap. 8) and the ability to overcome plant defenses seem to be commonly found in the genomes of the endophytic bacteria analyzed so far. QS systems allow bacteria to 'sense' their own concentration/abundance, thereby triggering the expression of specific target genes only at a high cell density. Considering that bacterial endophytes can reach high populations densities in defined sites, production of QS signals within plant tissues and how those signals operate in endophytic-mediated processes deserve investigation. In contrast, Type III (Chap. 7) secretion systems, which mediate, in gram-negative pathogens and rhizobial symbionts, the delivery to the host plant of effector proteins that suppress the host defense response, seem rare among endophytes. Instead, other types of secretion systems such as Type VI seem to be more frequent. Such findings can thus shed light on the potential differences between pathogenic and endophytic lifestyles displayed by plant-associated bacteria. Currently, however, the number of genomes and genomic information available from claimed endophytes is still too low to draw sound conclusions (Reinhold-Hurek and Hurek 2011; Mitter et al. 2013; Ali et al. 2014).

5.5 What Endophytes Do for the Plant and Their Potential to be Exploited in Agro-Ecosystems

The presence of non-deleterious endophytes can benefit the host plant in different ways, a scenario of utmost interest once these microbial communities and their multitrophic interactions are properly characterized, understood and harnessed. Endophytes are of increasing interest because of their potential biotechnological applications (see, for instance, Ryan et al. 2008; VV.AA. 2013; Brader et al. 2014; Mercado-Blanco and Lugtenberg 2014). Among the beneficial traits with potential to be exploited in agro-ecosystems, promotion of plant growth, and control of plant diseases are of particular significance. In most cases the mechanism(s) involved, particularly those related to biocontrol, remain to be elucidated. Nevertheless, it has been suggested that beneficial effects deployed by bacterial endophytes might operate through mechanisms similar to those described for rhizosphere bacteria (Kloepper and Ryu 2006). Additional applications of endophytes beyond agricultural biotechnology rely on the ability of the organisms to produce a broad range of bioactive metabolites which are relevant for other purposes, including human health (i.e. antibiotics, antitumor compounds, anti inflammatory agents, etc.) (see, for instance, Christina et al. 2013; Brader et al. 2014).

Plant growth promotion can be achieved either directly or indirectly. There is little knowledge available about mechanisms of growth promotion exerted by endophytes, and operating *in planta* (Hardoim et al. 2008). Nevertheless, considering that many endophytes are also free-living rhizosphere microorganisms, it is plausible to assume that mechanisms to stimulate plant growth deployed by the latter may also operate once endophytic growth is established. However, this assumption still needs to be confirmed.

Direct promotion of plant growth by endophytic bacteria and fungi can be achieved by the microbe providing (micro)nutrients (biofertilization) and/or phytohormones (phytostimulation) to the plant. Indirect plant growth promotion is a consequence of the suppression of plant diseases exerted by pathogenic microorganisms, and can be mediated by direct antagonism/antibiosis against the pathogens, advantageous out-competing for nutrients and/or space, or by triggering in the host plant enhanced defense capacities against pathogen attack. Growth can also be stimulated indirectly by alleviation of stress caused by environmental pollutants (rhizoremediation, see

Chap. 29) or other stressful abiotic (heavy metals, drought, salinated soils) conditions. For instance, synthesis of the enzyme ACC deaminase reduces the level of the stress hormome ethylene by converting ACC into α-ketobutyrate and ammonia. Production of this enzyme by plant-growth promoting bacteria, including those displaying endophytic lifestyles, can make host plants tolerant to a number of stresses (Chap. 27; Hardoim et al. 2008).

5.6 Concluding Remarks

Many questions on how, why and when any given endophyte(s)-plant consortium is established remain to be elucidated. A list of questions to be answered on this topic has been recently outlined by Mercado-Blanco and Lugtenberg (2014). For instance, little is known on how an 'endophytic candidate' is able to overcome or modulate defensive barriers/responses to successfully penetrate and establish within plant tissues. The identification of traits involved in the colonization and persistence within the plant is still incomplete. Similarly, understanding the influence of environmental, physiological, developmental stages and/or genetic factors on endophytes is essential if these organisms are expected to be further developed as biocontrol or (phyto)rhizoremediation tools. In summary, two main questions that should now be put forth are: (i) which driving forces are operating to build up an endophytic community, and (ii) what does the endophytic microbiome do for the plant. The implementation of currently available and powerful '-omics' and microscopy technologies (see Chap. 31) will undoubtedly provide some answer to these questions, as well as those aimed at unraveling the molecular processes that define endophytes (Mitter et al. 2013; Ali et al. 2014).

From a practical perspective, more studies are needed to understand whether the performance of any artificially-introduced endophyte can be affected by the native microbiome of the host plant; and *vice versa*, i.e. how the indigenous endophytic microbiota can be influenced by the introduction of a 'newcomer' into this delicately balanced microenvironment, and whether such introductions may alter the plant's development and fitness.

Acknowledgments Supported by grants P07-CVI-02624 from Junta de Andalucía (Spain) and AGL2009-07275 from Spanish MICINN/MINECO, both co-financed by ERDF of the EU. Thanks are due to Katherine Dobinson and Pilar Prieto for their critical reading and interesting suggestions.

References

Ali S, Duan J, Charles TC, Glick BR (2014) A bioinformatics approach to the determination of genes involved in endophytic behavior in *Burkholderia* spp. J Theor Biol 343:193–198
Ardanov P, Sessitsch A, Haggman H et al (2012) *Methylobacterium*-induced endophyte community changes correspond with protection of plants against pathogen attack. PLoS One 7(10):e46802

Bacon CW, Hinton DM (2006) Bacterial endophytes: the endophytic niche, its occupants, and its utility. In: Gnanamanickam SS (ed) Plant-associated bacteria, Springer, The Netherlands, pp 155–194

Brader G, Compant S, Mitter B et al (2014) Metabolic potential of endophytic bacteria. Curr Opin Biotechnol 27:30–37

Christina A, Christapher V, Bhore SJ (2013) Endophytic bacteria as a source of novel antibiotics. Pharmacogn Rev 7:11–16

Compant S, Duffy B, Nowak J et al (2005) Use of plant growth promoting bacteria for biocontrol of plant diseases: principles, mechanisms of action, and future prospects. Appl Environ Microbiol 71:4951–4959

Conn VM, Walker AR, Franco CMM (2008) Endophytic actinobacteria induce defense pathways in *Arabidopsis thaliana*. Mol Plant Microbe Interact 21:208–218

Hallmann J, Berg G (2006) Spectrum and population dynamics of bacterial root endophytes. In: Schulz B, Boyle C, Sieber T (eds) Microbial root endophytes, Springer-Verlag, Berlin, pp 15–31

Hardoim PR, van Overbeek LS, van Elsas JD (2008) Properties of bacterial endophytes and their proposed role in plant growth. Trends Microbiol 16:463–471

Kloepper JW, Ryu CM (2006) Bacterial endophytes as elicitors of induced systemic resistance. In: Schulz BJE, Boyle CJC, Sieber TN (eds) Microbial root endophytes, Springer-Verlag, Berlin Heidelberg, pp 33–52

Malfanova N, Kamilova F, Validov S et al (2013) Is l-arabinose important for the endophytic lifestyle of *Pseudomonas* spp.? Arch Microbiol 195:9–17

Mercado-Blanco J, Lugtenberg BJJ (2014) Biotechnological applications of bacterial endophytes. Curr Biotechnol 3:60–75

Mercado-Blanco J, Prieto P (2012) Bacterial endophytes and root hairs. Plant Soil 361:301–306

Mitter B, Petric A, Shin MW et al (2013) Comparative genome analysis of *Burkholderia phytofirmans* PsJN reveals a wide spectrum of endophytic lifestyles based on interaction strategies with host plants. Front Plant Sci 4:120

Prieto P, Schilirò E, Maldonado-González M et al (2011) Root hairs play a key role in the endophytic colonization of olive roots by *Pseudomonas* spp. with biocontrol activity. Microb Ecol 62:435–445

Reinhold-Hurek B, Hurek T (1998) Interactions of gramineous plants with *Azoarcus* spp. and other diazotrophs: identification, localization, and perspectives to study their function. Crit Rev Plant Sci 17:29–54

Reinhold-Hurek B, Hurek T (2011) Living inside plants: bacterial endophytes. Curr Opin Plant Biol 14:1–9

Rosenblueth M, Martínez-Romero E (2006) Bacterial endophytes and their interactions with hosts. Mol Plant Microbe Interact 19:827–837

Ryan RP, Germaine K, Franks A et al (2008) Bacterial endophytes: recent developments and applications. FEMS Microbiol Lett 278:1–9

Schilirò E, Ferrara M, Nigro F et al (2012) Genetic responses induced in olive roots upon colonization by the biocontrol endophytic bacterium *Pseudomonas fluorescens* PICF7. PLoS One 7:e48646

Schulz B, Boyle C, Sieber T (2006) Microbial root endophytes. Springer, Berlin

Van Overbeek L, van Elsas JD (2008) Effects of plant genotype and growth stage on the structure of bacterial communities associated with potato (*Solanum tuberosum* L.). FEMS Microbiol Ecol 64:283–296

VV.AA. (2013) Special issue: endophytes in biotechnology and agriculture. Fungal Divers 60:1–188

Zamioudis C, Pieterse CMJ (2012) Modulation of host immunity by beneficial microbes. Mol Plant Microbe Interact 25:139–150

Chapter 6
Microbial Cell Surfaces and Secretion Systems

Jan Tommassen and Han A. B. Wösten

Abstract Microbial cell surfaces, surface-exposed organelles, and secreted proteins are important for the interaction with the environment, including adhesion to hosts, protection against host defense mechanisms, nutrient acquisition, and intermicrobial competition. Here, we describe the structures of the cell envelopes of bacteria, fungi, and oomycetes, and the mechanisms they have evolved for the transport of proteins across these envelopes to the cell surface and into the extracellular milieu.

6.1 Basic Structure of Bacterial Cell Envelopes

Based on the Gram-staining method, bacteria are classically divided into two groups, Gram-positives and Gram-negatives, which have different cell envelope architecture. Gram-positives are enveloped by a cytoplasmic membrane (CM) and a thick cell wall. In Gram-negatives, the cell wall is thinner, but an additional membrane is present, the outer membrane (OM), which is located peripheral to the cell wall. The space in between the membranes is called the periplasm and because of the presence of the OM, the CM is also called the inner membrane (Fig. 6.1).

The Cytoplasmic Membrane The CM is a phospholipid bilayer with inserted proteins. The membrane-spanning segments of the integral membrane proteins are α-helical and consist of ~20 amino acids, the vast majority of them containing hydrophobic side chains (Fig. 6.1).

The Cell Wall The rigid cell wall consists of peptidoglycan and provides strength and shape to the bacteria. For example, it protects the bacteria against osmotic

J. Tommassen (✉) · H. A. B. Wösten
Section Molecular Microbiology, Department of Biology,
Utrecht University, Padualaan 8, 3584 CH Utrecht, The Netherlands
Tel.: +31-30-2532999
e-mail: j.p.m.tommassen@uu.nl

H. A. B. Wösten
Tel.: +31-30-2533448
e-mail: h.a.b.wosten@uu.nl

© Springer International Publishing Switzerland 2015
B. Lugtenberg (ed.), *Principles of Plant-Microbe Interactions,*
DOI 10.1007/978-3-319-08575-3_6

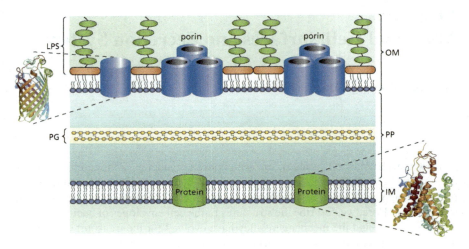

Fig. 6.1 Structure of the Gram-negative bacterial cell envelope. *OM* outer membrane containing LPS in its outer leaflet, *PP* periplasm containing a layer of peptidoglycan (*PG*), *IM* inner membrane. Examples of a typical β-barrel OMP and a typical α-helical inner-membrane protein are shown at the *left* and the *right*, respectively. Reproduced from Tommassen (2010) by permission of the publisher

pressure. Peptidoglycan consists of oligomers of a disaccharide composed of N-acetylglucosamine (GlcNAc) and N-acetylmuramic acid, which are covalently interconnected by small peptides, thus creating a network that enwraps the entire bacterial cell (Vollmer and Seligman 2010). In Gram-positive bacteria, the cell wall also contains large amounts of polymers known as (lipo)teichoic acids, which provide a net negative surface charge to the bacteria.

The Outer Membrane The OM protects Gram-negative bacteria from harmful compounds in the environment, including many antibiotics. This asymmetric bilayer contains phospholipids and lipopolysaccharides (LPS) in the inner and outer leaflets, respectively. LPS (Raetz and Whitfield 2002) consists of two or three moieties: lipid A, a core oligosaccharide, and a polysaccharide, the O-antigen, which is absent in several bacterial species. Lipid A is a signaling molecule for the innate immune system. It consists of a phosphorylated glucosamine disaccharide substituted with four 3-OH hydroxylated fatty acids, which can be esterified with secondary fatty acids. Repulsive forces between the negatively charged phosphate groups are compensated by divalent cations. The resulting network, together with the dense packing of the acyl chains, generates a barrier that is barely permeable to hydrophobic substances. Variations in the lipid A structure, affecting the acyl chains or the phosphate groups, are induced by environmental conditions and help bacteria to escape from the host's innate immune responses.

The core moiety of LPS is divided in an inner and an outer core. The inner core usually contains L-glycero-d-manno-heptose and 2-keto-3-deoxyoctonate with various substituents and is conserved among different strains of the same species. The outer core is more variable and the O-antigen, which is a polymer of repeating mono- or oligosaccharides, is highly variable. The O-antigen can form an effective barrier for bacteriophages, bacteriocins, and antibodies, targeting the underlying conserved parts of the OM.

Integral OM proteins (OMPs) structurally deviate from other membrane proteins. The membrane-spanning segments are not α-helices but β-strands, which form a β-barrel (Fig. 6.1). These β-strands are amphipathic with hydrophobic residues facing the lipids and hydrophilic ones directed toward the interior of the barrel. Some of these β-barrels form open channels through which small hydrophilic solutes, including nutrients, such as amino acids and small sugars, can pass by diffusion. Such pore-forming proteins are called porins. Thus, the OM functions as a molecular sieve, allowing the passage of small hydrophilic molecules but holding larger hydrophilic molecules and hydrophobic ones.

Besides integral OMPs, the OM also contains lipoproteins, which are bound to the membrane via an N-terminal lipid moiety. These lipoproteins can be very abundant, e.g. Braun's lipoprotein is, with $\sim 10^6$ copies per cell, the most abundant protein in *Escherichia coli*. This lipoprotein is also covalently bound to the peptidoglycan and functions to anchor the OM to the cell wall. Lipoproteins are also found in the CM.

Deviant Cell Envelope Architectures The general architecture of bacterial envelopes as described above was derived from studies on model organisms such as *E. coli* and *Bacillus subtilis*. However, deviations are now well known. For example, although not Gram-negative, Mycobacteria are covered with an OM, however, with a composition entirely different from the Gram-negative OM (Niederweis et al. 2010). It contains a large variety of lipids, including mycolic acids, which are covalently attached to the peptidoglycan via an arabinogalactan polymer. Also, bacteria that lack a cell wall have been described, such as mycoplasmas.

6.2   Additional Layers and Surface Appendices

Peripheral to the cell envelope, bacteria can be covered with additional layers, such as capsules and S-layers, and they can contain organelles, such as flagella, pili and fimbriae, which extend into the extracellular milieu.

Additional Layers Capsules (Bazaka et al. 2011) consist of polysaccharides with a highly variable composition. They have various functions, e.g. in protection against desiccation or against phagocytosis and other defense mechanisms of the host. In addition, they can have a role in the attachment of bacteria to biotic or abiotic surfaces. Bacteria can also produce extracellular polysaccharides (EPS), which do not form a capsule but are released in the environment (Bazaka et al. 2011). EPS can be important components of the extracellular matrix (ECM) of biofilms, which are surface-attached microbial communities embedded in a self-produced ECM (see also Chap. 7). S-layers are paracrystalline arrays of identical protein subunits with a large variety of functions (Fagan and Fairweather 2014). Amongst others, they may function as a molecular sieve, like the OM, or in determining the cell shape, and they may protect against bacteriophages, host defense mechanisms, or osmotic and mechanical stresses.

Flagella Flagella (Van Gerven et al. 2011) are long surface appendages that are used by bacteria to move towards favorable conditions or away from repellents in a process called chemotaxis. A flagellum consists of a long filament composed of multiple copies of a protein called flagellin. The filament is connected via a hook structure to a basal body that anchors the flagellum into the cell envelope and also forms the channel for the export of flagellin from the cytoplasm to the cell surface. The flagellum can rotate like a propeller to move the bacterium. Energy for this process is derived from the proton gradient across the CM. Flagella are also used for initial attachment of bacteria to a substratum during biofilm formation and, like LPS, flagellin is an important signaling molecule for defense mechanisms of animals and plants.

Pili and Fimbriae The names pili and fimbriae are often used interchangeably. These structures are built of subunits called pilins (Van Gerven et al. 2011). An abundant major pilin forms the filament, which usually exposes several minor pilins. Many pili/fimbriae have a role in adhesion, where one of the minor pilins functions as the adhesin that binds, for example, a eukaryotic target cell. The type IV pili form a special class of pili with multiple functions. These pili, which are based in the CM and cross the OM via a large oligomeric protein called secretin, are retractile. Extension and retraction of these pili, both at the expense of ATP, can be used by bacteria such as *Pseudomonas aeruginosa* to move over surfaces, a process called twitching motility. Some bacteria that are naturally transformable use type IV pili to take up DNA from the environment. Type IV pili can also function as nanowires that transfer electrons from the respiratory chain to extracellular electron acceptors. Also sex pili are retractile. They are produced by donor cells in the process of conjugation to establish contact with a recipient cell. Retraction of the pilus then results in the formation of a stable mating pair that allows for the transfer of DNA from donor to recipient. DNA transfer from *Agrobacterium tumefaciens* to plant cells requires similar machinery as in bacterial conjugation (see Chap. 37). A final class of pili is constituted by curli. Curli form amyloid fibers similar to the amyloids that cause neurodegenerative diseases in humans. Curli fibers are involved in adhesion, bacterial aggregation and biofilm formation.

6.3   Protein Export

Translocation Across the CM Proteins destined for export are synthesized as precursors with an N-terminal signal peptide of ∼25 amino-acid residues. Different signal peptides share a similar organization with three domains: an N-domain containing positively charged residues, an H-domain of ∼10–12 hydrophobic residues, and a C-domain containing the motif recognized by the enzyme that cleaves off the signal peptide after export. The signal peptide directs the precursor to the Sec (secretion) machinery, which mediates its transport across the CM (Kudva et al. 2013). The central component of this machinery is a complex of three integral CM proteins, the SecYEG translocon, which forms the protein-conducting channel. It is widely conserved in nature and corresponds to the Sec61 complex in the endoplasmic reticulum of eukaryotes. Energy for export is provided by the motor protein SecA, which hydrolyzes ATP, and the proton-motive force. The Sec machinery also inserts proteins into the CM. CM proteins are generally not produced with a cleavable signal peptide, but the most N-terminal membrane-spanning α-helix is recognized by the signal-recognition particle (SRP) and targeted via an SRP receptor (FtsY) to the Sec translocon. When such a long hydrophobic α-helix enters the translocon, the substrate is not released in the periplasm, but the translocon opens laterally to insert it into the CM. This is the reason why OMPs have a deviant structure: if OMPs would also consist of hydrophobic α-helices, they would never reach their destination but be inserted by the translocon into the CM. Besides the Sec translocon, YidC protein constitutes an alternate insertase for CM proteins.

The Sec translocon contains a narrow channel and exports delineated proteins. Some proteins need to be exported in a folded conformation, for example because they bind a co-factor in the cytoplasm. These proteins are exported via the Tat system (Kudva et al. 2013). Tat stands for twin-arginine translocation and refers to a characteristic twin-arginine motif in the signal peptides of the substrate proteins. The Tat system consists of three CM proteins, TatA, TatB, and TatC. Energy for transport is provided by the proton gradient.

OMP Assembly The Sec machinery releases OMPs into the periplasm, where they are bound by the chaperones Skp and/or SurA, which prevent their aggregation. Subsequently, they are folded and inserted into the OM by the Bam (β-barrel assembly machinery) complex. The central component of this complex, known as BamA or Omp85, is highly conserved and found in the OM of all Gram-negatives and even in mitochondria and chloroplasts. These eukaryotic cell organelles also contain β-barrel proteins in their OM, probably reflecting their endosymbiont origin. The Bam complex contains a variable number of accessory components, usually lipoproteins, which are less conserved (Tommassen 2010).

6.4 Protein Secretion

In Gram-positive bacteria, exported proteins are bound to the peptidoglycan or released in the environment. In Gram-negatives, secreted proteins have to pass another hurdle, the OM. These bacteria have developed six widely disseminated protein-secretion mechanisms, designated type 1–6 secretion systems (T1-6SS) (Chang et al. 2014). Many of these systems can be present in a single cell, often in multiple copies, each one dedicated to the secretion of a specific protein or set of proteins. For example, *P. aeruginosa* possesses five of the six secretion systems (Fig. 6.2).

Two-Step Mechanisms In two-step mechanisms, a periplasmic intermediate is translocated across the OM. The T2SS is dedicated to the secretion of folded proteins. The substrates, usually hydrolytic enzymes or toxins, are either folded in the cytoplasm and exported via the Tat system, or exported via the Sec system and folded in

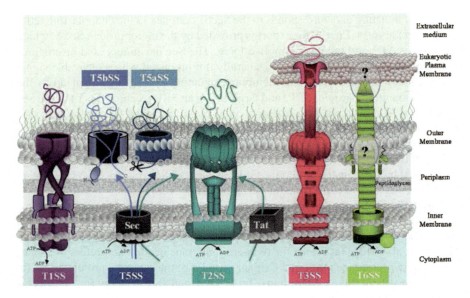

Fig. 6.2 Protein secretions systems in *P. aeruginosa*. Details are described in the text. Reproduced from Bleves et al. (2010) by permission of the publisher

the periplasm (Fig. 6.2). The T2SS consists of 12–16 proteins and resembles the machinery that builds type IV pili. It includes several pilin-like proteins and a secretin in the OM, which forms the protein-conducting channel. The model is that substrates bind the secretin, and a pilus-like structure that grows from the CM provides the mechanical force to push them through the secretin into the milieu.

The T5SS consists of five subtypes (a–e), including four variants of an autotransporter mechanism. Autotransporters consist of a signal peptide for export via the Sec system, a passenger, which is the secreted part, and a translocator domain, which forms a β-barrel that inserts into the OM via the Bam complex. During insertion, the connected passenger is translocated across the OM. Protein folding starts when the first part of the passenger appears at the external side and presumably provides the energy to thread the rest of the protein through the translocation channel. The passenger may stay associated with the translocator and function, for example, as an adhesin, or it may be released into the milieu often via autocatalytic proteolysis. In the fifth T5SS, known as two-partner secretion system or T5bSS (Fig. 6.2), the translocator is not connected to the secreted protein but is a separate protein. The secreted proteins are very large (up to > 6000 residues) β-helical proteins. Many of them contain a small toxic domain at the C terminus that inhibits the growth of related bacteria competing for the same niche (Ruhe et al. 2013).

One-Step Mechanisms The T1SS consists of three proteins, a CM-based ATPase, an OM-based tunnel protein, and a membrane-fusion protein that connects the other two. Together, they form a channel that translocates substrates directly from the cytoplasm into the external milieu. The substrates are very large proteins, belonging to the RTX (repeat-in-toxin) family, which refers to a glycine/aspartate-rich Ca^{2+}-binding nonapeptide repeat near the C terminus. Many substrates are toxins, but the family also includes adhesins, enzymes, and S-layer proteins.

The T3SS and T4SS translocate substrates directly from the cytoplasm into a eukaryotic target cell (Fig. 6.2), where they interfere with signal transduction and metabolism. The T4SS is very similar to the conjugation apparatus, which translocates DNA with associated proteins into other bacterial or eukaryotic cells. The T3SS contains a basal body resembling the structure that anchors the flagellum in the cell envelope and functions as the translocon for flagellin. In T3SS, the basal body is connected to an extracellular needle or pilus in animal or plant pathogens, respectively. These structures reach through any additional surface layers of the bacteria and the eukaryotic cell wall, if present. The T3SS first inserts a translocon into the eukaryotic membrane that serves to deliver effector proteins into these cells.

The T6SS delivers toxic proteins into competing bacteria or into eukaryotic cells. Several components of the T6SS resemble phage tail proteins, which serve to inject phage DNA into the bacterial cytoplasm. Thus, the T6SS appears to function as an inverted phage driving proteins out of the bacterial cell and delivering them straight into target cells.

6.5   Basic Structure of Cell Envelopes of Fungi and Oomycetes

The Fungal Cell Envelope The fungal cell envelope is the target of microbial control agents. Ergosterol is a main anti-fungal plasma-membrane target, while chitinases, glucanases and proteases attack the cell wall. These enzymes often work synergistically, thereby weakening or even killing the pathogen. The cell envelope has been best studied in *Saccharomyces cerevisiae*. This yeast functions as a model for the plasma membrane and cell walls of plant-pathogenic and plant-beneficial fungi such as *Fusarium* and *Trichoderma*, respectively.

The Fungal Plasma Membrane The plasma membrane of *S. cerevisiae* consists of the phospholipids phosphatidylcholine, phosphatidylethanolamine, phosphatidylinositol, phosphatidylserine as well as inositol sphingolipid and the sterol ergosterol (van Meer et al. 2008). The molar ratio between ergosterol and phospholipids is 0.5. Mechanical stress resistance is acquired by the relative dense packing of sphingolipids and sterols. Plasma membrane proteins are encoded by only ~4 % of the *S. cerevisiae* genome (i.e. ~250 proteins). Szopinska et al. (2011) identified > 100 of them and 68 % were integral membrane proteins. About one third were classified as transporters including, for example, seven glucose transporters. The plasma membrane also contains signaling proteins, e.g. of the cell-wall integrity pathway, and proteins involved in cell-wall synthesis, including two chitin synthases, one 1,3-β-D-glucan synthase, and three glucan elongases.

The Fungal Cell Wall The cell wall stabilizes internal osmotic conditions and turgor pressure and provides physical protection and shape to the cells. It represents a
considerable metabolic investment; in S. *cerevisiae*, it accounts for ~10–25 % of the
total cell mass (Klis et al. 2006). Cell walls of fungi generally consist of one or more
fibrillar components, one or more matrix components, and may have an outer protein layer. Often, the fibrillar components include both chitin (a 1,4-β-linked GlcNAc
polymer) and β-glucans, but occasionally mainly chitin (e.g. *Encephalitozoon cuni-
culi*) or 1,3-β-glucan (e.g. *Schizosaccharomyces pombe*) is present (Xie and Lipke
2010). Mannoproteins often form the matrix of cell walls, e.g. in *S. cerevisiae*, but
also galactomannoproteins and α-glucan are used, e.g. in *S. pombe*. The composition, molecular organization and thickness of the cell wall can vary depending on
environmental conditions. This adaption may be functional in an environment where
fungi are exposed to antibiotics, lytic enzymes secreted by other microorganisms or
to immune systems.

The cell walls of filamentous ascomycetes and even basidiomycetes have been
proposed to be very similar to the well-studied cell wall of *S. cerevisiae*, which
consists of an inner and outer layer (De Groot et al. 2005). The inner layer consists
of 1,3-β-glucan, 1,6-β-glucan and chitin (Fig. 6.3). The moderately branched 1,3-β-
glucan is the main polysaccharide. Its side-chains allow only local mutual association
by hydrogen bonding. Consequently, a three-dimensional network is formed that is
highly elastic and extended under normal osmotic conditions (Klis et al. 2006). Cells
of *S. cerevisiae* shrink when exposed to hypertonic conditions. The cell wall also
reduces in size then resulting in a surface loss of up to 50 % and a cell-wall porosity
of less than 1000 Da, while even medium-sized proteins can pass under normal
osmotic conditions. Chitin and 1,6-β-glucan are linked to the inner and outer parts
of the 1,3-β-glucan network, respectively. Growing buds have not yet formed chitin;
hence this polymer is not essential for cell-wall assembly and function. The 'alkali-
sensitive linkage' cell-wall proteins (ASL-CWPs) are linked to the 1,3-β-glucan in
the inner layer of the cell wall. Among the ASL-CWPs are the protein with internal
repeats (PIR)-CWPs, which interconnect two or even more 1,3-β-glucan chains,
thereby strengthening the cell wall (De Groot et al. 2005). Increased presence of
PIR-CWPs has been proposed to produce a less elastic cell wall as occurs during the
G1 phase of the cell cycle and during cell-wall stress (Klis et al. 2006).

The outer cell-wall layer is formed by mannoproteins, which are heavily glyco-
sylated; each glycochain may contain hundreds of mannose residues. At least 20
different proteins make up this layer. Their composition varies depending on the culture conditions (Klis et al. 2006). The glycosylphosphatidylinositol (GPI)-modified
CWPs form the largest group of CWPs within this outer layer. They are covalently
linked to 1,6-β-glucan through a truncated form of their original GPI-anchor (De
Groot et al. 2005). The genome of *S. cerevisiae* contains 66 GPI-CWP genes (De
Groot et al. 2003).

The cell wall of *S. cerevisiae* also contains proteins that are not covalently linked
to β-glucans. Collectively, the covalently linked and the non-covalently linked CWPs
have a wide range of functions. Both classes are involved in cell-wall synthesis and
remodeling. The GPI-CWPs also have a structural role by making the cell wall

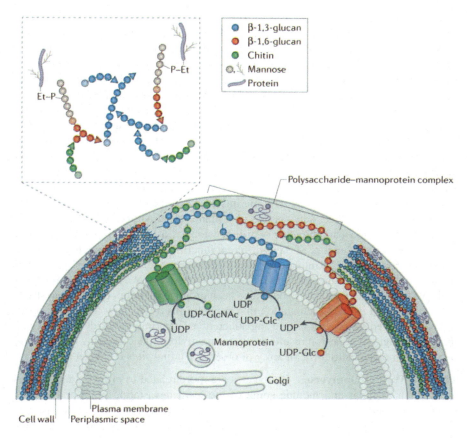

Fig. 6.3 Representation of the cell wall of *S. cerevisae*. Mannoproteins are delivered to the cell wall via secretory vesicles that fuse with the plasma membrane. The polysaccharides β-1,3-glucan and chitin are synthesized by synthases in the plasma membrane. The mechanism by which β-1,6-glucan is formed is not known. The cell wall polysaccharides and mannoproteins form a complex in the cell wall. The ASL-CWPs and proteins that are not covalently linked are not shown. *Et* ethanolamine, *Glc* Glucose, *P* phosphate. Reproduced from Cabib and Arroyo (2013) by permission of the publisher

less porous (Klis et al. 2006). Reduced porosity has been proposed to retain high-molecular-weight soluble proteins in the cell-wall matrix and may also protect the cell against lytic enzymes of competing microbes or plants or animals. In the latter case, it thus contributes to virulence. CWPs can also be involved in virulence by acquiring iron in the host or by inactivation of oxygen radicals released by the immune system (De Groot et al. 2005). Adhesins also contribute to the infection process. Adhesion of *S. cerevisiae* to foreign surfaces depends on the GPI-CWP Flo11 (Bojsen et al. 2012). This protein, which can also mediate mutual binding of yeast cells, is characterized by an A, B, and C domain. The A domain mediates cell-surface or cell-cell adherence, while the C domain contains the GPI anchor. Flo1, Flo5, Flo9 and Flo10 show high mutual homology and similarity to Flo11. These proteins also have A, B, and C domains. The A domain is a β-barrel involved in mutual binding of yeast cells. Binding to a cell surface or to each other is often accompanied by the formation of an ECM, which creates a micro-environment that may prevent dehydration or access of antibiotics to the cells. In the case of *S. cerevisiae,* glucose and mannose polysaccharides and proteins constitute the ECM (Beauvais et al. 2009).

As in bacteria, some surface-exposed fungal proteins can form amyloid-like structures (Gebbink et al. 2005). Amyloids are filamentous protein structures of ∼10 nm wide and 0.1–10 μm long that share a structural motif, the cross-beta structure, and have been associated with neurodegenerative diseases. One of the best studied amyloid-forming microbial proteins is the hydrophobin SC3 of *Schizophyllum commune.* The water-soluble form of this protein affects the polysaccharide cell-wall composition. When confronted with the interface between the cell wall and the air or a hydrophobic surface, such as that of a plant, the structure of SC3 changes. It self-assembles into an amphipathic two-dimensional mosaic film of parallel amyloid fibrils. In this conformation, SC3 has different functions. It allows fungi to escape the aqueous substrate to grow into the air, it confers hydrophobicity to aerial hyphae and mediates attachment of the fungus to a hydrophobic support (Wösten 2001). Adhesins in the yeast cell wall (see above) have also been proposed to adopt the amyloid structure. This would explain the paradox that the adhesins often show weak binding to ligands, yet mediate remarkably strong adherence. Experimental evidence indicates that the strength of adhesion results partly from amyloid-like clustering of hundreds of adhesin molecules to form an array of ordered binding sites (Lipke et al. 2012).

The Cell Wall of Oomycetes Oomycetes are more related to brown algae and diatoms than to fungi and include some of the most devastating plant and animal pathogens. Their cell envelopes represent an excellent target to control disease but little is known about their composition. The cell wall is classically described to consist of 1,3-β- and 1,6-β-glucans and 4–20 % of cellulose (Aronson et al. 1967). A recent detailed cell wall analysis of 10 species from two oomycete orders revealed high heterogeneity (Mélida et al. 2013). Three different cell wall types were distinguished primarily based on GlcNAc content. Types I, II and III contain 0 %, up to 5 %, and > 5 % of this sugar, respectively. Each type is also characterized by additional compositional features. For example, the type I cell walls of *Phytophtora* spp. contain glucuronic acid and mannose and have a cellulose content of 32–35 %. *Saprolegnia* has a type II cell wall. Its GlcNAc residues are contained in chitin. A unique feature of this type is the 1,3,4-linked glucosyl residues, which are indicative of cross-links between cellulose and 1,3-β-glucans. *Aphanomyces euteiches* has a type III cell wall and contains nearly 10 % GlcNAc including 1,6-linked polymers.

6.6   Protein Export in Fungi

Most fungal proteins are transported to the cell wall or beyond via the ER (Conesa et al. 2001). *S. cerevisiae* translocates proteins over the ER membrane via SRP-dependent and -independent pathways, in which translocation occurs co-

or post-translationally, respectively. Proteins with a less hydrophobic signal peptide are targeted through the SRP-independent route, which involves the ER chaperone BiP, whereas proteins with a more hydrophobic signal can take both routes. Both pathways make use of the same translocon, the Sec61 complex. Also in other fungi, both pathways are probably operational. In the ER lumen, proteins are modified and folded and then transported to the Golgi, where they are further processed, e.g. by proteolytic modification and/or by modifications of their glycan-chains. Subsequently, the proteins are packed in vesicles which fuse with the plasma membrane to release the proteins into the cell wall. Vesicle fusion does not occur uniformly along the plasma membrane but mainly at the growing bud of the daughter cell and later re-localizes at the mother-bud neck just prior to cytokinesis. This implies that proteins are mainly incorporated in the cell wall or released into the medium when the cell is formed and not once cell division has occurred (Sietsma and Wessels 2006). In filamentous fungi, which form hyphae that extend at their tips, vesicles also fuse at the growth site. Once extruded at the hyphal tip, proteins migrate together with the newly synthesized cell-wall polymers to the outer part of the wall. This migration is driven by the turgor pressure in the hyphae and the apposition of newly synthesized polymers at the inner part of the wall. At the outside, the proteins diffuse into the medium (Sietsma and de Vries 2006). The newly synthesized cell-wall polymers are initially not cross-linked, resulting in a deformable cell wall at the tip. Once deposited, the polymers start to crosslink, and the cell wall becomes more and more rigid and less porous in subapical direction. This implies that proteins that are secreted more subapically will be captured in the cell wall. Yeast buds grow over their whole surface and do not have an "ever" extending tip that creates a continuous flow of proteins to the medium. Therefore, protein release in this case depends more on pores in the cell wall.

References

Aronson JM, Barbara A, Cooper BA et al (1967) Glucans of oomycete cell walls. Science 155:332–335

Bazaka K, Crawford RJ, Nazarenko EL et al (2011) Bacterial extracellular polysaccharides. Adv Exp Med Biol 715:213–226

Beauvais A, Loussert C, Prevost MC et al (2009) Characterization of a biofilm-like extracellular matrix in FLO1-expressing *Saccharomyces cerevisiae* cells. FEMS Yeast Res 9:411–419

Bleves S, Viarre V, Salacha R et al (2010) Protein secretion systems in *Pseudomonas aeruginosa*: A wealth of pathogenic weapons. Int J Med Microbiol 300:534–543

Bojsen RK, Andersen KS, Regenberg B (2012) *Saccharomyces cerevisiae*–a model to uncover molecular mechanisms for yeast biofilm biology. FEMS Immunol Med Microbiol 65:169–182

Cabib E, Arroyo J (2013) How carbohydrates sculpt cells: chemical control of morphogenesis in the yeast cell wall. Nat Rev Microbiol 11:648–655

Chang JH, Desveaux D, Creason AL (2014) The ABCs and 123s of bacterial secretion systems in plant pathogenesis. Annu Rev Phytopathol. doi:10.1146/annurev-phyto-011014-015624

Conesa A, Punt PJ, van Luijk N et al (2001) The secretion pathway in filamentous fungi: a biotechnological view. Fungal Genet Biol 33:155–171

De Groot PW, Hellingwerf KJ, Klis FM (2003) Genome-wide identification of fungal GPI proteins. Yeast 20:781–796

De Groot PWJ, Ram A, Klis F (2005) Features and functions of covalently linked proteins in fungal cell walls. Fungal Genet Biol 42:657–675

Fagan RP, Fairweather NF (2014) Biogenesis and functions of bacterial S-layers. Nat Rev Microbiol 12:211–222

Gebbink MF, Claessen D, Bouma B et al (2005) Amyloids–a functional coat for microorganisms. Nat Rev Microbiol 3:333–341

Klis FM, Boorsma A, De Groot PW (2006) Cell wall construction in *Saccharomyces cerevisiae*. Yeast 23:185–202

Kudva R, Denks K, Kuhn P et al (2013) Protein translocation across the inner membrane of Gram-negative bacteria: the Sec and Tat dependent protein transport pathways. Res Microbiol 164:505–534

Lipke PN, Garcia MC, Alsteens D et al (2012) Strengthening relationships: amyloids create adhesion nanodomains in yeasts. Trends Microbiol 20:59–65

Mélida H, Sandoval-Sierra JV, Diéguez-Uribeondo J et al (2013) Analyses of extracellular carbohydrates in oomycetes unveil the existence of three different cell wall types. Eukaryot Cell 12:194–203

Niederweis M, Danilchanka O, Huff J et al (2010) Mycobacterial outer membranes: in search of proteins. Trends Microbiol 18:109–116

Raetz CRH, Whitfield C (2002) Lipopolysaccharide endotoxins. Annu Rev Biochem 71:635–700

Ruhe ZC, Low DA, Hayes CS (2013) Bacterial contact-dependent growth inhibition. Trends Microbiol 21:230–237

Sietsma JH, Wessels JGH (2006) Apical wall biogenesis. In: Kues U, Fisher R (eds) The mycota, part 1: growth, differentiation and sexuality, 2nd edn. Springer, Berlin, pp 53–73

Szopinska A, Degand H, Hochstenbach JF et al (2011) Rapid response of the yeast plasma membrane proteome to salt stress. Mol Cell Proteomics 10:M111.009589

Tommassen J (2010) Assembly of outer-membrane proteins in bacteria and mitochondria. Microbiology 156:2587–2596

Van Gerven N, Waksman G, Remaut H (2011) Pili and flagella: biology, structure, and biotechnological applications. Prog Mol Biol Transl Sci 103:21–72

van Meer G, Voelker DR, Feigenson GW (2008) Membrane lipids: where they are and how they behave. Nat Rev Mol Cell Biol 9:112–124

Vollmer W, Seligman SJ (2010) Architecture of peptidoglycan: more data and more models. Trends Microbiol 18:59–66

Wösten HAB (2001) Hydrophobins: multipurpose proteins. Annu Rev Microbiol 55:625–646

Xie X, Lipke PN (2010) On the evolution of fungal and yeast cell walls. Yeast 27:479–488

Chapter 7
Microbial Biofilms and Quorum Sensing

Aurelien Carlier, Gabriella Pessi and Leo Eberl

Abstract Many bacteria that form biofilms on various plant surfaces use small signal molecules for intra- and interspecies communication. In this chapter we will review the current knowledge on bacterial cell-to-cell signaling, referred to as quorum sensing (QS), in biofilms on the surfaces of plant roots and leaves. Particular focus will be laid on the role of QS in the formation of nitrogen-fixing root nodules and the expression of virulence factors in biofilms formed by plant pathogens. We will also discuss how plants can interfere with bacterial QS and thus manipulate microbial activity and persistence.

7.1 Quorum Sensing in Plant-Associated Biofilms

The term quorum sensing (QS) describes the phenomenon that bacteria are capable of perceiving and responding to self-generated signal molecules to coordinate their behavior in response to their population size (Fuqua et al. 1994). The general consensus is that bacteria trigger QS only when their cell density has reached a certain threshold (the "quorum"), upon which the expression of target genes is either activated or repressed. Among the various QS signal molecules identified to date, N-acyl-homoserine lactones (AHL) have been investigated to the greatest extent and have been shown to control the expression of various traits, including virulence, symbiosis, motility, biofilm formation, the production of antibiotics and toxins, and conjugation. Various AHL molecules have been described that all have a homoserine

L. Eberl (✉) · G. Pessi · A. Carlier
Institute of Plant Biology, University of Zurich, Zollikerstrasse 107,
CH-8008 Zurich, Switzerland
Tel.: + 41 44 634 8220
e-mail: leberl@botinst.uzh.ch

A. Carlier
Tel.: + 41 44 63 48227
e-mail: aurelien.carlier@gmail.com

G. Pessi
Tel.: + 41 44 63 52904
e-mail: gabriella.pessi@botinst.uzh.ch

© Springer International Publishing Switzerland 2015
B. Lugtenberg (ed.), *Principles of Plant-Microbe Interactions*,
DOI 10.1007/978-3-319-08575-3_7

45

lactone (HSL) moiety but differ in the length and structure of the acyl side chain. Cha et al. (1998) showed that the majority of plant-associated bacteria produce AHL signal molecules. All isolates of the genera *Agrobacterium*, *Erwinia*, *Pantoea*, and *Rhizobium*, and about half of the erwinias and pseudomonads tested, synthesize detectable levels of AHLs while only few AHL producers could be identified among *Xanthomonas* isolates. It is worthwhile to note that members of the latter genus are known to use another type of signal molecule, namely DSF (Diffusible Signal Factor; *cis*-11-methyl-2-dodecenoic acid). A structurally related molecule, BDSF (*Burkholderia* Diffusible Signal Factor; *cis*-2-dodecenoic acid), is also produced by many plant-associated *Burkholderia* species, which additionally produce AHL signal molecules (Suppiger et al. 2013). Elasri et al. (2001) screened 137 soil-borne and plant-associated *Pseudomonas* sp. strains using biosensors and identified 54 that were positive for AHL production. The authors of this study concluded that plant-associated and plant-pathogenic bacteria produce AHLs more frequently than soil borne strains and hypothesized that the more intimate the relationship of the bacteria with the host plant, the higher the probability that it produces AHLs.

QS is a particularly valuable regulatory mechanism when bacteria are living in close contact to each other. This is the case in biofilms, where the cells are embedded in a self-produced extracellular matrix, which consists of polysaccharides, proteins and DNA and acts as a diffusion barrier for signal molecules, thus creating an ideal environment for the induction of QS. Moreover, a direct role for AHL-mediated QS in biofilm formation has been demonstrated for many bacteria that are usually associated with plants, including members of the genera *Burkholderia, Pseudomonas* and *Serratia* (Aguilar et al. 2009). Employing a quorum quenching approach (i.e. the enzymatic degradation of AHL signal molecules), it was shown that AHL signaling regulates biofilm formation in the large majority of *Burkholderia* species. In several *Pseudomonas putida* strains AHL-dependent production of powerful cyclic lipopeptide biosurfactants (putisolvins) strongly affects biofilm formation. Putisolvins were found to inhibit biofilm formation and were shown to even break down existing biofilms. As a consequence it was observed that QS mutants produce denser biofilms than the wild-type strains. For *Serratia sp.* it has been shown that QS plays an important role in biofilm structural development, ultimately resulting in a highly porous biofilm composed of cell chains, filaments, and cell clusters.

Plants are known to support the growth of bacterial biofilms on and within their tissues, including aerial portions of the plant, the vascular network, and root tissues below ground (Fig. 7.1). These plant-associated biofilms may establish commensal, mutualistic and pathogenic interactions with plants, or simply grow saprophytically on the nutrients released.

QS in the Rhizosphere   The rhizosphere is the narrow soil compartment that is directly influenced by root secretions (see Chap. 3). Hence this niche is relatively rich in nutrients compared with the bulk soil and allows bacteria to form microcolonies on the root surface, consisting of multiple layers of cells that are embedded in a self-produced matrix. Steidle et al. (2001) screened over 300 bacterial strains isolated from the rhizosphere of tomato on standard laboratory media, and found that approximately 12 % of the isolates produced detectable AHLs. In this study, GFP-based AHL

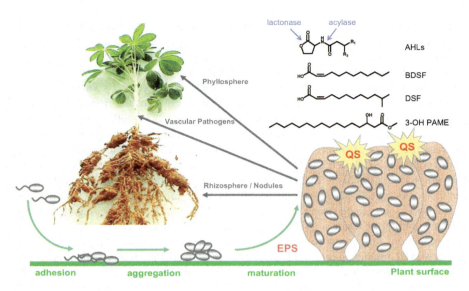

Fig. 7.1 Microbial biofilms on and within plants. In many plant-associated bacteria biofilm formation is controlled by QS. These bacterial communication systems utilize chemically diverse signal molecules. The site of action of two classes of enzymes that inactivate AHL signal molecules is indicated by blue *arrows*

monitor strains were used to visualize interspecies communication between bacteria colonizing the rhizosphere of tomato plants. These experiments provided evidence that AHL signal molecules serve as a universal language for communication between root-associated Gram-negative bacteria. Several subsequent studies provided further evidence that AHL-mediated QS is common among root-associated bacteria. *In situ* quantitative data on the spatial scale of AHL-mediated cell-to-cell communication of *P. putida* on tomato and wheat root surfaces were generated employing computer-assisted microscopy in combination with geostatistical modeling (Gantner et al. 2006). This analysis suggested that the effective "calling distance" on root surfaces was most frequent at 4–5 µm, extended to 37 µm in the root tip/elongation zone and further out to 78 µm in the root hair zone. This quantitative analysis also lent support to the hypothesis that the bacteria are able to use AHL gradients for sensing their positions relative to each other in the rhizosphere, an ability that may be particularly important for the formation of biofilms. An interesting link between AHL-dependent signaling and nitrogen cycling was noted for the bacterial consortium colonizing the *Avena* (wild oat) rhizosphere. In many rhizosphere isolates the expression of chitinolytic and proteolytic activity was found to be AHL-regulated, suggesting that QS could be a control point in the complex process of rhizosphere nitrogen mineralization (DeAngelis et al. 2008) .

QS in Aggregates and Biofilms of the Phyllosphere   The leaf surface is a stressful environment for colonizing bacteria, due to light, heat and desiccation (see Chap. 4).

Multicellular aggregates preferentially form around stomates and trichomes, or along leaf veins, responding to moisture and nutrient concentration. Such aggregates generally grow from one or a few cells that proliferate at the site of deposition to form a microcolony. Among the best studied epiphytes is *Pseudomonas syringae*, which causes brown spot disease on bean. Although single *P. syringae* cells can be observed on the leaf surface, the bulk of the adherent biomass that results from colonization is contained in large aggregates comprised of thousands of cells, which not only survive desiccation stress better than solitary cells but are also ideally suited for QS. In fact, the fitness of adherent *P. syringae* populations was shown to be dependent on a functional AHL-dependent QS system, as it controls various functions important for epiphytic persistence, including extracellular polysaccharide (EPS) production, oxidative stress tolerance, and motility (Quinones et al. 2005). QS of *P. syringae* on leaves was found to be strongly influenced both by the size of the aggregates and by the availability of water, as AHL transport on leaves appears to be a relatively local phenomenon, in contrast to that in the rhizosphere (see above). Moreover, a high level of AHL-mediated cross-talk as well as of QS quenching has been demonstrated for epiphytic communities and it has been suggested that these processes can be exploited for disease control (Dulla et al. 2009).

7.2   QS and Root Nodule Symbiosis

Rhizobia enter a symbiotic relationship with leguminous plants, resulting in differentiated bacteria that live enclosed in intracellular organelle-like structure within nodules on the root, called symbiosomes (see Chap. 23). Rhizobia are taxonomically diverse and polyphyletic; they include members of both the α-proteobacteria group (*Azorhizobium, Bradyrhizobium, Rhizobium, Mesorhizobium*, and *Sinorhizobium*) as well as *Burkholderia* and *Cupriavidus* strains, which belong to the β-proteobacteria. The Rhizobia-legume interaction begins with an initial attraction step involving various signal molecules, attachment to root hairs via formation of micro-colonies and biofilms, followed by root colonization and infection, and subsequent growth within legume roots, both in infection threads and in nodules where the bacteria fix atmospheric nitrogen (see Downie 2010).

The interaction between rhizobia and their leguminous host plants requires an exchange of molecular signals. In addition to well characterized flavonoids and Nod factor-mediated signalling, rhizobia also use EPS and QS molecules. The cell density of *Rhizobium* species is important for biofilm formation around the plant roots and for an efficient symbiotic life inside the root nodules. Most of the QS systems identified so far are based on AHLs. Compared to other genera, rhizobia produce the greatest diversity of AHLs, with strains producing from as few as one to as many as seven detectable AHL signals which range from short (C4) to very long acyl side chains. More recently, the soybean symbiont *B. japonicum* has been shown to produce low concentrations of isovaleryl-HSL that belongs to a new class of HSL signals, namely the aryl-HSLs. All the identified rhizobial QS regulation systems so far are based on acyl- or aryl-HSL synthesis and perception. Within a single rhizobial species, different isolates may have different AHL-dependent QS systems. The production of diverse, partially overlapping sets of AHLs by different strains suggests that QS regulation in each strain is likely to be complex and distinct.

QS affects many aspects of rhizobial physiology including EPS production, biofilm formation, legume nodulation and nitrogen fixation. Moreover, QS has been shown to induce the transfer of plasmids in several rhizobia and integrated symbiosis islands in *Mesorhizobium loti* as well as motility, growth inhibition and adaptation to abiotic stresses (Sanchez-Contreras et al. 2007). A conserved QS system is important for nitrogen fixation as well as growth and development of nitrogen-fixing symbiosomes in the pea symbiont *Rhizobium etli*. QS mutant bacteroids are always individually packed in the symbiosome membrane and are devoid of a large symbiosome space whereas wild-type symbiosomes usually contain multiple bacteroids. The *R. etli* QS system was shown to be expressed in infection threads and in differentiated bacteroids and AHLs could be extracted from *R. etli* (pea) bacteroids. In *Rhizobium leguminosarum* bv. *viciae* four QS systems have been identified which are hierarchically arranged. Interestingly, one of the AHL molecules produced by *R. leguminosarum*, *N*-(3-hydroxy-7-*cis*-tetradecenoyl) homoserine lactone, has been shown to be a small bacteriocin that inhibits growth of susceptible strains. This AHL-dependent inhibition of growth of *R. leguminosarum* is unusual and is mediated by two QS regulators that also regulate plasmid transfer. In *S. meliloti* and in several *Mesorhizobium* isolates, EPS production has been shown to be under the control of QS and is important for infection, attachment and biofilm formation on root hairs. EPS mutant strains form empty nodules that lack bacteroids, suggesting that EPS is essential for the development of nitrogen-fixing nodules.

Rhizobia belonging to the β-proteobacteria group, such as *Burkholderia phymatum*, use an AHL QS system that is highly conserved among plant-beneficial-environmental *Burkholderia* species. The *B. phymatum* QS system is responsible for the production and perception of several different AHL molecules. Although this system is not essential for legume nodulation, a QS mutant is impaired in EPS production (Coutinho et al. 2013).

7.3 QS in Plant Pathogens

The fact that QS can affect plant colonization at various stages is also reflected in the diversity of strategies adopted by phytopathogens for biofilm formation and dispersal as well as for infection.

Pathogens of the Plant Surface Biofilms are a high-cell density environment and thus not only a natural site for QS regulation but also for conjugative plasmid transfer. *Agrobacterium tumefaciens*, the agent that causes crown gall disease, regulates conjugation of the Ti virulence plasmid via an AHL-based QS system. The biofilm that *A. tumefaciens* forms on plant surfaces provides an ideal site for QS-dependent plasmid transfer (Danhorn and Fuqua 2007) .

Vascular Pathogens The role of biofilms in the later stages of infection is best studied in vascular pathogens, probably because of the well-established link between EPS production and virulence. *Pantoea stewartii* subsp. *stewartii* forms biofilms that clog the xylem vessels of susceptible varieties of sweet corn, causing water stress and the typical symptoms associated with Stewart's wilt. Dense bacterial mats are formed upon initial attachment to the protoxylem annular rings, structures that may provide nourishment to *P. stewartii*. Maturation of these biofilms is an essential process, as mutants blocked in EPS synthesis produce immature, flat biofilms and fail to colonize tissues beyond the infection site. An AHL-dependent QS system regulates the biosynthesis of EPS and biofilm maturation. At low cell density *P. stewartii* does not produce the EPS virulence factor, while at high cell density, the QS system activates EPS biosynthesis (Von Bodman et al. 2003). The importance of proper timing of the different stages of biofilm formation by QS is illustrated by the fact that mutants that express EPS constitutively form loose biofilms but display reduced virulence (Koutsoudis et al. 2006). *Ralstonia solanacearum* is the causative agent of wilts affecting over 200 plant species and can persist in soil or in water systems. When it encounters a susceptible host, *R. solanacearum* breaches the root cortex and spreads through the host's vasculature. Similar to *P. stewartii*, *R. solanacearum* biofilms lethally hinder the flow of nutrients through the xylem. Most of the traits required for infection, including EPS production and biofilm formation, are regulated by the Phc (phenotype conversion) regulatory system in a cell-density dependent manner. The Phc system is an atypical QS system that relies on the synthesis of the volatile 3-hydroxy-palmitic acid methyl ester (3-OH PAME) by a methyltransferase and its perception by a two-component sensor histidine kinase-response regulator pair. This QS system provides a regulatory switch from a free-living to a pathogenic lifestyle. This is illustrated by the fact that QS mutants produce low levels of EPS and plant cell wall degrading enzymes, while they display enhanced traits associated with the free-living stage in soil, e.g. motility and siderophore biosynthesis (Von Bodman et al. 2003) .

Xanthomonas campestris is a xylem-dwelling pathogen that is causing economically significant diseases. Pathovars of *X. campestris* cause black rot disease of virtually all cultivated brassicas, bacterial spot of pepper and tomato and angular leaf spot of cotton. The *rpf* genes of *X. campestris* are responsible for the synthesis and perception of the signal molecule DSF. DSF-dependent QS in *X. campestris* regulates the synthesis of extracellular hydrolytic enzymes and the polysaccharide xanthan, a structural component of biofilms. Mutations inactivating any of the *rpf* genes in *X. campestris* pv. *campestris* lead to reduced synthesis of xanthan EPS, extracellular enzymes and decreased virulence. Moreover, the DSF system positively controls the expression of an endo-β-(1,4)-mannase, involved in virulence and biofilm dispersal *in vitro* (Dow et al. 2003) .

7.4   Cross-Talk Between Bacteria and Plants

Interkingdom Signaling via Small Molecules Evidence has accumulated that plants can perceive and respond to bacterial signal molecules. Both *Medicago truncatula* and tomato have been shown to respond to exogenous AHLs by altering the global patterns of gene expression, including the activation of pathogenesis-related proteins that might contribute to disease resistance (Hartmann and Schikora 2012). AHLs with acyl side chain lengths smaller than C8 have been found to enter the roots readily and are transported up to the shoots whereas AHLs with side chains longer than C10 tend to stick to the root surface and are not transported substantially. The presence of AHL-producing bacteria in the tomato rhizosphere was shown to increase the salicylic acid levels in the leaves and enhance systemic resistance against the fungal leaf pathogen *Alternaria alternata*. Macro-array analyses showed that synthetic AHLs systemically induce salicylic acid- and ethylene-dependent defense genes, suggesting that AHLs can play an important role in the biocontrol activity of rhizobacteria. Several reports have demonstrated that AHLs can also influence root development of plants. Depending on the structure of the AHL molecule, (particularly the length of the acyl side chain), the thickness and length of the root but also root hair development can be affected. Recent research has also shown that a sub-family of bacterial AHL receptors responds specifically to yet unidentified small molecules produced by plants (Patel et al. 2013).

QS Inhibition and Quenching Plants also produce compounds that mimic or inhibit AHL-dependent QS of plant-associated bacteria (Kalia 2012). Signal antagonists, so-called QS inhibitors, have been isolated from a large number of plants and some of these compounds are currently considered for exploitation as antibacterial and antibiofilm agents. Many studies have demonstrated that AHL signals can be degraded by various bacteria by the production of lactonases or acylases and thereby "quench" QS (Fig. 7.1). Bacteria that are capable of quenching QS have been isolated from the rhizosphere as well as from the phyllosphere of various plants and have been shown to play an important role in the ecology of plant-associated consortia. Finally, some plants have also the capability to enzymatically inactivate AHL signal molecules.

References

Aguilar C, Carlier A, Riedel K et al (2009) Cell-to-cell communication in biofilms of gram-negative bacteria. In: Krämer R, Jung K (eds) Bacterial signaling. Wiley-VCH, Weinheim, pp 23–40

Cha C, Gao P, Chen Y et al (1998) Production of acyl-homoserine lactone quorum-sensing signals by gram-negative plant-associated bacteria. Mol Plant-Microbe Interact 11:1119–1129

Coutinho B, Mitter B, Talbi C et al (2013) Regulon studies and *in planta* role of the BraI/R quorum-sensing system in the plant-beneficial *Burkholderia* cluster. Appl Environ Microbiol 79:4421–4432

Danhorn T, Fuqua C (2007) Biofilm formation by plant-associated bacteria. Annu Rev Microbiol 61:401–422

DeAngelis KM, Lindow SE, Firestone MK (2008) Bacterial quorum sensing and nitrogen cycling in rhizosphere soil. FEMS Microbiol Ecol 66:197–207

Dow JM, Crossman L, Findlay K et al (2003) Biofilm dispersal in *Xanthomonas campestris* is controlled by cell-cell signaling and is required for full virulence to plants. Proc Natl Acad Sci U S A 100:10995–11000

Downie JA (2010) The roles of extracellular proteins, polysaccharides and signals in the interactions of rhizobia with legume roots. FEMS Microbiol Rev 34:150–170

Dulla G, Lindow S (2009) Acyl-homoserine lactone mediated cross talk among epiphytic bacteria modulates behavior of *Pseudomonas syringae* on leaves. ISME J 3:825–834

Elasri M, Delorme S, Lemanceau P et al (2001) Acyl-homoserine lactone production is more common among plant-associated *Pseudomonas* spp. than among soilborne *Pseudomonas* spp. Appl Environ Microbiol 67:1198–1209

Fuqua WC, Winans SC, Greenberg EP (1994) Quorum sensing in bacteria: the LuxR-LuxI family of cell density-responsive transcriptional regulators. J Bacteriol 176:269–275

Gantner S, Schmid M, Dürr C et al (2006) In situ quantitation of the spatial scale of calling distances and population density-independent N-acylhomoserine lactone-mediated communication by rhizobacteria colonized on plant roots. FEMS Microbiol Ecol 56:188–194

Hartmann A, Schikora A (2012) Quorum sensing of bacteria and trans-kingdom interactions of N-acyl homoserine lactones with eukaryotes. J Chem Ecol 38:704–713

Kalia VC (2012) Quorum sensing inhibitors: an overview. Biotechnol Adv 31:224–245

Patel HK, Suárez-Moreno ZR, Degrassi G et al (2013) Bacterial LuxR solos have evolved to respond to different molecules including signals from plants. Front Plant Sci 4:447

Quinones B, Dulla G, Lindow SE (2005) Quorum sensing regulates exopolysaccharide production, motility, and virulence in *Pseudomonas syringae*. Mol Plant Microbe Interact 18:682–693

Sanchez-Contreras M, Bauer WD, Gao M et al (2007) Quorum sensing regulation in rhizobia and its role in symbiotic interactions with legumes. Philos Trans R Soc Lond B Biol Sci 362:1149–1163

Steidle A, Siegl K, Schuhegger R, Ihring A et al (2001) Visualization of N-Acylhomoserine (AHL)-mediated cell-cell communication between bacteria colonizing the tomato rhizosphere. Appl Environ Microbiol 67:5761–5770

Suppiger A, Schmid N, Aguilar C et al (2013) Two quorum sensing systems control biofilm formation and virulence in members of the *Burkholderia cepacia* complex. Virulence 4:400–409

Von Bodman SB, Bauer WD, Coplin DL (2003) Quorum sensing in plant-pathogenic bacteria. Annu Rev Phytopathol 41:455–482

Chapter 8
Bacterial Volatiles as Airborne Signals for Plants and Bacteria

Choong-Min Ryu

Abstract Volatile compounds are found in all organisms. Bacteria communicate with their surrounding ecosystem using a diverse array of volatile metabolites. Some rhizosphere bacteria (rhizobacteria) emit volatile organic compounds (VOCs) that promote plant growth and elicit induced systemic resistance (ISR) and induced systemic tolerance (IST). This chapter reviews recent progress in understanding how VOCs mediate interactions between rhizobacteria and plants and among bacteria. Recent proteomics analysis of plant ISR and IST induced by rhizobacterial VOCs as well as the potential of VOCs for field applications will be discussed. These studies provide novel insights into the biological and ecological potential of rhizobacterial VOCs for modulating biotic and abiotic stress tolerance in modern agriculture.

8.1 Airborne Bacterial Signals

Bacterial volatile organic compounds (VOCs) are signaling molecules that are perceived by other bacteria, animals, insects, plants, and microorganisms (Farag et al. 2013). The first report of microbial VOCs was in 1921 (Zoller and Clark 1921). Bacterial volatiles are recognized as insect semiochemicals and inhibitors of fungal and plant growth (Baily and Weisskopf 2012; Davis et al. 2013). The study of bacterial volatiles traditionally focused on soil ecology and interactions with plants. In 2003, Ryu et al. identified bacterial VOCs that positively promote growth of *Arabidopsis thaliana* seedlings. These VOCs were emitted from a rhizobacterial class called plant-growth promoting rhizobacteria (PGPR). A year later, the same research group reported that PGPR trigger ISR (Ryu et al. 2004), which is a form of plant systemic defense. These two reports provided new information about novel bacterial compounds that elicit plant growth and ISR. There has been considerable progress in recent years in the understanding of bacterial VOC-mediated plant responses, including abiotic stress tolerance, and new evidence for the role of bacterial VOC in

C.-M. Ryu (✉)
Molecular Phytobacteriology Laboratory, KRIBB, Daejeon 305-806,
Republic of Korea
Tel.: + 82 42 879 8229
e-mail: cmryu@kribb.re.kr

© Springer International Publishing Switzerland 2015
B. Lugtenberg (ed.), *Principles of Plant-Microbe Interactions*,
DOI 10.1007/978-3-319-08575-3_8

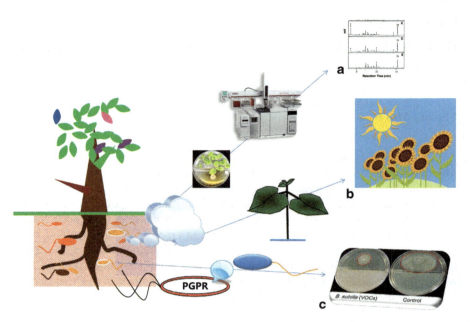

Fig. 8.1 Inter-kingdom and intra-kingdom communication-mediated by bacterial volatile organic compounds. **a** Bacterial volatiles are captured in fiber membranes, extracted with organic solvents, and identified using gas chromatography-mass spectrometry (*GC-MS*). The bacterial volatiles decoded by *GC-MS* analysis helps us to understand the roles of bacterial airborne signals in their interactions with plants and bacteria. **b** Plant protection by bacterial volatiles against biotic and abiotic stresses in *in vitro* as well as even under field conditions. **c** Bacterial interactions mediated by volatile compounds. The two plates indicate inhibition of the motility of a gram-negative bacterium (*E. coli*) by bacterial VOCs from the gram-positive bacterium *Bacillus subtilis* on the agar media

plants, animals, and bacteria. This chapter will focus on the role of bacterial VOCs in plants, bacteria, and other microbes (Fig. 8.1).

The major obstacle for early studies of the effect of bacterial VOCs on plants was the lack of suitable culture systems to perform the experiments, because natural culture conditions also contain VOCs produced by plants and microbes. It was necessary to develop a gnotobiotic system, in which plants and bacteria were spatially separated into a closed culture system that controls both biotic and abiotic variables. This problem was solved by using an I-plate system, which is a Petri dish that contains two physically separated compartments for growth of seedlings and bacteria and allows the free exchange of air. The I-plate system did not perform well in studies on fungal VOCs, due to fungal spore contamination of the seedlings. Therefore, microtiter plates (24- or 48-well) were used to study fungal VOCs, because the airspace between the plate and the lid prevented spore transmission but allowed VOC transmission.

The identification of VOCs from PGPR strains was necessary to isolate bioactive VOCs and study their effects on plants. Several analytical methods have been developed to capture, analyze, identify, and quantify airborne volatiles released from bacteria. Most of these techniques were adapted from those used to study plant VOCs. The most common method captures gasses in the headspace of culture systems and then employs gas chromatography (GC) for analysis of the extracted compounds. The volatile compounds contained in the dynamic air flow over a bacterial culture grown on solid Murashige-Skoog (MS) medium can be trapped by an absorbent filter from the headspace and then released by rinsing the filter with specific organic solvents. MS medium appeared to yield lower background signals compared to those produced by other microbial media (Ryu et al. 2003, 2004). Recently, a more advanced solid-phase micro extraction technique, coupled with software-driven extraction of compounds from overlapping peaks in GC analysis, was developed. Using this procedure, compounds are rapidly and directly released from bacterial cultures into a heated GC-injector in a no-flow, low-oxygen environment. The solid-phase micro extraction method can collect and identify more than 30 different volatile compounds and has been used to collect bacterial VOCs in several systems (Fig. 8.1a; Farag et al. 2006; Lee et al. 2012). A comparative review of different techniques used for volatile compound analysis is provided by Tholl et al. (2006).

8.2   Aromatic Therapy for Plant Stresses: Beneficial Effects of Bacterial VOCs on Plant Growth and Immunity

The I-plate enabled identification of specific airborne chemicals emitted from soil bacteria as signals that trigger plant growth and ISR (Ryu et al. 2003, 2004). A recent review by Bailly and Weisskopf (2012) describes the growth-promoting effects of bacterial VOCs. In the present chapter, I will focus on ISR/induced systemic tolerance (IST) triggered by bacterial VOCs (Fig. 8.1b).

Sweet Odor for Inducing Immunity Our early studies investigated the necrotrophic pathogen *Pectobacterium carotovorum* subsp. *carotovorum* (syn. *Erwinia carotovora* subsp. *carotovora*) because this pathogen causes visible soft-rot symptoms in many plant species within 24 h. In later studies we investigated bacterial speck disease caused by *Pseudomonas syringae* pv. tomato; however, the disease symptoms caused by this bacterium become visible only after 7 days (Lee et al. 2012). Maximum protection against necrotrophic disease was conferred by rhizobacterial strains *Bacillus subtilis* GB03 and *Bacillus amyloliquefaciens* IN937a, which also trigger ISR. Four other PGPR strains that trigger ISR when inoculated onto seeds failed to induce resistance against *P. carotovorum* subsp. *carotovorum* in the I-plate test. Continuous analysis for 24 h revealed that GB03 and IN937a consistently releases 2,3-butanediol and its precursor 3-hydroxy-2-butanone, whereas *E. coli* strain DH5α and *Pseudomonas putida* strain 89B61 did not release these VOCs and did not trigger ISR. Most bacterial species of the Proteobacteria and Firmicute groups produce 2,3-butanediol and acetoin under low-oxygen conditions; these VOCs provide an alternative electron sink for NAD^+ regeneration when aerobic respiration is limited (Ramos et al. 2000; Xiao and Xu 2007). Aerobic respiration may be limited in the rhizosphere due to low-oxygen conditions, and this is where PGPR naturally reside. The bioactivity of stereoisomers (*2R*, *3R* and *2S*, *3S*) should be compared to determine whether PGPR VOCs have stereoisomer specificity, and to identify the stereoisomer that triggers plant ISR and/or growth. Treatment with a synthetic 2,3-butanediol and a volatile extract collected from strain GB03 induce similar disease-resistance responses, which are comparable to those induced by direct inoculation of PGPR on plants (Ryu et al. 2004). Treatments with 0.2 pg/ml–0.2 μg/ml of the synthetic 2,3-butanediol (in increments of 1:100 dilutions) trigger similar levels of ISR in plants.

A recent proteomics study investigated *Arabidopsis* tissue exposed to VOCs derived from GB03. This study provided new insights into how plants perceive PGPR VOCs (Kwon et al. 2010). The study identified 95 peptides that differentially respond to GB03 VOCs, including 61 up-regulated and 34 down-regulated proteins. Of these, 20 spots correspond to twelve proteins involved in ethylene (ET) biosynthesis. Another proteomic study, exploring *in planta* effects of bacterial volatiles, confirmed that ISR triggered by *B. subtilis* FB17 against *P. syringae* pv. tomato DC3000 was mediated by the salicylic acid (SA) and ET signaling pathways, but was independent of the jasmonic acid (JA) pathway (Rudrappa et al. 2010). Bacterial mutants in the acetoin and 2,3-butanediol biosynthetic pathways fail to trigger ISR, which confirms that acetoin and 2,3-butanediol function as VOCs in eliciting ISR. In addition to these C4 volatile compounds, long-chain VOCs, such as tridecane released from *P. polymyxa* E681, prime transcriptional expression of marker genes for the SA, JA, and ethylene signaling pathways, including *PR1*, *ChiB*, and *VSP2*, respectively (Lee et al. 2012).

Treatment of maize seeds with the endophytic bacterium *Enterobacter aerogenes* led to the identification of 2,3-butanediol from soil-grown maize seedlings. Bacterial production of 2,3-butanediol resulted in maize plants that had greater resistance to the Northern corn leaf blight fungus *Setosphaeria turcica*. Treatment of seeds with *E. aerogenes* decreased plant attractiveness for the parasitoid *Cotesia marginiventris*, whereas treatment with the soil microbiome enhanced parasitoid recruitment. These contrasting observations indicate an indirect effect of bacterial VOCs on parasitoids, which depends on the microbiota composition (D'Alessandro et al. 2014).

Cool Scent for Thirsty Plants An increasing number of studies demonstrated that PGPR VOCs trigger plant tolerance to abiotic stress, including drought stress, salt stress, and nutrient deficiency. Our previous work proposed the term *induced systemic tolerance* for *"PGPR-induced physical and chemical changes in plants that result in enhanced tolerance to abiotic stress"*. *"Biotic stress is excluded from IST because conceptually it is part of biological control and induced resistance"* (Yang et al. 2009). Plants treated with GB03-derived VOCs had greater photosynthetic capacity and increased iron uptake (Zhang et al. 2009). GB03-derived VOCs also modulate *AtHKT1* function, which confers shoot-to-root Na^+ recirculation, possibly by loading Na^+ into phloem vessels. This result supports the role of AtHKT1 in controlling shoot-to-root Na^+ recirculation, and explains VOC-induced salt tolerance (Zhang et al. 2008). *Arabidopsis* plants grown with GB03 or with *Pseudomonas* *chlororaphis* O6 in the soil exhibit increased drought tolerance (Cho et al. 2008; Zhang et al. 2010). The SA signaling pathway may be involved in *P. chlororaphis* O6-mediated drought tolerance, because drought-stressed plants exposed to bacterial volatiles or 2,3-butanediol accumulate higher levels of SA compared to those in untreated plants (Cho et al. 2008).

8.3 Airborne Chemicals Mediate Communication Between Bacteria

Many bacterial species coexist in dynamic communities where they compete or cooperate with one another (see Chap. 43). Intercellular signaling systems enable bacteria to adapt to biotic and abiotic stresses. The most well-known phenomenon is quorum sensing (QS), which is mediated by a complex regulatory network (see Chap. 7). Many bacteria release a wide variety of bioactive VOCs that diffuse through heterogeneous mixtures of solids, liquids, and gasses (Kai et al. 2009). The mechanisms underlying VOC-mediated bacterial communication systems remain largely unknown.

Let's Speak with Odors Although many bacteria produce VOCs, little is known about their physiological functions. Several recent reports indicate that bacteria are responsive to ammonia, which is a volatile organic compound. Ammonia is a valuable nutrient for the production of proteins and nucleic acids. The Burgess group recently showed that the soil bacterium *Bacillus licheniformis* was sensitive to the presence of ammonium sulfate (Nijland and Burgess 2010), which was metabolized, converted to ammonia, and induced biofilm formation. The mechanisms underlying VOC-mediated bacterial communication are not known. Bernier et al. (2011) showed that ammonia released from *E. coli* influences antibiotic resistance in aerially-exposed Gram-positive and -negative bacteria. These authors discovered a previously uncharacterized antibiotic resistance mechanism, and showed that this contributes to the induced expression of genes involved in ammonia-polyamine metabolism during the stationary growth phase. This was the first evidence for a mechanism underlying VOC-modified gene expression in interacting bacteria. The same group recently reported that exposure to trimethylamine, a volatile compound produced by *E. coli* and other Gram-negative bacteria, altered antibiotic resistance in all tested bacteria. The study revealed that exposure to trimethylamine elevated the growth-medium pH, which modified antibiotic uptake and transiently altered antibiotic resistance.

  E. coli and *Bacillus subtilis* were used as representative models to study the roles of VOCs in Gram-negative and Gram-positive species, respectively (Kim et al. 2013). Microarray analysis was used to examine gene expression in *E. coli* exposed to *B. subtilis* VOCs. The results showed that VOCs dramatically affected bacterial phenotypes in receiver cells and altered the expression of more than one hundred genes. Changes in global gene expression related to motility and antibiotic resistance phenotypes were observed after treatment with 2,3-butanedione and glyoxylic acid from nanomolar to micromolar concentrations. VOC-mediated interactions were modulated by the previously uncharacterized *ypdB* gene product and its downstream transcription factors *soxS*, *rpoS*, and *yjhU*. These results strongly suggest that VOCs represent an interspecific signal for rapid sensing between bacteria, which can trigger variations in gene expression. VOC-mediated bacterial communication may induce modifications that affect defense and locomotion responses of physically separated bacteria (Fig. 8.1).

8.4 Volatile Compounds for Agricultural Applications: From Lab to Field

Many microbially produced VOCs function as signals that mediate interactions between microbes and plants or between different microbes. There have been few studies of field applications of VOCs. Treatment of aerial plant parts with the highly bioactive and inexpensive 2,3-butanediol to promote growth, induce ISR, and confer tolerance to drought and salinity offers a promising agricultural application. A selection of novel strategies utilizing bacterial VOCs is discussed.

Working in the Open Field? Although volatile compounds rapidly dissipate under natural conditions, VOCs may be useful for practical applications. Many VOCs (including 2,3-butanediol) are water soluble, inexpensive (< US \$/kg), function at extremely low concentrations (ng/ml to pg/ml), and are not overtly toxic to animals and humans. Under growth-chamber conditions, drenching roots with acetoin reduced the pathogen population on leaves (Rudrapa et al. 2010). Therefore, bacterial VOCs are promising candidates for triggering plant systemic defense and improving disease control. The primary requirement for field trials of bacterial VOCs is to develop an appropriate treatment protocol. Under natural conditions, bacteria emit the volatiles at a very low, steady level. This process is difficult to mimic, and the rapid and nonuniform evaporation of bacterial VOCs after application in an open field can cause inconsistent results.

Drench application of a bacterial volatile, 2-butanone, to pepper and cucumber plants has been successfully performed for four consecutive years under field conditions (Song and Ryu 2013). In another field trial, 4-week-old pepper plants were dip-treated with 1 mM 2-butanone before they were transplanted into the field. This successfully protected pepper crops against bacterial spot during a 2-year field trial (Fig. 7.2). Similarly drench treatment of cucumber with 2-butanone up-regulated the defense-related gene *CsLOX*, reduced the aphid (*Myzus persicae*) population, and increased the ladybird beetle population (the natural enemy of aphid). These results suggest that VOC-mediated induction of the oxylipin pathway, related to produce airborne signal molecules methyl jasmonate and green leaf volatiles, can help recruit a natural aphid enemy and may ultimately prevent plant disease and insect damage by eliciting induced resistance, even under open-field conditions (Song and Ryu 2013). Whether PGPR volatiles function to directly attract natural herbivore enemies, or

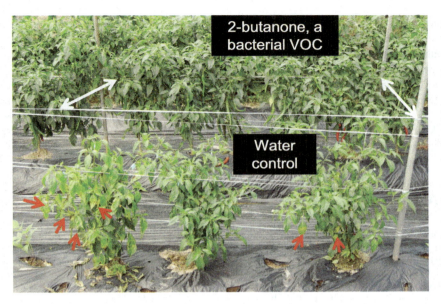

Fig. 8.2 Bacterial VOC-elicited induced resistance against naturally occurring bacterial spot caused by *Xanthomoans vesicatoria* and *Cucumber mosaic virus* in a pepper field. The picture was taken at 3 month after drench application of a bacterial volatile, 2-butanone when the pepper seedlings were transplanted. The upper and bottom rows indicate 2-butanone and water treatments respectively. The three plants in the first row show severe stunting by *Cucumber mosaic virus* infection. The white arrows indicate the same height of the two rows. The left and right plants on the first row demonstrate the beginning of chlorosis that is a typical symptom of bacterial spot caused by *X. vesicatoria*. The red arrows indicate the symptoms of bacterial spot: necrosis and chlorosis. The empty hole in the bottom row is caused by the removal of a plant because of infection by the late blight pathogen, *Phytophthora capsici*

whether bacterial VOCs induce volatile emissions from plants (Farag and Pare 2002; Arimura et al 2000) that function to attract herbivore enemies, has yet to be determined (Fig. 8.2).

8.5   Perspectives

The experiments in 2003 (Ryu et al. 2003), 2004 (Ryu et al. 2004), and 2013 (Kim et al. 2013) presented a new paradigm for interactions between bacteria and plants and among different bacteria. These studies revealed that bacterial VOCs can act as signal molecules that trigger plant growth and immunity. A detailed analysis of cellular and molecular mechanisms involved in plant and bacterial responses to VOCs is required. Potential agricultural applications of bacterial VOCs for promoting biotic and abiotic resistance in crop plants will be an important topic of future studies. The following topics remain to be addressed: (i) The mechanism of plant perception of bacterial VOCs is unknown. It is not known how plants differentiate between similar C4-alcohols such as 2,3-butanediol and acetoin. (ii) It is unknown whether plants possess a signaling pathway for bacterial VOCs that is distinct from known plant defense signaling pathways. (iii) It is unknown whether plant sensors for bacterial VOCs are tissue-specific or organ-specific. (iv) It is unknown how a single bacterial volatile compound promotes plant growth and elicits ISR. Theoretically, ISR occurs at the cost of plant fitness and may result in a growth penalty. (v) Future work will determine if bacterial volatiles can trigger the emission of plant green leaf volatiles.

References

Arimura GI, Ozawa R, Shimoda T et al (2000) Herbivory-induced volatiles elicit defense genes in lima bean leaves. Nature 406:512–515

Bailly A, Weisskopf L (2012) The modulating effect of bacterial volatiles on plant growth: current knowledge and future challenges. Plant Signal Behav 7:79–85

Bernier SP, Létoffé S, Delepierre M et al (2011) Biogenic ammonia modifies antibiotic resistance at a distance in physically separated bacteria. Mol Microbiol 81:705–716

Cho SM, Kang BR, Han SH et al (2008) 2R,3R-butanediol, a bacterial volatile produced by *Pseudomonas chlororaphis* O6, is involved in induction of systemic tolerance to drought in *Arabidopsis thaliana*. Mol Plant-Microbe Interact 21:1067–1075

D'Alessandro M, Erb M, Ton J et al (2014) Volatiles produced by soil-borne endophytic bacteria increase plant pathogen resistance and affect tritrophic interactions. Plant Cell Environ 37:813–826

Davis TS, Crippen TL, Hofstetter RW et al (2013) Microbial volatile emissions as insect semiochemicals. J Chem Ecol 39:840–859

Farag MA, Ryu CM, Sumner LW et al (2006) GC-MS SPME profiling of rhizobacterial volatiles reveals prospective inducers of growth promotion and induced systemic resistance in plants. Phytochemistry 67:2262–2268

Farag MA, Zhang H, Ryu CM (2013) Dynamic chemical communication between plants and bacteria through airborne signals: induced resistance by bacterial volatiles. J Chem Ecol 39:1007–1018

Kai M, Haustein M, Molina F et al (2009) Bacterial volatiles and their action potential. Appl Microbiol Biotechnol 81:1001–1012

Kim KS, Lee S, Ryu CM (2013) Interspecific bacterial sensing by airborne signals leads to modulated locomotion and drug-resistance. Nat Commun 4:1809

Kwon YS, Ryu CM, Lee S et al (2010) Proteome analysis of *Arabidopsis* seedlings exposed to bacterial volatiles. Planta 232:1355–1370

Lee BY, Farag MA, Park HB et al (2012) Induced resistance by a long-chain bacterial volatile: elicitation of plant systemic defense by a C13 volatile produced by *Paenibacillus polymyxa*. PLoS One 7:e48744

Létoffé S, Audrain B, Bernier SP et al (2014) Aerial exposure to the bacterial volatile compound trimethylamine modifies antibiotic resistance of physically separated bacteria by raising culture medium pH. MBio 5(1):e00944–13

Nijland R, Burgess JG (2010) Bacterial olfaction. Biotechnol J 5:974–977

Rudrappa T, Biedrzycki ML, Kunjeti SG et al (2010) The rhizobacterial elicitor acetoin induces systemic resistance in *Arabidopsis thaliana*. Commun Integr Biol 3:130–138

Ryu CM, Farag MA, Hu CH et al (2003) Bacterial volatiles promote growth in *Arabidopsis*. Proc Natl Acad Sci U S A 100:4927–4932

Ryu CM, Farag MA, Hu CH et al (2004) Bacterial volatiles induce systemic resistance in *Arabidopsis*. Plant Physiol 134:1017–1026

Song GC, Ryu CM (2013) Two volatile organic compounds trigger plant self-defense against a bacterial pathogen and a sucking insect in cucumber under open field conditions. Int J Mol Sci 14:9803–9819

Tholl D, Boland W, Hansel A et al (2006) Practical approaches to plant volatile analysis. Plant J 45:540–560

Xiao ZJ, Xu P (2007) Acetoin metabolism in bacteria. Crit Rev Microbiol 33:127–140

Yang J, Kloepper JW, Ryu CM (2009) Rhizosphere bacteria help plants tolerate abiotic stress. Trends Plant Sci 14:1–4

Zhang H, Xie X, Kim MS et al (2008) Soil bacteria augment *Arabidopsis* photosynthesis by decreasing glucose sensing and abscisic acid levels *in planta*. Plant J 56:264–273

Zhang H, Sun Y, Xie X et al (2009) A soil bacteria regulates plant acquisition of iron via deficiency-inducible mechanisms. Plant J 58:568–577

Zhang H, Murzello C, Sun Y et al (2010) Choline and osmotic-stress tolerance induced in *Arabidopsis* by the soil microbe *Bacillus subtilis* (GB03). Mol Plant-Microbe Interact 23:1097–1104

Zoller HF, Clark WM (1921) The production of volatile fatty acids by bacteria of the dysentery group. J Gen Physiol 3:325–330

Part II
Phytopathogens and Pest Insects

Chapter 9
Phytopathogenic Bacteria

Jan van der Wolf and Solke H. De Boer

Abstract A few hundred bacterial species, belonging to the Proteobacteria, Molle-
cutes and Actinomycetes cause a large number of different plant diseases, some of
which are devastating for agricultural crops. Symptoms of bacterial plant diseases are
diverse and include necrosis, tissue maceration, wilting, and hyperplasia. For successful infection to occur, the pathogen must overcome plant defense mechanisms,
which it often does by injecting effector molecules directly into plant cells to suppress
a host response. Virulence may also involve production of plant cell wall-degrading
enzymes, toxins and/or plant hormones often under control of quorum sensing mechanisms. Some phytopathogenic bacteria actively move to their host via chemotaxis
and enter the plant through natural openings such as stomata and lenticels or wounds
caused by insect feeding, fungal infection, or mechanical plant damage. Host plants
are internally colonized locally through intercellular spaces and systemically via the
vascular system. Control of bacterial plant diseases is achieved mainly by prevention
and exclusion of the pathogen since there are few effective chemical control agents
and sources of resistance against bacterial diseases are limited.

9.1 Introduction

While estimates of the number of bacterial species on earth vary widely from tens of
thousands to billions (Schloss and Handelsman 2004), only a few hundred species
cause significant damage to agricultural crops (Kado 2010). Phytopathogenic bacteria are a major threat to crop production due to (a) the lack of suitable agrochemicals
for their control, (b) the absence of resistance or immunity in host plants, and (c)
their inadvertent and undetected spread as contaminants or asymptomatic (latent)

J. van der Wolf (✉)
Plant Research International, PO BOX 69, 6700 AB Wageningen, The Netherlands
Tel.: +31.317.480598
e-mail: Jan.vanderWolf@wur.nl

S. H. De Boer
Emeritus Scientist, 29 Donald Drive, Charlottetown, PE C1E 1Z5, Canada
e-mail: Solkedb@gmail.com

© Springer International Publishing Switzerland 2015
B. Lugtenberg (ed.), *Principles of Plant-Microbe Interactions,*
DOI 10.1007/978-3-319-08575-3_9

infections in plant propagation materials. It is therefore not surprising that a relatively large number of phytopathogenic bacteria are quarantine organisms. In this chapter the taxonomy, host range, symptomatology and virulence factors are briefly discussed after which the fate of bacterial pathogens is followed from the first contact with their hosts to their dissemination and survival. The chapter closes with a description of disease management strategies in which exclusion and prevention play dominant roles.

9.2   Taxonomy, Host Range and Symptomatology

Plant disease-causing bacteria are a highly diverse group of prokaryotes. Although they are classified into many different taxa, the majority of them are found in the phylum Proteobacteria. A complete list of names of phytopathogenic bacteria has been published recently (Bull et al. 2010) as well as a general reference book on plant bacteriology (Kado 2010).

The current transition from phenotypic to genotypic methods for classification is expected to stabilize the taxonomy of plant pathogenic bacteria, which has been in flux for some time. In particular, the development of low-cost, next generation sequencing techniques enabling whole genome analysis will result in more accurate species delineation and the inclusion of virulence determinants as taxonomic criteria.

Some highly specialized bacterial species are pathogenic on only a single crop as illustrated by *Clavibacter michiganensis* subsp. *sepedonicus*, the causal agent of the potato bacterial ring rot disease. Other bacterial species are polyphagous and cause disease on multiple host plants as exemplified by *Ralstonia solanacearum* the causal agent of wilting diseases in various monocot and dicot crop plants. There are also a few phytopathogenic species that are so-called "kingdom crossers" which cause diseases in both plants and animals. *Dickeya dadantii* causes soft rot of vegetables and ornamental plants but is also an entomopathogen being highly virulent for pea aphid; *Serratia marcescens* causes onion bulb rot and clinical sepsis in humans. Bacterial plant diseases occur in all climatic zones, but tend to be less prevalent in arid climates because moisture is important for disease development and dissemination of the pathogens.

Infection of plants by phytopathogenic bacteria often, but not always, results in multiple symptoms of disease (Fig. 9.1) (Kado 2010). When seed, bulbs, tubers or rootstocks are infected, plants developing from them may succumb to disease before even emerging as sometimes occurs with *Clavibacter michiganensis* subsp. *michiganensis* in tomato and *Pectobacterium* spp. in potato. Many of the phytopathogenic *Xanthomonas* and *Pseudomonas* spp. initially cause water-soaked lesions in infected plant parts followed by necrosis and chlorosis of the tissues. Dieback of tree fruit branches, usually preceded by necroses of leaves, is caused by such bacteria as *Erwinia amylovora* and *Pseudomonas syringae* pv. *morsprunorum*. *Rhizobium rhizogenes*, *Rhodococcus fascians* and *Clavibacter michiganensis* subsp. *michiganensis* form cankers in their hosts, while *Streptomyces* spp. causes hyperplasia in the

Fig. 9.1 Examples of symptoms caused by plant pathogenic bacteria. **a** Chlorosis and necrosis on geranium leaves caused by *Xanthomonas hortorum* pv. *pelargonii*, **b** blackrot caused by *Xanthomonas campestris* pv. *campestris* on cauliflower leaves, characterized by V-shaped chlorotic and nectrotic lesions and black veins, **c** leaf cracking caused by *Curtobacterium flaccumfaciens* pv. *oortii* in tulip, **d** tumors caused by *Agrobacterium tumefaciens* on chrysanthemum stems, **e** wilting of potato plants caused by *Pectobacterium atrosepticum*, **f** potato tuber maceration caused by *Dickeya solani*. Figures 9.1a, **b**, **d** and **e** are from the collection of Jan van der Wolf. Figure 9.1c is a gift from Khanh Pham, PPO, Lisse, the Netherlands (*unpublished*). Figure 9.1f was reproduced from John Elphinstone, FERA, UK Crown copyright, by permission of the publisher

periderm of underground plant structures. *Dickeya, Pectobacterium, Burkholderia, Serratia* and *Ralstonia* spp. all cause soft rot in which plant tissue is macerated by the activity of microbial pectinolytic enzymes. Disease symptoms caused by phytoplasmas and spiroplasmas include stunting, phyllody (retrograde development of floral organs), witches' broom (development of multiple stalks), virescence (greening) and bolting (growth of elongated stalks) (Hogenhout and Loria 2008). Infection by some plant pathogenic bacteria such as *Erwinia amylovora* and *Ralstonia solanacearum* are characterized by oozing of bacterial slime from infection loci. Asymptomatic infections, while usually not causing direct disease losses, can still result in crop loss when detection of a quarantine pathogen dictates severe statutory measures including crop destruction.

9.3 Virulence Factors and Plant Defense

To successfully invade host plants, phytopathogenic bacteria must cope with a number of plant defense mechanism and have a means for acquiring water and nutrients for growth and colonization of plant tissues. Hence plant disease-causing bacteria display a broad variety of virulence factors such as cell-wall degrading enzymes, toxins, plant hormones, and effectors to overcome plant defense responses. Virulence factors are often associated with transposable elements or genomic islands and may be acquired from other microbes via lateral gene transfer which usually involve

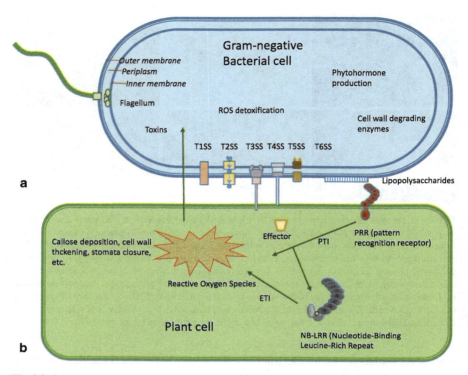

Fig. 9.2 Schematic representation of the interaction between a bacterial pathogen and a plant cell. Molecules are released from the pathogen into the host cell by different types of secretion systems (T1SS—T6SS). These molecules may be effectors that suppress plant defense mechanisms, but also cell wall degrading enzymes, toxins or phytohormones. Various bacterial components such as lipopolysaccharides and flagellar proteins, act as pathogen associated molecular patterns (*PAMPS*) that are recognized by cell surface pattern recognition receptors (*PRRs*) of the host, and elicit PAMP triggered immunity (*PTI*), which involve such responses as the production of reactive oxygen species (*ROS*), callose deposition, cell thickening etc. To suppress *PTI*, many plant pathogens release effector molecules. However, in some pathosystems effectors are recognized by nucleotide-binding leucine rich repeats (*NB-LRR*) of the host inducing effector triggered immunity (*ETI*). Some pathogens can also detoxify *ROS*. Figure from author Jan van der Wolf

mobile genetic elements, such as plasmids, bacteriophages and other integrative and conjugative elements. Major virulence factors will be discussed in the following and are indicated in bold.

Secretion Systems Phytopathogenic bacteria deliver pathogenicity factors directly into plant cells by various secretion systems (Chap. 6; Kado 2010) (Fig. 9.2). The type 1 secretion system (T1SS) is found in many proteobacteria and utilizes an ABC transporter cassette for ATP driven transmembrane delivery of biologically active molecules. It is involved in the translocation of metalloproteases, e.g. in *Dickeya* and *Pectobacterium* spp., and of lipases, sterols, toxins and drugs in other phytopathogenic bacteria. The type 2 secretion system (T2SS) is also found in many Gram-negative phytopathogenic bacteria. In T2SS, proteins sequentially pass through the cytoplasmic membrane and, after modification, through a pore in the outer membrane. This system is involved in the secretion of cell wall degrading enzymes by *Pectobacterium* sp., levansucrase by *Erwinia amylovora*, and a cysteine protease by *Xanthomonas oryzae* pv. *oryzae*. A functional type III secretion system (T3SS) is required for virulence of many important bacterial pathogens including *Pseudomonas syringae*, *Xanthomonas campestris*, *Ralstonia solanacearum* and *Erwinia amylovora* (Rosenberg et al. 2013). The expression of T3SS, which involves the injection of biologically active effector molecules into host cells through a pilus-like structure, is regulated by environmental factors such as pH and the balance between concentrations of calcium and magnesium ions. Effectors transported by T3SS generally have host defense-suppressing proteolytic activity as found in *Pseudomonas syringae* and *Xanthomonas campestris*, elicit water and nutrients from host cells into apoplasts, and modulate plant hormone physiology as evidenced by auxin regulation in *Xanthomonas campestris*. The chromosome or plasmid-encoded type 4 secretion system (T4SS) in which a pilus also forms an integral part of the secretion system, is yet another system by which various plant pathogenic bacteria transfer effectors, nucleoproteins and nucleic acids into host cells. *Agrobacterium tumefaciens*, for example, incites the crown gall disease by transferring its T-DNA into host cells via T4SS (Chap. 37). The type 5 secretion system (T5SS) is a so-called autotransporter protein system and is found in several plant pathogens including *Xylella fastidiosa*, *Xanthomonas oryzae* pv. *oryzae*, *Xanthomonas campestris* pv. *vesicatoria*, and *Dickeya* spp., where it plays a role in the adherence of the bacteria to surfaces of host tissues. Finally, the type 6 secretion system (T6SS), reminiscent of the injection machinery of phages, occurs in at least thirteen species of plant pathogenic bacteria. It appears to be involved in virulence, immunity responses, and possibly host specificity of *Pectobacterium atrosepticum*, *Pseudomonas syringae* and *Agrobacterium tumefaciens* (Records 2011).

PAMPS, DAMPS and Effectors Pathogen-associated molecular patterns (PAMPS) describe the immune response of host plants to bacterial infections that result in a basal level of resistance. PAMPS are triggered by conserved microbial molecules such as lipopolysaccharides, flagellar proteins, peptidoglycans, elongation factor peptides, activators of XA21-mediated immunity, double-stranded DNA, methylated DNA, and cell wall pieces (Li et al. 2013; Zeng et al. 2010) (Fig. 9.1). PAMP triggers are perceived by their cognate pattern-recognition receptors and their role in defence against infection is regulated by specific signal molecules that include ethylene, jasmonic acid and salicylic acid. Resulting defence mechanisms involve callose deposition, cell-wall thickening, stomata closure, and production of antimicrobial compounds and reactive oxygen species (ROS). Some pathogens are equipped with mechanisms to detoxify antimicrobial compounds and ROS. For example catalase and peroxidases are produced by *Xanthomonas campestris* pv. *campestris* and alkyl hydroperoxidase reductases are produced by *Xanthomonas campestris* pv. *phaseoli* to neutralize cell-damaging ROS. Damage-associated molecular patterns (DAMPS) are similar to PAMPS and are triggered by small endogenous peptides or cell wall fragments that are released from damaged or stressed cells and serve as signals to activate the immune response in infected host plants (Li et al. 2013).

A second layer of defence against phytopathogenic bacteria is effector-triggered immunity (ETI). All major groups of phytopathogenic bacteria studied to date produce effector proteins. Specific effector proteins, expressed by avirulence (Avr) genes, are injected into host cells via T3SS to suppress plant defences. In incompatible interactions, which are often host cultivar specific, the effectors are recognized by the product of corresponding resistance (R) genes and induce defence responses such as the hypersensitive response (HR). In some cases, however, the effector recognized by R gene products induces modifications to the effector mediated cascade so that instead of activating ETI, it suppresses it and results in effector-triggered susceptibility (ETS).

Quorum Sensing Many pathogenic bacteria use quorum sensing, a cell-to-cell communication mechanism, which allow them to respond to population density and delay production of virulence factors until a critical cell density has been achieved (Chap. 7; Venturi and Fuqua 2013). This mechanism postpones production of virulence factors, such as plant cell wall-degrading enzymes by soft rot *Enterobacteriaceae* and the transfer of the Ti-plasmid by *Agrobacterium tumefaciens*, until bacterial cell numbers are high enough to overwhelm the plant's defense response. Typically, these systems consists of a synthase enzyme belonging to the LuxI family and is responsible for the production of acyl homoserine lactone (AHL) signal molecules and a transcription factor of the LuxR family, which activates gene expression but can also be involved in gene repression. The luxR protein consists of an autoinducer-binding and a DNA-binding domain separated by a linker. Most LuxR proteins are inactive in the absence of the AHLs but some are inactivated in the presence of AHLs. Also LuxR solo systems have been described that respond to AHLs produced by other bacteria or by plant compounds. For instance the genes involved in the motility of *Xanthomonas oryzae* pv. *oryzae* are regulated during infection by host plant molecular activators. Conversely, AHLs can also influence expression of plant genes including those involved in plant defense mechanisms, stress responses and hormones as was found in *Medicago truncatula* and *Arabidopsis thaliana*. Moreover, these plants as well as rice and pea, produce compounds mimicking AHLs and thereby stimulate bacterial quorum sensing gene expression. These observations are indeed indicative of complex interkingdom communication.

Lipopolysaccharides The outer membrane of Gram-negative bacteria contains lipopolysacchides (LPS) that play an important role in their interaction with the environment and contact with the host (Newman et al. 2001). Some LPS mutants are also more sensitive to microbial compounds such as antibiotics, detergents, and antimicrobial peptides, possibly because the ability of their outer membrane to exclude them is impaired. An LPS mutant of *Ralstonia solanacearum*, for example, was unable to induce the hypersensitive response in its host and a mutant of *Xanthomonas oryzae* pv. *oryzicola,* causal agent of bacterial leaf streak of rice, was affected in the T3SS-mediated delivery of effector proteins.

Extracellular Polysaccharides (EPS) EPS can be either closely or loosely attached to the bacterial cell and serve different functions (Rosenberg et al. 2013). In *Ralstonia solanacearum* EPS is involved in the absorption of water, minerals, and nutrients whereas in *Xanthomonas campestris* EPS plays a role in adhesion and biofilm formation, protecting the pathogen against biotic and abiotic stresses such as desiccation. In *Erwinia amylovora*, EPS suppresses host recognition of bacterial colonization of plant tissues.

Phytotoxins Phytotoxins that cause direct damage to plant cells or modulate the metabolism and physiology of the host are produced by many plant pathogenic bacteria. One example is the production of taxtomin, a cyclic dipeptide that induces scab lesions, by *Streptomyces* spp. (Hogenhout and Loria 2008). The two lipodepsipeptide phytotoxins produced by *Pseudomonas syringae* pathovars, syringomycin and syringopeptin, induce necrosis by inserting into plant cell plasma membranes, thereby disrupting the membrane electrical potential (Rosenberg et al. 2013). These toxins also act as biosurfactants that enhance bacterial dispersion. The modified peptides tabtoxin and phaseolotoxin produced by other pseudomonads cause severe yellowing, possibly by inhibiting photosynthesis. Another pseudomonad phytoxin, coronatine, is a polyketide that mimics the plant hormones jasmonic acid and ethylene, upsetting plant defense mechanisms. Coronatine also promotes lesion formation, induces chlorosis, enhances pathogen multiplication, and inhibits root growth.

Plant Cell Wall Degrading Enzymes (PCWDE) PCDWE are important virulence factors of both Gram-positive and Gram-negative bacteria. Plant cell walls are composed of carbohydrate polymers in which pectins, hemicelluloses and celluloses are most abundant, but they also contain proteins and lignins. Plant cell wall degradation by invading pathogens allows the bacteria to spread in infected tissues, and makes water and nutrients available. Soft rot *Enterobacteriaceae* secrete multiple (iso)enzymes involved in pectin degradation as well as hemicellulases, cellulases and proteases which in concerted action very rapidly macerate host tissues (Charkowski et al. 2012). Plasmid encoded cellulases are also essential for virulence of the Gram-positive pathogens *Clavibacter michiganensis* subsp. *sepedonicus* and subsp. *michiganensis*.

Siderophores There is an important role for siderophores in plant infection for species of *Xanthomonas*, *Pseudomonas*, *Erwinia* and soft rot *Enterobacteriaceae*. Siderophores are ferric iron chelators essential for scavenging iron in iron-limited environments. For example, *Dickeya dadantii* produces two siderophores, achromobactin and chrysobactin, that are required for systemic progression of tissue maceration. Interestingly, chrysobactin not only supports the pathogen in its competition for iron but also modulates the defence mechanism of its host by interfering with salicylic and jasmonic acid signalling cascades (Dellagi et al. 2009).

Plant Hormones Several phytopathogenic bacteria produce hormones that enhance virulence by interfering in the physiology of host plants. *Rhodococcus fascians*, for example, produces a mixture of five cytokinins that directly affects plant development and indirectly stimulates the host to produce auxins and polyamines resulting in gall formation. Such galls serve as specific niches for survival of the bacterial pathogen. Similarly, *Pseudomonas syringae* subsp. *savastanoi* causes hyperplasia in several tree hosts by the production of indole-3-acetic acid inducing formation of knots that are typical of the disease. Ethylene production is also a virulence factor for *Ralstonia solanacearum* and *Pseudomonas syringae*.

9.4   Ecology and Epidemiology

Microbial population dynamics is a function of the ability to grow in an environment where water and nutrients may be limiting, and the ability to compete within a hostile environment. Competitive factors include such characteristics as possession of protective pigments and capsular materials, ability to form biofilms, production of antimicrobial metabolites, and ability to maintain cellular reserves. Plant pathogenic bacteria only cause disease if a suitable host plant is present and environmental conditions are conducive to infection.

The First Contact   The first step in pathogenesis is host recognition. Some phytopathogenic bacteria are chemotactic in that they are attracted by specific compounds released by host plants and move to suitable infection sites by swimming, swarming, or twitching types of motility. *Dickeya dadantii*, for example, is attracted by the phytohormone jasmonate, synthesized in plant wounds (Antunez-Lamas et al. 2009). Mutants affected in motility have reduced virulence. Similarly *Ralstonia solanacearum* is attracted to amino acids and organic acids in the root exudate of host tomato plants (Yao and Allen 2006). Some pathogenic bacteria are epiphytic on leaves and must often survive for considerable periods before conditions favor the infection process. Survival during periods of dessication is enhanced by osmoprotectants such as choline or glycine betaine of either microbial or host origin (Beattie 2011). *Xanthomonas* species on plant surfaces are protected from lethal ultraviolet light damage and host-produced photosensitizers by production of xanthomonadins, membrane-bound yellow pigments (Kado 2010).

Penetration   Plant pathogenic bacteria cannot penetrate plant cuticles directly and require openings to enter into their hosts. Bacteria invade plant tissue through wounds caused by cultural cropping practices and severe weather or naturally by lateral root formation, as well as by wounds caused by feeding insects and fungal growth. Other natural openings such as lenticels, hydathodes and in particular stomata are major routes of entry (Zeng et al. 2010). As part of their defense mechanism, some plants such as tomato, pea and grape close their stomata to prevent entry of bacteria. In response, some phytopathogenic bacteria including *Pseudomonas syringae* and *Xanthomonas campestris* produce compounds as virulence factors that reverse stomatal closure. Non-motile bacteria such as *Clavibacter michiganensis* subsp. *sepedonicus*, are introduced into susceptible vascular tissues of their potato hosts through crop management practices. *Candidatus* Liberibacter spp. and *Xylella fastidiosa* are among the few plant pathogenic bacteria that are specifically vectored by insects; host specific psyllids and leafhoppers, respectively, introduce these bacteria directly into the host. Both *Ca.* Liberibacter and *Xylella* spp. lack a T3SS apparatus, which is likely unnecessary for pathogenesis because the bacteria are directly injected into the xylem vessels bypassing many of the plant defense mechanisms. Some pathovars of *Pseudomonas syringae* possess unique ice-nucleating proteins that cause rapid crystallization of water molecules when air temperature approaches freezing. Frost damage to plants intensified by such ice crystals provides a means for epiphytically dwelling pathogens to invade host plants and take advantage of nutrients released from damaged tissue.

Colonization During colonization of host plants, many bacterial pathogens (e.g. *Xanthomonas fragariae* and *Clavibacter michiganensis* subsp. *michiganensis*) form biofilms in xylem or phloem vessels. For biofilm formation, attachment to internal plant surfaces is enhanced by extrapolysaccharides, several types of pili (fimbriae), non-fimbrial adhesions, and lipopolysaccharides. To move within the plant, bacterial cells must be released again from biofilms and in some disease scenarios rhamnolipids with biosurfactant proporties play a role in reestablishing planktonic cells. Within plants, bacteria often lose their flagella and move passively or by a twitching process through the xylem vessels.

Dissemination Phytopathogenic bacteria are disseminated via plant propagation material, including seeds, irrigation and other surface water sources, wind-blown rain and aerosols, and surface-contaminated harvesting, grading and cultivating equipment and the walls of storages and transportation vehicles. Several phytopathogenic bacteria are transmitted via insects (Nadarasah and Stavrinides 2011). Flying insects in particular can spread bacterial inoculum over relatively long distances, e.g. survival times of up to 12 days have been recorded for phytopathogenic bacteria on aphids. Most phytobacterial pathogens do not multiply within insect carriers but rather are dispersed by them as non-specific surface contaminants as found for *Erwinia amylovora*, the causal agent of the fireblight disease, on honey bees. Fruit flies and other insects transmit *Pectobacterium* and *Dickeya* spp. when attracted by soft rotted plant tissues and move with contaminated surface parts to adjacent susceptible plant material. Other phytopathogenic bacteria do, however, multiply in their insect vectors. For example *Xylella fastidiosa* and *Ca.* Liberibacter are plant pathogenic bacteria that replicate, in the gut and salivary glands of their leafhopper and psyllid vectors, respectively. In most cases of insect transmission of bacterial plant pathogens, bacterial cells are introduced directly into wounds generated during piercing, sucking or chewing activities of the insects.

Survival Generally plant pathogenic bacteria are hemi-biotrophs that do not survive very well outside their plant hosts. They do survive for various lengths of time as epiphytes, as root colonizers, in water streams, in plant rhizospheres, attached to soil particles, or in organic plant debris. There are exceptions, however, because some phytopathogens, such as *Streptomyces* spp., are classical soil-dwelling organisms and do not need plant hosts to survive. For some pathogens, populations of bacteria build up on perennial host plants or in the soil environment when the same crops are replanted year after year without crop rotation. There are also bacterial plant pathogens, such as *Ralstonia solanacearum* that survive well in aquatic environments and others that survive in association with insect vectors as described above.

9.5 Control

Control of bacterial plant diseases is challenging because few options are available. There are very few agrichemicals that are effective or practical to use, so control strategies need to be largely based on avoiding contamination of crop plants with pathogenic bacteria, or by growing disease-resistant plant cultivars. Initiation of crop plants from pathogen-tested seed or vegetative plant propagules such as cuttings, tubers, or bulbs is an effective strategy for avoiding bacterial pathogens. Elimination of overt inoculum sources by roguing and destruction of diseased and suspect plant material is another way to minimize disease spread. Cultural practices that minimize plant damage prevent creating entry points for bacteria, and meticulous sanitation on farm sites prevents the build-up of inoculum sources. To avoid pathogens, seed production is done in arid and semi-arid regions which reduces the risk of air-borne inoculum. Containment of seed production in glasshouses is another strategy that minimizes airborne pathogens. Planting resistant crop cultivars is effective when good varieties are available but care must be taken that disease tolerant plants do not play a role in maintaining and dispersing inoculum sources as latent infections.

Testing of Plant Propagation Material Control strategies that are based on the planting of material free from bacterial pathogens require a means to propagate such material and tests to ensure absence from pathogens. Most bacteria can be eliminated from plants by *in vitro* tissue culturing from growing tips and be maintained pathogen-free by axenic culture. Further multiplication of plant propagation material in a protected environment helps ensure that recontamination is kept to a minimum level. To ensure that propagation material remains free from pathogens, robust, specific and sensitive methods need to be applied to test for the possible presence of pathogens. Reliable serological and DNA-based amplification methods have been developed for detection of many bacterial plant pathogens. For some assays, sensitivity is enhanced by sample incubation in selective growth media to increase the population of target bacteria prior to application of the detection assay.

Cultivation based techniques provide the most definitive evidence for the presence of plant pathogenic bacteria but is challenging for many of them because of unavailability of good selective media or poor growth characteristics in the laboratory. Furthermore, some pathogens may persist in a viable but nonculturable (VBNC) state in response to environmental stress (Oliver 2010). The VBNC state has been described for a number of phytopathogenic bacteria, including *Agrobacterium tumefaciens*, *Erwinia amylovora*, *Pseudomonas syringae*, *Ralstonia solanacearum*, *Xanthomonas campestris* and *Xanthomonas axonopodis* pv. *citri*. The use of cultivation-independent detection methods for viable cells is therefore favored if they possess adequate detection sensitivity.

Disease Detection and Diagnosis Many bacterial plant diseases can be identified on the basis of symptomology although confirmation of the diagnosis by isolation of the pathogen is usually recommended. Techniques for non-destructive monitoring of bacterial plant diseases during crop production and post-harvest have been developed, although so far they have not been used widely in practice (Sankaran et al. 2010). Such methods include spectroscopic and imaging techniques, including fluorescence and spectral imaging, infrared-, fluorescence-, multiband- and nuclear magnetic resonance spectroscopy. Fluorescence spectroscopy has been used specifically to detect *Xanthomonas axonopodis* pv. *citri* and GC-MS analysis has been used to detect volatiles in onion bulbs infected with *Pectobacterium carotovorum* subsp. *carotovorum*. Such indirect techniques detect the response of plants to infections rather than the pathogen. Consequently, latent infections cannot be detected and in particular the spectroscopic and imaging techniques cannot always distinguish the response of plants to a pathogen from that caused by other biotic or abiotic stress factors.

Chemical and Biocontrol Treatments In some cases, chemical treatments or bio-control agents can decrease populations of phytopathogenic bacteria in and on plant material to reduce damage and yield loss. The efficacy of chemical treatments, including antibiotics, copper compounds and general biocides—such as peroxides and chlorine—are limited. However in some pathosystems acceptable control levels can be achieved with streptomycin or copper sprays, such as for control of fireblight caused by *Erwinia amylovora*. However, with the high mutation rate of bacteria, typically one per million bacterial cells, development and selection of resistant mutants is a considerable problem for chemical control strategies. True seed, but also bulbs and tubers, can be treated with warm water, aerated steam or hot air to reduce populations of bacterial pathogens associated with such propagation materials. The temperature window between control and plant damage however, is very small.

The first example of successful biocontrol of a phytobacterial disease is the use of *Rhizobium rhizogenes* K84 against *Agrobacterium tumefaciens*, the causative agent of crown gall. Selective control of *Agrobacterium tumefaciens* occurs because its Ti plasmid codes for production of agrocinopine permease that is inserted into its cell membrane for uptake of nutrients, as well as plasmid-induced agrocinopines produced by the host plant. The agrocinopine permease also allows specific uptake of agrocin produced by the biocontrol agent; agrocin, a nucleotide analogue, blocks DNA synthesis in the pathogen. Another biocontrol agent is *Pantoea herbicola* used to control fireblight caused by *Erwinia amylovora*. Phage therapy has been applied commercially for control of the foliar tomato pathogens *Xanthomonas vesicatoria* and *Pseudomonas syringae* pv. *tomato* but the rapid development of resistance against the phages requires a constant selection of new hypervirulent strains.

Resistance Breeding Knowledge of molecular plant-pathogen interactions, such as pattern and effector—triggered immune systems (PTI and ETI) is useful in resistance breeding, often via marker-assisted selection. One typical example is how knowledge of the receptor-like kinase involved in PTI signaling controlled by the rice disease resistance gene Xa21 was exploited for developing broad spectrum resistance against

Xanthomonas oryzae pv. *oryzae*, causative agent of bacterial blight (Zhang et al. 2013). ETI, particularly the NBS-LRR R genes that contain a nucleotide binding site and a leucine rich repeat region are also useful targets for resistance breeding. To avoid pathogens overcoming ETI-based resistance, a strategy is used by which several R genes, each recognizing a specific range of pathogen strains, are stacked.

Disease Resistant Transgenic Crops Development of transgenic disease resistant crops is controversial because of consumers' concern and skepticism (Chaps. 15 and 16; Collinge et al. 2010). Currently no GM crops with resistance to bacterial diseases are grown commercially but release of such crops is imminent. Recently, in the US, 127 applications representing 12 % of those submitted for field testing, concerned transgenic crops with resistance against bacterial pathogens. In most cases, resistance was achieved by insertion of genes for antimicrobial proteins or metabolites. For example, transgenic potatoes were developed with resistance to *Pectobacterium carotovorum* using a synthetic antimicrobial magainin peptide originating from the amphibian *Xenopus*. Other traits introduced to achieve bacterial disease resistance include enzymes involved in plant resistance such as protein kinases, transcription factor proteins, and disruptors of quorum sensing. The use of cisgenic crops in which recipient plants are modified with homologous R-genes from the same or related species may be more acceptable to consumers and opponents of GM crops.

References

Antunez-Lamas M, Cabrera E, Lopez-Solanilla E et al (2009) Bacterial chemoattraction towards jasmonate plays a role in the entry of *Dickeya dadantii* through wounded tissues. Mol Microbiol 74:662–671

Beattie GA (2011) Water relations in the interaction of foliar bacterial pathogens with plants. Annu Rev Phytopathol 49:533–555

Bull CT, De Boer SH, Denny TP et al (2010) Comprehensive list of names of plant pathogenic bacteria, 1980–2007. J Plant Pathol 92:551–592

Charkowski A, Blanco C, Condemine G et al (2012) The role of secretion systems and small molecules in soft-rot *Enterobacteriaceae* pathogenicity. Annu Rev Phytopathol 50:425–449

Collinge DB, Jørgensen HJL, Lund OS et al (2010) Engineering pathogen resistance in crop plants: current trends and future prospects. Annu Rev Phytopathol 48:269–291

Dellagi A, Segond D, Rigault M et al (2009) Microbial siderophores exert a subtle role in Arabidopsis during infection by manipulating the immune response and the iron status. Plant Physiol 150:1687–1696

Hogenhout SA, Loria R (2008) Virulence mechanisms of Gram-positive plant pathogenic bacteria. Curr Opin Plant Biol 11:449–456

Kado CI (2010) Plant bacteriology. APS Press, St. Paul, p 336

Li Y, Huang F, Lu Y et al (2013) Mechanism of plant–microbe interaction and its utilization in disease-resistance breeding for modern agriculture. Physiol Mol Plant Pathol 83:51–58

Nadarasah G, Stavrinides J (2011) Insects as alternative hosts for phytopathogenic bacteria. FEMS Microbiol Rev 35:555–575

Newman MA, Dow JM, Daniels MJ (2001) Bacterial lipopolysaccharides. Mol Plant Pathogen Interact 107:95–102

Oliver JD (2010) Recent findings on the viable but nonculturable state in pathogenic bacteria. FEMS Microbiol Rev 34:415–425

Records AR (2011) The Type VI secretion system: a multipurpose delivery system with a phage-like machinery. Mol Plant-Microbe Interact 24:751–757

Rosenberg E, DeLong E, Lory S et al (2013) Virulence strategies of plant pathogenic bacteria. In: The prokaryotes. Springer, Berlin, pp 61–82

Sankaran S, Mishra A, Ehsani R et al (2010) A review of advanced techniques for detecting plant diseases. Comput Electron Agric 72:1–13

Schloss PD, Handelsman J (2004) Status of the microbial census. Microbiol Mol Biol Rev 68:686–691

Venturi V, Fuqua C (2013) Chemical signaling between plants and plant-pathogenic bacteria. Annu Rev Phytopathol 51:17–37

Yao J, Allen C (2006) Chemotaxis is required for virulence and competitive fitness of the bacterial wilt pathogen *Ralstonia solanacearum*. J Bacteriol 188:3697–3708

Zeng W, Melotto M, He SY (2010) Plant stomata: a checkpoint of host immunity and pathogen virulence. Curr Opin Biotechnol 21:599–603

Zhang Q, Wing RA, Chen H et al (2013) Transformation and transgenic breeding. In: Genetics and genomics of rice, vol 5. Springer, New York, pp 363–386

Chapter 10
Plant Pathogenic Fungi and Oomycetes

Pierre J. G. M. de Wit

Abstract Fungi and Oomycetes are notorious plant pathogens and use similar strategies to infect plants. The majority of plants, however, is not infected by pathogens as they recognize pathogen-associated molecular patterns (PAMPs) by pattern recognition receptors that mediate PAMP-triggered immunity (PTI), a basal defense response effective against potential pathogens. Successful pathogens secrete effectors to suppress PTI and alter host plant physiology. In turn, plants have evolved immune receptors that recognize effectors, resulting in effector-triggered immunity (ETI). ETI includes the hypersensitive response which is effective against biotrophic plant pathogens that require living cells to feed on. Other pathogens are hemi-biotrophic, which start infection as a biotroph, but after having colonized the host tissue can also feed on death tissue. Necrotrophic pathogens kill host tissue before they start to feed on it. Co-evolution between pathogens and their hosts had led to the development of numerous effectors produced by pathogens and corresponding resistance proteins in host plants, which has generated an arms race genetically described by the gene-for-gene concept. Resistance genes can now successfully be transferred to crop plants by classical breeding or as transgenes stapled into one cultivar.

10.1 Fungal and Oomycetous Pathogens and Their Life Styles

Fungi and Oomycetes are uni- or multicellular eukaryotic heterotrophic organisms producing filamentous structures, designated hyphae, that show tip growth, except for yeasts that multiply by budding. Hyphae are usually 1–2 μm in diameter, but may reach more than 100 μm for some fungi. Cells of hyphae contain one, two or multiple nuclei often divided by septa. Until the 1990s, Oomycetes were considered true fungi, but based on genome analyses they are now classified among the Chromista (Chap. 39). Pathogenic Oomycetes are still treated as fungi because they have many properties in common, including their filamentous growth and their mode of plant

P. J. G. M. de Wit (✉)
Laboratory of Phytopathology, Wageningen University,
Droevendaalsesteeg 1, 6708 PB Wageningen, The Netherlands
Tel.: + 31 317 48 31 30
e-mail: Pierre.dewit@wur.nl

© Springer International Publishing Switzerland 2015
B. Lugtenberg (ed.), *Principles of Plant-Microbe Interactions,*
DOI 10.1007/978-3-319-08575-3_10

infection. Fungal cell walls contain chitin, α- and β- glucans and (glyco)proteins, but no cellulose, while those of Oomycetes contain cellulose and glucans but lack chitin. Several fungi and Oomycetes are important for industrial production of enzymes, bread, cheese, alcohol and organic acids. As producers of antibiotics they can have important medical applications. On the other hand, many fungi and Oomycetes can cause disease to humans, animals and plants. Around 150,000 fungal species have now been described, which represent around 10 % all fungi estimated to be present on earth (Hawksworth 1991). The most important fungal plant pathogens belong to (i) Ascomycetes, producing sexual spores (ascospores) in a sac-like structure, the ascus, and asexual spores (conidia), (ii) Basidiomycetes producing sexual spores (basidiospores) on a basidium, dikaryotic vegetative mycelium and asexual spores, (iii) Oomycetes producing sexual spores (oospores) and asexual spores (sporangia that can germinate directly or produce zoospores) (Agrios 2005). In the remainder of this chapter I will jointly discuss fungi and Oomycetes as fungi, unless phenomena are specific to Oomycetes only. Fungi are one of the largest living organisms; the soil-borne fungus *Armillaria ostoyae* is estimated to be 2400 years old, covering 8.4 km^2 of soil in Oregon, USA (Burdsall and Volk 2008). Pathogenic fungi enter plants via natural openings (e.g. stomata) or penetrate directly by a penetration peg produced by an appressorium on a host cell (see Chap. 25). Many fungi produce haustoria in plant cells, specialized feeding organs for retrieval of nutrients. Extracellular pathogens grow as epiphytes on the outside of plants or in the apoplastic space between cells without producing haustoria. Ascomytous fungi are haploid for the major part of their life and produce haploid spores (1N). During the sexual stage haploid hyphae of Ascomycetes may fuse to produce a dikaryon inside the ascogenous hyphae where the nuclei soon fuse to produce a zygote (2N) that quickly divides meiotically to produce eight haploid ascospores per ascus. Oomycetes are diploid for the major part of their life and have a life cycle that is very similar to that of algae. During the sexual stage Oomycetes produce gametangia in which meiosis occurs, followed by fertilization and production of the diploid zygote, the oospore, that produces a sporangium that geminates directly or indirectly by producing zoospores. Basidiomycetes are dikaryotic (N + N) for the major part of their life. During the sexual stage in the basidium the paired nuclei fuse and form a zygote (2N) that quickly divides meiotically to produce four haploid basidiospores (Agrios 2005). In most fungi the asexual cycle repeats multiple times during the growth season and is most damaging to plants, whereas the sexual cycle usually occurs only once a year at the end of the growth season when host plants senesce and nutrients become limiting.

Diseases Caused by Oomycetes The most important oomycetous plant pathogens comprising species belonging to the genera *Pythium, Phytophthora, Peronospora* and *Plasmopara*. The late blight disease of potato is an important oomycetous pathogen discussed in Chap. 39. Downy mildew of grape is another important disease caused by *Plasmopara viticola* (Fig. 10.1a), that almost completely destroyed the grape and wine industry in France soon after it was imported into Europe from the United States around 1875; the first fungicide, Bordeaux mixture, effective against *P. viticola* was

Fig. 10.1 Symptoms of economically important plant diseases caused by fungi and Oomycetes. **a** Plasmopara viticola, downy mildew of grape (http://www.biolib.cz/cz/image/id100651/). **b** *Blumeria graminis* f.sp. *tritici,* powdery mildew of wheat (http://www.apsnet.org/publications/imageresources/Pages/fi00189.aspx). **c** *Nectria galligena*, apple canker (http://www.downgardenservices.org.uk/cankerapp.htm). **d** *Venturia inaequalis*, apple scab (http://www.nature.com/news/us-regulation-misses-some-gm-crops-1.13580). **e** *Ophiostoma ulmi*, Dutch elm disease (http://commons.wikimedia.org/wiki/File:Ceratocystis_ulmi_1_beentree.jpg). **f** *Melampsora lini*, flax rust (http://www.sciencearchive.org.au/events/frontiers/frontiers2008/dodds.html). **g** *Puccinia graminis* f.sp. *tritici,* wheat leaf rust (http://www.mississippi-crops.com/2012/03/02/wheat-leaf-rust-and-stripe-rust-update-march-2-2012/). **h** *Ustilago tritici*, loose smut of wheat (http://www.bayercropscience.cl/soluciones/fichaproblema.asp?id=27). **i** *Ustilago maydis*, corn smut (http://commons.wikimedia.org/wiki/File:Maisbrand,_Maisbeulenbrand_(Ustilago_maydis)_-_hms(1).jpg)

discovered by accident in 1885. Downy mildews are obligate biotrophic pathogens that overwinter as oospores. Many resistance genes against downy mildews have been cloned (Takken and Goverse 2012). They belong to the cytoplasmic nucleotide binding, leucine-rich repeat protein receptors (NB-LRRs) (discussed later).

Diseases Caused by Ascomycetes Powdery mildews can infect virtually all plant species and are easy recognizable by the white powdery overlay on infected plant leaves (Fig. 10.1b). Powdery mildews are obligate biotrophic pathogens that can only infect epidermis cells from which they retrieve nutrients by multiple haustoria. Short conidiophores are produced on the plant surface producing chains of conidia.

At the end of the growth season the fungus may produce cleistothecia with asci and ascospores. Many resistance genes against powdery mildews have been cloned. They belong to the cytoplasmic nucleotide binding, NB-LRRs (Takken and Goverse 2012) (discussed later).

Tree Cankers These are caused by ascomycetous fungi. Cankers generally begin at a wound from which they expand in all directions, but the host may survive the disease by producing callus tissue around the dead areas thereby limiting the canker. In subsequent years the fungus invades additional healthy tissue, and new concentric ridges of callus tissue are produced every year, resulting in a typical canker. Canker of apple is caused by *Nectria galligena* (Fig. 10.1c), one of the most important diseases of apple worldwide. Apple scab is caused by *Venturia inaequalis* (Fig. 10.1d), and occurs in areas with cool, moist springs and summers. Infected fruits develop scab lesions and the cuticle is ruptured at the margin of these lesions. The mycelium in living tissues is located between the cuticle and epidermal cells, and produces short conidiophores bearing conidia. The fungus is a hemi-biotroph starting infection as a biotroph, but at the end of the growth season, the mycelium grows through dead leaf tissues and produces pseudothecia with asci and ascospores. In spring, when pseudothecia become thoroughly wet, the asci forcibly discharge the ascospores that infect young apple leaves. Apple varieties resistant against apple scab exist, and the resistance genes encode typical receptor-like proteins (RLPs) (discussed later).

Vascular Wilts These are widespread, very destructive plant diseases and are caused by fungal pathogen residing in the xylem vessels of plants that may be clogged with mycelium, spores, or polysaccharides produced by the fungus and gels and gums produced by plant cells upon attack by the fungus. In some hosts, tyloses are produced by parenchyma cells adjoining xylem vessels. The disease can sometimes be controlled by using disease-resistant cultivars.

Dutch Elm Disease This disease is caused by *Ophiostoma ulmi* (Fig. 10.1e). It is a very destructive wilt disease that affects all elm species. Usually trees that become infected in spring or early summer die quickly, whereas those infected in late summer are less seriously affected and may recover. Perithecia with asci and ascospores may be produced. The spread of the Dutch elm disease fungus depends on bark beetles belonging to the genus *Scolytus* that carry fungal spores from infected wood to healthy elm trees. The fungus overwinters in the bark of dead elm trees as mycelium or spores. Adult female beetles tunnel through the bark and lay eggs that, after hatching, develop into adult beetles that carry thousands of spores. The beetles feed on healthy elms and carry fungal spores that infect healthy xylem vessels. Control of Dutch elm disease depends primarily on removal and destruction of diseased elm trees. Inoculation of the xylem vessels with a spore suspension of the wilt fungus *Verticillium dahliae* offers protection through induced resistance (see Chap. 14), whereas protection can also be achieved by inoculating trees with particular strains of *Pseudomonas* bacteria. Resistance genes against wilt diseases have been cloned; they encoded either NB-LRRs or RLPs (discussed later).

Diseases Caused by Basidiomycetes   Basidiomycetes produce their sexual spores, basidiospores, on a club shaped basidium. Most Basidiomycetes are either saprobes or fungi causing wood decay including root and stem rots of trees. However, they also include plant pathogens that cause rust and smut diseases.

Rusts   Rusts are obligate biotrophic pathogens that have caused many famines in the history of many countries by destroying cereal crops. Presently the rust strain Ug99 causes a dramatic epidemic on wheat in eastern Africa from which it has spread into Arabia (Ward 2007). There are about 5000 species of rusts, mostly belonging to the genus *Puccinia*. They are very specialized pathogens attacking only particular genera or only certain species or even only some varieties of particular cereals. In the latter case they are called races that can be identified only by a set of differential varieties carrying different resistance genes. The gene-for-gene system proposed by Flor in the USA was based on research performed on the rust fungus of flax caused by *Melampsora lini* (Fig. 10.1f), (Flor 1971). Rust fungi can produce up to five different types of spores: uredospores, teliospores and basidiospores (on wheat), spermatia and aeciospores (on barbary) (Agrios 2005).

Stem Rust of Wheat   This disease is caused by *Puccinia graminis* f.sp. *trititici* (Fig. 10.1g), and affects wheat wherever it is grown. It attacks all aboveground parts of wheat and the alternate host barberry *(Berberis vulgaris)*. Symptoms on wheat appear as pustules and uredia. The latter produce red-colored uredospores. Later in the season two-celled black teliospores are produced in telia. On barberry, spermagonia with spermatia and receptive hyphae are produced on the upper side of leaves, and on the lower side orange cup-like aecia are produced containing aeciospores. The most effective means of control of the rust pathogen is by growing resistant wheat varieties. The resistance proteins are typical NB-LRR receptor proteins (Takken and Goverse 2012) (discussed later).

Smut Fungi   More than 1200 species of smut fungi exist that occur throughout the world. Most smut fungi attack the ovaries of grains and grasses and develop in the kernels, which they destroy completely. Some smuts infect seeds or seedlings before they emerge from the soil, in which they grow systemically until they reach the inflorescence. Cells in affected tissues are either destroyed and replaced by black smut spores, or they are first stimulated to divide and enlarge to produce a swelling or gall that is subsequently destroyed and replaced by the black smut spores. Most smut fungi produce only teliospores and basidiospores. When haploid basidiospores germinate, the germ tubes fuse to produce dikaryotic infectious mycelium. Important smut fungi are *Ustilago tritici* (Fig. 10.1h), causing smut on wheat and *Ustilago maydis* (Fig. 10.1i), causing smut galls on corn. Together with Flor in the USA, Oort in the Netherlands proposed the gene-for-gene hypothesis based on research performed on *Ustilago tritici* and wheat (Oort 1944).

Root Rots   Root rots of trees are often caused by species of *Armillaria* that occur worldwide and affect hundreds of species of trees. The pathogen *Armillaria mellea* is one of the most common fungi in forest soils. Diagnostic characteristics of *Armillaria* root rot appear at decayed areas in the bark, at the root-stem junction, and on the roots.

White mycelial mats are formed between the bark and wood. Another characteristic sign of the disease is the formation of rhizomorphs or "shoe strings", cordlike threads of mycelium 1–3 mm in diameter that can grow long distances and infect healthy trees (Agrios 2005).

10.2 Characteristcis of Fungal and Oomycetous Plant Pathogens

These organisms can be obligate biotrophic, biotrophic, hemi-biotrophic or necrotrophic pathogens. Biotrophic pathogens thrive on living host cells. Some biotrophs cannot be cultured on synthetic media and are called obligate biotrophic pathogens like rusts, downy mildews and powdery mildews. However, many biotrophic pathogens grow on host plants under natural conditions but can still be cultured on synthetic media like the tomato pathogen *Cladosporium fulvum* (Joosten and De Wit 1999). Hemi-biotrophic pathogens start infection as a biotroph, but later in the growth season they can live as a saprophyte on death host tissue. A defense response that is very efficient against obligate biotrophic and biotrophic pathogens is the hypersensitive response (HR), death of a few plant cells at the site of infection. In contrast, necrotrophic fungal pathogens kill host tissue before they retrieve nutrients, and the HR is not effective against them.

Extracellular and Intracellular Pathogens   Many fungi live on plants as epiphytes of which some have developed into pathogens. They usually enter plants through stomata and thrive in the apoplastic space surrounding cells without producing haustoria. Most extracellular pathogens are slow growers and show long latent periods as the apoplast often contains antimicrobial compounds and antimicrobial enzymes, and is poor in nutrients. Intracellular pathogen often produce haustoria.

10.3 Fungal and Oomycetous Infection Strategies and Host Defense Mechanisms

Infection Strategies   Infection strategies employed by fungal pathogens depend strongly on where they thrive in their host plants. The cell wall is a major obstacle for plant pathogens which contains physical and chemical barriers like the plant waxy cuticle, (lignified) cell walls and antimicrobial metabolites and proteins (Balmer et al. 2013). Many fungal genomes have been sequenced showing that pathogenic fungi contain many different classes of cell-wall degrading enzymes. The expression of these genes is highly up-regulated during infection by necrotrophic plant pathogens (Zhao et al. 2013). Biotrophic pathogens often contain similar numbers of genes encoding cell-wall degrading enzymes as necrotrophs, but they are often lowly expressed by these pathogens or only at specific sites and phases of infection. Necrotrophic fungi often also produce secondary metabolites that are toxic to plant cells thereby facilitating their necrotrophic lifestyle, while those metabolites are often absent in (obligate) biotrophic pathogens. For example, the powdery mildew *Blumeria graminis* contains hardly any genes encoding secondary metabolites (Spanu 2012) and in the tomato leaf pathogen *C. fulvum* these genes are down-regulated during infection (De Wit et al. 2012). Apart from producing enzymes that enable pathogens to degrade plant cell walls and to retrieve the released mono/oligosaccharides, fungi also need to protect themselves against antifungal proteins that are present in plant-cell walls and the apoplast by detoxifying enzymes like tomatinase produced by *Cladosporium fulvum* that detoxifies the toxic saponin, α-tomatine occurring at high concentrations in tomato (Okmen et al. 2013).

Basal Defense Strategies   Plants have developed sophisticated defense strategies to recognize pathogens and to defend themselves against fungal pathogens. All plants can recognize pathogen-associated molecular patterns (PAMPs), like chitin from fungi, by pattern recognition receptors (PRRs) that mediate PAMP-triggered immunity (PTI), a response that protects plants against potential microbial pathogens (Jones and Dangl 2006; Liebrand et al. 2014). PRRs are extracellular LRR-containing receptor-like transmembrane proteins with a cytoplasmic kinase signaling domain known as RLKs. They mediate PAMP-triggered basal structural and chemical defense responses including callose deposition, accumulation of reactive oxygen species (ROS), cell wall enforcements and accumulation of pathogenesis-related (PR) proteins including chitinases, proteases and glucanases.

Effector-Triggered Susceptibility   Although plants have developed basal defense strategies against microbes, successful pathogens have found ways to overcome basal defense responses. They can suppress PTI by secreting different types of effectors that target various components of PTI (Stergiopoulos and de Wit 2009). By suppressing PTI they cause effector-triggered susceptibility (ETS). Various types of effectors have been described in various pathogenic fungi (Stergiopoulos and De Wit 2009). Overall they manipulate host defenses and host physiology to facilitate virulence in various ways. Here I discuss the intrinsic functions of two *Cladosporium fulvum* effectors that can serve as an example of many other fungal effectors. *C. fulvum* secretes the Avr2 effector protein which inhibits plant cysteine proteases including Rcr3[pim] (required for *C. fulvum* resistance 3) (Rooney et al. 2005). Heterologous expression of *Avr2* in tomato and Arabidopsis resulted in increased susceptibility to different fungal pathogens including *Botrytis cinerea* and *Verticillium dahliae* (Van Esse et al. 2008). Mutation studies on Rcr3[pim] showed it also to be active against *Phytophthora infestans*, which secretes EPIC1 and EPIC2B proteins that also bind and inhibit Rcr3[pim]. Also the root parasitic nematode *Globodera rostochiensis* secretes a venom allergen-like effector protein, Gr-VAP1, that targets and inhibits Rcr3 perturbing its active site (Lozano-Torres et al. 2012).

   C. fulvum also secretes the cysteine-rich Avr4 protein which is a chitin-binding lectin that protects fungal cell walls against plant chitinases, providing a defensive role during infection (Van den Burg et al. 2006). In addition, silencing of the *Avr4* gene in *C. fulvum* reduces virulence on tomato plants (Van Esse et al. 2007). Protection against plant chitinases is important for virulence of most plant pathogenic fungi

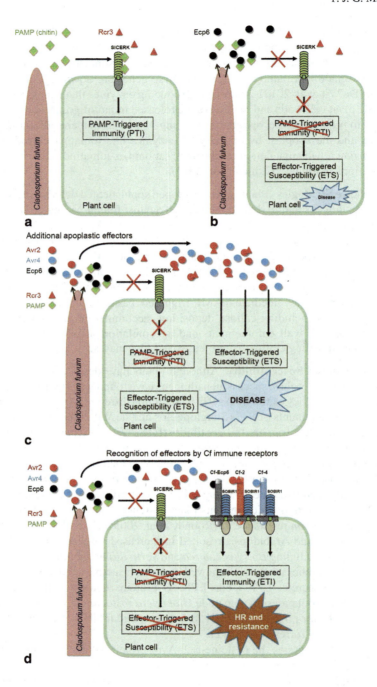

Fig. 10.2 Schematic overview of an evolutionary scenario for adaptation of the leaf mould pathogen *Cladosporium fulvum* to tomato. **a** Chitin fragments (pattern-associated molecular pattern; *PAMP*) are recognized by the tomato chitin receptor (SlCERK) that triggers PAMP-triggered immunity

because chitin is a basic component of fungal cell walls. Indeed homologs of *Avr4* have been identified in several dothideomycetous fungi, including *Mycosphaerella fijiensis* and *Dothistroma septosporum* (Stergiopoulos et al. 2010). Interestingly many fungi, including *C. fulvum*, also secrete proteins that contain LysM carbohydrate-binding domain proteins (Kombrink and Thomma 2013). The LysM protein Ecp6 of *C. fulvum* also binds to chitin, but does not protect the fungus against basic plant chitinases. Ecp6 mainly binds small chitin fragments (PAMPs) that are released from the fungal cell wall *in planta* and prevents them from being recognized by chitin receptors present in plants to induce PTI (De Jonge et al. 2010). Ecp6 is a virulence factor because silencing of *Ecp6* results in reduced virulence (Kombrink and Thomma 2013). Many homologs of *Ecp6* are identified in various fungal species (De Jonge and Thomma 2009; Kombrink and Thomma 2013) , suggesting an important function in preventing chitin-triggered immunity in many plant-fungus interactions. Recently, it was shown that *Mycosphaerella graminicola* secretes three LysM effectors called Mg1LysM, Mg3LysM and MgxLysM (Marshall et al. 2011) of which both Mg1LysM and Mg3LysM bind chitin, whereas only Mg3LysM prevents chitin-triggered immunity. However, only the Mg3LysM deletion mutant showed significantly reduced virulence. *Magnaporthe oryzae* LysM effector Slp1 also binds and scavenges small chitin fragments. In rice, the chitin elicitor binding protein, CEBiP, recognizes chitin and activates PTI. Indeed it was found that Slp1 prevents chitin-triggered immunity by competing with the CEBiP receptor for chitin binding (Mentlak et al. 2012).

Effector-Triggered Immunity   In turn, plants have evolved sophisticated ways to recognize and respond to effectors. In addition to PRRs that recognize PAMPs, plants have developed immune receptors that recognize effectors or host plant targets manipulated by effectors resulting in effector-triggered immunity (ETI) (Jones and Dangl 2006). One of the most typical characteristics of ETI is the HR. At the host species and cultivar level, co-evolution between hosts and their pathogens has caused an arms race that has led to the development of numerous novel effectors and corresponding resistance proteins, which are described by the gene-for-gene concept (De Wit et al. 2009) (Fig. 10.2a, 10.2b, 10.2c, 10.2d).

Fig. 10.2 (continued) (*PTI*) providing basal defense against the pathogen. **b** To become pathogenic, *C. fulvum* secretes the Ecp6 effector that scavenges chitin fragments to prevent PTI. **c** *C. fulvum* secretes additional effectors to increase virulence including Avr2 inhibiting apoplastic cysteine protease Rcr3, and Avr4 protecting chitin present in the cell wall of the fungus against tomato chitinases. **d** Tomato responds by developing RLP immune receptors Cf-Ecp6, Cf-2 and Cf-4 to recognize Ecp6, Avr2 and Avr4, respectively, and to mediate effector-triggered immunity (*ETI*) leading to the hypersensitive response (*HR*) and resistance against fungal strains secreting these effectors. *C. fulvum* is supposed to produce numerous effectors and tomato a comparable number of immune receptors

10.4   Resistance Against Fungi and Oomycetes

The cloning of effector genes speeded up the cloning of the matching resistance genes that encode either cell surface-localized receptor-like proteins known as RLPs (Liebrand et al. 2014) or cytoplasmic nucleotide binding, leucine-rich repeat (NB-LRR) or NLR proteins (Takken and Goverse 2012). RLPs are integral plant membrane proteins containing an extracellular leucine-rich repeat or LRR domain, a membrane spanning domain, and a short cytoplasmic tail without signaling domain. RLPs recognize effectors of extracellular fungal pathogens and mediate ETI. However, cytoplasmic pathogens exploit the cytoplasm of plant cells by injecting effectors into host cells that interact with cytoplasmic targets to suppress PTI. The cytoplasmic effectors are usually recognized by cytoplasmic NLR immune receptors. Pathogens can develop mutations in effectors or simply loose them to escape recognition by RLP and NRL immune receptors or develop new effectors for compensation. Some immune receptors can work in concert in receptor complexes active against more than one pathogen (Macho and Zipfel 2014). Defense systems of plants are able to respond to different types of PAMPs and effectors, but they cannot always be distinguished easily and downstream defense pathways in plants activated during PTI and ETI often overlap and operate against a broad spectrum of pathogens (Thomma et al. 2011). These responses include the generation of reactive oxygen species (ROS), antimicrobial phytoalexins, chitinases, glucanases, proteases, and often the HR. Many immune receptors encoded by resistance genes have been cloned in the last decade. They can now be introduced in many copies by breeders in crops by classical breeding or by cis/transgenesis (Zhu et al. 2012). It is important to introduce multiple immune receptor genes in plants against multiple pathogen effectors in order to obtain durable plant resistance (Brunner et al. 2010; Vleeshouwers and Oliver 2014) . Overcoming multiple immune receptors by a pathogen by modification or loss of effectors is expected to cause a fitness cost and decrease in virulence.

References

Agrios GN (2005) Plant pathology handbook, 5th edn. Elsevier, Amsterdam, 922 p

Balmer D, Planchamp C, Mauch-Mani B (2013) On the move: induced resistance in monocots. J Exp Bot 64:1249–1261

Brunner S, Hurni S, Streckeisen P et al (2010) Intragenic allele pyramiding combines different specificities of wheat Pm3 resistance alleles. Plant J 64:433–445

Burdsall HH, Volk TJ (2008) *Armillaria solidipes*, an older name for the fungus called *Armillaria ostoyae*. N Am Fungi 3:261–267

De Jonge R, Thomma BPHJ (2009) Fungal LysM effectors: extinguishers of host immunity? Trends Microbiol 17:151–157

De Jonge R, Van Esse HP, Kombrink A et al (2010) Conserved fungal LysM effector Ecp6 prevents chitin-triggered immunity in plants. Science 329:953–955

De Wit PJGM, Mehrabi R, Van den Burg HA et al (2009) Fungal effector proteins: past, present and future. Mol Plant Pathol 10:735–747

De Wit PJGM, Van der Burgt A, Ökmen B et al (2012) The genomes of the fungal plant pathogens *Cladosporium fulvum* and *Dothistroma septosporum* reveal adaptation to different hosts and lifestyles but also signatures of common ancestry. PLoS Genet. doi:10.1371/journal.pgen.1003088

Flor HH (1971) Current status of the gene-for-gene concept. Annu Rev Phytopathol 9:275–296

Hawksworth DL (1991) The fungal dimension of biodiversity -magnitude, significance, and conservation. Mycol Res 95:641–655

Jones JDG, Dangl JL (2006) The plant immune system. Nature 444:323–329

Joosten MHAJ, De Wit PJGM (1999) The tomato-*Cladosporium fulvum* interaction: a versatile experimental system to study plant-pathogen interactions. Annu Rev Phytopathol 37:335–367

Kombrink A, Thomma BPHJ (2013) LysM effectors: secreted proteins supporting fungal life. Plos Pathog 9:e1003769. doi:10.1371/journal.ppat.1003769

Liebrand TWH, van den Burg HA, Joosten MHAJ (2014) Two for all: receptor-associated kinases SOBIR1 and BAK1. Trends Plant Sci 19:123–132

Lozano-Torres JL, Wilbers RHP, Gawronski P et al (2012) Dual disease resistance mediated by the immune receptor Cf-2 in tomato requires a common virulence target of a fungus and a nematode. Proc Natl Acad Sci U S A 109:10119–10124

Macho AP, Zipfel C (2014) Plant PRRs and the activation of innate immune signaling. Mol Cell 54:263–272. doi:10.1016/j.molcel.2014.03.028

Marshall R, Kombrink A, Motteram J et al (2011) Analysis of two in planta expressed LysM effector homologs from the fungus *Mycosphaerella graminicola* reveals novel functional properties and varying contributions to virulence on wheat. Plant Physiol 156:756–769

Mentlak TA, Kombrink A, Shinya T et al (2012) Effector-mediated suppression of chitin-triggered immunity by *Magnaporthe oryzae* is necessary for rice blast disease. Plant Cell 24:322–335

Okmen B, Etalo DW, Joosten MHAJ et al (2013) Detoxification of α-tomatine by *Cladosporium fulvum* is required for full virulence on tomato. New Phytol 198:1203–1214

Oort AJP (1944) Onderzoekingen over stuifbrand. II. Overgevoeligheid van tarwe voor stuif-brand (*Ustilago tritici*) with a summary: hypersensitiviness of wheat to loose smut. Tijdschr Plantezieekten 50:73–106

Rooney HCE, Van 't Klooster JW, Van der Hoorn RAL et al (2005) Cladosporium Avr2 inhibits tomato Rcr3 protease required for Cf-2-dependent disease resistance. Science 308:1783–1786

Spanu PD (2012) The genomics of obligate (and nonobligate) biotrophs. Annu Rev Phytopathol 50:91–109

Stergiopoulos I, De Wit PJGM (2009) Fungal effector proteins. Annu Rev Phytopathol 47:233–263

Stergiopoulos I, van den Burg HA, Okmen B et al (2010) Tomato Cf resistance proteins mediate recognition of cognate homologous effectors from fungi pathogenic on dicots and monocots. Proc Natl Acad Sci U S A 107:7610–7615

Takken FLW, Goverse A (2012) How to build a pathogen detector: structural basis of NB-LRR function. Curr Opin Plant Biol 15:375–384

Thomma B, Nurnberger T, Joosten M (2011) Of PAMPs and effectors: the blurred PTI-ETI dichotomy. Plant Cell 23:4–15

Van den Burg HA, Harrison SJ, Joosten MHAJ et al (2006) *Cladosporium fulvum* Avr4 protects fungal cell walls against hydrolysis by plant chitinases accumulating during infection. Mol Plant Microbe Interact 19:1420–1430

Van Esse HP, Bolton MD, Stergiopoulos I et al (2007) The chitin-binding *Cladosporium fulvum* effector protein Avr4 is a virulence factor. Mol Plant Microbe Interact 20:1092–1101

Van Esse HP, Van 't Klooster JW, Bolton MD et al (2008) The *Cladosporium fulvum* virulence protein Avr2 inhibits host proteases required for basal defense. Plant Cell 20:1948–1963

Vleeshouwers VGAA, Oliver RP (2014) Effectors as tools in disease resistance breeding against biotrophic, hemibiotrophic, and necrotrophic plant pathogens. Mol Plant Microbe Interact 27:196–206

Ward R (2007) The global threat posed by Ug99. Phytopathol 97: S136

Zhao ZT, Liu HQ, Wang CF et al (2013) Comparative analysis of fungal genomes reveals different plant cell wall degrading capacity in fungi. BMC Genomics 14:274. doi:10.1186/1471-2164-14-274

Zhu SX, Li Y, Vossen JH et al (2012) Functional stacking of three resistance genes against *Phytophthora infestans* in potato. Transgenic Res 21:89–99

Chapter 11
Phytopathogenic Nematodes

Johannes Helder, Mariëtte Vervoort, Hanny van Megen, Katarzyna
Rybarczyk-Mydłowska, Casper Quist, Geert Smant and Jaap Bakker

Abstract Soil is teeming with life, and rhizosphere soil is even more densely in-
habited than bulk soil. In terms of biomass, bacteria and fungi are dominant groups,
whereas nematodes (roundworms) are the most abundant Metazoans. Bulk soil, soil
not directly affected by living plant roots, typically harbours around 2000–4000 nematodes per 100 g, while in the rhizosphere these numbers should be multiplied by
a factor 3–5. This difference is not only explained by a higher density of plant parasites, as also bacterivorous and fungivorous nematodes benefit from the local boost
of the bacterial and fungal community. Most nematodes feeding on higher plants
are obligatory parasites. In this chapter four independent lineages of plant-parasitic
nematodes are discussed. Facultative plant parasites often occupy basal positions

J. Helder (✉) · M. Vervoort · H. van Megen · C. Quist · G. Smant · J. Bakker
Laboratory of Nematology, Wageningen University, Droevendaalsesteeg 1,
6708 PB Wageningen, The Netherlands
Tel.: +31 317 483 136
e-mail: Hans.Helder@wur.nl

K. Rybarczyk-Mydłowska
Museum and Institute of Zoology PAS, Wilcza 64, 00-679 Warsaw, Poland
Tel.: +48 888981281
e-mail: katarzynar@miiz.waw.pl

M. Vervoort
Tel.: +31 317 483136
e-mail: Jet.Vervoort@wur.nl

H. van Megen
Tel.: +31 33 2864283
e-mail: Hannyvanmegen@kpnmail.nl

C. Quist
Tel.: +31 317 483136
e-mail: Casper.Quist@wur.nl

G. Smant
Tel.: +31 317 485154
e-mail: Geert.Smant@wur.nl

J. Bakker
Tel.: +31 317 483136
e-mail: Jaap.Bakker@wur.nl

© Springer International Publishing Switzerland 2015
B. Lugtenberg (ed.), *Principles of Plant-Microbe Interactions*,
DOI 10.1007/978-3-319-08575-3_11

91

Fig. 11.1 Schematic overview of a plant-parasitic nematode. Amphids and phasmids are chemosensory organs. The stylet is a protrusible, hollow puncturing device, that is used to penetrate plant cell walls. Figure courtesy of Shinya et al. (2013)

within a lineage. Most, but not all, economically high impact plant parasites such as root knot, cyst and lesion nematodes belong to the most distal nematode clade (Clade 12; Holterman et al. Mol Biol Evol 23:1792–1800, 2006). In this chapter, some of the latest insights on the evolution, the ecology and the biology of phytopathogenic nematodes will be covered.

11.1 Nematodes

Nematodes are small (mostly between 0.2 and 2.5 mm in length), worm-shaped animals that together constitute the phylum Nematoda. This phylum, which is thought to have arisen during early phases of the Cambrian explosion (≈ 550 million years ago) in marine habitats, belongs to the superphylum Ecdysozoa that encompasses all moulting animals (Aguinaldo et al. 1997). Nematodes are not only present in terrestrial systems, but also in freshwater and marine habitats (Bongers and Ferris 1999). Next to their ubiquity, nematodes are abundant and can reach densities of up to millions per square meter in soil, residing in the water films attached to soil particles or e.g. in or around plant roots. Due to their size and their mobility, nematodes are easily extractable from soil as compared to protozoans, fungi and bacteria. The phylum Nematoda show a high trophic diversity; nematodes may feed upon bacteria, fungi, protozoa, algae, other nematodes, a combination of the aforementioned food sources (omnivores), or they may be facultative or obligate parasites of plants or animals (Yeates et al. 1993). Due to this diversity in feeding habits, nematodes can be found at all three levels of the soil food web (Ferris et al. 2001). Plant parasites, in fact herbivores, reside in the first trophic level as they feed directly on roots of higher plants. Bacterivores and fungivores are found in the second trophic level as they feed on primary decomposers. Carnivorous nematodes, also referred to as predators, feed mainly on other nematodes and are therefore positioned at the third trophic level. It should be noted that plant-parasites (see Fig. 11.1) usually constitute a minority within the nematode community, both in terms of number of individuals as well as of species.

11.2   Use of an SSU rDNA Framework to Reconstruct the Evolution of Nematodes

Nematodes are morphologically highly conserved; even for experts, the ability to identify certain species depends on the life stage or sex of the present individuals (Floyd et al. 2002). Considering the age of this phylum, the saying 'never change a winning team' is certainly applicable. As the number of informative morphological characters is limited, identification of nematodes on the basis of such characters is challenging and requires a considerable amount of time, experience and expertise.

Keeping the economical relevance of this animal phylum in mind, it is remarkable to see that nematode systematics is far from established. It has a long history of constant revision, and over a dozen general schemes for nematode classification have been proposed. One of the first phylum-wide classifications was proposed by Chitwood and Chitwood (1933). They divided the phylum into two classes, the Phasmidia and Aphasmidia, later renamed to Secernentea and Adenophorea respectively. This was based mainly on the fact that the Secernentea share several characters, including the presence of phasmids, small sensory organs on the tail. Although it was already recognized at the time that the Adenophorea did not form a natural group, the division of the Nematoda into these two groups persisted for a long time. The first person to apply cladistic principles to nematode systematics was Lorenzen (Lorenzen 1981). He also recognized that the Adenophorea were not a monophyletic group, but could not provide an alternative. Also at lower taxonomic levels (order, family and genus level), systematics were far from stable (De Ley and Blaxter 2002). Despite their ecological and physiological diversity, their conserved morphology and small size resulted in a paucity of observable, phylogenetically informative characters. Furthermore, many characters show a convergent evolution. In recent years DNA sequence data have brought a revival to the field of systematics. The first major classification to incorporate both morphological and molecular phylogenetic information was presented by De Ley and Blaxter (2002).

Appreciating that nematodes arose early in animal evolution, it would be conceivable that even relatively conserved genes such as the ribosomal RNA-encoding genes could offer us insights in the evolution of plant parasitism within this phylum. Especially among invertebrates, the small subunit ribosomal DNA (SSU rDNA) gene (coding for SSU rRNA) is frequently used to deduce deep phylogenetic relationships. Because of their vital role in the assembly of proteins in the ribosomes, there is a strong selection on the SSU and LSU ribosomal DNA genes (LSU: large subunit). As a consequence these genes—at least parts thereof—are very conserved. Among the ribosomal RNA encoding genes, the SSU rDNA is most conserved. Ribosomal DNA genes are usually present in multiple copies (the *Caenorhabditis elegans* genome harbours $\approx$ 55 copies; Ellis et al. 1986) and this implies that a relatively small quantity of starting material (e.g. a single nematode corresponding to ≈ 0.2 ng DNA) is sufficient for a polymerase chain reaction (PCR)-based amplification. Normally it would not be advisable to use a multicopy gene in phylogenetics because there could be the risk of comparing paralogs instead of orthologous gene copies.

However, intrachromosomal homogenization ensures that mutations in a copy of a SSU rDNA gene are either removed or taken over in the other SSU rDNA copies (Dover et al. 1993).

For some time, SSU rDNA in nematodes was suggested to have a substitution rate 2–3 times higher than in most other Metazoa. In a study on the phylogenetic position of the arthropods, Aguinaldo and co-workers carefully selected nematode species with relatively low SSU rDNA substitution rates (Aguinaldo et al. 1997). The SSU rDNA substitution rates within the phylum Nematoda appeared to be relatively variable. The model organism *Caenorhabditis elegans* and the animal parasite *Strongyloides* showed remarkably high substitution rates; respectively 0.192 and 0.187 substitutions per site, whereas another, more basal animal parasite—*Trichinella*—was shown to have a significantly lower rate of change (0.110 substitutions per site). Related Arthropods have values ranging from 0.040 to 0.080 (Aguinaldo et al. 1997). Higher rDNA substitution rates as observed for nematodes with a short life cycle and/or an advanced parasitic life style could result in unexpected resolution at lower phylogenetic levels.

The small subunit ribosomal DNA gene ($\approx$ 1700 bp) has been used successfully for the generation of molecular frameworks throughout the animal kingdom. The SSU rDNA harbours a few regions that are almost fully conserved among eukaryotes and this allowed the design of so-called "universal primers" for the amplification of this particular gene from a wide range of Metazoans. In fact SSU rDNA has—in terms of diversity—a mosaic structure of very conserved regions interspersed by more variable elements. Amplifiability from unknown samples in combination with the presence of multiple regions with ample phylogenetic signals explains why this gene became popular in nematode phylogenetics. A first key paper by Blaxter et al. (1998) (53 nematode taxa), was followed by Holterman et al. (2006) (360 taxa), and Van Megen et al. (2009) (1250 taxa). Currently, the SSU rDNA molecular framework of nematodes at the Laboratory of Nematology, Wageningen University, includes approximately 2800 taxa. Bayesian inference based analysis of 360 nematode taxa by Holterman et al. (2006) gave rise to a phylogenetic framework characterized by a backbone consisting of 12 consecutive bifurcations. Based on these bifurcations, 12 clades were defined within the phylum Nematoda. From this analysis it was clear that trophic specialisations such as plant, mammal or insect parasitism all have arisen multiple times during evolution. It should be noted that due to the absence of good fossil records it is not possible to label nodes with a proper assessment of the era in which such a hypothetical common ancestor has lived.

11.3   The Evolution of Plant Parasitism Among Nematodes

From molecular analyses of a wide diversity of representatives of the phylum Nematoda, it became clear that plant parasitism has arisen at least four times independently. In all lineages plant parasites are equipped with a protrusible, hollow, injection

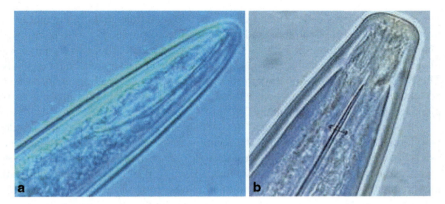

Fig. 11.2 Head regions of representatives of two distinct lineages of plant-parasitic nematodes. *Left*: *Trichodorus primitivus* (Triplonchida, Clade 1) and *right*: *Longidorus intermedius* (Dorylaimida, Clade 2). Pictures were taken at a 1000 times magnification. (Source: Hanny van Megen, Wageningen University, Laboratory of Nematology, The Netherlands)

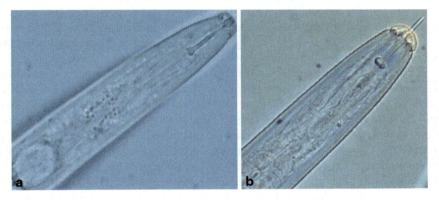

Fig. 11.3 Head regions of representatives of two distinct lineages of plant-parasitic nematodes. *Left*: *Aphelenchoides subtenuis* (Aphelenchida, Clade 10) and *right*: *Globodera rostochiensis* (Tylenchida, Clade 12). Pictures taken at a 1000 times magnification. (Source: Hanny van Megen, Wageningen University, Laboratory of Nematology, The Netherlands)

needle-like device that is used to puncture the plant cell wall, to deliver pathogenicity-related secretions in the plan root, and to take up nutrients from the plant (see Figs. 11.1, 11.2, and 11.3). Another common denominator among plant parasitic nematodes is the large size of the pharyngeal gland cells (usually a single dorsal and multiple subventral glands (Fig. 11.1), and the high activity of these glands just before and during plant parasitism.

Within Clade 1 (most basal clade of this phylum), obligatory plant parasites can be found within the order Triplonchida, in the family Trichodoridae. Plant-parasitic representatives within this family are generally known as 'stubby root nematodes'. These nematodes are ectoparasites, and they use a curved protrusible onchiostyle to puncture rhizodermis cells (Fig. 11.2). Nematode secretions that are thought to be involved in root cell penetration and manipulation are produced in the pharyngeal glands located in the basal bulb. Stubby root nematodes harbour five pharyngeal glands cells: one dorsal gland, two posterior and two anterior ventrosublateral glands. These glands open into the lumen of the nematode, and produce pathogenicity related-proteins. It is so far unclear whether these five glands act during distinct phases of plant feeding (Karanastasi et al. 2003).

The second lineage can be found in Clade 2, in a family embedded in the order Dorylaimida called Longidoridae. The common name of this category of plant parasites is 'dagger nematodes'. Dagger nematodes are equipped with an odontostyle, a protrusible hollow spear (Fig. 11.2) that is used to puncture the plant cell wall, to secrete pathogenicity-related proteins and to take up food from the plant cell. As compared to stubby root nematodes, members of the family Longidoridae tend to feed on deeper cell layers (e.g. cortical instead of rhizodermal cells). The terminal bulb in the Dorylaimida contains five pharyncheal gland cells, but in case of the family Longidoridae, two ventrosublateral glands seems to be degenerated (or fused) as the nuclei of this second pair have disappeared (although the five gland orifices are still intact) (Loof and Coomans 1972). Also in this respect, the plant parasitic Longidoridae are distinct from the more basal plant-parasitic members of the order Triplonchida. It might be worth noting that both a number of representatives of both the Trichodoridae and the Longidoridae are transmitters of plant viruses.

The plant-parasitic members of the third lineage are mainly (or even exclusively) facultative plant parasites; as an alternative food source they can feed on fungi as well. They are members of the families Aphelenchoididae and Parasitaphelenchidae, and the most notorious representatives of these groups are the causal agent of white tip in rice, *Aphelenchoides besseyi*, and the pine wood nematode *Bursaphelenchus xylophilus*. Currently these two families are positioned in Clade 10, but this clade positioning is not robust, as the GC contents of SSU rDNAs of these families is relatively low as compared to related families (Holterman et al. 2006). Members of these two families are equipped with a stomatostylet (Fig. 11.3), a device functionally comparable with the onchiostyle and the odontostyle mentioned above. However, as suggested by their names, the origins of these injection needle-like devices are fundamentally distinct.

The fourth, and economically most important group of plant parasitic nematodes can be found in the most distal nematode clade (Clade 12). This clade roughly corresponds to the order Tylenchida, with the exception of the suborder Hexatylina, a branch that mainly comprises insect parasitic nematodes. The other three suborders, the Hoplolaimina, the Criconematina, and the Tylenchina, harbour plant-parasitic nematode species. Contrary to the first two suborders, numerous members of the Tylenchina are facultative parasites of higher plants, being able to use lower plants, fungi and oomycetes as alternative food sources. The most well-known and best-studied plant parasites such as the root knot (*Meloidogyne* sp.), cyst (*Heterodera* and *Globodera* sp.) and lesion (*Pratylenchus* sp.) nematodes all belong to the suborder Hoplolaimina. Tylenchida are equipped with a stomatostylet (Fig. 11.3), and at least of part of the secretory components involved in plant parasitism are produced in the dorsal and in the two subventral pharyngeal glands. It should be noted that the subventral glands are most active in pre-parasitic second stage juveniles, and their activity slows down rapidly in parasitic second and third stage juveniles. The peak of dorsal gland activity is observed in parasitic life stages, and thus lags behind the peak of subventral gland activity.

11.4   Some Biological Characteristics of Migratory and Endoparasitic Plant Parasites

Root-knot, lesion and cyst nematodes pose a serious threat to main agricultural crops such as potato, sugar beet, and soybean. As such these distal representatives of the order Tylenchida constitute the economically most detrimental group of plant parasitic nematodes. Among the three genera mentioned above, root-knot nematodes such as *Meloidogyne incognita*, *M. hapla*, and *M. chitwoodi*, are most polyphagous, being able to infect almost all domesticated plants worldwide (Trudgill and Blok 2001). Lesion nematodes are members of a species-rich genus named *Pratylenchus* (*ca* 70 valid species), and species such as *P. penetrans*, *P. neglectus*, and *P. thornei* are notorious parasites in crops such as potato and tomat as well as a range of cereals and legumes. Lesion nematodes are migratory endoparasites (see below), and as such provide other opportunistic soil bacteria and fungi access to the plant root. The invasion of plant roots by root-knot and cyst nematodes leads to the formation of nematode feeding sites. In case of sedentary endoparasites, establishment of so called 'giant cells' or 'syncytia' (root-knot and cyst nematodes, respectively) is considered as one of the most sophisticated adaptations of plant parasitism. The syncytium, a conglomerate of plant cells fused by partial degradation of plant cell walls, functions as a metabolic sink that transfers plant assimilates from the conductive tissues in the vascular cylinder to the sedentary nematode. In case of rot knot nematodes, the formation of five to seven giant-cells grouped around the head region is induced. Upon the injection of nematode secretions, parasitized cells rapidly become larger, hypertrophied and multinucleate as nuclear division occurs in the absence of cell wall formation. Feeding sites, giant cells or syncytia, are used till the end of the nematode life cycle and serve as sink tissues to which nutrients are imported in a symplastic and/or apoplastic manner (Hoth et al. 2008).

On the other hand, migratory endoparasites, such as *Pratylenchus* species, exhibit feeding strategies that may be considered as less refined, but they are no less successful keeping their proliferation and host range in mind. In various life stages, lesion nematodes move freely through the root to feed and reproduce, creating numerous local tissue lesions, which are used as an entrance by the secondary pathogens such as bacteria or fungi. The feeding takes place mostly in the root cortex, but root hair feeding is observed for younger life stages as these are unable to perforate thicker epidermal cell wall (Zunke 1990). The parasitic success of the mentioned groups of nematodes is undoubtedly a result of their unusual ability to overcome the barrier of the plant cell wall. This biological obstacle is mechanically and chemically weakened by the stylet, spear or onchiostyle-driven puncturing of the cell wall in combination with the local release of cell wall-degrading and modifying proteins.

11.5 The Role of Lateral Gene Transfer in the Evolution of Plant Parasitism

Contrary to vertical gene transfer, the transmission of genes from parents to their offspring, lateral gene transfer (LGT) is about the stable acquisition of (parts of) genes in a manner other than traditional reproduction. Such non-conventional transfer of genetic material requires, among other things, close physical contact between the between donor and recipient. LGT could for instance occur between organisms with a trophic relationship. As an example, Doolittle described the transfer of genes to early eukaryotes from the bacteria taken by them as food (Doolittle 1998). In the same year, Smant and co-workers (Smant et al. 1998) discovered that the potato and soybean cyst nematodes (*G. rostochiensis and H. glycines* (order Tylenchida) produce and secrete β-1,4-endoglucanase (cellulases). Till this discovery, animals were thought to be incapable degrading plant cell walls; herbivores use microbial endosymbionts for the degradation of these recalcitrant polymers.

This finding constituted the starting point of a series of papers reporting a range of cell wall-degrading enzymes (CWDE) from plant-parasitic nematodes, including pectate lyases (Popeijus et al. 2000), exo-polygalacturonase (Jaubert et al. 2002), xylanases (Mitreva-Dautova et al. 2006) and expansins (Qin et al. 2004). It should be noted that CWDEs are produced in the subventral glands of plant parasitic nematodes during the early phases of infection. These enzymes are unlikely to be involved in feeding site formation in the plant root.

The discovery of a set of cell wall-degrading enzymes (CWDE) is remarkable. Nematodes are devoid of plant cell wall-like structures, and hence it seems safe to state that plant cell wall penetration by parasitic nematodes is the result of mechanical weakening and local depolymerization. Remarkably these plant cell wall-degrading enzymes produced by nematodes were far more similar to their bacterial equivalents than to orthologs from other eukaryotes such as higher plants, fungi or oomycetes. This prompted Keen and Roberts (1998) in a commentary paper to the hypothesize that ancestral bacterivorous nematodes could have acquired a pathogenicity island with multiple plant parasitism-related genes by the ingestion of (plant-parasitic) soil bacteria. Such an acquisition could have enabled them to penetrate a plant cell wall, and exploit a food source that was till that time inaccessible.

If it is true that the evolution of plant parasitism among nematodes was facilitated by the acquisition of CWDEs of bacterial origin, one might wonder whether the four independent lineages as described in 10.3 harbour distinct or similar core sets of cell wall-degrading enzymes. Here we will use cellulases as an example as this is so far the best-characterized CWDE among plant parasitic nematodes. As indicated above, β-1,4-endoglucanases (cellulases) were first discovered in two cyst nematode species *Globodera rostochiensis* and *Heterodera glycines* (Smant et al. 1998).

They are encoded by a multi-copy gene family, which is expressed in the subventral esophageal glands of the infective second stage juveniles (J_2). The biochemical activity of these enzymes is determined as a hydrolysis of β-1,4 glycosidic bonds of cellulose microfibrils. According to their biochemical characteristics, cellulases are represented in various glycoside hydrolase (GH) families. All members of the order Tylenchida (lineage 4 in this chapter) investigated so far (even the insect parasite *Delandenus siridicola*) harbour cellulases belonging to family 5 (GHF5; glutamic acid (Glu) residues essential for catalysis).

The GHF5 genes in Tylenchida comprise at least a catalytic domain, and occasionally this is connected to a linker and/or a type II cellulose-binding domain (CBDII). Due to the slight differences in intron-exon composition and noticeable phylogenetic distance, those cellulases are thought to belong to at least two distinct lineages (Kyndt et al. 2008). These lineages are referred to as *CelI* and *CelII* by (Rehman et al. 2009a), and in a more recent study these are similar to catalytic domains type B and type C respectively (Rybarczyk-Mydłowska et al. 2012). Phylogenetic analysis of nematode GHF5 cellulases suggests for an early acquisition of this category of plant cell wall-degrading enzymes, maybe even by the common ancestor of the order Tylenchida and the family Aphelenchidae (subfamilies Aphelenchinae and Paraphelenchinae) (Rybarczyk-Mydłowska et al. 2012).

A facultative plant parasite belonging to lineage 3, the pinewood nematode *Bursaphelenchus xylophilus*, was shown to produce cellulases from glycoside hydrolase family 45 (GHF 45; aspartic acid (Asp) residues essential for catalysis) (Kikuchi et al. 2004). *Bursaphelenchus* belongs to the family Parasitaphelenchidae, and recently a cellulase from its sister family, Aphelenchoididae, was isolated and characterized. A small-scale analysis of the transcriptome of the foliar nematode Aphelenchoides besseyi resulted in the discovery of another GHF45 cellulase that is produced and presumably secreted (as the core protein is preceded by a predicted signal peptide for secretion) by this nematode species (Kikuchi et al. 2014). So far each plant parasite lineage was thought to be characterized by the production of cellulases from a single GH family. Therefore the finding of a GHF5 cellulase from another facultative plant parasite within this lineage, *Aphelenchoides fragariae*, by Fu et al. (2012) was remarkable Notably, this finding was the result of a directed search as the authors used degenerated GHF5 primers to screen for the presence of cellulases in A. *fragariae*. The GHF5 cellulases described from A. *fragariae* seemed to have some unusual features. For example, it was not possible to detect genomic copies of this sequence in nematodes reared on fungi. Until the finding of this GHF5 cellulase in this foliar nematodes species is confirmed by more detailed research, we tend to state that the presence of GHF45 cellulases is a typical and unique characteristic of this lineage of plant parasitic nematodes.

As compared to the two most distal lineages of plant parasites, the two remaining families within the more basal clade 1 and 2 are poorly characterized. Within the family Longidoridae, a cellulase was identified from *Xiphinema index*, and this enzyme belonged to yet another GH family (Jones and Helder, unpublished results). Based on this fragmentary information, we hypothesize that individual lineages of plant-parasitic nematodes are characterized by the presence of a single, lineage-dependent kind of cellulase.

11.6   Plant Manipulation by Parasitic Nematodes

Even plant-parasitic nematode species residing in the most basal lineage, ectoparasites belonging the family Trichodoridae, do not show a hit-and-run strategy, i.e. a strategy by which the nematode would just insert its puncturing device into a plant cell, and take up the cytosol, and move on to the next plant cell. In case of the stubby root nematode species *Paratrichodorus anemones* the delivery of nematode secretion into the rhizodermis cell resulted in the redistribution of cytoplasm from all areas of the cell towards the penetration site (Karanastasi et al. 2003). In the interaction of *Ficus carica* seedlings with a representative of the second lineages of plant parasites, the dagger nematode *Xiphinema index,* hypertrophied, multinucleate cells were induced. Most likely the re-differentiation of plant cells is induced by saliva proteins produced by this ectoparasite (Wyss et al. 1980). Hence, plant cell re-differentiation as an essential process linked to parasitism is not a unique characteristic for the most distal nematode taxa in Clade 12, and also this trait seems to have arisen multiple times. Most likely proteins produced in de dorsal glands are responsible for plant cell re-differentiation, but so far the underlying mechanism is unknown. Recently, a number of effector proteins has been identified that are produced in these glands in infectious life stages. It should be noted that the examples given below are all from cyst and root knot nematodes, representatives of the most distal lineages. Indications for local auxin manipulation as an early step in feeding site formation (Goverse et al. 2000) were recently confirmed by the identification of an effector protein from the soybean cyst nematode *Heterodera glycines*, called 19C07, that was shown to interact with an auxin influx transporter (LAX3) in Arabidopsis (Lee et al. 2011). Another category of dorsal gland proteins, the so-called SPRYSECs (secreted proteins containing a SPRY domain) have been identified from the potato cyst nematodes *Globodera rostochiensis* and *G. pallida,* close relatives of the soybean cyst nematode. As compared to e.g. *Caenorhabditis elegans,* the pinewood nematode *Bursaphelenchus xylophilus* and the tropical root knot nematode *Meloidogyne incognita*, this protein family has expanded enormously in cyst nematodes; both potato cyst nematodes were shown to harbor dozens of secreted SPRY proteins (Rehman et al. 2009b, Cotton et al. 2014). Although the function of these proteins in the interaction with the host plant is unknown for most members of this family, one member, SPRYSEC-19, was demonstrated to suppress the CC-NB-LRR disease resistance response in host plants (Postma et al. 2012).

Hence, plant parasitism within the phylum Nematoda is characterized by ample convergent evolution. As we have seen, all of the plant-parasitic lineages developed similar morphological adaptations that allow them to overcome the plant cell wall, a major physical barrier. A similar picture starts to arise for the non-morphological characteristics of these lineages. Although the origin and the exact blend of cell wall-degrading enzymes might be unique for individual taxa, a roughly similar pallet of enzymes is likely to be produced by all obligatory plant parasites. It will probably become clear in the next years whether or not convergent evolution can be observed in the mechanisms underlying host cell manipulation and the suppression of resistance responses.

References

Aguinaldo AMA, Turbeville JM, Linford LS et al (1997) Evidence for a clade of nematodes, arthropods and other moulting animals. Nature 387:489–493

Blaxter ML, De Ley P, Garey JR et al (1998) A molecular evolutionary framework for the phylum Nematoda. Nature 392:71–75

Bongers T, Ferris H (1999) Nematode community structure as a bioindicator in environmental monitoring. Trends Ecol Evol 14:224–228

Chitwood BG, Chitwood MB (1933) The characters of a protonematode. J Parasitol 20:130

Cotton JA, Lilley CJ, Jones LM et al (2014) The genome and life-stage specific transcriptomes of *Globodera pallida* elucidate key aspects of plant parasitism by a cyst nematode. Genome Biol 15:R43

De Ley P, Blaxter ML (2002) Systematic position and phylogeny. In: Lee DL (ed) The biology of nematodes. Taylor & Francis, London, pp 1–30

Doolittle WF (1998) You are what you eat: a gene transfer ratchet could account for bacterial genes in eukaryotic nuclear genomes. Trends Genet 14:307–311

Dover GA, Linares AR, Bowen T et al (1993) Detection and quantification of concerted evolution and molecular drive. Method Enzymol 224:525–541

Ellis RE, Sulston JE, Coulson AR (1986) The rDNA of *C. elegans*: sequence and structure. Nucleic Acids Res 14:2345–2364

Ferris H, Bongers T, De Goede RGM (2001) A framework for soil food web diagnostics: extension of the nematode faunal analysis concept. Appl Soil Ecol 18:13–29

Floyd R, Abebe E, Papert A, Blaxter M (2002) Molecular barcodes for soil nematode identification. Mol Ecol 11:839–850

Fu Z, Agudelo P, Wells CE (2012) Differential expression of a beta-1,4-endoglucanase induced by diet change in the foliar nematode *Aphelenchoides fragariae*. Phytopathology 102:804–811

Goverse A, Overmars H, Engelbertink J et al (2000) Both induction and morphogenesis of cyst nematode feeding cells are mediated by auxin. Mol Plant Microbe Interact 13:1121–1129

Holterman M, van der Wurff A, van den Elsen S et al (2006) Phylum-wide analysis of SSU rDNA reveals deep phylogenetic relationships among nematodes and accelerated evolution toward crown clades. Mol Biol Evol 23:1792–1800

Hoth S, Stadler R, Sauer N et al (2008) Differential vascularization of nematode-induced feeding sites. Proc Natl Acad Sci U S A 105:12617–12622

Jaubert S, Laffaire JB, Abad P et al (2002) A polygalacturonase of animal origin isolated from the root- knot nematode *Meloidogyne incognita*. FEBS Lett 522:109–112

Karanastasi E, Wyss U, Brown DJF (2003) An *in vitro* examination of the feeding behaviour of *Paratrichodorus anemones* (Nematoda: Trichodoridae), with comments on the ability of the nematode to acquire and transmit Tobravirus particles. Nematology 5:421–434

Keen NT, Roberts PA (1998) Plant parasitic nematodes: digesting a page from the microbe book. Proc Natl Acad Sci U S A 95:4789–4790

Kikuchi T, Jones JT, Aikawa T et al (2004) A family of glycosyl hydrolase family 45 cellulases from the pine wood nematode *Bursaphelenchus xylophilus*. FEBS Lett 572:201–205

Kikuchi T, Cock PJA, Helder J et al (2014) Characterisation of the transcriptome of *Aphelenchoides besseyi* and identification of a GHF 45 cellulase. Nematology 16:99–107

Kyndt T, Haegeman A, Gheysen G (2008) Evolution of GHF5 endoglucanase gene structure in plant-parasitic nematodes: no evidence for an early domain shuffling event. BMC Evol Biol 8:305

Lee C, Chronis D, Kenning C et al (2011) The novel cyst nematode effector protein 19C07 interacts with the Arabidopsis auxin influx transporter LAX3 to control feeding site development. Plant Physiol 155:866–880

Loof PAA, Coomans A (1972) The oesophageal gland nuclei of Longidoridae (Dorylaimida). Nematologica 18:213–233

Lorenzen S (1981) Entwurf eines phylogenetischen systems der freilebenden Nematoden. Veröff Inst Meeresforsch Breme, Supplement 7:1–472

Mitreva-Dautova M, Roze E, Overmars H et al (2006) A symbiont-independent endo-1,4-bata-xylanase from the plant-parasitic nematode *Meloidogyne incognita*. Mol Plant Microbe Interact 19:521–529

Popeijus H, Overmars H, Jones J et al (2000) Enzymology: degradation of plant cell walls by a nematode. Nature 406:36–37

Postma WJ, Slootweg EJ, Rehman S et al (2012) The effector SPRYSEC-19 of *Globodera rostochiensis* suppresses CC-NB-LRR-mediated disease resistance in plants. Plant Physiol 160:944–954

Qin L, Kudla U, Roze EHA et al (2004) A nematode expansin acting on plants. Nature 427:30.

Rehman S, Butterbach P, Popeijus H et al (2009a) Identification and characterization of the most abundant cellulases in stylet secretions from *Globodera rostochiensis*. Phytopathology 99:194–202

Rehman S, Postma W, Tytgat T et al (2009b) A secreted SPRY domain-containing protein (SPRYSEC) from the plant-parasitic nematode *Globodera rostochiensis* interacts with a CC-NB-LRR protein from a susceptible tomato. Mol Plant Microbe Interact 22:330–340

Rybarczyk-Mydłowska K, Ruvimbo Maboreke H, Van Megen H et al (2012) Rather than by direct acquisition via lateral gene transfer, GHF5 cellulases were passed on from early Pratylenchidae to root-knot and cyst nematodes. BMC Evol Biol 12:221

Shinya R, Morisaka H, Kikuchi T (2013) Secretome analysis of the pine wood nematode *Bursaphelenchus xylophilus* reveals the tangled roots of parasitism and its potential for molecular mimicry. PLoS One 8(6):e67377

Smant G, Stokkermans J, Yan YT et al (1998) Endogenous cellulases in animals: isolation of beta-1,4-endoglucanase genes from two species of plant-parasitic cyst nematodes. Proc Natl Acad Sci U S A 95:4906–4911

Trudgill DL, Blok VC (2001) Apomictic, polyphagous root-knot nematodes: exceptionally successful and damaging biotrophic root pathogens. Annu Rev Phytopathol 39:53–77

Van Megen H, Van Den Elsen S, Holterman M et al (2009) A phylogenetic tree of nematodes based on about 1200 full-length small subunit ribosomal DNA sequences. Nematology 11:927–950

Wyss U, Lehmann H, Jank-Ladwig R (1980) Ultrastructure of modified root-tip cells in *Ficus carica*, induced by the ectoparasitic nematode *Xiphinema index*. J Cell Sc 41:193–208

Yeates GW, Bongers T, De Goede RGM et al (1993) Feeding-habits in soil Nematode families and genera-an outline for soil ecologists. J Nematol 25:315–331

Zunke U (1990) Ectoparasitic feeding behaviour of the root lesion nematode, *Pratylenchus penetrans*, on root hairs of different host plants. Rev Nématol 13:331–337

Chapter 12
Herbivorous Insects—A Threat for Crop Production

Eddy van der Meijden

Abstract It is estimated that, in spite of plant breeding and pest control efforts, 15 % of crop yield is worldwide lost to herbivory by insects. Examples demonstrate how insect pests have developed in the past and why they will develop in the future. The evolutionary potential of insects to become new pests is considered for traditionally and genetically modified crop varieties. The immune system of plants is presented step by step. Generalist herbivores can be effectively repelled, but specialist herbivores are much harder to repel. They use plant defenses as cues for host plant recognition. Next to direct defense, indirect defense by attracting natural enemies of (specialist) herbivores is explained. Finally, the interactions of plants and insect herbivores with microbial symbionts—and their consequences—are discussed.

12.1 Introduction

Probably not a single wild plant will complete its life cycle without being victim of herbivory by one or usually many more insect species during some stage of its life. Without special treatment, like the application of insecticides, the release of natural enemies and/or modification of their immune system, crop plants are even more vulnerable than their wild relatives. The worldwide loss of crop yield to insects is estimated to be 15 % (Maxmen 2013). Loss of crops has been familiar to man as long as he has been growing them. Locust pests were already well known in ancient Egypt. Especially the Migratory locust (*Locusta migratoria*) belongs to the most voracious pests and at the same time to the most difficult pests to control. A high local juvenile population density stimulates individuals to adapt their physiological development. They grow into adults with effective wings and lightweight energy reserves. These individuals start migrating (Fig. 12.1) in search for new food sources. During their flight they mix with other groups of the same species and eventually these swarms may become incredibly large. Swarms have been observed with an estimated number of 70 billion individuals, ten times as many individuals as the total human world

E. van der Meijden (✉)
Institute of Biology, Leiden University, P.O. Box 9505, RA 2300 Leiden, The Netherlands
Tel.: +31715275119
e-mail: e.van.der.meijden@biology.leidenuniv.nl

© Springer International Publishing Switzerland 2015
B. Lugtenberg (ed.), *Principles of Plant-Microbe Interactions*,
DOI 10.1007/978-3-319-08575-3_12

Fig. 12.1 A swarm of locusts is landing, an insect pest develops. (Reproduced with permission by FAO (©FAO/Yasuyoshi Chibam))

population. Locusts feed on *Sorghum*, maize and wheat and other grass species. A recent outbreak led to 42 % reduction in farmland grass production of 36,000 ha in Northern China in 2003 (Tanaka and Zhu 2005).

Interestingly, already long ago (2500 BC) Sumerians used sulphur compounds to control insects and mites (Dent 2000). Recent control methods include warning systems with drones, insecticides, resistant plant varieties and biological control with predators, parasitoids or insect pathogens. Despite immense scientific efforts, plagues continue to affect plants and consequently human food supply. It may be argued that we will probably never get completely rid of insect (and other) pests (Chaps. 9, 10, 11, and 13). What makes insects such outstanding guerilleros? Their short generation time (relative to that of their food sources) and their extremely high fecundity (their potential number of offspring), make them evolutionarily extremely successful. They can often adapt to insecticides and to new resistance genes within a few generations. Moreover, for each plant species probably several hundreds of insect herbivore species exist that may grow into effective pest species.

To demonstrate how insect pests work, how they develop and how some pests are in the waiting room to develop, I will give some examples. This continuous battle is probably best illustrated by one of the most important crop plants worldwide, maize. Maize (*Zea mays*) originates from Central America and has been grown for more than 6000 years. Today maize is grown all over the world. By far the largest part of the maize crop is fed to livestock; smaller amounts serve as human food or are turned into ethanol. Economically, maize is extremely important but it is vulnerable to insect feeding. In the U.S. more than 100 different insect species were reported to cause important pest damage. Some of these can be dealt with by using insecticides. Others—like the European corn borer (*Ostrinia nubilalis*)—feed inside the stem, in immature kernels, or inside the roots. In these cases insecticides are less effective. Alternative measures are biological control by insect parasitoids, viruses or microorganisms. Another alternative is the development of resistant varieties by genetic modification. Products containing the soil bacterium *Bacillus thuringiensis* (Bt) have been known to be effective insecticides since the first half of the last century. More recently, several toxic proteins were detected and the coding genes were identified (Chap. 40). Maize is now one of the crop plants for which varieties have been developed which harbour Bt-genes. They produce the Bt-proteins, which make them resistant to several butterfly, moth and beetle pests such as the European corn borer. These genetically modified varieties have become very popular in some countries, like the USA and Spain. It must be stressed however, that similar to natural varieties of wild plants and varieties of crop plants that have been developed with traditional breeding techniques, these phenotypes are prone to loss of their resistance by natural selection and evolution of their insect herbivores. Counter resistance is especially expected to develop when host plants are grown in large monocultures over longer periods of time. In 2012, Gassmann reported the first examples in which the European corn borer broke through the resistance of Bt-maize in fields in Iowa that had been grown with this variety for 3 to 6 years. Right now it has become an urgent question how to deal with these economically extremely important plant resistance characteristics. This makes future pest control one of the most urgent and at the same time most exciting fields of the biological sciences.

A very ancient but yet illustrative example of loss of resistance comes from the grape vine. Wine production has been of great economic importance for ages, probably even before the Greek and Roman cultures. In 1864 a disease was observed among grape vines in Southern France that eventually affected most European vineyards. It took quite some time before it was realized that the disease was in fact caused by sap sucking on leaves and roots by the Grape phylloxera. This small, aphid-related, insect may even kill vines. In France the total production of wine between 1875 and 1889 was reduced by three quarters. The insect, which is native to North America, must have been accidentally introduced in France. Native vines in North America, however, were not seriously affected by this herbivore. The solution to control the Grape phylloxera in Europe has been to graft native varieties onto phylloxera-resistant North American rootstocks (Campbell 2006). However, be careful. Also the grape phylloxera is continuously evolving and is presently causing disasters in Northern California.

Each year crop failures due to insect herbivores take place worldwide, sometimes only locally, sometimes on a very large scale. All major and minor crops are vulnerable. Rice, the second most important crop plant with its main areas of production in China and Thailand, is frequently subject to crop losses of 30 % due to pest insects like the Rice brown plant hopper. Cassava, an important crop of Africa, is frequently subject to large-scale destruction by the Cassava mealy bug, the Cassava green mite, the whitefly *Bemisia tabaci*, locusts and other insects. *Bemisia* is not only a serious threat itself, it is also a vector for the Cassava mosaic virus (CMV). Yield losses in Africa by CMV have been estimated up to 50 %.

The climate of our planet will change in the future as it did in the past. There are several indications that global warming takes place. Will warming affect the influence of insects on human crops? Population dynamics of insects are usually affected to a great extent by weather conditions. Weather can have direct and indirect effects. Especially drought conditions may lead to insect pest development (a. o. White 1969). Under drought conditions that are limited in duration, sap-feeders may benefit from stress-induced increases in plant nitrogen. If our climate continues to change towards higher temperatures and periods of drought, it is very likely that we will be faced with more insect outbreaks all over the world in all kinds of crops from timber to food crops like wheat, maize and potato (Maxmen 2013).

12.2 Pest and Beneficial Insects

The overwhelming variety of insect species that feed on plants may be challenging for an entomologist, but is an absolute threat for plant growers. There are more than a million different insect species and it is estimated that more than 360,000 species are plant feeders.

Insects of some orders undergo complete metamorphosis. Larvae turn into pupae, and pupae into adult insects. The larval stage is usually the most voracious life stage. This is typical for beetles (Coleoptera), butterflies and moths (Lepidoptera) and flies (Diptera). In other groups, like the bugs, whiteflies, leafhoppers and aphids (Hemiptera), locusts and grasshoppers (Orthoptera) and thrips (Thysanoptera) there is so-called incomplete metamorphosis during which immature stages (called nymphs) already resemble the mature stage.

Beetle larvae, with chewing mouthparts, often feed on or inside plant roots or stems, or behave as leaf-miners. Adult beetles feed on leaves of trees, shrubs or herbs. Some species are specialized on seeds. Lepidopteran larvae are mainly leaf chewers. Leaf chewers can have very different feeding patterns; some start eating at the leaf rim, some make feeding holes, and others skeletonise leaves. Hemiptera have piercing mouthparts that enable them to feed from the vascular system of plants or even from individual cells. In this way they can avoid feeding contact with particular (less palatable) tissues. Locusts and grasshoppers are leaf chewers; thrips suck from epidermis cells. Although all insects may be vectors of plant diseases by transmission of bacteria, fungi or viruses from one plant to another, especially the groups with piercing mouthparts bring along this extra threat for plants. Because they feed with their mouthparts inside plants, they may be difficult to control with insecticides.

Beneficial insects are found in the orders of Diptera (flies) and Hymenoptera (bees, wasps and ants). Both orders have pollinators and parasitoids of herbivorous insects. These parasitoids may be extremely important in controlling pest species. Several companies, (like Koppert: http://www.koppert.com/) are specialized in breeding parasitoids for crop protection and pollinators for glasshouse environments. Among the beetles, the bugs and the ants, several species are specialized as predator of herbivorous insects. Because they immediately kill their prey upon finding it, their herbivory-reducing effect may be considerable.

12.3 Which Characteristics Make Insects Dangerous?

For each individual plant, the immune system provides protection against by far the largest majority of 360,000 potential herbivore species. However, they are quite vulnerable to a much smaller group, the so-called specialists. These specialists (a few to more than a hundred different insect species per plant species) have apparently penetrated the plant's immune system during their (co)evolution. These particular plants have become their specific food plants. Specialist insects often use the defense substances of their food plants to recognize and locate these plants. This phenomenon will be illustrated with an example of *Brassica* species and their herbivores. *Brassica* species like Cabbage and Oilseed rape (Fig. 12.2) contain a large group of specific chemical substances, the glucosinolates. We are all familiar with these substances because of their distinct "cabbage smell". Specialist insect herbivores like the Diamondback moth (*Plutella xylostella*), the Cabbage white butterflies (*Pieris* spec.) and the Crucifer flea beetle (*Psylliodes chrysocephala*), all important pest species of *Brassica* on a worldwide scale, use these glucosinolates to find their food plants and to start feeding. Generalist feeders (other insects, birds and slugs, etc) on the other hand, are effectively repelled by the same substances (Fig. 12.2). Alkaloids constitute another group of plant substances that is highly toxic to generalist herbivores (like horses and cattle) but not to specialists. Small amounts of alkaloids spread on pieces of filter paper are sufficient to attract individuals of the specialist Cinnabar moth (Macel and Vrieling 2003). This leads to an awkward dilemma for plant breeders and plant protection in general. The use of these plant substances as insecticides may increase the level of defense of crop plants towards generalist herbivores. At the same time it makes them more attractive to the specialists.

All insects of crop plants are potential pest species. If they can multiply fast, they will soon cause damage. Under natural circumstances the low number of food plants available and the presence of natural enemies, like parasitoids, will (often) keep numbers low. However, under agricultural circumstances large monocultures provide them with excess of food. Development of populations of natural enemies large enough for control can only follow after a time lag of at least one generation. That means after at least one season of crop growth. A special group of insects that may cause crop pests are the so-called invaders. We have seen examples of the Grape phylloxera (*Daktulosphaira vitifoliae*), colonizing Europe from the United States, and the European corn borer colonizing the United States. These species enter a new continent without their natural enemies which under natural circumstances might control them. This clearly gives them a head start in their new environment.

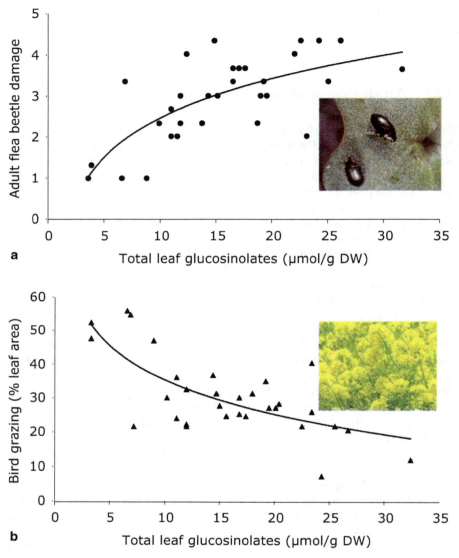

Fig. 12.2 a Relationship between (specialist) Crucifer flea beetle damage and total leaf glucosi-nolates in Oilseed rape (*Brassica rapa*). (Inserted photograph of Crucifer flea beetle by Richard Mithen). **b** Relationship between (generalist) bird feeding and total leaf glucosinolates in Oilseed rape. (Inserted photograph of oilseed rape by Eddy van der Meijden. Figures redrawn from Giamoustaris and Mithen (1995) by permission of the publisher)

12.4 The Plant's Immune System, Step by Step

Firstly, each plant produces a blend of volatile chemicals. These substances pass through the stomata and cuticle and surround the whole plant. Such a cloud may contain a few dozens to several hundreds of different compounds. Undamaged plants emit leaf volatiles. Upon damage the blend of compounds may change considerably. The production of some new substances is induced by herbivore damage. They are plant-species specific and their composition sometimes also depends on the particular herbivore. Leaf volatiles constitute the outer layer of a plant's immune system. They provide information to the majority of herbivores that can smell that a particular plant is not their host plant and is consequently unsuitable to feed or lay eggs upon. As was mentioned earlier, each plant has also some specialist herbivores that are not repelled by the plant's immune system, and these use these volatiles to find their particular host plant. Insects have advanced olfactory organs in their antennae that enable them to sense particular volatile substances and blends in extremely low concentrations (Schoonhoven et al. 2005).

The second component of the plant's immune system that an insect has to deal with is the outer surface, the epidermis. It provides protection by its toughness caused by lignin and cellulose. Grasses in general, are three times tougher than herbs (Schoonhoven et al. 2005). The epidermis is covered with a wax layer which contains a great variety of molecules that play an essential role in plant defense. Insects can sense these substances with their antennae and with the sense organs in their tarsae. The maintenance of the chemical composition of these compounds in the wax layer is an active process. Trichomes on the leaf surface are penetrating through this layer. They may be glandular or non glandular. In the first case they may secrete repelling or even toxic substances.

The insects that have not been stopped by leaf volatiles and other external defenses are subsequently confronted by a world that is dominated by an incredible variety of complicated chemical substances within the plant. These chemicals do not play an important role in the primary activity of growth, and are therefore called secondary metabolites. Up till now about 200,000 of these substances have been detected. Many of them reduce herbivory by particular insect species. This is the *constitutive defense system* of plants. These metabolites may be toxic or just repelling. Some act as digestibility reducers. For instance, saponins inhibit enzymes in the gut of insects that digest proteins. Some even act as attractors of pollinators. Some of these compounds clearly have several different functions within a plant. Most of the substances in the cloud of leaf volatiles are also secondary metabolites.

The glucosinolates or mustard oil glucosides are characteristic for the Brassicaceae, a plant family with more than 3500 species. All the cabbage varieties—Black mustard, Indian mustard and Oilseed rape—belong to this family, but also the model species for molecular research, *Arabidopsis thaliana*. About 120 different glucosinolates have been identified and each different species usually contains twenty or more of them, providing a specific fingerprint for that particular species. Typical is that they contain at least one glucose residue, one sulphur and one nitrogen atom.

Some other groups of secondary metabolites are the alkaloids (16,000) that occur in, for instance, the Solanaceae (a. o. potato, tomato) and the sesquiterpenes (6500) of the Asteraceae (a. o. sunflower and artichoke).

Not all plant parts contain the same concentration of secondary metabolites. Hound's tongue (*Cynoglossum officinale*) is a poisonous plant with a high concentration of pyrrolizidine alkaloids. However, young leaves may have a ten- to fifty-fold higher concentration than old leaves. Specialist herbivores and generalist herbivores that were fed on these leaves demonstrated a totally different preference. Specialists fed predominantly on the younger leaves; generalists avoided these leaves and fed on the older leaves with much lower concentrations (Van Dam et al. 1995; Fig. 12.3).

In general, much higher concentrations of secondary metabolites are found in the reproductive organs of plants and in younger leaves that are important for future photosynthesis, than in older leaves. The former plant parts are thus better protected (against generalist herbivores). The Hound's tongue example demonstrates that less important plant parts (from the plant's point of view) with low concentrations may be attacked by generalist herbivores. Significantly different patterns in secondary compounds were even detected among cell layers, like epidermis and mesophyll (Nuringtyas et al. 2012).

Insect herbivore-challenged plants do not only act passively with their constitutive defenses, but also respond to herbivory with the production of toxins and defensive proteins that target physiological processes in insects (Howe and Jander 2008). This is the extremely important *inducible defense system*. Contrary to the constitutive defenses, induced responses follow upon particular damage cues and may thus be more directed towards defense against the specific herbivore that is causing the damage. To respond in this specific way, plants should be able to recognize herbivore species. Such recognition has been earlier found in several plant-pathogen studies (Chap. 14). During the past 20 years many experimental studies have demonstrated differences in plant physiological responses to mechanical damage and insect herbivory. These are the result of differences in induced gene expression patterns and transcriptional responses.

There is strong evidence from experimental studies that plant defense is induced by the oral secretions of insects (Howe and Jander 2008). The presence of fatty acid-amino acid conjugates (FACs) in insect oral secretions was found to be an important induction elicitor. These substances are derived from moieties from both insect and host plant. There are indications that plants have specific FAC receptors. Other insect and plant derived substances with similar functions are being studied. This field of research is very actively developing right now. Signal transduction pathways from insect signals to induced plant responses are still relatively unknown. What we do know, is that especially the jasmonates play a crucial role in signalling and regulating defense responses to insect herbivory. Many studies have demonstrated that jasmonate mutants lack the ability to induce defenses against a wide variety of insect species, whereas experimentally application of jasmonate to leaves increases the level of defenses (Howe and Jander 2008).

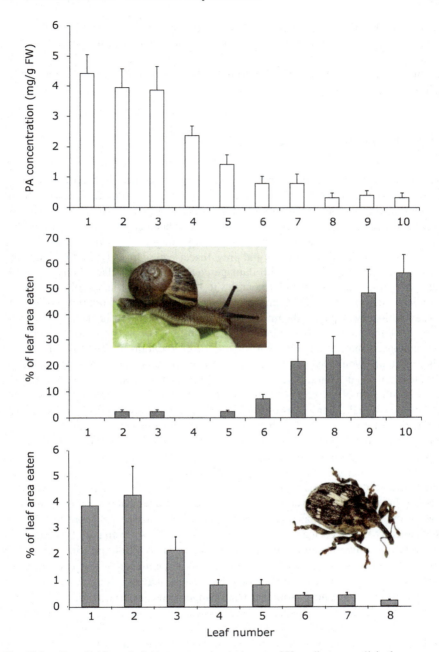

Fig. 12.3 **a** Pyrrolizidine alkaloid concentration in leaves of Hound's tongue (1 is the youngest leaf). **b** Fraction of leaves eaten by the generalist *Helix Aspersa*; (Inserted photograph of *Helix aspersa* by Eddy van der Meijden.) **c** Fraction of leaves eaten by the specialist *Mogulones cruciger*. (Inserted photograph of *Mogulones cruciger* by Henri Goulet, Agriculture and Agri-Food Canada. Figures redrawn from Van Dam et al. (1995) by permission of the publisher)

How Specific is the Plant Immunity System? The impression that one gets today from a wide variety of studies on different species from feeding guilds like leaf chewing caterpillars to phloem feeding aphids, is that the type of defense reaction is more related to feeding guild than to individual insect species (Howe and Jander 2008).

12.5 Tritrophic Systems

Why would a plant change its production of leaf volatiles after herbivory? One possibility is that it follows from damage without any particular function. Studies demonstrate an incredible variety of organisms that can perceive these volatile signals. During the last few decades, evidence has been collected that natural enemies of herbivores—parasitoids, predators and pathogens—use this induced blend of compounds to locate their hosts and prey. Insect-eating birds (great tits) are being attracted by volatiles of trees with damage caused by caterpillars. Predatory soil-living nematodes are attracted by volatiles to feeding sites of larvae of the Western corn borer in maize roots. It is tempting to believe that the induction process was the next evolutionary step after the immune system of the plant was hacked by specialist herbivores. Induction of plant volatiles provides information to parasitoids, predators and pathogens that may lead to parasitization or predation of insect herbivores. If this eventually would lead to a reduction of herbivore damage and a higher fitness of individual plants, natural selection would favour this type of induced response. Another function of the production of volatiles by herbivore-damaged leaves might be that they signal other leaves of the same plant to become induced so that further damage can be restricted. An alternative would be to signal through the phloem. Future research will give the answer, but it is clear that rapid reactions are extremely important for limiting herbivore damage.

How fast inducible responses through volatile emission work, is beautifully illustrated by a study of Allmann and Baldwin (2010). Coyote tobacco in the Western U.S. produces nicotine as a constitutive defense substance. Nicotine is toxic to most insects, but not to the Tobacco hornworm, a specialist insect herbivore of tobacco. Mechanically damaged leaves produce a. o. the volatiles hexenal and hexenol. So do Tobacco hornworm-damaged leaves. But there is an important difference. The cloud of volatiles of mechanically damaged leaves is dominated by z-isomers of these substances, whereas the tobacco hornworm-damaged leaves have more or less equal concentrations of z- and e-isomers. The change from the production of e-isomer dominated volatiles to z-e balanced blends takes place within a few minutes after the onset of damage, and is induced by the oral secretion of the specialist herbivore. Oral secretions of two generalist insect herbivores had no such effect. Blends with predominantly e-isomers were much more attractive for an important predator of the tobacco hornworm, the predatory bug *Geocoris*. By producing the induced blend locally in an extremely fast way, predation may start immediately and the search activity of the predator is guided by signals of the damaged leaf to its prey.

12.6 Insect Symbiotic Microbes Hijack Plant Defense Signalling

There are relatively few studies on the role of microorganisms on insect-plant interactions, but that is not a reliable indicator for their impact. Both insects and plants carry microbial pathogens and symbionts. One of the most important functions of this symbiosis is that microbial symbionts provide insects with (essential) amino acids that are not available in their food plants. They also play a role in resistance against natural enemies. Microbial symbionts of plants fulfil the same roles: the uptake of nutrients and the protection against pathogens.

Recent studies suggest that microbial symbionts of herbivorous insects play a crucial role in modifying food plant defenses. Larvae of the Colorado potato beetle produce oral secretions that suppress the induced defenses in tomato and potato. To unravel the mechanism, larvae were fed either antibiotic-treated leaves or non-treated leaves. The first group was not able to suppress the jasmonate regulated defenses. The second group did suppress these defenses. The bacteria in the insect's oral secretion (belonging to the genera *Stenotrophomonas*, *Pseudomonas* and *Enterobacter*) elicit salicylic acid-regulated defenses. This negatively cross-talks with jasmonate signalling, which in turn disables the plant to fully activate its jasmonate-mediated resistance. Apparently the food plant does not recognize the insect herbivore any more, but instead it defends itself against a microbial attack (Chung et al. 2013). Several more or less similar experiments have been published now, which gives support to the idea that we are dealing with an important phenomenon. Further study in this particular field is expected to give information on why plant defense sometimes fails.

References

Allmann S, Baldwin IT (2010) Insets betray themselves in nature to predators by rapid isomeration of green leaf volatiles. Science 329:1075–1077
Campbell C (2006) The Botanist and the Vintner: how wine was saved for the world. Algonquin Books, Chapel Hill
Chung SH, Rosa C, Scully ED et al (2013) Herbivore exploits orally secreted bacteria to suppress plant defenses. Proc Natl Acad Sci U S A 10:15278–15733
Dent D (2000) Insect pest management, 2nd edn. CABI Wallingford
Gassmann AJ (2012) Field-evolved resistance to Bt maize by western corn rootworm: predictions from the laboratory and effects in the field. J Invertebr Pathol 110:287–293
Giamoustaris A, Mithen R (1995) The effect of modifying the glucosinolate content of leaves of oilseed rape (*Brassica napus* ssp oleifera) on its interaction with specialist and generalist pests. Ann Appl Biol 126:347–363
Howe GA, Jander G (2008) Plant immunity to insect herbivores. Annu Rev Plant Biol 59:41–66
Macel M, Vrieling K (2003) Pyrrolizidine alkaloids as oviposion stimulants for the cinnabar moth, *Tyria jacobaeae*. J Chem Ecol 29:1435–1446
Maxmen A (2013) Under attack. The threat of insects to agriculture is set to increase as the planet warms. What action can we take to safeguard our crops? Nature 501:15–17
Nuringtyas TR, Choi YH, Verpoorte R et al (2012) Differential tissue distribution of metabolites in *Jacobaea vulgaris*, *Jacobaea aquatica* and their crosses. Phytochemistry 78:89–97

Schoonhoven LM, van Loon JJA, Dicke M (2005) Insect-plant biology, 2nd edn. Oxford: Oxford University Press

Tanaka S, Zhu DH (2005) Outbreaks of the migratory locust *Locusta migratoria* (Orthoptera: Acrididae) and control in China. Appl Entomol Zool 40:257–263

Van Dam NM, Vuister LWM, Bergshoeff C et al (1995) The 'raison d'ê tre' of pyrrolizidine alkaloids in *Cynoglossum officinale*—Deterrent effects against generalist herbivores. J Chem Ecol 21: 507–523

White TCR (1969) An index to measure weather-induced stress of trees associated with outbreaks of psyllids in Australia. Ecology 50:905–909

Chapter 13
Phytopathogenic Viruses

Carmen Büttner, Susanne von Bargen and Martina Bandte

Abstract Plant viruses are small sized plant pathogens. They are obligate parasites and among the major limiting factors to modern agriculture. The incidence of plant viruses has been shown in woody and herbaceous plants, soil and surface waters. Many of them have a wide host range and are characterized by efficient virus transmission. Since curative plant protection measures are lacking, it is important to focus on preventive measures according to phytosanitary practices, interruption of transmission pathways and vector control to combat plant viruses in practical crop production. Viral diseases require a constant vigil. The suitability and efficacy of different measures depend on the specific characteristics of the virus and the biology of the plants, potential vectors, and the environment.

13.1 Characteristics of Plant Viruses

Viruses are responsible for severe losses in crop plants worldwide. Annual crop losses caused by plant viruses were estimated in billions of US$. To date, more than 400 different plant virus species have been described and are summarized in the Ninth Report of the International Committee on Taxonomy of Viruses (King et al. 2012).

Among plant viruses, *Tomato spotted wilt virus* (TSWV) is responsible for numerous epidemics in different regions of the world and considered to have one of the largest host ranges including at least 1090 plants species. It causes reductions in yield of 1 billion USDs annually while losses due to *African cassava mosaic virus*

C. Büttner (✉) · S. von Bargen · M. Bandte
Division Phytomedicine, Faculty of Life Sciences, Humboldt-Universität zu Berlin,
Lentzeallee 55/57, 14195 Berlin, Germany
Tel.: +49-30-2093-46445
e-mail: carmen.buettner@agrar.hu-berlin.de

S. von Bargen
Tel.: +49-30-2093-46447
e-mail: susanne.von.bargen@agrar.hu-berlin.de

M. Bandte
Tel.: +49-30-2093-46447
e-mail: martina.bandte@agrar.hu-berlin.de

© Springer International Publishing Switzerland 2015
B. Lugtenberg (ed.), *Principles of Plant-Microbe Interactions,*
DOI 10.1007/978-3-319-08575-3_13

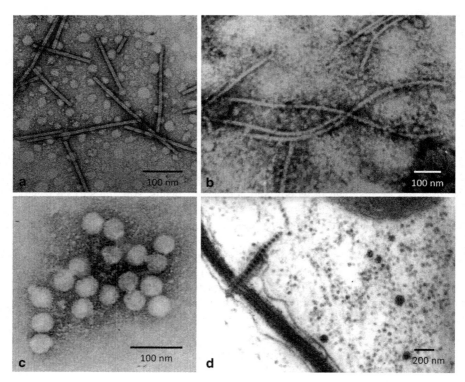

Fig. 13.1 Electron micrographs of different particle morphologies of plant viruses and intercellular movement via plasmodesmata. *Rod shaped* particles of *Tobacco mosaic virus* (*TMV*) 300 nm in length (**a**). *Flexible* particles of *Potato virus Y* (*PVY*) with an estimated length of 800 nm (**b**). *Isometric* particles with a diameter of approx. 28 nm of *Cherry leaf roll virus* (*CLRV*) (**c**) and cell-to-cell transport of the virus via tubules induced at plasmodesmata of *Sambucus nigra* cells (**d**). (Figures 13.1**a, b, c** are from the authors. Figure 13.1**d** has not been published previously and was kindly provided by J. Hamacher, Agro-Horti Testlabor, Bonn, Germany)

(ACMV) and related species, involved in the cassava disease pandemic in East and Central Africa, are now estimated at US\$ 1.9–2.7 billion (Scholthof et al. 2011). Depending on the virus-host interaction and environmental conditions, the response of the plant to virus infection may range from symptomless to severe disease or even plant death. The induced symptoms include deformations in shape, undesirable color, changes in taste or a reduction in keeping quality of the product.

Virus Properties Viruses are of small size, ranging within nanometers. They vary from simple helical and icosahedral shapes to more complex structures (Fig. 13.1a, b, c). Their morphology is only visible by electron microscopy. Viruses carry the genetic information encoded in one type of nucleic acid, DNA or RNA. Transcription, translation of viral proteins as well as replication of the viral genome is completely dependent upon the cellular metabolism of the host. Therefore, plant viruses are often considered replicators rather than forms of life. Some scientists classify viruses as microorganisms, but others consider them as nonliving. Rybicki (1990) described viruses as "organisms at the edge of life". They replicate by creating multiple copies of themselves through self-assembly processes, and evolve by natural selection but do not metabolize.

Plant viruses have at least three genes encoding the capsid protein, the movement protein and a protein accounting for nucleic acid replication.

The current state of taxonomy that follows a Linnaean-like classification of order, family, subfamily, genus, and species is summarized by King et al. 2012. As more sequence data have accumulated, more viruses have been placed in newly described or existing taxa. The highest level of virus classification recognizes six major taxa, based on the nature of the genome. In contrast to human and animal viruses the majority of plant viruses have a positive sense single-stranded RNA (ssRNA (+)) which functions directly as messenger RNA after entering the host cell and release from the virus particle. Only very few plant viruses have negative sense single-stranded RNA (ssRNA (−)), single stranded DNA (ssDNA), or double-stranded RNA (dsRNA).

The infection process of plants by viruses differs from that of other host's such as fungi, bacteria and animals. The plant has a robust cellulose cell wall covered by a waxy epidermis and viruses cannot penetrate them unaided. The entry into the cell succeeds in most cases through wounds by mechanical damage and/or vector assistance. In contrast entry of viruses infecting bacteria or animals is usually mediated by cell surface receptors. There are also differences in the distribution of viruses within the plant. Plasmodesmata enable virus movement by cytoplasmatic connections between cells and vascular tissue throughout most parts of the plant (Fig. 13.1d).

Transmission The mode of transmission is characteristic for each virus and often serves as a starting-point for plant protection management. Few viruses, such as tobamoviruses, rely on passive mechanical transmission from plant to plant. These viruses belong to the stable ones which can be easily transmitted also through water and soil, and they were detected even in clouds (Büttner and Koenig 2014). Other viruses can be transmitted by pollen and seeds (Sastry 2013). The major types of vector organisms of plant viruses are plant-feeding insects, nematodes and plant-parasitic fungi (Bragard et al. 2013). The economically most important insect vectors are restricted to a few hemipteran families such as aphids (Aphididae), whiteflies (Aleyrodidae), leafhoppers (Cicadellidae) and delphacid planthoppers (Delphacidae) (Hogenhout et al. 2008). For instance, aphids transmit about 200 plant virus species of which 110 belong to the genus *Potyvirus*. Whiteflies transmit 128 virus species; 115 of them are solely transmitted by these vectors and belong to the genus *Begomovirus*. Four different mechanisms of insect transmission of plant viruses have been described to date. If the vector retains the virus for only a few seconds/minutes after acquisition of the virus from plants and the virus is lost after molting (ecdysis), the transmission mode is called non-persistent. In contrast for persistent viruses, the vector retains the virus for long periods (days to weeks), often throughout the vector's lifespan, and keeping it even after molting. Semipersistent viruses display an intermediate category whereupon the vector retains the virus for a few hours to a few days and vectors lose it upon molting. Propagative viruses not only infect plants but also use the vectors as hosts as they are able to invade and replicate in various tissues of their vectors. Plant viruses can induce changes in vector behavior, survival and performance that promotes transmission efficiency (Ingwell et al. 2012; de Oliveira et al. 2014).

To date, 14 different viruses are known to be transmitted by ectoparasitic nematodes. These viruses belong to two genera (ii) nepoviruses which are transmitted by nematodes belonging to the family *Longidoridae* (genera *Longidorus, Paralongidorus,* and *Xiphinema*) and (ii) tobraviruses which are transmitted by nematodes from the family *Trichodoridae* (genera *Trichodorus* and *Paratrichodorus*). Ectoparasitic nematodes feed from the outside of the root, most often targeting the region at or near the root tip.

Six fungal species vector plant viruses: all are soilborne zoosporic obligate endoparasites that belong either to the protists (plasmodiophorids) or to the chytrid fungi (*Olpidium*). For instance, *Polymyxa betae* transmits four viruses to sugar beet, *Polymyxa graminis* transmits 14 viruses to cereals and groundnut, and the chytrid fungi *Olpidium brassicae, Olpidium bornovanus,* and *Olpidium virulentus* vector 15 viruses, all grouped in the *Ophioviridae* and *Tombusviridae* families as well as in the unassigned genus *Varicosavirus*.

13.2   Virus Interference with Host Plants

The presence of plant viruses has been shown in herbaceous and woody plants. Many of them have a wide host range and are characterized by efficient virus transmission, the rapidity with which new variants arise, and difficulties in vector control. Some hosts are more studied than others, e.g. investigations of viruses in forest and urban trees are rare (Gonthier and Nicolotti 2013).

Symptoms of viral diseases vary according to the virus and its host. The effect may range from severe and easily perceived to negligible depending on the plants' predisposition. The most evident symptoms are caused by a systemic infection. Deformation and coloring appear in fruits, leaves, stems, roots, or other parts of the plant, leading directly to crop yield reductions or losses of quality and/or quantity. One common symptom is hyperplasia: the abnormal proliferation of cells that causes the appearance of plant tumors known as galls. Other viruses induce hypoplasia, or decreased cell growth, in the leaves of plants, causing thin, yellow areas.

Virus-Host Interactions One of the most effective defense mechanisms of plants is the presence of resistance genes. Each of these genes confers resistance to a particular virus by triggering localized areas of cell death in the proximity to the infected cell, which is often visible as necrotic lesions. This phenomenon prevents the viral infection from spreading within the plant. Local lesions are often induced under experimental conditions when inoculating indicator plants within the scope of bioassays. Latent viruses do not cause symptoms in one crop, but may do so in another crop.

Another central part of the plants resistance response to virus infection and spread is called RNA silencing. It is a conserved intrinsic mechanism in eukaryotes,

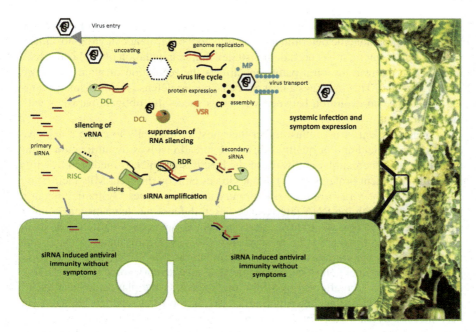

Fig. 13.2 Schematic representation of plant-virus interactions in host plant tissue. Infection process of a plant virus (virus life cycle), RNA-based antiviral immune response of the host plant cell (silencing of viral RNA, amplification and systemic spread of small interfering RNA-based signal), and counter strategy of the virus (suppression of RNA silencing) leading to mosaic symptoms. *DCL* dicer like protein, *RISC* RNA-induced silencing complex, *RDR* plant RNA-dependent RNA polymerase, *VSR* viral suppressor of RNA silencing, *CP* coat protein, *MP* movement protein. (This figure was created by the authors)

mediating antiviral defense via sequence-specific small RNA molecules. Virus-induced RNA silencing and regulation of endogenous plant genes are the major components of the mechanism called RNA interference (RNAi). In plants it is a key regulator of gene expression and involved in many developmental processes. In the RNA-based antiviral immunity, viral double-stranded RNAs, produced as replicative intermediates or present in viral genomic RNAs with extensive hairpin structures, are recognized by a host specific pathway as pathogen-associated molecules (Fig. 13.2). This silencing pathway leads either to post-transcriptional gene silencing (PTGS) by translational repression or to RNA degradation by endonucleolytic cleavage (slicing) of targeted cognate vRNAs. The aberrant RNAs generated by slicing can be amplified by host RNA-dependent RNA polymerases (RDRs) and may accumulate to high levels of secondary vsiRNAs serving as a systemic signal which triggers specific antiviral immunity throughout the plant.

Plant viruses are efficient pathogens having evolved various strategies to avoid or overcome silencing. Most notably, many viruses encode one or more proteins functioning as viral suppressors of RNA silencing (VSRs). The functions of VSRs are very diverse as they target different sites of the silencing pathway (Burgyan and

Havelda 2011). Many VSRs are also responsible for virus-induced symptoms and are important pathogenic determinants and effectors of virulence in virus-host interactions. Furthermore, it has been demonstrated that manipulation of the host's RNA interference pathway by viral pathogens affects the regulation of plant resistance genes (R genes). This indicates a host counter-counter defense mechanism accounting for a close co-evolution between viruses as successful obligate parasites with their host plants.

In general, plant viruses must infect their host systemically to cause a negative effect on crop yield and/or quality. This viral spread involves the utilization of intracellular and intercellular pathways of the host tissue which is mediated by virus encoded movement proteins (MPs). Plant viruses employ different MPs facilitating distribution of the virus from the site of entry into new tissues. Systemic infection in combination with the RNA interference-based-defense mechanisms and counter strategies employed by plant viruses described above are considered to be the main causes of symptom induction by these plant pathogens. For instance, mosaic is the most characteristic symptom of plant virus infections (Fig. 13.2). Mosaic is comprised of light green or yellow cell areas interspersed with dark green islands (DGIs). DGI formation can be affected by VSRs thus indicating the plant's defense mechanism through RNA silencing which are active in the tissues which have not yet been invaded by the virus. The formation of mosaic symptoms are therefore a good example of the multiple sites a plant virus interferes within its host. The underlying molecular mechanisms of these fundamental processes involved in the RNA interference-based immune response of plants to viral attack and its relevance to symptom induction are thoroughly discussed in Pumplin and Voinnet (2013), Wang et al. (2012), and Palukaitis (2011).

Beneficial Aspects of Plant Viruses Some plant viruses have been discovered to be beneficial and used by horticulturists to enhance the aesthetics of ornamental plants. Desirable aesthetic value is realized through flower color breaking, vein discolorations and foliar or flower variegations. Often this is the product of mutations that affect plastid development or transposable genetic elements that result in anomalous production of pigments but plant viruses can cause symptoms that mimic genetic variegations. One of the oldest virus diseases recorded is tulip breaking disease. *Tulip breaking virus* (TBV) causes the petals to variegate due to the irregular distribution of pigments, instead of being uniformly colored. Such an appearance may be considered a benefit due to its economic value, but most virus-infected plants do indicate less vitality in the long term.

13.3 Management of Viral Diseases

Introduction of viruses to new areas, often a result of human-aided transport, and change in vector dynamics are the main factors underlying the global emergence of viral diseases. Higher levels of human management result in higher disease and virus infection risk, which is associated with decreased habitat species diversity and host genetic diversity, and with increased host plant density (Rodelo-Urrego et al. 2013).

Phytosanitary Measures Virus disease management begins with a precise diagnosis of the pathogen. The most important tools of integrated disease management are the use of virus free planting or sowing material, the removal of vectors that transmit the virus as well as of weed populations serving as secondary hosts and the interruption of transmission paths while being aware of sources of contamination. The quality of water and soil is a condition precedent to virus-free crop production. If crops have to be irrigated, all potentially contaminated nutrient solutions have to be controlled. Such contamination often occurs in surface water collected from intensively cropped landscapes and in nutrient solutions of recirculating systems (Hong et al. 2014). In order to identify possible sources of contamination threatening the production a risk assessment shall be carried out.

Several procedures have been developed to deliver water that does not pose a risk to infect crop plants. Just recently, Hong et al. (2014) summarized the knowledge on biology, detection and management of plant pathogens in irrigation water.

Development of Virus Resistant Plants The above described measures focus primarily on avoiding the (i) introduction of plant viruses into an area and (ii) dispersion of plant viruses by controlling the vector (iii) interruption of all possible paths of virus transmission. Removal of weedy hosts and volunteer crops can reduce the virus reservoir but may contribute to a more homogenous agricultural landscape, which in turn decrease populations of predators and parasitoids that may regulate vector populations. With low cost-benefit ratio, environmental concerns and concomitant development of insecticide resistance, other strategies should be taken into consideration. An alternative control strategy for viruses is the use of resistant crop cultivars or varieties. These resistant genotypes carry heritable traits that are responsible for the suppression of virus multiplication or/and spread even under environmental conditions favoring infection. Thus no additional costs arise for the grower during the growing season. Virus resistance traits are available so that cultivars with varying degrees of resistance have been obtained by conventional breeding and gained commercial status. Technical solutions exist to provide transgenic virus resistance even for those crop species where virus resistance traits are lacking. For instance coat protein genes of many plant viruses have been transformed into a wide range of many plant species to obtain virus protection. Today, papaya, potato, squash, bean and plum have been introduced to the global market with transgenic virus resistance. For instance in 1998, two transgenic papaya (*Carica papaya* L.) with resistance to *Papaya ringspot virus* (PRSV) were released for commercial cultivation in Hawaii (Gonsalves 2014). The development and commercialization of these transgenic cultivars controlled the devastating aphid-vectored disease and saved the papaya industry in Hawaii. Currently, the cultivar Rainbow accounts for about 70 % of Hawaii's papaya acreage.

General Strategies Several national programs have been developed for the management of viruses that affect vegetatively propagated perennial plants including fruit trees, small fruits, grapevines and hop plants. These programs are based on virus-free stock providing certified virus-free propagation or growing material to be used in nurseries and vineyards. Prevention and control of vector transmission of plant viruses is imperative to promote sustainable agricultural practices and to reduce species invasion. Vector transmission of plant viruses is multi-scale and driven by characteristics of the specific virus, crop, and field properties. Therefore a good understanding of plant viruses, their ecology and vectors is required to develop and implement reliable prediction models for timely intervention. Intervention can be conducted using chemical or biological methods, particularly in greenhouse production.

References

Bragard C, Caciagli P, Lemaire O et al (2013) Status and prospects of plant virus control through interference with vector transmission. Annu Rev Phytopathol 51:1–25

Burgyan J, Havelda Z (2011) Viral suppressors of RNA silencing. Trends Plant Sci 16:265–272

Büttner C, Koenig R (2014) Viruses in water. In: Hong C, Moorman GW, Wohanka W, Büttner C (eds) Biology, detection and management of plant pathogens in irrigation water. ISBN 978-0-89054-426-6. St. Paul, Minnesota, USA

de Oliveira CF, Long EY, Finke DL (2014) A negative effect of a pathogen on its vector? A plant pathogen increases the vulnerability of its vector to attack by natural enemies. Oecologia 174:1169–1177

Gonsalves D (2014) Hawaii's transgenic papaya story 1978–2012: a personal account. In: Ming R, Moore P (eds) Genetics and genomics of papaya. Springer, New York, pp 115–142

Gonthier P, Nicolotti G (2013) Infectious forest diseases. Oxfordshire, UK

Hogenhout SA, Ammar ED, Whitfield AE et al (2008) Insect vector interactions with persistently transmitted viruses. Annu Rev Phytopathol 46:327–359

Hong C, Moorman GW, Wohanka W et al (2014) Biology, detection and management of plant pathogens in irrigation water. St. Paul, Minnesota, USA

Ingwell LL, Eigenbrode SD, Bosque-Perez NA (2012) Plant viruses alter insect behavior to enhance their spread. Sci Rep 2, Article number 578. doi:10.1038/srep00578

King AMQ, Adams MJ, Carstens EB et al (2012) Virus taxonomy: classification and nomenclature of viruses. Ninth report of the International Committee on Taxonomy of viruses. London, UK

Palukaitis P (2011) The road to RNA silencing is paved with plant-virus interactions. Plant Pathol J 27:197–206

Pumplin N, Voinnet O (2013) RNA silencing suppression by plant pathogens: defence, counter-defence and counter-counter-defence. Nat Rev Microbiol 11:745–760

Rodelo-Urrego M, Pagán I, González-Jara P et al (2013) Landscape heterogeneity shapes host-parasite interactions and results in apparent plant virus codivergence. Mol Ecol 22:2325–2340

Rybicki EP (1990) The classification of organisms at the edge of life, or problems with virus systematics. S Afr J Sci 86:182–186

Sastry KS (2013) Seed-borne plant virus diseases. New Delhi, India

Scholthof KB, Adkins S, Czosnek H et al (2011) Top 10 plant viruses in molecular plant pathology. Mol Plant Pathol 12:938–954

Wang MB, Masuta C, Smith NA et al (2012) RNA silencing and plant viral diseases. Mol Plant Microbe Interact 25:1275–1285

Chapter 14
Induced Disease Resistance

Corné M. J. Pieterse and Saskia C. M. Van Wees

Abstract During the co-evolutionary arms race between plants and pathogens, plants evolved a sophisticated defense system to ward off their enemies. In this plant immune system, plant receptor proteins recognize non-self molecules of microbial origin, which leads to the activation of a basal level of disease resistance. The onset of these local plant immune reactions often triggers a systemic acquired resistance (SAR) in tissues distal from the site of infection. Beneficial microbes in the rhizosphere microbiome stimulate a phenotypically similar induced systemic resistance (ISR) that, like SAR, is effective against a broad spectrum of pathogens. There are differences and similarities in the SAR and ISR signaling pathways. The plant defense hormone salicylic acid is a major regulator of SAR, whereas jasmonic acid and ethylene play important roles in ISR. Priming of systemic tissue to express an accelerated defense response upon attack by a pathogen is a common phenomenon in both SAR and ISR. This chapter will outline the current concept of the plant immune system, with special emphasis on mechanisms of systemically induced disease resistance and priming for enhanced defense.

14.1 The Plant Immune System

In the past decade, ground-breaking conceptual advances have been made in the understanding of the evolutionary development and functioning of the plant immune system (Jones and Dangl 2006). In the current concept of the plant immune system, pattern-recognition receptors (PRRs) have evolved to recognize pathogen- or microbe-associated molecular patterns (PAMPs or MAMPs), such as bacterial flagellin or fungal chitin (Boller and Felix 2009). MAMP recognition is translated into

C. M. J. Pieterse (✉) · S. C. M. Van Wees
Plant-Microbe Interactions, Department of Biology, Faculty of Science,
Utrecht University, PO BOX 800.56, 3508 TB Utrecht, The Netherlands
Tel.: +31302536887
e-mail: C.M.J.Pieterse@uu.nl

S. C. M. Van Wees
Tel.: +31302536861
e-mail: S.VanWees@uu.nl

© Springer International Publishing Switzerland 2015
B. Lugtenberg (ed.), *Principles of Plant-Microbe Interactions*,
DOI 10.1007/978-3-319-08575-3_14

a basal defense called pattern-triggered immunity (PTI) (Dodds and Rathjen 2010). Successful pathogens evolved virulence effector molecules to bypass this first line of defense, either by preventing detection by the host, or by suppressing PTI signaling (Dodds and Rathjen 2010; Pel and Pieterse 2013). To fight these successful pathogens, plants developed a second line of defense in which resistance (R) proteins mediate recognition of attacker-specific effectors (formerly known as avirulence factors), resulting in highly powerful effector-triggered immunity (ETI) (Dodds and Rathjen 2010). ETI is a manifestation of the classic gene-for-gene resistance that is accompanied by a hypersensitive response that prevents biotrophic pathogens from further entry (Chap. 10).

Activation of PTI and ETI in locally infected tissues often triggers an induced resistance in tissues distal from the site of infection and involves one or more long-distance signals that propagate an enhanced defensive capacity in still undamaged plant parts. This pathogen-induced systemic resistance is known as systemic acquired resistance (SAR) (Fu and Dong 2013). While PTI and ETI are activated rapidly and act locally to limit growth of the specific invader at the site of infection, SAR takes more time to develop but confers an enhanced defensive capacity that is typically effective against a broad spectrum of pathogens (Walters et al. 2013; Fu and Dong 2013).

Besides pathogen infection, also colonization of plant roots by beneficial microbes has been shown to stimulate the plant immune system, resulting in a phenotypically similar type of broad-spectrum disease resistance, commonly referred to as induced systemic resistance (ISR) (Pieterse et al. 2014). Moreover, insect herbivory and specific chemicals can also induce resistance (Howe and Jander 2008; Pastor et al. 2013). After more than three decades of research, the picture is emerging that the different forms of induced resistance are regulated by a complex network of interconnecting signaling pathways in which plant hormones play an important regulatory role (Pieterse et al. 2012). Induced resistance signaling pathways that are triggered by pathogens, beneficial microbes, and insects partly overlap and share common signaling components (Pieterse et al. 2014). This provides plants with an enormous regulatory potential to rapidly adapt to their biotic environment and to utilize their limited resources for growth and survival in a cost-efficient manner. Intriguingly, successful pathogens evolved mechanisms to rewire the plant's hormone signaling network to suppress or evade the host immune system (Robert-Seilaniantz et al. 2011; Pieterse et al. 2012), highlighting the central role of plant hormones in the regulation of immunity.

The concepts of PTI and ETI that act locally in the plant immune system will be discussed in more depth elsewhere in this issue (see Chap. 10). In this chapter we will focus on the important principles and recent findings of induced disease resistance that acts systemically throughout the plant.

14.2 Pathogen-Induced Systemic Acquired Resistance (SAR)

Hallmarks of SAR The term SAR was first coined by Ross for the phenomenon that uninfected systemic plant parts become more resistant in response to a prior infection elsewhere in the plant (Ross 1961). SAR is typically triggered upon local activation of a PTI or ETI response (Shah and Zeier 2013). In systemic tissues, SAR is characterized by increased levels of the hormone salicylic acid (SA), one of the hallmarks of SAR (Vlot et al. 2009) (Fig. 14.1). Early genetic studies in tobacco showed that SA accumulation and signaling is essential for the establishment of SAR (Vernooij et al. 1994). Another hallmark of SAR is the coordinate activation of *PATHOGENESIS-RELATED* (PR) genes, several of which encode PR proteins with antimicrobial activity (Van Loon et al. 2006). *PR −1* is amongst the best characterized *PR* genes and is in many plant species used as a marker for SAR (Van Loon et al. 2006; Fu and Dong 2013).

Long-Distance Signals Because the expression of SAR occurs in plant parts that are distant from the site of induction, a long-distance mobile signal is required that is produced locally and is responsible for the systemic onset SAR in still healthy tissues. The identity of the mobile SAR signal(s) has been a subject of controversy for many years. The lipid-transfer protein DEFECTIVE IN INDUCED RESISTANCE1 (DIR1) was shown to act as a chaperone for an unknown mobile SAR signal in the vascular tissue (Maldonado et al. 2002; Champigny et al. 2011). Despite the fact that SA accumulates in the phloem sap of SAR-expressing plants, grafting experiments with tobacco showed that SA itself is not the mobile SAR signal (Vernooij et al. 1994). Recent genetic and biochemical studies uncovered several plant metabolites involved in long-distance SAR signaling. These include the methyl ester of SA (MeSA), the diterpenoid (DA), a glycerol -3-phosphate (G3P)-dependent factor, azelaic acid (AzA), and pipecolic acid (Pip) (Fig. 14.1). From these findings a more comprehensive view on the identity and functioning of the long-distance SAR signals started to emerge in which different signals may be operative under different environmental conditions (Shah and Zeier 2013; Dempsey and Klessig 2012; Kachroo and Robin 2013). In systemic tissues, the onset of SAR requires the function of FLAVIN-DEPENDENT MONOOXYGENASE 1 (FMO1) (Mishina and Zeier 2006), possibly to transduce or amplify long-distance signals originating from primary leaves, which then results in enhanced SA biosynthesis in still healthy tissues.

SAR Signaling Upon activation of SAR, the SA signal is transduced by the redox-regulated protein NONEXPRESSOR OF *PR* GENES1 (NPR1), which functions as a transcriptional co-activator of a large set of *PR* genes (Fu and Dong 2013). In non-stimulated cells, NPR1 is sequestered in the cytoplasm as an oligomer through intermolecular disulfide bonds. Upon SA accumulation, changes in the cellular redox state mediate monomerization of NPR1, which allows translocation of NPR1 into the nucleus. In the nucleus, NPR1 interacts with TGA transcription factors that together with WRKY transcription factors activate SA-responsive *PR* genes. Proper functioning of NPR1 requires that the protein is broken down by the proteasome,

126   C. M. J. Pieterse and S. C. M. Van Wees

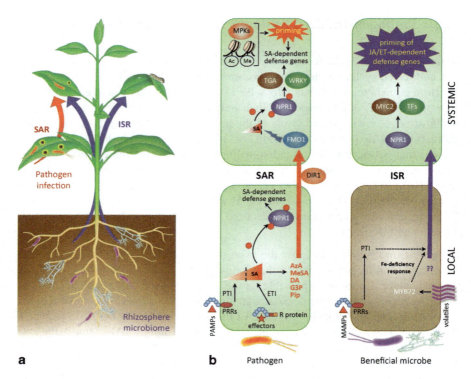

Fig. 14.1 a Schematic representation of biologically induced disease resistance triggered by pathogen infection (*SAR; red arrow*) and colonization of the roots by beneficial microbes (*ISR; purple arrow*). Induced resistance involves long-distance signals that are transported through the vasculature or as airborne signals, and systemically propagate an enhanced defensive capacity against a broad spectrum of attackers in still healthy plant parts. **b** Schematic representation of molecular components and mechanisms involved in pathogen-induced SAR and rhizobacteria-mediated ISR. Solid black lines indicate established interactions; dashed black lines indicate hypothetical interactions. Colored arrows indicate systemic translocation of long-distance signals (indicated in the same color at the base of the arrows). *Ac* acetylation, *ET* ethylene, *ETI* effector-triggered immunity, *Fe* iron, *ISR* induced systemic resistance, *JA* jasmonic acid, *MAMP* microbe-associated molecular pattern, *Me* methylation, *PAMP* pathogen-associated molecular pattern, *PRR* pattern-recognition receptor, *PTI* PAMP-triggered immunity, *R* protein Resistance protein, *SA* salicylic acid, *SAR* systemic acquired resistance, *TF* transcription factor

possibly to allow new NPR1 proteins to reinitiate the *PR* transcription cycle (Spoel et al. 2009). Recently, NPR1 and its paralogues NPR3 and NPR4 were identified as SA receptors that bind to SA with different affinity thereby influencing the stability of NPR1 (Fu et al. 2012; Wu et al. 2012).

14.3   Induced Systemic Resistance (ISR) by Beneficial Microbes

Besides microbial pathogens, also large communities of commensal and mutualistic microbes interact with plants providing them with essential services, such as enhanced mineral uptake, nitrogen fixation, growth promotion, and protection from pathogens (Chap. 20; Lugtenberg and Kamilova 2009; Zamioudis and Pieterse 2012; Pieterse et al. 2014). This community of microbes is predominantly hosted by the root system and is also referred to as the rhizosphere microbiome (Chap. 28; Berendsen et al. 2012; Mendes et al. 2011). In 1991, it was demonstrated that colonization of plant roots by selected strains of plant growth-promoting rhizobacteria (PGPR) can stimulate the plant immune system in above-ground plant parts, resulting in a broad-spectrum disease resistance called rhizobacteria-ISR (Fig. 14.1) (Van Peer et al. 1991; Wei et al. 1991; Alström 1991). Since then, hundreds of studies in dicots and monocots have reported on the ability of PGPR to promote plant health via ISR. These studies mainly involved *Bacillus*, *Pseudomonas,* and *Serratia* PGPR strains. In addition, non-pathogenic plant growth-promoting fungi (PGPF) strains from species like *Fusarium oxysporum*, *Trichoderma* spp., and *Piriformospora indica* strains, but also symbiotic arbuscular mycorrhizal fungi have been shown to trigger ISR (Pieterse et al. 2014).

Microbial Elicitors of ISR In order to stimulate ISR, beneficial microbes must produce elicitors that are responsible for the onset of systemic immunity. Early research on MAMPs and other elicitors of ISR-inducing *Pseudomonas* and *Bacillus* PGPR focused on the involvement of lipopolysaccharides (LPS) and the iron-regulated metabolites pyoverdin and SA (De Vleesschauwer and Höfte 2009). Other microbial ISR elicitors include antibiotics, like 2,4-diacetylphloroglucinol (DAPG) and pyocyanin, flagella, N-acyl homoserine lactones, siderophores, and biosurfactants (De Vleesschauwer and Höfte 2009). Also specific volatile organic compounds produced by beneficial microbes were demonstrated to elicit ISR (Ryu et al. 2004; Lee et al. 2012). Several of these ISR elicitors were shown to act redundantly, indicating that multiple microbial elicitors can trigger common signaling pathway leading to systemic immunity (Bakker et al. 2003). This resembles PTI in plant-pathogen interactions, where recognition of multiple PAMPs is channeled into the same PTI signaling pathway (Boller and Felix 2009).

Rhizobacteria-ISR Signaling Pathways Because of its broad spectrum effectiveness, rhizobacteria-ISR was initially thought to be mechanistically similar to pathogen-induced SAR. However, in radish it was shown that *Pseudomonas fluorescens* WCS417r (hereafter called WCS417r) triggered ISR without stimulating the accumulation of the PR proteins that are characteristic for SAR (Hoffland et al. 1995). Also in *Arabidopsis thaliana* (Arabidopsis) WCS417r-ISR developed without the activation of *PR* genes (Pieterse et al. 1996). Transgenic SA-nonaccumulating Arabidopsis NahG plants mounted wild-type levels of ISR upon colonization of the roots by WCS417r, providing genetic evidence that ISR can be mediated via an SA-independent signaling pathway (Pieterse et al. 1996). Hence, rhizobacteria-mediated

ISR and pathogen-induced SAR are regulated by distinct signaling pathways (Fig. 14.1). Analysis of a large number of ISR-triggering plant-beneficial microbe interactions in which a role for SA had been functionally tested, revealed that the ability to activate an SA-independent ISR pathway is common for beneficial microbes and occurs in a broad range of plant species (Van Loon and Bakker 2005; Van Wees et al. 2008).

Although ISR by beneficial microbes is often regulated through SA-independent mechanisms, certain strains of beneficial microbes have been reported to trigger ISR in an SA-dependent fashion, which resembles pathogen-induced SAR (De Vleesschauwer and Höfte 2009; Van de Mortel et al. 2012). In these cases, reactive oxygen species that accumulate at the site of tissue colonization seem to act as important elicitors (De Vleesschauwer and Höfte 2009). Since SA-dependent signaling triggered by beneficial microbes is likely to follow the SAR signaling pathway, we refer to the above section on pathogen-induced SAR for information on mechanisms underlying this phenomenon.

Role of Jasmonic Acid and Ethylene in ISR After the discovery of SA as an important defense hormone, also the plant hormones jasmonic acid (JA) and ethylene (ET) emerged as important regulators of plant immunity (Pieterse et al. 2012). By using JA or ET signaling mutants of Arabidopsis, it was shown that not SA, but JA and ET are central regulators of WCS417r-ISR (Pieterse et al. 1998). For many other PGPR and PGPF genetic evidence pointed to a role for JA and/or ET in the regulation of ISR (Pieterse et al. 2014), supporting the notion that JA and ET are dominant players in the regulation of SA-independent systemic immunity conferred by beneficial soil-borne microbes.

Master Regulators of ISR The first regulatory protein identified as being essential for rhizobacteria-ISR was NPR1 (Pieterse et al. 1998). While in SAR, NPR1 functions as a transcriptional co-activator of SA-responsive *PR* genes, JA/ET-dependent ISR typically functions without *PR* gene activation. Hence, the role of NPR1 in ISR seems to be different from that in SAR. In SA signaling, NPR1 is clearly connected to a nuclear function (Fu and Dong 2013), while in JA/ET signaling and ISR evidence is accumulating for a cytosolic function of NPR1 (Spoel et al. 2003; Stein et al. 2008; Pieterse et al. 2012). Interestingly, simultaneous activation of SAR and ISR leads to an additively enhanced defensive capacity (Van Wees et al. 2000). Whether this is based on the notion that SAR and ISR do not seem to compete for the same subcellular pool of NPR1 is unknown, as the exact molecular mechanism by which NPR1 functions in JA/ET-dependent ISR remains to be investigated.

Although ISR involves long-distance root-to-shoot signaling, only few studies have investigated the signaling components of the plant root that are involved in the onset of ISR. Analysis of the transcriptome of WCS417-colonized Arabidopsis roots revealed the R2R3 type MYB transcription factor gene *MYB72* as one of the significantly induced genes (Verhagen et al. 2004). In non-stimulated plants, *MYB72* is lowly expressed in the root vascular bundle, but becomes highly expressed in root epidermis and cortical cells upon colonization by ISR-inducing PGPR or their volatiles (Zamioudis et al. 2014b). Knockout *myb72* mutants are impaired in their ability to express ISR, indicating that this root-specific transcription factor is essential for the onset of ISR (Van der Ent et al. 2008). *MYB72* is also induced in *Trichoderma*-colonized Arabidopsis roots and shown to be crucial for *Trichoderma*-ISR (Segarra et al. 2009), suggesting that MYB72 is a node of convergence in the ISR signaling pathway triggered by different beneficial microbes. Being a transcriptional regulator, it was postulated that MYB72 plays an important role in the generation and/or translocation of a long-distance ISR signal. Besides its crucial role in the onset of ISR, MYB72 is also implicated in the iron-deficiency response of plant roots (Zamioudis et al. 2014a; Zamioudis et al. 2014b). How ISR and the iron-deficiency response are interconnected is currently unknown.

14.4 Induced Disease Resistance: Priming for Enhanced Defense

While SA accumulation and *PR* gene expression are hallmarks of SAR, ISR triggered by beneficial microbes is lacking such universal characteristics associated with the onset of systemic immunity. In many cases, colonization of plant roots by beneficial microbes does not lead to major changes in defense-related gene expression in the above-ground plant parts. Instead, pathogen infection or insect herbivory on ISR-expressing plants often leads to an accelerated expression of defense-related gene expression in comparison to similarly attacked control plants (Van Wees et al. 1999; Van Oosten et al. 2008). Large-scale analysis of the WCS417r-ISR transcriptome of Arabidopsis before and after pathogen challenge showed that ISR is associated with potentiated expression of a large set of JA/ET-regulated defense genes that are induced upon pathogen challenge (Fig. 14.1) (Verhagen et al. 2004). This preparation of the whole plant to better combat pathogen or insect attack is called 'priming' and is characterized by a faster and/or stronger activation of cellular defenses upon invasion, resulting in an enhanced level of resistance (Conrath 2011). To date, a large number of studies with PGPR and PGPF have supported the notion that ISR by beneficial microbes is commonly based on defense priming (Pieterse et al. 2014).

Priming for enhanced defense emerged as an important cellular process in many types of biologically and chemically induced systemic immunity, including SAR, ISR, and herbivore-induced resistance (Frost et al. 2008; Luna et al. 2014; Pastor et al. 2013; Conrath 2011). For instance, low doses of SAR-inducing agents do not directly activate *PR* gene expression, but prime systemic tissues for enhanced *PR* gene expression after pathogen challenge, indicating that priming is also an important component of this type of induced resistance (Conrath 2011). By studying the costs and benefits of defense priming, it was shown that the fitness costs of priming are lower than those of constitutively activated defenses (Van Hulten et al. 2006; Walters et al. 2008; Vos et al. 2013). The fitness benefit of priming was shown to outweigh its cost when under pathogen pressure, suggesting that priming functions as an ecological adaptation of the plant to respond faster to its hostile environment.

Priming: A Molecular Memory of Immunization Because defense priming is clearly expressed at the transcriptional level, research on the mechanisms underlying the primed state has focused on the expression of signaling intermediates in transcriptional networks. These factors are thought to remain inactive in the absence of an attacker, but their accumulation can provide the plant with the capacity to react with an accelerated defense response upon perception of a pathogen- or insect-derived stress signal. In Arabidopsis, the ISR-primed state was shown to be associated with elevated transcript levels of genes that encode transcription factors of the AP2/ERF family and MYC2, both of which have been implicated in the regulation of JA- and/or ET-dependent defenses (Van der Ent et al. 2009; Pozo et al. 2008). This is in agreement with the observation that in particular JA/ET-regulated genes show a primed expression pattern in challenged ISR-expressing plants (Verhagen et al. 2004).

Mitogen-activated protein kinases (MAPKs) have also been implicated in defense priming. Inactive forms of the MPK3 and MPK6 were shown to accumulate after treatment of plants with low concentrations of the SAR-inducing SA-analogue benzothiadiazole (BTH) (Beckers et al. 2009). After pathogen challenge, these latent signaling molecules were activated, resulting in accelerated *PR -1* gene expression and the development of enhanced disease resistance. Priming is also associated with chromatin modifications in the promoters of WRKY transcription factor genes that regulate SA-dependent defenses, thereby facilitating potentiated expression of these regulatory genes upon pathogen attack (Jaskiewicz et al. 2011). Recently, epigenetic regulation of pathogen- and chemically-induced priming for SA-dependent defenses and herbivore-induced priming for JA-dependent defenses was shown to be inherited to the offspring via chromatin remodeling (Slaughter et al. 2012; Rasmann et al. 2012; Luna et al. 2012). Hence, plants seem to have the capacity to "memorize" a stressful situation and subsequently immunize not only themselves, but also their next generation against future attacks (Pastor et al. 2013).

14.5   Induced Resistance: Shaping the Plant's Social Network

Exciting developments in induced disease resistance research provided a wealth of information on the molecular details of how this adaptive defense system functions. In nature, plants are attacked by a multitude of pathogens and pests. However, beneficial associations between plants and mutualistic microbes are abundant in nature as well, improving plant growth and health. Hormone-regulated plant defense signaling networks finely balance plant responses to beneficial microbes, pathogens, and insects to maximize both profitable and protective functions. Defense signaling pathways that are recruited in response to parasitic and beneficial organisms can overlap, indicating that the regulation of the plant's adaptive response to its biotic environment is finely balanced between protection against aggressors and acquisition of benefits. Plant hormones play pivotal roles in the regulation of the defense signaling network. Their signaling pathways interact in a synergistic or antagonistic manner, providing the plant with the capacity to tailor its immune response to the attacker encountered

(Pieterse et al. 2012). In agricultural and ecological settings, plants often interact with a whole suite of other organisms that range from beneficial microbes on their root system to foliar pathogens and insect herbivores. Detailed mechanistic knowledge on how the plant immune signaling network functions during multi-organisms interactions is fundamental to develop novel strategies for sustainable protection of our future crops that need to produce more with less input of pesticides and fertilizers.

References

Alström S (1991) Induction of disease resistance in common bean susceptible to halo blight bacterial pathogen after seed bacterization with rhizosphere pseudomonads. J Gen Appl Microbiol 37:495–501

Bakker PAHM, Ran LX, Pieterse CMJ et al (2003) Understanding the involvement of rhizobacteria-mediated induction of systemic resistance in biocontrol of plant diseases. Can J Plant Pathol 25:5–9

Beckers GJM, Jaskiewicz M, Liu Y et al (2009) Mitogen-activated protein kinases 3 and 6 are required for full priming of stress responses in *Arabidopsis thaliana*. Plant Cell 21:944–953

Berendsen RL, Pieterse CMJ, Bakker PAHM (2012) The rhizosphere microbiome and plant health. Trends Plant Sci 17:478–486

Boller T, Felix G (2009) A renaissance of elicitors: perception of microbe-associated molecular patterns and danger signals by pattern-recognition receptors. Annu Rev Plant Biol 60:379–406

Champigny M, Shearer H, Mohammad A et al (2011) Localization of DIR1 at the tissue, cellular and subcellular levels during systemic acquired resistance in Arabidopsis using DIR1:GUS and DIR1:EGFP reporters. BMC Plant Biol 11:125

Conrath U (2011) Molecular aspects of defence priming. Trends Plant Sci 16:524–531

De Vleesschauwer D, Höfte M (2009) Rhizobacteria-induced systemic resistance. In: Van Loon LC (ed) Plant Innate Immunity, vol 51. Advances in botanical research. Academic Press Ltd-Elsevier Science Ltd, London, pp 223–281

Dempsey DA, Klessig DF (2012) SOS—too many signals for systemic acquired resistance? Trends Plant Sci 17:538–545

Dodds PN, Rathjen JP (2010) Plant immunity: towards an integrated view of plant-pathogen interactions. Nat Rev Genet 11:539–548

Frost CJ, Mescher MC, Carlson JE et al (2008) Plant defense priming against herbivores: getting ready for a different battle. Plant Physiol 146:818–824

Fu ZQ, Dong X (2013) Systemic acquired resistance: turning local infection into global defense. Annu Rev Plant Biol 64:839–863

Fu ZQ, Yan S, Saleh A et al (2012) NPR3 and NPR4 are receptors for the immune signal salicylic acid in plants. Nature 486:228–232

Hoffland E, Pieterse CMJ, Bik L et al (1995) Induced systemic resistance in radish is not associated with accumulation of pathogenesis-related proteins. Physiol Mol Plant Pathol 46:309–320

Howe GA, Jander G (2008) Plant immunity to insect herbivores. Annu Rev Plant Biol 59:41–66

Jaskiewicz M, Conrath U, Peterhansel C (2011) Chromatin modification acts as a memory for systemic acquired resistance in the plant stress response. EMBO Rep 12:50–55

Jones JDG, Dangl JL (2006) The plant immune system. Nature 444:323–329

Kachroo A, Robin GP (2013) Systemic signaling during plant defense. Curr Opin Plant Biol 16:527–533

Lee B, Farag MA, Park HB et al (2012) Induced resistance by a long-chain bacterial volatile: elicitation of plant systemic defense by a C13 volatile produced by *Paenibacillus polymyxa*. PLoS ONE 7:e48744

Lugtenberg B, Kamilova F (2009) Plant-growth-promoting rhizobacteria. Annu Rev Microbiol 63:541–556

Luna E, Bruce TJA, Roberts MR et al (2012) Next-generation systemic acquired resistance. Plant Physiol 158:844–853

Luna E, Van Hulten M, Zhang Y et al (2014) Plant perception of β-aminobutyric acid is mediated by an aspartyl-tRNA synthetase. Nat Chem Biol 10:450–456

Maldonado AM, Doerner P, Dixon RA et al (2002) A putative lipid transfer protein involved in systemic resistance signalling in *Arabidopsis*. Nature 419:399–403

Mendes R, Kruijt M, De Bruijn I et al (2011) Deciphering the rhizosphere microbiome for disease-suppressive bacteria. Science 332:1097–1100

Mishina TE, Zeier J (2006) The Arabidopsis flavin-dependent monooxygenase FMO1 is an essential component of biologically induced systemic acquired resistance. Plant Physiol 141:1666–1675

Pastor V, Luna E, Mauch-Mani B et al (2013) Primed plants do not forget. Environ Exp Bot 94:46–56

Pel MJC, Pieterse CMJ (2013) Microbial recognition and evasion of host immunity. J Exp Bot 64:1237–1248

Pieterse CMJ, Van Wees SCM, Hoffland E et al (1996) Systemic resistance in Arabidopsis induced by biocontrol bacteria is independent of salicylic acid accumulation and pathogenesis-related gene expression. Plant Cell 8:1225–1237

Pieterse CMJ, Van Wees SCM, Van Pelt JA et al (1998) A novel signaling pathway controlling induced systemic resistance in Arabidopsis. Plant Cell 10:1571–1580

Pieterse CMJ, Van der Does D, Zamioudis C et al (2012) Hormonal modulation of plant immunity. Annu Rev Cell Dev Biol 28:489–521

Pieterse CMJ, Zamioudis C, Berendsen RL et al (2014) Induced systemic resistance by beneficial microbes. Annu Rev Phytopathol 52:347–375

Pozo MJ, Van der Ent S, Van Loon LC et al (2008) Transcription factor MYC2 is involved in priming for enhanced defense during rhizobacteria-induced systemic resistance in *Arabidopsis thaliana*. New Phytol 180:511–523

Rasmann S, De Vos M, Casteel CL et al (2012) Herbivory in the previous generation primes plants for enhanced insect resistance. Plant Physiol 158:854–863

Robert-Seilaniantz A, Grant M, Jones JDG (2011) Hormone crosstalk in plant disease and defense: more than just jasmonate-salicylate antagonism. Annu Rev Phytopathol 49:317–343

Ross AF (1961) Systemic acquired resistance induced by localized virus infections in plants. Virology 14:340–358

Ryu C-M, Farag MA, Hu CH et al (2004) Bacterial volatiles induce systemic resistance in Arabidopsis. Plant Physiol 134:1017–1026

Segarra G, Van der Ent S, Trillas I et al (2009) MYB72, a node of convergence in induced systemic resistance triggered by a fungal and a bacterial beneficial microbe. Plant Biol 11:90–96

Shah J, Zeier J (2013) Long-distance communication and signal amplification in systemic acquired resistance. Front Plant Sci 4:30

Slaughter A, Daniel X, Flors V et al (2012) Descendants of primed Arabidopsis plants exhibit resistance to biotic stress. Plant Physiol 158:835–843

Spoel SH, Koornneef A, Claessens SMC et al (2003) NPR1 modulates cross-talk between salicylate- and jasmonate-dependent defense pathways through a novel function in the cytosol. Plant Cell 15:760–770

Spoel SH, Mou ZL, Tada Y et al (2009) Proteasome-mediated turnover of the transcription coactivator NPR1 plays dual roles in regulating plant immunity. Cell 137:860–872

Stein E, Molitor A, Kogel KH et al (2008) Systemic resistance in Arabidopsis conferred by the mycorrhizal fungus *Piriformospora indica* requires jasmonic acid signaling and the cytoplasmic function of NPR1. Plant Cell Physiol 49:1747–1751

Van de Mortel JE, De Vos RCH, Dekkers E et al (2012) Metabolic and transcriptomic changes induced in Arabidopsis by the rhizobacterium *Pseudomonas fluorescens* SS101. Plant Physiol 160:2173–2188

Van der Ent S, Verhagen BWM, Van Doorn R et al (2008) MYB72 is required in early signaling steps of rhizobacteria-induced systemic resistance in Arabidopsis. Plant Physiol 146:1293–1304

Van der Ent S, Van Hulten MHA, Pozo MJ et al (2009) Priming of plant innate immunity by rhizobacteria and ß-aminobutyric acid: differences and similarities in regulation. New Phytol 183:419–431

Van Hulten M, Pelser M, Van Loon LC et al (2006) Costs and benefits of priming for defense in Arabidopsis. Proc Natl Acad Sci U S A 103:5602–5607

Van Loon LC, Bakker PAHM (2005) Induced systemic resistance as a mechanism of disease suppression by rhizobacteria. In: Siddiqui ZA (ed) PGPR: biocontrol and biofertilization. Springer, Dordrecht, pp 39–66

Van Loon LC, Rep M, Pieterse CMJ (2006) Significance of inducible defense-related proteins in infected plants. Annu Rev Phytopathol 44:135–162

Van Oosten VR, Bodenhausen N, Reymond P et al (2008) Differential effectiveness of microbially induced resistance against herbivorous insects in Arabidopsis. Mol Plant-Microbe Interact 21:919–930

Van Peer R, Niemann GJ, Schippers B (1991) Induced resistance and phytoalexin accumulation in biological control of fusarium wilt of carnation by *Pseudomonas* sp. strain WCS417r. Phytopathology 81:728–734

Van Wees SCM, Luijendijk M, Smoorenburg I et al (1999) Rhizobacteria-mediated induced systemic resistance (ISR) in Arabidopsis is not associated with a direct effect on expression of known defense-related genes but stimulates the expression of the jasmonate-inducible gene *Atvsp* upon challenge. Plant Mol Biol 41:537–549

Van Wees SCM, De Swart EAM, Van Pelt JA et al (2000) Enhancement of induced disease resistance by simultaneous activation of salicylate- and jasmonate-dependent defense pathways in *Arabidopsis thaliana*. Proc Natl Acad Sci U S A 97:8711–8716

Van Wees SCM, Van der Ent S, Pieterse CMJ (2008) Plant immune responses triggered by beneficial microbes. Curr Opin Plant Biol 11:443–448

Verhagen BWM, Glazebrook J, Zhu T et al (2004) The transcriptome of rhizobacteria-induced systemic resistance in *Arabidopsis*. Mol Plant-Microbe Interact 17:895–908

Vernooij B, Friedrich L, Morse A et al (1994) Salicylic acid is not the translocated signal responsible for inducing systemic acquired resistance but is required in signal transduction. Plant Cell 6:959–965

Vlot AC, Dempsey DA, Klessig DF (2009) Salicylic acid, a multifaceted hormone to combat disease. Annu Rev Phytopathol 47:177–206

Vos IA, Pieterse CMJ, Van Wees SCM (2013) Costs and benefits of hormone-regulated plant defences. Plant Pathol 62:43–55

Walters DR, Paterson L, Walsh DJ et al (2008) Priming for plant defense in barley provides benefits only under high disease pressure. Physiol Mol Plant Pathol 73:95–100

Walters DR, Ratsep J, Havis ND (2013) Controlling crop diseases using induced resistance: challenges for the future. J Exp Bot 64:1263–1280

Wei G, Kloepper JW, Tuzun S (1991) Induction of systemic resistance of cucumber to *Colletrotichum orbiculare* by select strains of plant-growth promoting rhizobacteria. Phytopathology 81:1508–1512

Wu Y, Zhang D, Chu JY et al (2012) The Arabidopsis NPR1 protein is a receptor for the plant defense hormone salicylic acid. Cell Rep 1:639–647

Zamioudis C, Pieterse CMJ (2012) Modulation of host immunity by beneficial microbes. Mol Plant-Microbe Interact 25:139–150

Zamioudis C, Hanson J, Pieterse CMJ (2014a) β-Glucosidase BGLU42 is a MYB72-dependent key regulator of rhizobacteria-induced systemic resistance and modulates iron uptake responses in Arabidopsis roots. New Phytol 204:368–379

Zamioudis C, Korteland J, Van Pelt JA et al (2014b) Root bacteria stimulate iron uptake in plants via a novel photosynthesis-dependent iron sensing system. *Submitted*

Chapter 15
Apologies to the Planet—Can We Restore the Damage?

Dulce Eleonora de Oliveira and Marc Van Montagu

Abstract Many of us believe that human ingenuity can promote the innovations required to face the challenge of demographic pressure. But twentieth century experience has shown that pollution by human beings (through increased population), by manufacturing industries and by agricultural practices have severely damaged the environment, requiring continuous mitigation efforts. And still we are unable to bring an acceptable standard of living to half of the world population.

How to render intensive agriculture and small farming more sustainable? Good governance and innovative science are essential, but we can no longer delay applying the knowledge generated in the last decennia by the plant scientists. Intensive cooperation among agronomists, agro-ecologists and biotechnologists is urgently needed, together with communication to society on the value of applying science to agriculture, to achieve global food security and improved environment.

Why not doubt about the ingenuity of this *Homo sapiens sapiens*, if he is not able or willing to make birth control acceptable; unable or unwilling to develop an economy with better profit sharing; unable or unwilling to apply science for developing sustainable agriculture and industry.

15.1 Death in this Garden

Humanity's path was not a wandering around a Garden of Eden. For most of our history human beings survived as hunters and gatherers, depleting locally the wild food resources. Extensive land areas were necessary to sustain small groups. Our possessions were limited to what we could transport when moving to a new site. Depending on the local environment this could be very hard and it is probably the

M. V. Montagu (✉) · D. E. de Oliveira
VIB—Institute of Plant Biotechnology Outreach, Department of Biotechnology
and Bioinformatics, Ghent University, IIC/UGent Technologiepark 3,
B-9052 Gent-Zwijnaarde, Belgium
Tel.: + 32 9 473 78 47 79
e-mail: marc.vanmontagu@vib-ugent.be

D. E. de Oliveira
Tel.: + 32 9 264 87 27
e-mail: dulce.deoliveira@vib-ugent.be

© Springer International Publishing Switzerland 2015
B. Lugtenberg (ed.), *Principles of Plant-Microbe Interactions*,
DOI 10.1007/978-3-319-08575-3_15

reason why humans developed environmental engineering skills to alter terrestrial habitats.

Clearing vegetation, tilling the soil and domesticating wild animals and plants to meet specific requirements resulted in the advent of organized agriculture, which played an enormous role in the development of humanity. The battle was not won at the first round though. Early farmers have paid a price for their new way of life as they traded quality for quantity. Paleophatology studies reveal that with the adoption of agriculture the average height of individuals of post-agricultural communities in Greece and Turkey reduced drastically. Skeletons from different archaeological sites also indicated childhood malnutrition, anemia, and diseases such as tuberculosis and leprosy, as well as significant decrease in life expectancy (Diamond 1987). Although the hunter and gatherer diet was probably superior to that of new agricultural societies, there is little question that the advent of the agriculture supported the increase of populations.

Successive cultural revolutions have led to surges in population. Together with medical advances, the increase in agricultural productivity is responsible for the phenomenal population growth of the last century. Already in 1798 Thomas Malthus studied the population growth in Europe and claimed that the population was increasing faster than food production. He feared global starvation, which did not happen thanks to the development of new technologies that expanded food production.

When the Club of Rome asked the world to pay attention to overpopulation (Meadows et al. 1972), stressing that food security would become a problem, many thoughtful scientists were wondering if Thomas Malthus had not been right with his predictions. The majority however continued to deride these scientists who didn't trust human ingenuity. The successes of diligent innovative plant breeders such as Norman Borlaug, the "Father of the Green Revolution", brought overconfidence. It took a period of time before it became accepted that intensive agriculture, with its need for a high input of synthetic fertilisers, crop protection chemicals and irrigation caused environmental problems. Today it is too late to argue against intensive agriculture; world food production cannot go on without this high yielding agriculture; we are too numerous. We were 2 billion when Norman was a student. We are now well over the 7.2 billion and demographers predict we will be about 10 or 11 billion within 40 years. But at this very moment there are 1.2 billion people who are underfed amongst which 0.8 billion enter the FAO statistics of being underfed for a whole year

15.2 From Enlightenment to Darwin and Back to Superstition

The Enlightenment displaced beliefs and revelation by the scientific method, including empiricism and rational thought. Hand in hand with it came a new concept of nature. The nineteenth century gave us Darwin's Origin of Species with its revolutionary theory of evolution and Gregor Mendel's laws of genetic inheritance, two wonderful outcomes of the newly-born biological science.

Life Sciences evolved at an exponential rate in the twentieth century and became one of the most relevant fields of research and innovation to the benefit of mankind, and hopefully in the future to the entire ecosystem. In the last decade we have witnessed an eruption of information generated by the 'omics', computational models, genome engineering, and bio-imagining. These large datasets are progressively generating knowledge on the biological interactions behind the phenotype, allowing biology to evolve from the classical reductionist mode to the more holistic approach of systems biology. We can now appreciate how genomes within a metaorganism interact and affect one another. It will soon be possible to unravel the complex network of interactions that link a host with its associate microbiome and how this can influence the host's performance and even *evolution* (Guerrero et al. 2013)

Agrobacterium had already given us a hint on how it can be relevant to plants. The holistic approach will bring us new insights on other types of plant-microbe interactions such as symbiosis with endophytes. Although plant-growth promoting endophytes have been identified for a long time, the predictive success at positively influencing plant growth under field conditions has been limited. We can now tackle this issue through the extensive examination of endophyte community dynamics and how they are influenced by abiotic and biotic factors such as soil conditions, biogeography, plant species, microbe–microbe interactions and plant–microbe interactions (Gaiero et al. 2013)

High-throughput deep sequencing has shown that many non-coding sequences produce RNA molecules that regulate gene expression. Several studies have indicated that small RNAs participate in plant resistance to bacterial pathogens (Seo et al. 2013). We are tantalized by the future discoveries on the role of small RNAs in the intra-holobiont communication, leading to better understanding of the mechanisms underlying the new Hologenome Theory of Evolution (Zilber-Rosenberg and Rosenberg 2008),

All these advancements in our knowledge of how biology works are sources of major innovations and bring us hope that we can improve agricultural productivity while reducing its environmental footprint. And yet, ironically, despite all the scientific progress, from the standpoint of public acceptability it seems that we have returned to medieval times. We face now a kind of fundamentalism which deems all that is "natural" to be "sacred". The "natural" or "back to nature" viewpoint is opposed to human intervention in the natural world, and therefore to the biotechnologies, as if what is "natural" can only be good and science can only be bad.

To a certain extent this is nothing new. Human beings have always had mixed feelings in their relationship to Nature. Sometimes nature is perceived as hostile to the persons who try to transform it, such as farmers or engineers. Sometimes it is perceived as a sacred "Mother Earth" by more contemplative people such as poets, philosophers, naturalists and traditional cultures. From a Science point of view, Nature is neither good nor bad. It is our living environment. Ecology shows the interdependence between the natural world and human beings, who have shaped the world and environment as we see it and experience it today over thousands of years. Modern biology shows the way an organism works to interact with other organisms in a habitat. It is also an invaluable tool to find strategies to improve human life while preserving the environment. Again the example of how *Agrobacterium* survives is an excellent illustration.

Our society rightfully wants to have a bottom-up voice on every aspect of social life like education, technology, ecology, arts, kinship, or sexuality. Such laudable search for self-management can only be successful in a well-informed society. We have to cultivate discernment based on intellectual honesty and not on individual passions. Human passions, especially self-esteem and fear, are at the origin of our emotion-driven myth-making tendency (Russel 1943). These traits were useful to our hunters and gatherers ancestors, but are a threat to modern society. Irrational fear is nourished by ignorance and is an easy prey to intellectual dishonesty.

15.3 *Agrobacterium* has Opened Our Eyes: The Living World is a Continuous Genetic Experiment

The TIP Story In the beginning of the twentieth century, several new scientific disciplines such as virology, epidemiology, internal medicine and public health developed. Some discoveries, like the demonstration (in 1911 by Peyton Rous) that a cancer on chicken wings could be caused by an infectious agent, were so revolutionary that the medical world could not be convinced. At the Rockefeller Institute, the quest for the "Tumour Inducing Principle" (TIP) started. A Belgian scientist, Albert Claude, joined the team in the late 1920s. He stressed that new enabling technologies would be needed and introduced ultracentrifugation and electron microscopy to the life sciences. This opened the field for "cell biology" and allowed the identification of TIP as a virus, later called Rous Sarcoma Virus. It brought a Nobel Prize to Peyton Rous (in 1966) and to Albert Claude (together with Christian de Duve and George Palade in 1974). Another astonishing infectious agent was a bacterium inducing tumorous growth (called crown galls) on a wide variety of plants. Amin Braun (also at Rockefeller) initiated the quest for the TIP present in *Agrobacterium tumefaciens* with the postulate that this "principle" was transferred by the bacterium to the plant cell to induce transformation. Again it was ultracentrifugation and electron microscopy that showed that TIP was a product of a large plasmid, which was then called Ti-plasmid (Tumour inducing plasmid) (Zaenen et al. 1974). Later in the 1970s it was demonstrated that a particular DNA segment of the bacterial Ti plasmid, called the transferred (T)-DNA, was integrated into the plant cell genome upon *Agrobacterium* infection (Depicker et al. 1978; Chilton et al. 1980; Zambryski et al. 1980).

The First Transgenic Plants As soon as it was established that gene transfer from *Agrobacterium* to plant cells occurred, there was a rush to use *Agrobacterium* as a vector for plant transformation. It was soon employed worldwide for a systematic and refined analysis of the impact of single genes on all aspects of plant biology, giving a new life to plant science.

At Plant Genetic Systems (PGS), a spin-off company from the plant science lab at Ghent University, this tool was employed to develop transgenic plants with different useful agronomic traits: tolerance to the herbicide glufosinate ammonium (De Block et al. 1987); insect resistance using the insecticidal protein genes from *Bacillus thuringiensis* (Vaeck et al. 1987; see Chap. 20); and nuclear male sterility, a trait that formed the basis for an efficient hybrid production system (Mariani et al. 1990, 1992).

Scientific Hit, Public Denial Since the first transgenic plants, scientists, mostly from the public sector, have identified numerous genes for introducing new traits into plants, including resistance to pests and disease-causing agents, enhanced stability or shelf-life, increased yield, environmental tolerances including salt resistance, nutritional enhancements (especially for vitamin A deficiency), and pharmaceutical and industrial value-added traits. Sadly today, in contrast to the scientific achievements, 20 years after their first commercialization, new GM varieties with novel genes are being introduced very slowly worldwide. The commercial development of transgenic germplasm has advanced mostly in stacking the few traits commercially available in essentially four crops (maize, soybean, canola and cotton). The cost of global regulation coupled with the lack of consumer acceptance is stopping the spread of the technology beyond the most profitable seed crops previously mentioned, with R&D (but no commercialization) occurring in rice and wheat. Specialty crops and traits with high environmental or social value but low economic value for commercial seed companies have not been introduced or considered or have been abandoned in the face of NGO opposition. Today only a handful of companies with global experience in deregulation have the capability of developing and launching a GM plant product.

Much of the frustrating delay on the development and use of GM crops can be attributed to the significant and successful opposition deployed by NGOs such as GreenPeace and Friends of the Earth through highly sophisticated marketing campaigns. These campaigns raise the specter of adverse social implications, as well as health and environmental risks. The latter have been debated at length and no serious study has concluded that GMOs are unsafe for health or the environment (EASAC report 2013). But fear persists at consumer level and is mainly linked to the tool of genetic engineering with which these crops are constructed. The fact that one can transfer a gene from a particular species into a completely different species has triggered people's irrational fears.

Scientific thinking is alien to irrational gut feelings. The scientific community did not expect to face this challenge. The gene splicing technology used to generate a transgenic organism is a minute genetic alteration compared to the genomic changes induced during all crosses and breeding events traditionally practiced in agriculture and husbandry. Our planet is one large natural genetic laboratory, where all the living organisms continuously activate and silence part of their genomes, as a reaction to all the environmental stresses endured (Pigliucci 2005; Heger and Wiehe 2014; Bateson et al. 2014). The natural gene engineering of *Agrobacterium* is, as indicated above, one example of very large phenomena. Biological evolution cannot be achieved by single point mutations in a static genome. It is essential to adaptation and survival that living organisms can alter their genomes through transposition of movable elements, accumulation of deletions, insertions, gene amplifications, and point mutations. Taxonomy is a useful tool to group and categorize organisms according to important homologous attributes, but in reality species often have indistinct boundaries. Genomic studies of the last decade have well documented that a genome is a dynamic structure continuously refining its gene pool. We are convinced that, when the metagenomic studies will be comprehensibly analysed, we will find that horizontal gene transfer is a frequent and essential event for evolution (Fuentes et al. 2014). Once again the example of *Agrobacterium* will have shown us the way.

15.4   Give Science a Chance

According to calculations by the FAO, the world already produces enough food to feed 12 billion people (da Silva 2012). But global estimates can be misleading because access to food varies enormously around the world (Vermeulen et al. 2012). Many of the world's poorest rural populations continue to rely on locally produced food distributed in economies that are poorly integrated into global markets. Therefore improvement of the quality and quantity of the agricultural production of the smallholder farmer is central to eradicating hunger and poverty in the world. Some 3 billion poor people still live in rural areas of low-income countries and derive the major part of their income from the agricultural sector and related activities (Dethier and Effenberger 2012).

Several empirical studies support the highly positive impact of agricultural progress on poverty alleviation. Agricultural growth generates income and employment in rural areas and provides cheaper food for urban areas. A labour-intensive sector such as agriculture has a larger impact on poverty reduction than less labour-intensive activities. The difference is greatest among the extremely poor, who live on less than $ 1 a day (Christiaensen et al. 2010).

It is undisputed that solutions to improve the agriculture in low-income countries pass through good governance, political will and concerted actions of different segments of society. But there is a broad consensus that agricultural technologies are at the heart of long-run agricultural growth. The Green Revolution in Asia is an example of how R&D for seeds, fertilizers, pesticides and irrigation was able to significantly expand agricultural production. However, any *New* Green Revolution for Asia and developing countries will have to be sustainable over time, economically, environmentally and from a societal dimension, as explained above. Smallholder farmers in low-income countries face this challenge now. Drought, low-yielding crop varieties, pests and diseases, poor soils, low fertilizer use, limited irrigation and lack of modern technologies are among the problems that plague tropical agriculture.

Modern biotechnology is part of the answer to achieve an agricultural revolution in low-income countries. Already, the technology has brought significant improvement to earned income, quality of life and per acre productivity. The crops produced by biotechnology and molecular-assisted breeding currently on the market are helping agriculture to achieve higher yields in a more sustainable way. One remarkable example is *Bt* cotton, which is grown by over 15 million smallholder farmers in India, China, Pakistan, and a few other developing countries. This technology has reduced food insecurity by 15–20 % among cotton-producing households (Qaim and Kouser 2013). Another example from the Philippines shows that the adoption of *Bt* corn has a positive yield impact especially to farmers at the lower end of the yield distribution (Sanglestsawai et al. 2014). Other studies indicate that if restrictions on the present GM-crops were lifted in all countries, including sub-Saharan Africa, the region might have significant gains in income generation and improved public health (Dethier and Effenberger 2012). These insights can help to develop policies that encourage the use of GM technology among smallholder farmers as the high costs restrict its adoption. Indeed, the capacity to adopt modern technologies is directly related to the health of the economy of the developing country.

However, other constraints have prevented the adoption of GM technology where is it needed most. One major obstacle is the fact that the other three major GM crops—soybean, maize, and canola—do not match the interests of most of smallholder farmers of the least developed countries. Although more than half of the global GM crop area is located in developing countries, much of this relates to large farms in South America. Relevant development country crops such as sorghum, millet, groundnut, cowpea, common bean, chickpea, pigeon pea, cassava, yam and sweet potato are grown in niche geographical areas by small farmers. These key staple crops are often referred to as 'orphan crops'. Because they are not extensively traded, except for sorghum these crops present little economic interest for commercial seed companies and therefore have been mostly ignored by GM agriculture.

Public research funded plant science has produced numerous breakthroughs that can help to alleviate many of the entrenched problems of impoverished nations. But most of these are proof-of-concept in model plants. For downstream applications relevant to low-income countries, it is essential to develop efficient transformation protocols for niche geographical crops and that breeding programs are funded for each crop.

Plant transformation is a daring research topic because, although *Agrobacterium* has a wide host range in plants, the external conditions that permit or prohibit transformation are diverse and the balance needs to be determined empirically. Despite intensive research it is still not established why some plant species can be transformed easily, while others are recalcitrant to *Agrobacterium*-mediated transformation. There is a continuous striving for simpler, more robust and more efficient transformation protocols for model plants and crop species for intensive agriculture (Pitzschke 2013). The advances are driven by the demand for higher yields and improved stress tolerance. Unfortunately major crops of marginal environments of Africa, Asia and South America have not received the same attention; generating stable transformed plants is still a challenge for most orphan crops.

Poor farmers might benefit especially from transgenic food crops that are more nutritious or particularly tolerant to biotic and abiotic stress. Research is well advanced in the development of more nutritious maize and rice (Pérez-Massot et al. 2013). Still a lot has yet to be done for staple crops of geographical niches, many of them requiring not only nutrient improvements but also the elimination of anti-nutritional factors. *Lathyrus sativa* is one such example (Van Moorhem et al. 2011).

As for biotic stress, development of resistant phenotypes via classical breeding is no longer an option. Pests and pathogens easily break down resistance based on single genes. Crop breeding is limited by the time taken to move resistance traits into elite crop genetic backgrounds and the limited gene pools in which to search for novel resistance (Bruce 2011). The pyramiding of resistance genes from different wild relatives using gene engineering is a strategy that deserves to be applied to orphan crops

Systems biology is showing that biotic and abiotic regulatory pathways are integrated in multidimensional ways that include genetic variation, abiotic environment and plant life history (Kliebenstein 2014). The understanding on how a plant modulates defence, symbiosis and growth as well as the signalling communication between individual plants will help to find the best ways to develop agroecology. This approach is widely recognized as a way to enhance the sustainability of intensive agricultural systems.

Modern agroecology cannot ignore the needs of low-income countries. The paradigm shift proposed in agroecology is not an option but a prerequisite to poor farmers. There is no other choice but to adapt the plant genotypes to the environment, since the inverse is not affordable. Actions must be taken to immediately apply the discoveries of this research to orphan crops.

This is even more necessary in view of the predicted impacts of climate change, particularly to tropical countries. Rain-fed crops account for nearly 60 % of cropland area (Vermeulen et al. 2012). Climate instability will have large negative impacts on the productivity of these agricultural systems, which do not have the required adaptive capacity, with major implications for rural poverty and both rural and urban food security.

Food aid can be of help in emergency situations but is quite questionable in a longer term development strategy. It has been shown that this type of assistance can have adverse effects such as price fluctuations, disincentives to agricultural production and market development, and can cause a cycle of dependency in farming communities (Dethier and Effenberger 2012). The solution will have to come from innovations in agriculture. We have to develop better-adapted crops faster. It can be that, even using GM technology, the development of new varieties will not fast enough respond to our needs. It is predicted that new races of pathogens will evolve rapidly under elevated temperature and CO_2. Another implication will be a changing pathogen geographic distribution (Chakraborty 2013). The unpredictability and the speed of the events will demand more flexible and speedier responses. Novel biotechnology tools will have to come. We dare to bet that the rapid evolution of miRNA technology will soon allow the production of these pathogen-specific molecules in quantities sufficient to apply on the field at the moment of a disease outburst. More human solidarity will be necessary for this tool to be used by the poor farmers.

15.5 Ite Missa Est

It has been well documented that the unprecedented population growth and the inequity in resource availability have generated huge famines, massive deforestation and intensive industrial pollution in the last century. To achieve global food security both intensive and subsistence agriculture are necessary. We head straight for disaster if we do not find ways to enhance land use productivity while preventing worldwide ecosystem degradation and fragmentation. This can only happen when the best technologies available are being used.

GM plant technology is a mature technology that has proved its social and economic benefits as well as its safety. After 26 years of field trials and 18 years of regular consumption, not a single health or environmental incident that could be attributed to GMO has been reported. Academies of Sciences worldwide endorse GM technology.

Science evolves with the development of new tools and all knowledge is tentative and may be adapted in the light of future knowledge. The Scientific Method cannot prove or disprove ideas and all scientific ideas are subject to change if warranted by the evidence. This normal and healthy process of Science is often misused by NGOs to serve political projects or simply to raise money. Mediatised "experts" abuse people's fears of the unknown by proclaiming that something new is more dangerous than something old. More alarming is that certain governments embrace this anti-science and anti-progress ideology.

It is time to learn not to surrender to fear and the images that fear engenders. We will have to learn to share emotions without fooling ourselves by blurring the real world with mythical dreams. Time is pressing, we have to act now and fear is a breeding ground for inactivity.

References

Bateson P, Peter G, Hanson M (2014) The biology of developmental plasticity and the Predictive Adaptive Response hypothesis. J Physiol 592:2357–2368

Bruce T (2011) GM as a route for delivery of sustainable crop protection. J Exp. Botany 63:537–541

Chakraborty S (2013) Migrate or evolve: options for plant pathogens under climate change. Global Change Biol 19:1985–2000

Chilton MD, Saiki RK, Yadav N et al (1980) T-DNA from Agrobacterium Ti plasmid is in the nuclear-DNA fraction of Crown gall tumor-cells. Proc Natl Acad Sc USA 77:4060–4064

Christiaensen LJ, Demery L, Kuhl J (2010) The (evolving) role of agriculture in poverty reduction-an empirical perspective. J Dev Econ 96:239–254

da Silva JG (2012) The Economist Conference: feeding the world in 2050, Geneva, 28 February 2012

De Block MB, Vandewiele M et al (1987) Engineering herbicide resistance in plants by expression of a detoxifying enzyme. EMBO J 6:2513–2518

Depicker A, Van Montagu M, Schell J (1978) Homologous DNA sequences in different Ti plasmids are essential for oncogenicity. Nature 275:150–153

Dethier JJ, Effenberger A (2012) Agriculture and development: a brief review of the literature. Econ Syst 36:175–205

Diamond J (1987) The worst mistake in the history of the human race. Discover Magazine, May 1987, pp 64–66

European Academies Science Advisory Council (2013) Planting the future: opportunities and challenges for using crop genetic improvement technologies for sustainable agriculture. http://www.easac.eu. Accessed 10 June 2014

Fuentes I, Stegemann S, Golczyk H et al (2014) Horizontal genome transfer as an asexual path to the formation of new species. Nature 511:232–235

Gaiero JR, McCall CA, Thompson KA et al (2013) Inside the root microbiome: bacterial root endophytes and plant growth promotion. Am J Bot 100:1730–1750

Guerrero R, Margulis L, Berlanga M (2013) Symbiogenesis: the holobiont as a unit of evolution. Int Microbiol 16:133–143

Heger P, Wiehe T (2014) New tools in the box: an evolutionary synopsis of chromatin insulators. Trends Genet 30:161–171

Kliebenstein DJ (2014) Orchestration of plant defence systems: genes to populations. Trends Plant Sci 19:250–255

Malthus TR (1798) An essay on the principle of population. Oxford World's Classics reprint, Chapter 1. p 13

Mariani C, De Beuckeleer M, Truettner J et al (1990) Induction of male sterility in plants by a chimaeric ribonuclease gene. Nature 347:737–741

Mariani C, Gossele V, De Beuckeleer M et al (1992) A chimaeric ribonuclease-inhibitor gene restores fertility to male sterile plants. Nature 357:384–387

Meadows DH, Dennis L. Meadows DL et al (1972) Limits to growth. New American Library, New York

Pérez-Massot BR, Gómez-Galera S et al (2013) The contribution of transgenic plants to better health through improved nutrition: opportunities and constraints. Genes Nutr 8:29–41

Pigliucci M (2005) Evolution of phenotypic plasticity: where are we going now? Trends Ecol Evol 20:481–486

Pitzschke A (2013) Agrobacterium infection and plant defence-transformation success hangs by a thread. Front Plant Sci 4:1–12

Qaim M, Kouser S (2013) Genetically modified crops and food security. PLoS ONE 8:e64879

Russel B (1943) An outline of intellectual rubbish: a hilarious catalogue of organized and individual stupidity. Haldeman-Julius, Girard

Sanglestsawai S, Rejesus R, Yorobe J (2014) Do lower yielding farmers benefit from Bt corn? Evidence from instrumental variable quantile regressions. Food Pol 44:285–296

Seo JK, Wu J, Lii Y et al (2013) Contribution of small RNA pathway components in plant immunity. Mol Plant Microbe Interact 26:617–625

Vaeck M, Reynaerts A, Höfte H et al (1987) Insect resistance in transgenic plants expressing modified Bacillus thuringiensis toxin genes. Nature 328:33–37

Van Moorhem M, Lambein F, Leybaert L (2011) Unraveling the mechanism of β-N-oxalyl-α,β-diaminopropanoic acid (β-ODAP) induced excitotoxicity and oxidative stress, relevance for neurolathyrism prevention. Food Chem Toxicol 49:550–555

Vermeulen S, Zougmore R, Wollenberg E et al (2012) Climate change, agriculture and food security: a global partnership to link research and action for low-income agricultural producers and consumers. Curr Opinion Environ Sustain 4:128–133

Zaenen I, Van Larebeke N, Teuchy H, Van Montagu M, Schell J (1974) Supercoiled circular DNA in crown-gall inducing Agrobacterium strains. J Mol Biol 86:109–127

Zambryski P, Holsters M, Kruger K et al (1980) Tumor DNA structure in plant cells transformed by A. tumefaciens. Science 209:1385–1391

Zilber-Rosenberg I, Rosenberg E (2008) Role of microorganisms in the evolution of animals and plants: the hologenome theory of evolution. FEMS Microbiol Rev 32:723–735

Chapter 16
Will the Public Ever Accept Genetically Engineered Plants?

Inge Broer

Abstract Compared to transgenic herbicide- or insect-resistant plants, pathogen-resistant ones are rarely found in the field. There are several economic reasons for this, but an additional cause is the scientific and public debate about potential risks and a low attractiveness for consumers to buy these plants and their products. This paper gives a short overview of traits theoretically available or already on the market and describes the concerns raised that reduce their market opportunity. Finally it proposes solutions on how to proceed in the future to allow a rational dealing with the technology.

16.1 Transgenic Pathogen Resistant Plants

Since the beginning of their existence plants and pathogens survive by ongoing development of sophisticated strategies to either attack or defend. In plant breeding, humans try to pyramid as many defense strategies as possible in crops in order to combine a healthy growth with a minimal input of pesticides and with optimal yield. Theoretically, this can be achieved using pathogen resistant plants. Nevertheless, since in classical breeding complete genomes are mixed, stacking of many beneficial genes without retaining deleterious ones is complicated and analysis of high numbers of crossings and descendants is unavoidable. In spite of the development of molecular markers, which reduce these numbers drastically, it is still an enormous challenge. In addition, classical breeding is restricted to defense genes present in sexually compatible cultivars. Hence several attempts have been made to enhance crop resistance to pathogens via genetic engineering which enables scientists to exploit the whole gene pool of the world.

Most of the defense genes transferred up to now originate from plants, but in many cases donor and recipient are not sexually compatible like for instance rice and maize. According to Collinge et al. (2008) out of twenty eight examples for transgene encoded disease resistance only 8 of the donors were sexually compatible

I. Broer (✉)
Agricultural and Environmental Faculty, University of Rostock,
18059 Rostock, Germany
Tel.: + 49 381 498 3080
e-mail: Inge.Broer@uni-rostock.de

© Springer International Publishing Switzerland 2015
B. Lugtenberg (ed.), *Principles of Plant-Microbe Interactions,*
DOI 10.1007/978-3-319-08575-3_16

to the recipient. Only for these a transfer via conventional breeding would be possible. All other donors were non compatible plants, viruses, insects, fungi or bacteria. The establishment of plant resistance using these resources is therefore strictly dependent on gene transfer.

So far forty five different cultivars have been successfully transformed to receive pathogen resistance. However, only virus resistant squash and papaya are cultivated in the USA and some other countries and virus resistant pepper and tomato in China (James 2014). The transgenic papaya was developed and commercialized by the University of Hawaii and is now approved for consumption in Japan (Dec 2011) and China (Dangl et al. 2013). This seems surprisingly little considering that herbicide and insect resistant plants are grown worldwide with great success. In 2013, transgenic plants, most of them herbicide or insect resistant, covered 175 million ha worldwide (ISAAA 2014). In the USA, about half of the total land was used for GM crops (USDA 2014). E.g. no pathogen resistant soybean variety is on the market (Hudson et al. 2013) although more than 81 % of the worldwide production is transgenic. So up till now, transgenic pathogen resistant plants play no visible role. This is especially surprising considering that crop losses due to bacteria, fungi and viruses account for about 16 % yield loss over all major crops (wheat, rice, maize, potato and soybean for the period 2001–2003) (Oerke 2006). An overview of the status of transgenic pathogen-resistant plants is shown in Table 16.1. Numbers of field trials for pathogen resistant plants are also much lower than those of herbicide and insecticide tolerance. 2616 total field releases of transgenic plants with virus/fungal resistance have been conducted in the USA until September 2013. This has to be compared to 6772 releases of plants with herbicide tolerance, 4809 with insect resistance, 4896 with improved product quality and 5190 with agronomic properties (USDA 2014). One interesting example of a commercially potent, virus resistant plant is a virus resistant bean, created by the Brazilian Agricultural Research Corporation (EMBRAPA). Beans are very important crops in South and Middle America and are mostly produced by small farmers. The golden mosaic virus causes severe yield loses (40–100 %). Descendants of one single transformant (event) showed resistance up to 93 %, hence this might be a real breakthrough for the small farmers (Bonfim et al. 2007). The bean is approved in Brazil und expected to appear on the market in 2014/2015.

16.2 Reasons of Why so Few Transgenic Pathogen Resistant Plants Exist in Agriculture

There seem to be several reasons for the minimal use of transgenic pathogen resistant plants in the world. In many countries, economic concerns might be of most importance. In Europe, political and public concerns are dominant while scientific concerns seem to matter less.

Table 16.1 Approval and cultivation of transgenic, pathogen resistant plants worldwide. (Source: ISAAA/TransGEN/Hudson et al. 2013)

Culture	Trait	Approval	Cultivation
Apple	Fungal/bacterial resistance	–	–
Apricot	Virus resistance	–	–
Avocado	Fungal resistance	–	–
Banana	Fungal/bacterial/virus resistance	–	–
Barley	Fungal resistance	–	–
Broad bean (*Vicia faba*)	Fungal resistance	–	–
Carrot	Fungal resistance	–	–
Cassava	Virus/bacterial resistance	–	–
Chestnut	Fungal resistance	–	–
citrus fruits	Fungal/virus/bacterial resistance	–	–
Cocoa	Fungal/virus resistance	–	–
Coconut	Virus resistance	–	–
cowpea (*Vigna unguiculata*)	Fungal resistance	–	–
Eggplant	Fungal resistance	–	–
Grape vine	Fungal/virus/bacterial resistance	–	–
Grapefruit	Fungal/virus/bacterial resistance	–	–
Hop plant	Fungal resistance	–	–
Kidney bean (*Phaseolus vulgaris*)	Virus/fungal resistance	Brazil	Starting 2014
Kiwi	Fungal resistance	–	–
Lettuce	Fungal/virus resistance	–	–
Lupine	Virus resistance	–	–
Mango	Fungal/bacterial resistance	–	–
Melon	Virus resistance	–	–
Oat	Virus resistance	–	–
Olive	Fungal resistance	–	–
Papaya	Virus resistance	USA 2, China 1, Canada 1, Japan 1	USA, China
Pea (*Pisum sativum*)	Fungal/virus resistance	–	–
Peanut	Fungal/virus resistance	–	–
Pear	Bacterial resistance	–	–
Pepper	Virus resistance	China	Cultivated In China

Table 16.1 (continued)

Culture	Trait	Approval	Cultivation
Persimmon	Fungal resistance	–	–
Pineapple	Virus resistance	–	–
Plum	Virus resistance	USA 1	USA 1
Potato	Fungal/virus/bacterial resistance	USA/ Canada	Not now
Raspberry	Virus resistance	–	–
Soybean	Virus/fungal resistance	–	–
Squash	Virus resistance	USA/ Canada	USA
Strawberry	Fungal resistance	–	–
Sugar beet	Virus resistance	–	–
Sunflowers	Fungal resistance	–	–
Sweet potato	Virus/fungal resistance	–	–
Taro	Fungal resistance	–	–
Tobacco	Virus resistance	–	–
Tomato	Virus resistance	–	Cultivated in China
Water melon	Virus resistance	–	–
Wheat	Fungal resistance	–	–

Economic Causes The approval process required to develop permitted pathogen resistant plants is often too costly to make it viable for any business to pursue. Unfortunately, pathogen resistance is much more complex compared to herbicide or insect resistance. Since there are several kinds of organisms causing disease (bacteria, fungi, viruses, oomycetes) no single gene is expected to protect against all of them. In addition, the pathogen's strategies to exploit the plants resources are quite divergent. Hence the diversity of defense genes is huge, ranging from chitinases, virus repressor genes and viral coat proteins, bacterial (Bt-toxin) and viral toxins (KP4) to RNAi (Collinge et al. 2008). Unfortunately, most plants suffer from the constitutive expression of defense genes that quite often causes cell death. Consequently, in order to sustain yield, inducible expression is often necessary, complicating the process further.

The high divergence of pathogens and defense mechanisms also complicates their administrative approval. In Europe 10–50 million € are estimated as final costs for approval of a single event. Compared to a herbicide or insect resistance, the risk assessment of a pathogen resistant plant is much more complex and expensive. Due to the high capability of pathogens to adapt to the plant defense, several different defense mechanisms should be introduced. In addition to the location of these transgenes in the plants genome, potential side effects of each transgene have to be analyzed. Although from a scientific point of view it might be assumed that the analysis of each combination of transgene and cultivar is sufficient (trait specific) the European legislation requests the analysis of each event for approval (event specific). Taking into account that the market for pathogen resistant plants is much smaller compared to insect- or herbicide resistant plants, the chance for a sufficient return of investment seems for the time being too low.

Scientific Causes   The most obvious scientific concern is the spread of the resistance gene to wild relatives, giving these hybrids a better chance to survive. Although this could also happen to the resistance genes in classical breeding, and is not regulated there, in transgenic plants it is required to be analyzed when cultivars are used that have sexual compatible wild relatives in the area where the transgenic plants are likely to grow. This is for instance important for pathogen resistant maize in South America, for transgenic canola at river beets where *Brassica oleracea* is found, or for sugar beets at the coastline of Europe where wild beets are growing. Hence intensive out crossing studies have been conducted (www.transgen.de).

Secondly, unwanted effects on non-target organisms have to be taken into account. Stefani and Hamelin (2010) summarize eighty six studies investigating the impact of transgenic fungal resistant crops on target and non-target fungi. No significant changes in the populations of non-target fungi could be detected. Similar results were obtained for fungal resistant trees. Nevertheless, this result cannot be generalized since the studies were mostly carried out under controlled conditions; only eleven were conducted in the field.

The uncontrolled spread of the transgenic plants is quite often named as one of the biggest problems. Nevertheless, as long as crop plants are used that are not able to survive without human support, this risk is restricted to the agricultural area. Here, strict seed thresholds defining the maximum content of transgenic seeds in a conventional seed lot and corresponding thresholds for the labeling of products that contain small amounts of transgenic plants in combination with guidelines that define the distance between fields where conventional or transgenic sexual compatible plants are grown, is sufficient to secure the availability of non GMP products on the market, as long as the demand is high enough to justify the costs of the separation. In case of plants that are able to spread into the wild, more caution has to be taken. It has to be assured that the expression of the transgene does not lead to any advantage compared to the near isogenic lines.

Big advantages of certain varieties lead to an increase of these varieties in the field. Hence it might be assumed that transgenic, pathogen resistant varieties will be preferred by the farmers and thereby reduce the number of different varieties available on the market. However, this holds true for any interesting variety independent on the method with which it was created. Actually, it could be much easier to integrate an interesting trait into several varieties via genetic engineering instead of crossings, but this is only possible if the legislation changes from event specific to a trait specific approval.

In contrast to insect or herbicide resistant plants, the creation of resistance is not of great concern. Resistance of pathogens will occur at some point, no matter which strategy is used to create the resistant plant; this is just a matter of time and evolution. Nevertheless, the pyramiding of resistance genes, which is much easier and more effective using genetic engineering, should prolong the utilization phase of specific traits drastically.

Political and Social Concerns As shown in Brazil, pathogen resistant plants are of importance for small farmers and could support a sustainable agriculture due to the reduction of pesticides. This is of public interest and should therefore—as happened in Brazil—be supported by public money. For several years the development as well as risk assessment of transgenic plants was funded by governments in the whole world. Nevertheless, opposition against this policy arose quite early and with increasing intensity. Several non-governmental organizations and even political parties declared the opposition to genetically modified plants as a primary goal, leading to increasing financial and political support by the public. Unjustified dispraise of scientists conducting risk assessment analysis led to a loss of trust in the independence of science. Politicians expressing their doubts on the governmental assessment system and countless scandals in food and feed production undermined the public faith in security systems. Contradictive scientific studies also caused confusion. For example, a study by Séralini et al. concluded that a herbicide-tolerant maize caused tumors and early death in rats (Séralini et al. 2012). This was widely dismissed by scientists including EFSA (European Food Safety Authority 2012). In November 2013, this article was retracted by the journal concerned (Elsevier 2013). All this supported attacks on field trials, scientists and watchmen by activists without disapproval from the public. Understandably, neither the business nor the governments are now very keen to support the production and cultivation of GMP in Europe.

In addition to the scientific reasons, which are unfortunately less and less part of the debate, ethical and moral concerns influence the public opinion. The integrity of a being and of its genome is often mentioned as an important value, ignoring the fact that genomes always undergo changes due to the mixing in sexual mating, recombination, mutation through environmental factors and jumping genes. These changes are the prerequisite of evolution. The potential dependence on multinational companies is another, and very relevant, concern. Only very few big multinational companies are able to finance and support the approval process necessary, especially in Europe. All applications for approval in Europe have been made by only nine companies. This is much different for instance in the United States, where approval is cheaper and easier and therefore also smaller companies are able to bring their events on the market. Due to the high cost in Europe, the companies join to apply for the approval of one event that is then inbreed into the different varieties.

One last important concern is the fact that high licensing costs make it impossible for small farmers to use transgenic, pathogen resistant plants. The example of the golden rice, where license holders agreed to waive the fee if the income of the farmer is below a certain level, shows that it is possible to solve this problem. In addition, the support of governmental research independent of companies, like in Brazil, would allow more innovations to come to the market with lower licensing fees.

16.3 What has to Happen?

Pathogen resistant transgenic plants have the potential to make agriculture more cost-effective and with less adverse impact upon the natural environment. The following suggestions might contribute to use this potential and to ensure that all people, not just multinationals, can benefit from the technology.

Worldwide, mainly plants with small market potential are problematic. Here, first of all the costs for approval and licenses have to go down while keeping the relevant safety level. In addition, the trait has to reduce the costs for the farmer in order to justify a technology fee. Even then, it might also be necessary to refund the farmer for the environmentally friendly cultivation in order to set enough incentives to saw these seeds. In addition, public acceptance is a prerequisite for the cultivation. This seems only achievable when there is an advantage for the consumer that is directly obvious, which unfortunately seems to be not the case for ecological advantages.

Even then, taking the huge opposition into account, it is unrealistic to think that there will be transgenic pathogen resistant plants on the market in Europe in 10 or even 20 years. Here, scientists should prepare for the time when cultivation of transgenic plants will be needed and possible too, since it takes more than 20 years from the first idea to a safety assessed transgenic variety ready for the market. This is only possible if field trials can be conducted. For instance in 2014, not a single field trial is planned in Germany. This is also due to the high costs of field protection (for fences and watchmen) that exceed the actual research costs. The Swiss government is funding a protected site where all field trails can be conducted (Romeis et al. 2013). This is a good start; nevertheless the site is much too small. Similar attempts have to be undertaken in other European countries or even by the EU.

References

Bonfim K et al (2007) RNAi-mediated resistance to Bean golden mosaic virus in genetically engineered common bean (*Phaseolus vulgaris*). Mol Plant Microbe Interact 20:717–726

Collinge DB, Lund OS, Thordal-Christensen H (2008) What are the prospects for genetically engineered, disease resistant plants? Eur J Plant Pathol 121:217–231

Dangl JL, Horvath DM, Staskawicz BJ (2013) Pivoting the plant immune system from dissection to deployment. Science 341(2013):746–751

Elsevier (2013) Elsevier announces article retraction from Journal Food and Chemical Toxicology November 28, 2013. http://www.elsevier.com/about/press-releases/research-and-journals/elsevier-announces-article-retraction-from-journal-food-and-chemical-toxicology

European FoodSafety Authority(2012) Final review of the Séralini et al. (2012) publication on a 2-year rodent feeding study with glyphosate formulations and GM maize NK603 as published online on 19 September 2012 in Food and Chemical Toxicology, 1. EFSA J 10(11):1–10

Hudson LC et al (2013) Advancements in transgenic soy: from field to bedside. In: Board J (ed) A comprehensive survey of international soybean research—genetics, physiology, agronomy and nitrogen relationships. InTech. doi:10.5772/52467

James C (2014) Status of Commercialized Biotech/GM Crops: 2013 ISAAA Brief 47, Ithaca, NY

Oerke EC (2006) Crop losses to pests. J Agric Sci 144:31–43

Romeis J et al (2013) Plant biotechnology: research behind fences. Trends Biotechnol 31(4): 222–224

Séralini GE et al (2012) Long term toxicity of a Roundup herbicide and a Roundup-tolerant genetically modified maize. Food Chem Toxicol 50(11):4221–4231

Stefani FOP, Hamelin RC (2010) Current state of genetically modified plant impact on target and non-target fungi. Environ Rev 18:441–475

USDA (United States Department of Agriculture) (2014) Report Summary from the Economic Research Service 2014

Part III
Control of Plant Diseases and Pests using Beneficial Microbes

Chapter 17
Microbial Control of Phytopathogenic Nematodes

Xiaowei Huang, Keqin Zhang, Zefen Yu and Guohong Li

Abstract Phytopathogenic nematodes, mainly comprised of plant parasitic nematodes, cause serious losses in a variety of agricultural crops worldwide. However, because traditional nematicides are associated with major environmental and health concerns, developing safe and effective nematicides is urgently needed. Among the recent developments, biocontrol measures using nematophagous microorganisms have shown significant promise and attracted much attention. Nematophagous microorganisms are the most important natural enemies of phytopathogenic nematodes and these microorganisms employ a variety of physical, chemical, and biochemical mechanisms to attack nematodes. This chapter introduces nematophagous microorganisms as well as their virulence factors against nematodes.

17.1 Introduction to Nematodiasis and Biocontrol

Nematodes are among the most abundant multi-cellular organisms on earth and can reach densities of up to 10 million individuals/m^3 in soil. Plant parasitic nematodes (Chap. 11), a major group of worms that feed and reproduce on living plants, widely exist as one of the main pathogens for global agriculture and cause crop damages greater than those by bacteria and viruses (Cai et al. 1997). The significant impact of specific nematodes on world agriculture is the result of their wide distribution,

X. Huang (✉) · K. Zhang · Z. Yu · G. Li
Yunnan University, 650091 Kunming, People's Republic of China
Tel.: +13888865821
e-mail: xwhuang@ynu.edu.cn

K. Zhang
Tel.: +86-871-6503487
e-mail: kqzhang@ynu.edu.cn

Z. Yu
Tel.: +86-871-65033805
e-mail: zfyuqm@hotmail.com

G. Li
Tel.: +86-871-65031092
e-mail: ligh@ynu.edu.cn

© Springer International Publishing Switzerland 2015
B. Lugtenberg (ed.), *Principles of Plant-Microbe Interactions*,
DOI 10.1007/978-3-319-08575-3_17

their ability to attack a variety of cultivated plants, as well as the multiple types of damages to individual plants. Take Mel*oidogyne.* spp as an example. The root-knot nematodes (i.e. *Meloidogyne* spp.) and the cyst nematodes (i.e. *Heterodera* and *Globodera* spp.) are the two main groups of plant parasitic nematodes responsible for the majority of crop losses (Molinari 2011). *Meloidogyne.* spp. infects about 3000 flowering plant species belonging to 114 different families, resulting in 10–20 % crop failure per year (Wang et al. 2002). They preferentially invade the roots of plants and cause the formation of giant cells and root knots in plants. Such damages lead to the fracture in catheter, epidermis and cortical tissue, which consequently endangers the plants by slowing their respiration, photosynthesis, and transpiration. Furthermore, infection by *Meloidogyne* can also weaken the immunity or protective enzyme system in plants, which makes the damaged plants easier to become infected by pathogenic fungi or/and bacteria. Through the different modes described above, plant parasitic nematodes cause damages to plants ranging from negligible injury to serious destruction. It has been estimated that globally, plant parasitic nematodes bring direct economic losses worth about US$ 157 billion per year (Abad et al. 2008).

The management of nematodes is more difficult than that for other pests because nematodes mostly inhabit the soil and usually attack the underground parts of the plants. For a long time, due to its effectiveness, rapid effects, and ease of use, the application of chemical products has been the most commonly used method to control nematodes. However, public health and environmental concerns have focused scientific interest on the development of environmentally acceptable alternatives to chemical nematocides. Biological control, exploiting the interaction between nematode-antagonistic microorganisms and their hosts, represents an alternative or complementary approach to chemical controls.

17.2 Microbial Populations Antagonizing Phytopathogenic Nematodes

A diversity of microorganisms exist in natural environments that function as antagonists of nematodes. These microorganisms include fungi, bacteria, and viruses. Those nematocidal microorganisms have evolved a variety of physical, chemical, and biochemical mechanisms to kill nematodes, and they have been the most important resources used in biocontrol of phytopathogenic nematodes. Among the nematocidal microorganisms, nematophagous fungi and nematophagous bacteria have been the most frequently investigated.

Nematophagous Fungi More than 700 species of nematophagous fungi have been described. These fungi belong to widely divergent Orders and Families. The reported fungal taxa of nematophagous fungi include Ascomycetes (e.g. asexual *Orbiliaceae* and *Clavicipitaceae*), Basidiomycetes (e.g. *Pleurotaceae*), Chytridiomycetes, Oomycetes and Zygomycetes (*Zoopagales*). They are broadly distributed in terrestrial and aquatic ecosystems (Hao et al. 2005)

According to the modes of nematicidal action, the nematophagous fungi can also be divided into four groups, consisting of endoparasitic fungi, nematode-trapping fungi, opportunistic fungi and toxic fungi respectively. Endoparasitic fungi infect nematodes via special spores and subsequently mycelia grow from spores within the nematodes. Endoparasitic fungi are typically regarded as obligate parasites since they are poor saprotrophic competitors in soil. So far, about 120 nematode-endoparasitic species have been reported, and the typical genera include *Drechmeria* W. Gams & H.-B. Jansson, *Myzocytium* Schenk, *Harposporium* Lohde, *Hirsutella* Pat. and *Nematoctonus* Drechsler (Zhang et al. 2011). The second group is the nematode-trapping fungi and they capture nematodes by trapping devices produced from the vegetative mycelia. The typical traps mainly include adhesive networks, adhesive knobs, constricting rings, non-constricting rings, and adhesive branches (Fig. 17.1). The majority of nematode-trapping fungi are asexual taxa, mostly known as hyphomycetes. Three hundred and forty seven nematode-trapping species have been described, and they are mainly in Ascomycota, Basidiomycota and Zygomycota. The representative genera include *Arthrobotrys* Corda, *Cystopage* Drechsler, *Dactylella* Grove, *Dactylellina* M. Morelet, *Drechslerella* Subram, *Hyphoderma* Wallr., *Hohenbuehelia* Schulzer, *Monacrosporium* Oudem, *Nematoctonus* Drechsler, *Orbilia* Fr., *Stylopage* Drechsler, *Triposporina* Höhnel, *Tridentaria* Preuss and *Zoophagus* Sommerst (Zhang et al. 2011). The third group is the opportunistic nematophagous fungi. These are also called egg and cyst parasitizing fungi because they can colonize nematode reproductive structures and affect their reproductive capabilities. These fungi commonly use appressoria or zoospores to infect their hosts. This group of fungi includes species in genera *Pochonia*, *Paecilomyces*, *Lecanicillium* and *Nematophthora* (Siddiqui and Mahmood 1996). The fourth group is the toxic fungi and they produce low molecular metabolites that are toxic to nematodes. Generally, these fungi first immobilize nematodes by secreting toxins and then their hyphae penetrate the nematode cuticle. About 270 species of toxin-producing fungi have been reported and they belong to widely divergent orders and families. Most of these fungi are Basidiomycota (e.g., *Pleurotus*, *Coprinus*) although several ascomycete species (e.g., *Lecanicillium*, *Paecilomyces*, *Pochonia*) also produce nematocidal compounds (Lòpez-Llorca et al. 2008).

Nematophagous Bacteria   Since nematophagous bacteria have several advantages over fungi, such as their fast multiplication, ease of cultivation and mass production, they have been more commonly applied as biocontrol agents in the field. Several bacterial agents have been shown similar effectiveness in controlling nematodes as chemical pesticides (Zhou et al. 2002). Strains of the following bacterial genera have shown the capabilities to infect nematodes: *Actinomycetes*, *Agrobacterium*, *Arthrobacter*, *Alcaligenes*, *Aureobacterium*, *Azotobacter*, *Bacillus, Beijerinckia*, *Chromobacterium*, *Clavibacter*, *Clostridium*, *Comamonas*, *Corynebacterium*, *Curtobacterium*, *Desulforibtio*, *Enterobacter*, *Flavobacterium*, *Gluconobacter*, *Hydrogenophaga*, *Klebsiella*, *Methylobacterium*, *Pasteuria*, *Pseudomonas*, *Phyllobacterium*, *Phingobacterium*, *Rhizobium*, *Stenotrotrophomonas*, and *Variovorax*. Additionally, several human bacterial pathogens, such as species of *Burkholderia*, *Enterococcus*, *Staphylococcus*, *Serratia* and *Streptococcus* have also been reported to have antagonistic effects against nematodes (Liu et al. 2013).

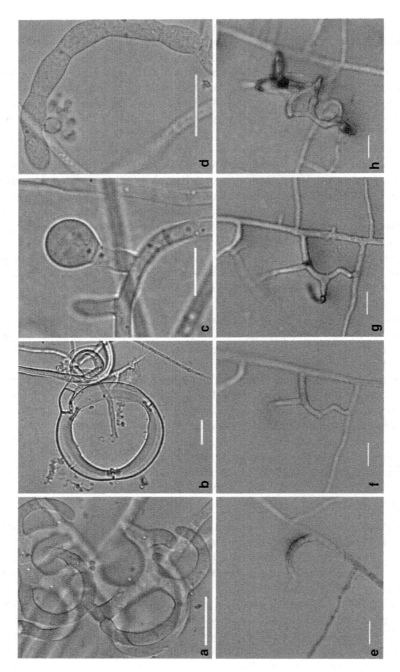

Fig. 17.1 The typical trapping devices in nematode-trapping fungi and the formation process of the adhesive networks. **a** adhesive networks, **b** constricting rings, **c** adhesive knob, **d** adhesive branch, and **e–h** adhesive networks after adding nematode extracts. **e** 8 h, **f** 16 h, **g** 24 h, **h** 32 h. Bars are 20 μm in length

Based on the modes of nematicidal action, nematophagous bacteria can be divided into three groups. These groups are obligate parasitic bacteria, parasporal crystal-forming bacteria, and opportunistic parasitic bacteria. Four *Pasteuria* species (*Pasteuria ramosa, P. penetrans, P. thornei,* and *P. nishizawae*) are obligate, mycelial and endospore-forming bacterial parasites of nematodes. They parasitize 323 nematode species belonging to 116 genera, including main phytopathogenic nematode genera *Meloidogyne* spp, *Heterodera* and *Globodera* (Atibalentja et al. 2000). During their pathogenesis, the spores of *Pasteuria* attach to the cuticles of the second-stage juveniles, and germinate after the worms enter plant roots and begin feeding. Parasporal crystal-forming bacteria, represented by *Bacillus thuringiensis* (Chap. 20), can produce one or more parasporal crystal inclusions (Cry proteins) toxic to the larvae of free-living or plant parasitic nematodes (Wei et al. 2003; Kotze et al. 2005). Most of nematophagous bacteria, except for obligate parasitic bacteria, are opportunistic parasites because they usually live a saprophytic life. But once the conditions become suitable for parasitism, the bacteria will infect the nematode hosts as a possible nutrient source. Most of the opportunistic parasitic bacteria are also rhizobacteria, represented by *Bacillus* and *Pseudomonas* (Table 17.1). The rhizobacteria antagonize plant-parasitic nematodes by a variety of strategies to directly or indirectly kill nematodes.

17.3   Virulence Factors as well as Molecular Mechanisms Used by Nematophagous Microorganisms

Nematophagous microorganisms produce a series of virulence factors or employ different strategies to reduce the populations of plant parasitic nematodes. They antagonize nematodes mainly by competing for essential nutrients, promoting plant growth, inducing systemic resistance, interfering with plant-nematode recognition, regulating nematode behavior and/or directly killing the nematodes by means of enzymes, small molecular metabolic products, protein toxins, and trapping devices. In the next section, the microbial virulence factors that are directly used for killing nematodes will be described in detail.

Hydrolytic Enzymes The cuticle of nematodes, which completely surrounds the animal except for small openings, is a very rigid but flexible multilayered extracellular exoskeleton structure and an effective barrel preventing nematodes from being environmentally damaged. The cuticle of adult nematodes consists mainly of proteins including keratin, collagen and fibers running diagonally in opposite directions from each other. Therefore, penetration to the nematode cuticle requires the activities of hydrolytic enzymes. Indeed, both nematophagous fungi and nematophagous bacteria have been reported to secrete different types of extracellular enzymes to help the infectious process of pathogens (Huang et al. 2004; Tian et al. 2007).

Subtilisin-like serine protease is a virulence enzyme characterized by its catalytic domain of aspartic acid-histidine-serine, and its roles in infection have been most

Table 17.1 The main rhizobacterial species and their target nematodes

Genus	Species	Target nematodes
Bacillus	B. subtilis, B. thuringiensis	Meloidogyn javanica
	B. cereus, B. circulans	Meloidogyn incognita
	B. mycoides, B. polymyxa	Meloidogyn hapla
	B. pumilus, B. sphaericus	Heterodera glycines
	B. stearothermophilus	Trichodorus primitivus
		Paratrichodorus pachydermus
		Globodera pallida
Pseudomonas	P. aeruginosa, P. putida	Meloidogyn javanica
	P. chlororaphis, P. cepacia	Meloidogyn incognita
	P. fluorescens, P. gladioli	Criconemella xenoplax
	P. mendocina, P. picketti	Heterodera glycines
	P. aureofaciens	Pratylenchus penetrans
		Radopholus similes
		Tylenchulus semipenetrans
		Globodera pallida
		Rotylenchulus reniformis
Burkholderia	B. cepacia, B. ambifaria	Meloidogyn incognita
	B. glathel	Heterodera glycines
Streptomyces	S. violaceus niger	Meloidogyn incognita
	S. lavendulae	Heterodera glycines
Hydrogenophaga	H. flava	Meloidogyn incognita
	H. pseudofalva	Heterodera glycines

intensively studied in nematophagous microbes. Currently, more than 20 serine proteases from nematophagous fungi and four from nematophagous bacteria have been identified respectively. These serine proteases share several biochemical traits: most of them are highly sensitive to the serine protease inhibitor phenylmethylsulfonyl fluoride (PMSF); their maximal activities can generally be obtained under alkaline conditions; the purified proteases can degrade the proteins from the nematode cuticle and even destroy the adult nematode cuticle or nematode eggshells. Additionally, comparisons of the deduced amino acid sequences showed relatively low similarities between those from nematophagous fungi and nematophagous bacteria except for the active center of catalytic domains and the substrate-binding pockets. However, higher degrees of similarity can be found among members within either fungal serine proteases or bacterial serine proteases. Besides the alkaline serine proteases, a few neutral metalloproteinase have also been identified as a virulence factor during bacterial infection.

A chitinous layer, the thickest component in the shell of nematode eggs, is also the major barrier against infections by nematophagous microorganisms. Therefore chitinases, a group of inducible enzymes capable of degrading chitin, have been assumed to be required for egg infection. Since the first extracellular chitinase (CHI43) was identified from *Pochonia chlamydosporia* and *P. rubescens*, several chintinases were subsequently investigated in the opportunistic nematophagous fungi. The experimental data from microscopic observations demonstrated a series of changes in the chintinase-treated eggs: large vacuoles in the chitin layer of nematode eggs; swollen, deformed and even degraded eggs; abnormal development or hatch of eggs (Gan et al. 2007; Mi et al. 2010).

Certainly, other extracellular enzymes, such as collagenase, lipases and elastases, are also reported to be involved in the infection of worms or eggs. In conclusion, those hydrolytic enzymes are believed to be involved in several steps of infection, such as facilitating microbial penetration by degrading proteins of the cuticle, causing nematode deaths by targeting intestinal epithelium, and releasing nutrients to further support the growth of nematophagous microbes.

Cry Proteins Cry proteins, called δ-endotoxins, are generally produced by *B. thuringiensis* during sporulation. Those Cry proteins are encoded by *cry* genes that are commonly located on a plasmid in most strains of *B. thuringiensis*. Cry proteins were first shown to be strictly insecticidal, but later some authors described them as also toxic to other invertebrates, including nematodes. To date, six Cry proteins (Cry5, Cry6, Cry12, Cry13, Cry14, and Cry21) have been described to have nematotoxic activities to free-living and phytopathogenic nematodes (Kotze et al. 2005). Based on their amino acid sequences, the Cry proteins with nematocidal activity can be divided into two clusters. Among them, Cry5, Cry12, Cry13, Cry14, Cry21 were arranged together in a branch; but Cry6 was assigned to a separate cluster that had relatively low similarity in either amino acid sequence or protein crystal structure to the first cluster. The molecular mechanism underlying the Cry5 cluster has been elucidated in the model nematode *Caenorhabditis elegans*. Once the Cry5 toxin has been ingested by the target nematode larvae, the crystals dissolve within the gut and then the toxin molecules bind to a receptor in the epithelial cell. This binding leads to vacuole and pore formation, pitting, and eventual degradation of the intestine (Crickmore 2005; Marroquin et al. 2000).

Small Toxic Metabolites Nematophagous microorganisms can produce a large number of small molecular metabolites to reduce nematode reproduction, egg hatching and juvenile survival, and even kill nematodes directly. So far, more than 200 compounds with nematicidal activities have been reported from about 280 fungal species in 150 genera of Ascomycota and Basidiomycota. These nematocidal compounds are very diverse in structure and include alkaloids, quinones, pyrans, benzofuran, furan, peptides, macrolides, lactones, terpenoids, fatty acids, diketopiperazines, simple aromatics, as well as alkynes etc (Li et al. 2007). Fewer metabolites have been obtained from nematophagous bacteria, likely due to the much simpler metabolic pathway present in bacteria than in fungi.

It is particularly worth mentioning that avermectin compounds, which have potent insecticidal and nematocidal properties, are extensively used in agriculture worldwide. The avermectins consist of a series of 16-membered macrocyclic lactone derivatives. These naturally occurring compounds are generated as fermentation products by the soil actinomycete *Streptomyces avermitilis* (Omura and Shiomi 2007) and have shown significant antagonistic effects to some plant parasitic nematodes such as *Ditylenchus dipsaei* and *Meloidogyne hapla*. Experiments with the model worm *C. elegans* suggested that avermectins mediate their nematocidal effects via interacting with a common receptor molecule, glutamate-gated chloride channels (Arena et al. 1995).

Trap Formation Mechanical activities, especially trapping devices in nematode-trapping fungi mentioned above, is another important factor involved in the infection against nematodes. Though the trapping devices are developed from vegetative hyphae, they show a large variation in morphology and a distinguishable ultrastructure from regular hyphae. For example, the dense bodies of peroxisomes and the extra-cellular polymers are much more prominent in traps than in regular hyhae. Because the development of traps is a complex process involving many genes, a few high throughput methods of genomics, microarray, proteomics, and transcriptomics have been applied to investigate the mechanisms of trap formation. Recently, a model of trap formation in *A. oligospora* (three-dimensional networks) was suggested to include active cytoskeleton assembling, enhanced cell wall biosynthesis, increased glycerol synthesis and accumulation, as well as peroxisome biogenesis (Yang et al. 2011). Additionally, a few novel mechanical structures with nematicidal activity have been reported in our lab, including the spiny ball from *Coprinus comatus* and acanthocytes from *Stropharia rugosoannulata*, which can damage the cuticle of the nematode with the mechanical force provided by their sharp projections (Luo et al. 2006, 2007).

17.4   Commercial Biocontrol Nematicides

Despite its long history and recent developments, relatively few commercial biocontrol products based on nematophagous fungi and bacteria have been developed. However, the pace of discovery and commercialization is accelerating. For instance, several commercial products based on the bacteria *P. penetrans*, *Bacillus firmus*, *Burkholderia cepacia* and *Bacillus* spp., as well as on the fungi *Purpureocillium lilacinus*, *Pochonia chlamydosporia* and *Myrothecium verrucaria* have been developed to control the root-knot nematodes *Meloidogyne* spp. (LamovŠek et al. 2013). However, for more effective biocontrol, a more thorough understanding of the complexity of soil ecology in agricultural fields is required.

References

Abad P, Gouzy J, Aury JM et al (2008) Genome sequence of the metazoan plant-parasitic nematode *Meloidogyne incognita*. Nat Biotechnol 8:909–915

Arena JP, Liu KK, Paress PS et al (1995) The mechanism of action of avermectins in *Caenorhabditis elegans*: correlation between activation of glutamate-sensitive chloride current, membrane binding, and biological activity. J Parasitol 81:286–294

Atibalentja N, Noel GR, Domier LL (2000) Phylogenetic position of the North American isolates of *Pasteuria* that parasitizes the soybean cyst nematodes, *Heterodera glycines*, as inferred from 16S rDNA sequence analysis. Int J Syst Evol Microbiol 50:605–613

Cai DG, Kleine M, Kifle S et al (1997) Positional cloning of a gene for nematode resistance in sugar beet. Science 275:832–834

Crickmore N (2005) Using worms to better understand how *Bacillus thuringiensis* kills insects. Trends Microbiol 13:347–350

Gan ZW, Yang JK, Tao N et al (2007) Cloning of the gene *Lecanicillium psalliotae* chitinase Lpchi1 and identification of its potential role in the biocontrol of root-knot nematode *Meloidogyne incognita*. Appl Microbiol Biotechnol 76:1309–1317

Hao YE, Mo MH, Su HY et al (2005) Ecology of aquatic nematode-trapping hyphomycetes in southwestern China. Aquat Microb Ecol 40:175–181

Huang XW, Zhao NH, Zhang KQ (2004) The extracellular enzymes serve as virulent factors in nematophagous fungi involved in the infection of host. Res Microbiol 155:811–816

Kotze AC, O'Grady J, Gough JM et al (2005) Toxicity of *Bacillus thuringiensis* to parasitic and free-living life stages of nematodes parasites of livestock. Int J Parasitol 35:1013–1022

Lamovšek J, Urek G, Trdan S (2013) Biological control of root-knot nematodes (*Meloidogyne* spp.): microbes against the pests. Acta Agric Slov 101:263–275

Li GH, Zhang KQ, Xu JP et al (2007) Nematicidal substances from fungi. Recent Pat Biotechnol 1:212–233

Liu Z, Budiharjo A, Wang P et al (2013) The highly modified microcin peptide plantazolicin is associated with nematicidal activity of *Bacillus amyloliquefaciens* FZB42. Appl Microbiol Biotechnol 97:10081–10090

Lòpez-Llorca LV, Macia-Vicente JG, Jansson HB (2008) Mode of action and interactions of nematophagous fungi. In: Ciancio A, Mukerji KG (eds) Integrated management and biocontrol of vegetable and grain crops nematodes. Springer, Dordrecht, pp 13–16

Luo H, Li X, Li GH et al (2006) Acanthocytes of stropharia rugosoannulata function as a nematode-attacking device. Appl Environ Microbiol 72:2982–2987

Luo H, Liu YJ, Fang L et al (2007) *Coprinus comatus* damages nematode cuticles mechanically with spiny balls and produces potent toxins to immobilize nematodes. Appl Environ Microbiol 73:3916–3923

Marroquin LD, Elyassnia D, Griffitts JS et al (2000) *Bacillus thuringiensis* (Bt) toxin susceptibility and isolation of resistance mutants in the nematode *Caenorhabditis elegans*. Genetics 155:1693–1699

Mi QL, Yang JK, Ye FP et al (2010) Cloning and overexpression of *Pochonia chlamydosporia* chitinase gene *pcchi44*, a potential virulence factor in infection against nematodes. Process Biochem 45:810–814

Molinari S (2011) Natural genetic and induced plant resistance, as a control strategy to plant-parasitic nematodes alternative to pesticides. Plant Cell Rep 30:311–323

Omura S, Shiomi K (2007) Discovery, chemistry, and chemical biology of microbial products. Pure Appl Chem 79:581–591

Siddiqui ZA, Mahmood I (1996) Biological control of plant parasitic nematodes by fungi: a review. Bioresour Technol 58:229–239

Tian BY, Yang JK, Zhang KQ (2007) Bacteria used in the biological control of plant-parasitic nematodes: populations, mechanisms of action, and future prospects. FEMS Microbiol Ecol 61:197–213

Wang LF, Yang BJ, Li CD (2002) A review of biological control of biological nematodes. J Nanjing For Univ 26:64–68

Wei JZ, Hale K, Carta L et al (2003) *Bacillus thuringiensis* crystal proteins that target nematodes. Proc Natl Acad Sci U S A 100:2760–2765

Yang JK, Wang L, Ji XL et al (2011) Genomic and proteomic analyses of the fungus *Arthrobotrys oligospora* provide insights into nematode-trap formation. PLoS Pathog 7:e1002179

Zhang Y, Li GH, Zhang KQ (2011) A review on the research of nematophagous fungal species (In Chinese). Mycosystema 30:836–845

Zhou XS, Kaya HK, Heungens K et al (2002) Response of ants to a deterrent factor(s) produced by the symbiotic bacteria of entomopathogenic nematodes. Appl Environ Microbiol 68:6202–6209

Chapter 18
Microbial Control of Root-Pathogenic Fungi and Oomycetes

Linda Thomashow and Peter A. H. M. Bakker

Abstract The rhizosphere is a complex and dynamic environment in which microbes introduced to control root pathogens must establish and maintain populations of sufficient size and activity to antagonize pathogens directly or by manipulating the host plant's own defenses. Genetic and physiological studies of rhizobacteria with the capacity to control root pathogens have given considerable insight into the microbial side of these interactions, but much remains to be learned about the physical conditions and the chemical and biological activities that take place at the root-microbe interface. This chapter focuses on advances in our understanding of the constraints to the successful introduction of microbial agents for the control of soil-borne root pathogens and the mechanisms involved in pathogen suppression. Chapters elsewhere in this volume address related topics including plant growth promotion, stress control, the activation of the plant's own defense mechanisms by introduced microbes, and powerful new biotechnological advances available to gain insight into rhizosphere processes.

18.1 Colonization: A Necessary Requisite for Biological Control

Root colonization, a challenge to both indigenous microbes and those introduced to control root pathogens, is the process by which bacteria become distributed along the root, propagate, and persist for weeks or longer in the presence of the indigenous rhizosphere microflora. How many bacteria are needed to achieve pathogen control? The minimum population size required will depend in part on the disease pressure,

L. Thomashow (✉)
USDA-ARS, Root Disease and Biological Control Research Unit,
Washington State University, PO Box 646430, Pullman, WA, USA
Tel.: +1-509-335-0930
e-mail: thomashow@wsu.edu

P. A. H. M. Bakker
Plant-Microbe Interactions, Institute of Environmental Biology,
Utrecht University, 3508 TB Utrecht, The Netherlands
Tel.: +31 30 253 6861
e-mail: P.A.H.M.Bakker@uu.nl

© Springer International Publishing Switzerland 2015
B. Lugtenberg (ed.), *Principles of Plant-Microbe Interactions*,
DOI 10.1007/978-3-319-08575-3_18

the distribution of the bacteria along the root, and the mechanism of control, but several lines of evidence suggest a minimum average population size of 10^5 per gram of root, at least for antibiosis and the induction of systemic resistance (Pieterse et al. 2014).

Bacteria on root surfaces are not uniformly distributed; they reside in discrete mucigel-enclosed aggregates termed biofilms where nutrients are available. Such sites include wounds along the root, root tips, the junctions between epidermal cells, and regions where root hairs and lateral roots emerge. Exudates from these sites are a dominant source of nutrients for rhizobacteria, and there is increasing evidence that the sugars, organic and amino acids, phenolics, and other signal molecules in exudates maintain a complex chemical dialog between the plant and its associated microflora (Zolla et al. 2013). The quality, quantity, and composition of root exudates vary widely with the plant species and the biotic and abiotic stresses acting upon it, and these factors have a significant effect on the structure and composition of the associated microbial communities. There has been considerable effort in recent years to isolate and characterize exudates, but the best methods now available still identify only a fraction of the compounds present (Zolla et al. 2013).

Biofilms The extracellular matrix in which bacterial cells on roots are embedded in biofilms is composed mainly of proteins and exopolysaccharides, of which the latter vary in composition among strains but are key structural components (Martínez-Gil et al. 2013). In nature, biofilms provide microbes with a stable protective barrier against chemical stresses and protozoal grazing. In the model strain *Pseudomonas putida* KT2440, an efficient colonizer of seeds and roots, two very large extracellular proteins, LapA and LapF, have sequential roles in biofilm development. LapA first facilitates a cell-surface interaction resulting in irreversible bacterial attachment, and LapF mediates subsequent cell-cell interactions, providing support for expansion and maturation of the biofilm. Complex interactions modulated by the two-component global regulators GacS/GacA and the bacterial universal second messenger cyclic dimeric guanosine phosphate (c-di-GMP) have integral roles in the balance of protein and polysaccharide constituents within the biofilm (Martínez-Gil et al. 2014) and therefore its structural integrity and characteristics. c-di-GMP also has an important role in the transition of cells in biofilms to the planktonic form associated with motility. Biosurfactants such as cyclic lipopeptides (cLPs) and rhamnolipids also can influence the formation, as well as the stability and dissolution, of biofilms, and it has been postulated that the contrasting roles of this diverse family of compounds may be due to differences in their chemical structures and physicochemical properties, as well the ionic conditions and pH of their environment (Raaijmakers et al. 2010).

Motility Bacterial movement along roots may occur passively with root elongation or redistribution with water. Alternatively, dispersal mediated by flagellar swimming or swarming can be active. Numerous studies have demonstrated that bacterial mutants defective in motility and chemotaxis, the ability to detect and move towards nutrients, also are impaired in rhizosphere colonization (Lugtenberg and Kamilova 2009; Pliego et al. 2011). Recent studies suggest that these mechanisms are likely to require relatively hydrated root surfaces (Dechesne et al. 2010; Dechesne and

Smets 2012) and/or the presence of biosurfactants that can modulate the viscosity and surface tension of the thin water films that are thought to be present on roots under common agricultural conditions. Biosurfactants clearly contribute to microbial surface motility *in vitro,* but whether they have a similar role in the environment remains uncertain (Raaijmakers et al. 1999). Collectively, however, it is apparent that introduced bacteria can and do move along the root, that movement can be facilitated by one or more different mechanisms, and that a greater understanding is needed of the physicochemical and hydrodynamic forces that prevail in the rhizosphere.

18.2   Mechanisms of Biological Control

Biological control of plant root pathogens may be mediated indirectly, by competition for nutrients and niches on the root or by induction of resistance in the host plant. Alternatively, biocontrol agents may directly antagonize the pathogen via the production of biosurfactants, antibiotics, or enzymes that hydrolyze the pathogen cell wall. Biocontrol agents often express more than one mechanism of action, and metabolites such as the antibiotics and iron-sequestering siderophores produced by many biocontrol agents may have multiple activities, affecting not only the target pathogen but also the host plant and other members of the microbial community.

Competition for Niches and Nutrients (CNN)   Kamilova et al. (2005) used a classic selective enrichment procedure to recover novel biocontrol agents enhanced in competitive colonization of the root. A mixture of rhizosphere strains was introduced onto surface-sterilized tomato seeds and the seeds were allowed to germinate and grow for one week in a gnotobiotic (germ-free) system. Bacteria that had moved along the root to the tip were then collected, introduced onto sterile seeds, and the cycle was repeated three times. Many of the isolates recovered from root tips after this procedure were found to protect tomato against foot and root rot caused by *Fusarium oxysporum* f. sp. *radicis-lysopersici,* and studies with mutants showed that both motility and the ability to grow efficiently on nutrients in tomato root exudate were important for biocontrol (Lugtenberg and Kamilova 2009; Pliego et al. 2011). Not all of the recovered isolates were effective against the pathogen, however, and results of another study (Pliego et al. 2011) involving the white root rot pathogen of avocado indicated that not only aggressive colonization of the root, but also colonization of specific target sites of pathogen attack on the root, or even colonization of the surface of the pathogen itself, may be required to achieve disease control.

Induced Systemic Resistance (ISR)   In induced systemic resistance, a plant's first line of defense against pathogens, selected beneficial microbe-associated inducers on roots activate and sensitize (*i.e., prime*) the entire plant, including the aboveground parts, for defense against diverse pathogens (Pieterse et al. 2014). Like immunization in animals, *priming* sensitizes the immune system to an enhanced state of readiness such that defenses can be mobilized rapidly and systemically upon subsequent pathogen attack. However, priming differs from immunization in that immunization targets specific pathogens whereas priming by beneficial root-associated microbes enhances defenses against diverse pathogens and even herbivorous insects. The exposure of roots to certain common microbial structural elements such as flagella, lipopolysaccharides, and chitin, and microbially-produced compounds such as siderophores, certain antibiotics, biosurfactants, and volatiles, can result in priming in a wide variety of plant species.

Because the induction of defense and the enhanced defense reaction in ISR take place in spatially separated locations, plants must employ a complex long-distance signaling pathway that starts at the root-microbe interface. Signaling is modulated by the hormones ethylene and jasmonic acid (and less commonly by salicylic acid). In all cases, however, a complex array of signaling proteins and transcriptional regulators is involved that accumulates after induction of the primed state. This pathway, and the identity of ISR long-distance signals, are currently subjects of active investigation.

Siderophore-Mediated Competition for Iron and Induced Systemic Resistance
Under conditions of low iron availability bacteria produce siderophores, low molecular weight compounds that sequester iron in the environment and facilitate its uptake by bacterial cells (Höfte 1993). Since basically all organisms require iron to grow and function, and iron availability in soil is extremely low, competition for iron between microorganisms in the rhizosphere is expected to be a common phenomenon. Siderophores produced by different microorganisms can belong to different classes and they have different structures (Höfte 1993). Accordingly, specific receptors are required to utilize ferric siderophore complexes (Hartney et al. 2011). Thus it was postulated that bacteria that produce siderophores with a high affinity for iron and that are specific, that is they can be utilized by the producer but not by other microbes, could be effective biological control agents. Indeed competition for iron between microorganisms in the rhizosphere has been demonstrated using reporter gene constructs that respond to bioavailability of iron (Loper and Henkels 1999). In several studies on fluorescent pseudomonads it was demonstrated that mutants unable to produce their fluorescent siderophore also (partially) lost their ability to control disease (see for example De Boer et al. 2003). Based on such studies, siderophore-mediated competition for iron is considered an important mechanism in biological control of soil-borne diseases.

Siderophores also have been implicated as effective elicitors of ISR in plants (Meziane et al. 2005). Thus, when plants are protected from disease by siderophores produced by biological control agents, it is difficult to assess if this is due to competition for iron, to ISR, or to both. In radish, *Pseudomonas putida* WCS358 effectively controls fusarium wilt, caused by *Fusarium oxysporum* f. sp. *raphani,* through the production of its pyoverdin siderophore (De Boer et al. 2003). Strain WCS358 cannot elicit ISR in radish and thus in this case the suppression of disease most likely depends on competition for iron between the biological control agent and the pathogen. Evidence that the fluorescent siderophore produced by strain WCS358 can elicit ISR comes from studies in *Arabidopsis thaliana,* bean and tomato. In all three plant species, application of the purified pyoverdin of WCS358 did elicit ISR.

In *A. thaliana* and bean, a pyoverdin mutant was as effective as the wild-type, but in tomato the pyoverdin mutant did not elicit ISR (Meziane et al. 2005). For tomato the role of pyoverdin in ISR is straightforward but for *A. thaliana* and bean there seems to be redundancy of bacterial elicitors of ISR and next to the pyoverdin, both lipopolysaccharides and flagella play a role. The observed redundancy may give robustness to biological control, since if one of the factors involved in biocontrol is not produced by the bacterium the additional determinants can still be effective. This situation may however further complicate studies on siderophore-mediated competition for iron. In several studies it was suggested that siderophores do not play a role in disease suppression since knock out mutants were as effective as the wild-type. When we consider that there can be redundancy in biological control traits in a single biological control agent, the mutant approach to study possible involvement of siderophores is not waterproof and in several cases, rejecting their role may have been unjustified. Nevertheless, siderophores have been demonstrated to play a significant role in biocontrol and both competition for iron and triggering ISR can be involved. If both mechanisms were active simultaneously this would first weaken the pathogen in the rhizosphere by iron depletion and then confront it with an enhanced plant defense resulting in effective biocontrol. Such a scenario may explain why siderophores were early on revealed as being important effectors of biological control of soil-borne diseases.

Antibiosis Antibiotics are small organic molecules produced by microorganisms that are deleterious to the growth or metabolic activities of other microorganisms. Most bacteria involved in biocontrol have the capacity to produce antibiotics, and many produce multiple such metabolites with broad-spectrum activity against a wide range of plant pathogens (Raaijmakers and Mazzola 2012). Antibiotic-producing strains typically are detected by their ability to inhibit the growth of target pathogens *in vitro* and then are tested for biocontrol in soil, but this approach often fails due to the producer strain's inability to compete successfully in the rhizosphere or because conditions there do not support the synthesis of inhibitory levels of the active agent in the sites where pathogens attack. While the spatiotemporal and quantitative aspects of antibiotic production in nature are only now beginning to be explored, it has become apparent in recent years that antibiotics play multiple roles in natural systems. At subinhibitory concentrations these molecules can function as molecular signals in such diverse activities as biofilm formation and cellular differentiation, motility and dispersal, and defense against predators and competitors (Raaijmakers and Mazzola 2012), all of which may impact upon biological control, and some antibiotics are produced in the environment in sufficient quantities to inhibit pathogens (Mavrodi et al. 2012). The availability since 2005 of genomic DNA sequences not only for model biocontrol strains, but also for environmental isolates, has greatly facilitated the identification of a repertoire of novel metabolites and gene clusters with the potential to exhibit antibiotic activity. *Pseudomonas* and *Bacillus* spp. are the most widely studied biocontrol agents to date, and *Bacillus* spp. are the most frequently commercialized because they are more readily formulated.

Biosurfactant Antibiotics Produced by _Bacillus_ spp Cyclic lipopeptides (cLP) produced by root-associated bacilli are composed of a lipid tail linked to a short cyclic oligopeptide and include members of three main families, the surfactins, iturins and fengycins, which differ in the type, number and sequence of amino acid residues and the nature of peptide cyclization. Given their structural diversity, it is not surprising that these antibiotics also differ in their modes of action and their contributions to the biocontrol activity of the cells that produce them. For example, surfactins are powerful biosurfactants that can interfere with biological membrane integrity, but they have no marked toxicity to fungi. In contrast, both iturins and fengycins exhibit antifungal activity which is thought to be due to their ability to form ion-conducting pores in target cell membranes, leading to an imbalance in transmembrane ion fluxes and cell death. These latter classes of cLPs have been implicated directly in antagonism of soil-borne and foliar fungal phytopathogens including _Fusarium graminearum_, _Rhizoctonia solani_, _Botrytis cinerea_, _Penicillium roqueforti_, and _Aspergillus flavus_ (Raaijmakers et al. 2010).

Rhizosphere factors profoundly influence the production of cLPs and the induction of systemic resistance by _Bacillus_ spp. Members of the surfactin family, which are inducers of ISR in a broad range of plants, are the main cLPs secreted by _B. amyloliquefaciens_ S499 in biofilms and on the roots of tomato, whereas iturins were detected at much lower concentrations and fengycins were not produced in measurable quantities. This is in marked contrast to production _in vitro_, where iturins and fengycins are produced in similar quantities and each was more than half the amount of surfactin present. A combination of elegant electrospray and imaging mass spectrometry-based approaches showed that cLPs, mainly surfactin, were formed by microcolonies directly on tomato roots and root hairs, and that surfactin diffused into the surrounding medium, reaching biologically relevant concentrations within the diffusion zone close to the colonized roots (Nihorimbere et al. 2012). Moreover, a strong correlation was demonstrated between the amount of surfactin produced and the ability to elicit ISR, suggesting that screening for strong producers among members of the _B. subtilis/amyloliquefaciens_ complex could facilitate the selection of effective biocontrol agents (Cawoy et al. 2014).

Antibiotics produced by Pseudomonas spp Lipopeptide-producing strains of _Pseudomonas_ spp., like those of _Bacillus_, are widely distributed in nature. Many have significant impacts on oomycetes such as _Phytophthora_ and _Pythium_ spp., with members of the viscosin group of compounds including massetolide A of particular note for their effects (ranging with increasing concentration from inducement of encystment to immobilization and lysis) on zoospores. Effects of these cLPs on mycelial morphology and physiology not only of oomycetes, but also of _Rhizoctonia_, typically are less pronounced than on zoospores and include increased hyphal branching and swelling (Raaijmakers et al. 2010; D'aes et al. 2011). cLPs produced by _Pseudomonas_ spp. have been implicated in the control of pathogens such as _Pythium ultimum_ on sugar beet, _Phytophthora infestans_ on tomato, _Pythium_ and _Rhizoctonia_ spp. on bean, and _R. solani_ and _Gaeumannomyces graminis_ var. _tritici_ on wheat (Yang et al. 2014).

Among the most commonly detected small metabolite antibiotics produced by *Pseudomonas* spp. are phenazine compounds, 2, 4-diacetylphloroglucinol (DAPG), pyrrolnitrin and pyoluteorin. The genetics, regulation of synthesis, and mode(s) of action of these compounds have been studied intensively, and the compounds themselves have, in most cases, been isolated from roots colonized by introduced or indigenous strains of the producing bacteria (Raaijmakers and Mazzola 2012). Of particular note are DAPG, produced by indigenous strains responsible for control of the wheat root disease take-all throughout much if not all of the USA and Europe (Chap. 38) and phenazine-1-carboxylic acid (PCA), present on the roots of wheat grown in dryland regions throughout the American Pacific Northwest (Mavrodi et al. 2012). The polyketide antibiotic DAPG is active against a wide range of phytopathogens, and also can elicit ISR in plants (Weller et al. 2012). The phenazines encompass a large family of pigmented, heterocyclic redox-active compounds that function not only in biological control, but also as microbial fitness determinants, virulence factors in plant and animal disease, and in microbial community dynamics and physiology. The phenazine-producing strain *P. aureofaciens* 1393 is marketed as an active ingredient in the biopesticide "Pseudobacterin-2" in the Russian Federation for the control of a wide range of phytopathogens as well as for the induction of resistance to plant diseases in organic and conventional crops (Thomashow 2013). PCA is the major ingredient in Shenqinmeisu, a microbial pesticide produced in China and applied commercially to protect rice and vegetables against important phytopathogens (Xu 2013).

18.3 Conclusions

In this chapter we have touched briefly upon recent insights towards understanding interactions among microorganisms and their plant hosts, as well as knowledge gaps and needs for future research. Technological advances that enable sensitive detection of metabolites including root exudate components produced *in situ* will continue to be critical to unraveling the complex molecular and organismal interrelationships in the rhizosphere habitat. Better knowledge of the microbe-plant dialogue is essential given the need for increased agricultural productivity to provide food and biofuel feedstocks in the face of climate change, the increasing world population and the loss of arable lands.

Acknowledgements Parts of this work were supported by USDA-NRI Grant No. 2011-67019-30212 from the USDA-NIFA Soil Processes program. Mention of trade names or commercial products in this publication is solely for the purpose of providing specific information and does not imply recommendation or endorsement by the U.S. Department of Agriculture. USDA is an equal opportunity provider and employer. We thank Dr. David Weller for review comments.

References

Cawoy H, Mariutto M, Henry G et al (2014) Plant defense stimulation by natural isolates of *bacillus* depends on efficient surfactin production. Mol Plant Microbe Interact 27:87–100

D'aes J, Hua GK, De Maeyer K et al (2011) Biological control of root rot of bean by phenazines and cyclic lipopeptides-producing *Pseudomonas* CMR12a. Phytopathology 101:996–1004

De Boer M, Bom P, Kindt F, Keurentjes JJB et al (2003) Control of *Fusarium* wilt of radish by combining *Pseudomonas putida* strains that have different disease-suppressive mechanisms. Phytopathology 93:626–632

Dechesne A, Smets BF (2012) Pseudomonad swarming motility is restricted to a narrow range of high matric water potentials. Appl Environ Microbiol 78:2936–2940

Dechesne A, Wang G, Gülez G et al (2010) Hydration-controlled bacterial motility and dispersal on surfaces. Proc Natl Acad Sci U S A 107:14369–14372

Hartney SL, Mazurier S, Kidarsa TA et al (2011) TonB-dependent outer-membrane proteins and siderophore utilization in *Pseudomonas fluorescens* Pf-5. BioMetals 24:193–213

Höfte M (1993) Classes of microbial siderophores. In: Barton LL, Hemming BC (eds) Iron chelation in plants and soil microorganisms. Academic Press, San Diego, pp 3–26

Kamilova F, Validov S, Azarova T et al (2005) Enrichment for enhanced competitive plant root tip colonizers selects for a new class of biocontrol bacteria. Environ Microbiol 7:1809–1817

Loper JE, Henkels MD (1999) Utilization of heterologous siderophores enhances levels of iron available to *Pseudomonas putida* in the rhizosphere. Appl Environ Microbiol 65:5357–5363

Lugtenberg B, Kamilova F (2009) Plant growth-promoting rhizobacteria. Annu Rev Microbiol 63:541–556

Martínez-Gil M, Quesada JM, Ramos-González M et al (2013) Interplay between extracellular matrix components of *Pseudomonas putida* biofilms. Res Microbiol 164:382–389

Martínez-Gil M, Ramos-González M, Espinosa-Urgel M (2014) Roles of cyclic di-GMP and the Gac system in transcriptional control of the genes coding for the *Pseudomonas putida* adhesins LapA and LapF. J Bacteriol 196:1484–1495

Mavrodi DV, Mavrodi OV, Parejko JA et al (2012) Accumulation of the antibiotic phenazine-1-carboxylic acid in the rhizosphere of dryland cereals. Appl Environ Microbiol 78:804–812

Meziane H, Van der Sluis I, Van Loon LC et al (2005) Determinants of *Pseudomonas putida* WCS358 involved in inducing systemic resistance in plants. Mol Plant Pathol 6:177–185

Nihorimbere V, Cawoy H, Seyer A et al (2012) Impact of rhizosphere factors on cyclic lipopeptide signature from the plant beneficial strain *Bacillus amyloliquefaciens* S499. FEMS Microbiol Ecol 79:176–191

Pieterse C, Zamioudis C, Berendsen RL et al (2014) Induced systemic resistance by beneficial microbes. Annu Rev Phytopathol 52:347–375

Pliego C, Kamilova F, Lugtenberg B (2011) Plant growth-promoting bacteria: fundamentals and exploitation. In: Maheshwari DK (ed) Bacteria in agrobiology: crop ecosystems. Springer, Germany, pp 295–343

Raaijmakers JM, Mazzola M (2012) Diversity and natural functions of antibiotics produced by beneficial and plant pathogenic bacteria. Annu Rev Phytopathol 50:403–424

Raaijmakers JM, de Bruijn I, Nybroe O et al (2010) Natural functions of lipopeptides from *Bacillus* and *Pseudomonas*: more than surfactants and antibiotics. FEMS Microbiol Rev 24:1037–1062

Thomashow LS (2013) Phenazines in the environment: microbes, habitats, and ecological relevance. In: Chincholkar S, Thomashow L (eds) Microbial phenazines. Springer, Berlin, pp 199–216

Weller DM, Mavrodi DV, van Pelt JA et al (2012) Induced systemic resistance in *Arabidopsis thaliana* against *Pseudomonas syringae* pv. *tomato* by 2,4-diacetylphloroglucinol-producing *Pseudomonas fluorescens*. Phytopathology 102:403–412

Xu Y (2013) Genomic features and regulation of phenazine biosynthesis in the rhizosphere strain *Pseudomonas aeruginosa* M18. In: Chincholkar S, Thomashow L (eds) Microbial phenazines. Springer, Berlin, pp 177–198

Yang M-M, Wen S-S, Mavrodi DV et al (2014) Biological control of wheat root diseases by the CLP-producing strain *Pseudomonas fluorescens* HC1-07. Phytopathology 104:248–256

Zolla G, Bakker MG, Badri D et al (2013) Understanding root-microbe interactions. In: de Bruijn F (ed) Molecular microbial ecology of the rhizosphere, vol 2. Wiley Blackwell, Hoboken, New Jersey, pp 745–754

Chapter 19
Control of Insect Pests by Entomopathogenic Nematodes

Vladimír Půža

Abstract Entomopathogenic nematodes (EPNs) of the families Steinernematidae and Heterorhabditidae (Rhabditida, Nematoda) are obligate pathogens of insects and are associated with specific symbiotic bacteria of the genera *Xenorhabdus* and *Photorhabdus*, respectively. Due to their ability to infect various insects, the possibility of mass production by industrial techniques, and safety to non-target organisms and the environment, these nematode-bacteria complexes are attractive agents for the biological control of many insect pests. In this chapter, a brief characteristic of these organisms is given, together with information on their use in the biological control of insect pests.

19.1 Entomopathogenic Nematodes and their Bacterial Symbionts

Entomopathogenic nematodes of the families Stenernematidae and Heterorhabditidae do not form a monophyletic group, but evolved independently from a free-living bacterivorous nematode in the Devonian period (Poinar 1993). Both *Xenorhabdus* and *Photorhabdus* are gram-negative rods belonging to the family of *Enterobacteriaceae*, within the gamma subdivision of Proteobacteria. Unlike their nematode associates, both genera form a monophyletic group and *Proteus* bacteria are the closest sister taxon.

The family Steinernematidae comprises over 80 species, and the genus *Xenorhabdus* over 20 described species. The family Heterorhabditidae, probably a phylogenetically younger group of nematodes, has 18 described species, and four species of associated *Photorhabdus* bacteria. Single species of *Steinernema* may be associated with only one species of *Xenorhabdus*. The same applies to *Heterorhabditis* with the exception of *H. bacteriophora* that is associated either with *P. luminescens* or

V. Půža (✉)
Institute of Entomology, Biology Centre of the AS CR, Branišovská 31,
37005 České Budějovice, Czech Republic
Tel.: +420732561579
e-mail: vpuza@seznam.cz

© Springer International Publishing Switzerland 2015
B. Lugtenberg (ed.), *Principles of Plant-Microbe Interactions,*
DOI 10.1007/978-3-319-08575-3_19

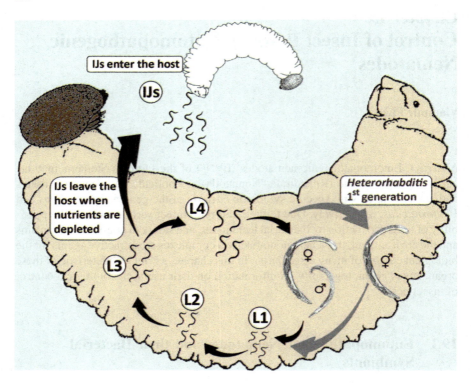

Fig. 19.1 Generalized life cycle of entomopathogenic nematodes

P. temperata. On the other hand, species of *Photorhabdus* and certain species of *Xenorhabdus* are hosted by several species of *Heterorhabditis* and *Steinernema*.

Over 80 % of EPNs and bacteria have been described since the year 2000 thanks to the use of molecular methods, and further descriptions of new EPN and bacterial species are to be expected in the future. New EPN species present a unique combination of characteristics and may also have a great potential for the biological control of particular insect pests.

Life Cycle (Fig. 19.1) The only free-living stage is a non-feeding and developmentally arrested third stage larva, called infective juvenile (IJ). Infective juveniles actively seek insect host, using various chemical cues. Once the host is found, they penetrate into the haemocoel via natural openings. Upon entering an insect host, the nematode expels its bacterial symbionts, which produce extracellular toxins and kill the host within 1–3 days. The insect carcasses with *Steinernema-Xenorhabdus* turn yellowish-brown in color, while those killed by *Heterorhabditis-Photorhabdus* turn red. The bacteria produce degradative enzymes, and proliferate, providing a food source that supports the reproduction of an increasing population of nematodes. Larvae develop into amphimictic adults in the genus *Steinernema* or hermaphrodites in

Heterorhabditis. The first generation is followed by several amphimictic generations in both genera. When the insect carcass is depleted of nutrients, the nematodes revert back to the resting form, and their receptacle is colonized by bacterial cells. This usually happens within 7–14 days after infection, depending on the host size, temperature, and other factors. At the end of the cycle, thousands of infective juveniles emerge from the carcass in search of a new insect host (Fig. 19.1).

Symbiosis The relationship between *Steinernema* and *Xenorhabdus* and *Heterorhabditis* and *Photorhabdus* is obligate in the natural environment (Akhurst and Boemare 1990). For an excellent review on this subject, see Ciche et al. (2006). Under experimental conditions, the nematodes may also consume other bacteria, but the IJs can associate only with their own symbionts. Thus, under natural conditions, the nematodes do not live without their symbiont, and, with the exception of *P. asymbiotica*, the bacteria do not live freely without the nematodes.

The location of the symbiont within the IJ nematode differs between the two genera. In the genus *Steinernema* the IJs carry dozens of symbionts in a specialized structure called the receptacle. IJs of *Heterorhabditis* species transmit the symbionts in the anterior part of the intestine. The bacteria persist for many weeks within the free-living IJs in a quiescent state until becoming pathogenic within the insect host.

Besides killing the host and providing a food source to the nematodes, the bacteria also protect the cadaver against other microorganisms and scavengers by the production of bacteriocins, antibiotics, antimicrobials, and scavenger deterrent compounds. On the other hand, the nematodes protect the bacteria in the external environment and vector them into the insect hemocoel. In some associations, the nematode also inhibits the insect immune response by producing proteases and immune suppressive factors.

Typical for symbionts of both genera, *Xenorhabdus* and *Photorhabdus*, is the phenomenon of phase variation. The primary phase is isolated from IJ nematodes and in EPN-infected insects, while the secondary phase sometimes occurs during the late stages of nematode development in insect cadavers and during the sub-culturing of the primary form. Secondary stage cells are smaller, produce a lower number of secondary metabolites, and have lower nutritional value. When the bacteria are cultured under standard conditions, they are in the primary phase, however, stress conditions can evoke a shift to the secondary stage. The shift to the secondary phase has a detrimental effect on nematode development, particularly in liquid culture (Ehlers et al. 1990). Thus all measures should be taken during production to avoid the occurrence of phase variation.

Host Range The host range of EPNs is still not fully understood. In laboratory experiments, they are capable of invading and killing a large number of insects and even other arthropods, e.g. spiders, ticks and millipedes (Poinar 1979). However, under natural condition, the host spectrum is much narrower and can include only insects that spend at least some time within the soil environment. Insects naturally infected with EPNs are rarely found. The available data show that the most frequently infected larval stages of Diptera, Coleoptera, Lepidoptera, and Hymenoptera prevail (Peters 1996). Peters (1996) concluded that some species like *S. carpocapsae* and *S.*

feltiae are generalists with a broader host range, whereas specialists like *S. glaseri* and *S. scapterisci* have a more restricted host range. On the other hand Půža and Mráček (2010) found no difference in the natural host range of *S. affine* and *S. kraussei.*

Many insects, such as elaterid larvae (wireworms) possess various morphological barriers that may protect them from invasion. Other insects, e.g. scarab larvae, display aggressive and evasive behaviour, or other responses such as frequent defecation to expel nematodes from the gut, as observed in some fly larvae. Alternatively, some insects are able to detect EPNs, and move to an EPN-free place to prevent invasion, as observed in fire-ant colonies treated with *Steinernema carpocapsae*. Once in the host haemocoel, the nematodes may encounter a strong immune response. Often, these defense strategies are powerful enough to make the insect resistant to infection by EPNs.

Foraging Strategies From the biocontrol point of view, the foraging strategies of entomopathogenic nematodes are of particular interest, because they make the nematodes differentially suitable for the control of different insect pests. The strategies range from sit and wait (ambusher) to actively foraging (cruiser) (Lewis et al. 1992). Cruisers move through the soil and look for the sedentary host whereas ambushers remain near the soil surface preferably attacking passing moving insects. Moreover, the foraging strategy is not a two-class model, but it is a continuum with two poles (Campbell and Gaugler 1997) and the majority of species possess an intermediate strategy. Recently, Wilson et al. (2012) have demonstrated, that a typical 'ambush' forager, *Steinernema carpocapsae*, can use cruiser strategy in habitats other than mineral soils and successfully control sedentary or cryptic pests in organic habitats. The authors concluded that the classification of *S. carpocapsae* as an ambush forager cannot be sustained.

19.2   Entomopathogenic Nematodes in Biocontrol

Collection For the use in biocontrol, EPN collection should be made in areas with the occurrence of target insects, where EPN strains adapted to this pest are to be expected. As was mentioned above, naturally infected insects are rarely found. Thus the most frequently used technique used for EPN isolation is Galleria baiting (Bedding and Akhurst 1975) using larvae of the greater wax moth, *Galleria mellonella*, as bait. The insects caged in a wire mesh are placed in the soil (either in a soil sample, or in situ). A few days later, the infected insect cadavers are removed, washed, and placed on a White trap (White 1927) (top of a Petri dish with moist filter paper placed in a larger Petri dish with water). One to two weeks later, thousands of fresh IJs are collected for further use.

Mass Production EPNs can be mass produced using *in vivo* and *in vitro* methods. In vivo production uses living insects, mostly the greater wax moth (*Galleria mellonella*) larvae, or mealworms (*Tenebrio molitor*). The method is simple and cheap, but is labor and cost effective only at a small scale. This method is thus appropriate for laboratory use or small scale applications. It may also be used by small growers, where large investments into in vitro culture technology cannot be made.

Solid or liquid fermentation *in vitro* technologies are used when large scale production is needed at a reasonable quality and cost. In both these methods the nematodes are cultured monoxenically, to ensure quality consistency and predictability (Lunau et al. 1993). A symbiont is extracted from the nematodes and subsequently sterile nematode eggs are prepared and applied to the medium pre-inoculated with bacterial symbiont. A comprehensive review of the current situation regarding the *in vitro* mass production of EPNs was published by Shapiro-Ilan et al. (2012).

Solid culture was first performed in two dimensional arenas (e.g, Petri dishes), using various media. A substantial improvement was achieved by adopting a three-dimensional rearing system with the liquid medium mixed with an inert carrier (e.g. pieces of polyurethane foam). Media were initially based on animal products (e.g. pig kidney) but were later improved by including various other ingredients (e.g. peptone, yeast extract, eggs, soy flour). The culture starts with the inoculation of the sterilized medium with bacteria followed by the nematodes. Nematodes are then harvested within 2–5 weeks by placing the foam onto sieves immersed in water. Only a few companies currently use this approach.

The *in vitro* liquid culture method is a complex process requiring medium development, understanding of the biology of the nematode-bacteria complex, the development of bioreactors, and understanding and control of the process parameters. The process takes place in large fermentors (up to 80.000 l). It is necessary to supply enough oxygen, and prevent excessive shearing of the nematodes. Once the culture is completed, nematodes are removed from the medium through centrifugation. This method is currently the most cost-effective and thus the majority of EPN products result from liquid culture. Major producers using this method are E-Nema GmbH, Koppert B.V., Becker Underwood, Andermatt Biocontrol, BioLogic, etc. Commercialized EPNs are listed in Table 19.1.

The repeated culturing of nematodes can result in the reduction of beneficial traits such as infectivity, environmental tolerance, or fecundity (Shapiro-Ilan et al. 2012). Thus precautions against trait deterioration have been proposed consisting of the minimization of serial passages, the introduction of fresh genetic material, the cryopreservation of stock cultures, and the creation of homozygous inbred lines which are more resistant to trait deterioration.

Formulation and Application Methods Entomopathogenic nematodes are always applied as infective juveniles, and are mainly used for controlling the larval or pupal stages of insect pests in the soil or cryptic habitats. Application against foliar pests may be successful under specific conditions. Application technologies for entomopathogenic nematodes were thoroughly reviewed by Wright et al. (2005). According to these authors, EPNs can be applied using the equipment for other control agents, but EPNs are of the most expensive control agents, and thus application techniques should be optimized in order to achieve a cost-effective control.

At present, besides a classical aqueous suspension, EPNs are formulated in water-dispersible granules, nematode wool, gels, vermiculite, clay, peat, sponge, etc. The shelf life of the EPN based products depends on the formulation and nematode species. Actively moving nematodes can remain alive and infective for 1–6 months

Table 19.1 Commercialized EPN species, their bacteria symbionts, target pests and production status

Nematode	Symbiont	Target insects	On market
H. bacteriophora	*P. luminescens, P. temperata*	White grubs, weevils	Widely
H. indica	*P. luminescens*	White grubs, weevils, small hive beetle	In the USA
H. megidis	*P. temperata*	Weevils, white grubs	-[a]
H. zealandica	*P. zealandica*	White grubs	-[a]
S. carpocapsae	*X. nematophila*	Armyworms, cutworms, leatherjackets	Widely
S. feltiae	*X. bovienii*	Fungus gnats, thrips, weevils, moths	Widely
S. glaseri	*X. poinarii*	White grubs	-[a]
S. kraussei	*X. bovienii*	Black vine weevil	Widely
S. kushidai	*X. japonica*	Grubs	In Asia
S. longicaudum	*X. bedingii*	Grubs	-[a]
S. riobrave	*X. cabanillassii*	Weevils	In the USA
S. scapterisci	*X. innexi*	Mole crickets	-[a]

[a] **Currently unavailable**

under refrigeration ranges. EPNs with reduced mobility (formulations in gels) are still infective after up to 9 months of storage while EPNs formulated in partial anhydrobiosis (formulations in water soluble powders) remain so for up to 1 year.

After mixing with water, the nematodes are applied using sprayers, mist blowers, or irrigation systems. The application of infected insect host cadavers can enhance EPN persistence (Shapiro-Ilan et al. 2001). The recently proposed "lure and kill" approach based on the application of the nematodes in capsules with insect attractant may reduce the number of nematodes necessary to control the insect pest (Hiltpold et al. 2012).

Several recommendations can be made for EPN applications. The application should target the most vulnerable insect pest stage. Moderate pre-irrigation few hours before an EPN application will enable the nematodes to move easily through the soil pores. Adequate moisture should be maintained several days after application. For any spray application, the nozzle openings should not be smaller than 50 microns, and operating pressures within a system should not exceed 1000–2000 kPa, depending on the EPN species.

Genetic Improvement Entomopathogenic nematode strains with improvements of their pathogenicity, host range, environmental tolerances, and shelf-life are highly desirable. Such strain enhancement might be achieved through genetic improvement approaches. These methods were comprehensively reviewed by Burnell (2002).

Hybridization and selective breeding is the most promising approach to enhance specific survival attributes, such as desiccation and heat tolerance, cold activity, host-finding and nematicide resistance. Thanks to recent advances in EPN and bacteria genomics it will be possible to identify genes from the whole genome that are being expressed, in order to detect those that are involved in a particular process and target them via genetic engineering methods. However, at present, it is unlikely that a transgenic EPN strain would meet regulatory and public acceptance as a control agent (Perry et al. 2012).

Safety In the past four decades, numerous studies have assessed the effect of entomopathogenic nematodes and their bacterial symbionts on non-target organisms. A thorough review of this subject was given by Ehlers (2005). The available data show that EPNs are safe to humans and animals. In laboratory and field tests, non-target arthropods were less susceptible to EPNs. Also, no adverse effect on the non-target soil arthropods has been observed after EPN application. According to Bathon (1996), the mortality of non-target animals in the field may occur, but will be temporal, spatially restricted, affecting only part of the population, and its impact can be considered negligible.

A few decades ago, a close relative to symbiotic *Photorhabdus* species, *P. asymbiotica* was identified as a facultative human pathogen. It was later revealed that strains of this species are associated with *Heterorhabditis indica* and *H. gerrardi*. This fact raises concerns about the safety of EPNs to humans. However, commercially produced *H. bacteriophora* and *H. megidis* have never been found in association with this bacterium. According to Grewal (2012), more research is needed on these bacteria, but as a precaution it is advisable to avoid contact between these bacteria and human wounds.

Regulation In general, regulations form a major obstacle in the commercialization of biopesticides. Luckily, EPNs have been exempted from the registration process in many countries, e.g. in the USA, while other countries require registration for the exotic or non-indigenous species/strains. In the EU, EPNs are not covered by the legislation of plant protection microbial products but are usually covered as macroorganisms together with beneficial arthropods (Ehlers 2005). Thus in most EU countries, no registration is required. Those countries that require registration usually ask for information that is freely available in the scientific literature. Some countries further require the data of each new product from field trials performed within their borders. The recent EU directive 2009/128/EC on the sustainable use of pesticides states that non-chemical methods should be given priority wherever possible. Demand for EPN-based products is therefore likely to increase in the future.

Control of Target Pests Entomopathogenic nematodes are used for the control of a variety of insect pests in soil, cryptic, and foliar habitats. EPNs are native to soil and thus many soil insects have received attention regarding control efforts with EPNs. The most important beetles targeted by EPNs in soil are weevils (e.g. the black vine weevil *Ottiorhynchus sulcatus*, the root weevil *Diaprepes abbreviatus*, and the sweet potato weevil *Cylas formicarius*), and grubs (e.g. the japanese beetle

Popillia japonica and the garden chafer *Phyloperta horticola*). EPNs are further used to control soil-borne dipterans such as fungus gnats (e.g. *Bradysia* spp., *Lycoriella* spp.) and maggots (e.g. *Delia radicum*). Among lepidopterans, cutworms (*Agrotis* spp.) and earworms (e.g. the corn earworm *Helicoverpa zea*) spend some or all of their feeding stages in contact with the soil and thus can be controlled by the use of EPNs.

Cryptic habitats are also considered to be favorable for EPN survival, as UV radiation and desiccation are minimized. In these habitats, the bark and wood boring moths (*Synanthedon* spp., *Euzophera semifuneralis*), or codling moth (*Cydia pomonella*) represent the most important target pests.

Foliar habitats are less suitable for EPN use due to the adverse conditions associated with them. Thus EPN-based control in these habitats tends to be less efficient. However, some foliar pests, e.g. dipteran leafminers (e.g. *Liriomyza trifolii, Tuta absoluta*), or diamondback moth larvae (*Plutella xylostella*), can be successfully controlled with EPNs.

References

Akhurst RJ, Boemare NE (1990) Biology and taxonomy of *Xenorhabdus*. In: Gaugler R, Kaya HK (eds) Entomopathogenic nematodes in biological control. CRC, Boca Raton, pp 75–90

Bathon H (1996) Impact of entomopathogenic nematodes on non-target hosts. Biocontr Sci Techn 6:421–434

Bedding RA, Akhurst RJ (1975) A simple technique for the detection of insect parasitic rhabditid in soil. Nematologica 21:109

Burnell A (2002) Genetics and genetic improvement. In: Gaugler R (ed) Entomopathogenic nematology.CABI, New York, pp 333–356

Campbell JF, Gaugler R (1997) Inter-specific variation in entomopathogenic nematode foraging strategy: dichotomy or variation along continuum? Fundam Appl Nematol 20:393–398

Ciche TA, Darby C, Ehlers RU et al (2006) Dangerous liaisons: the symbiosis of entomopathogenic nematodes and bacteria. Biol Control 38:22–46

Ehlers RU (2005) Forum on safety and regulation. In: Grewal PS, Ehlers RU, Shapiro-Ilan DI (eds) Nematodes as biocontrol agents. CABI, Oxford, pp 107–115

Ehlers RU, Stoessel S, Wyss U (1990) The influence of phase variants of *Xenorhabdus* spp. and *Escherichia coli* (Enterobacteriaceae) on the propagation of entomopathogenic nematodes of the genera *Steinernema* and *Heterorhabditis*. Rev Nematol 13:417–424

Grewal PS (2012) Entomopathogenic nematodes as tools in integrated pest management. In: Abrol DP, Shankar U (eds) Integrated pest management: principles and practice. CABI, Wallingford, pp 162–236

Hiltpold I, Hibbard BE, French BW et al (2012) Capsules containing entomopathogenic nematodes as a Trojan horse approach to control the western corn root-worm. Plant Soil 358:11–25

Lewis EE, Gaugler R, Harrison R (1992) Entomopathogenic nematode host finding—response to host contact cues by cruise and ambush foragers. Parasitology 105:309–315

Lunau S, Stoessel S, Schmidt-Peisker AJ et al (1993) Establishment of monoxenic inocula for scaling up in vitro cultures of the entomopathogenic nematodes *Steinernema* spp. and *Heterorhabditis* spp. Nematologica 39:385–399

Perry RN, Ehlers RU, Glazer I (2012) A realistic appraisal of methods to enhance desiccation tolerance of entomopathogenic nematodes. J Nematol 44:185–190

Peters A (1996) The natural host range of *Steinernema* and *Heterorhabditis* spp. and their impact on insect populations. Biocontrol Sci Techn 6:389–402

Poinar GO Jr (1979) Nematodes for biological control of insects. CRC, Boca Raton, p 249

Poinar GO Jr (1993) Origins and phylogenetic relationships of the entomophilic rhabditids, *Heterorhabditis* and *Steinernema*. Fudam Appl Nematol 16:333–338

Půža V, Mráček Z (2010) Mechanisms of coexistence of two sympatric entomopathogenic nematodes, *Steinernema affine* and *S. kraussei* (Nematoda: Steinernematidae), in a central European oak woodland soil. Appl Soil Ecol 45:65–70

Shapiro-Ilan DI, Lewis EE, Behle RW et al (2001) Formulation of entomopathogenic nematode-infected-cadavers. J Invertebr Pathol 78:17–23

Shapiro-Ilan DI, Han R, Dolinksi C (2012) Entomopathogenic nematode production and application technology. J Nematol 44:206–217

White GF (1927) A method for obtaining infective juvenile nematode larvae from cultures. Science 66:302–303

Wilson MJ, Ehlers RU, Glazer I (2012) Entomopathogenic nematode foraging strategies—is *Steinernema carpocapsae* really an ambush forager? Nematology 14:389–394

Wright DJ, Peters A, Schroer S et al (2005) Application technology. In: Grewal PS, Ehlers RU, Shapiro-Ilan DI (eds) Nematodes as biocontrol agents. CABI, New York, pp 91–106

Chapter 20
Bacillus thuringiensis-Based Products for Insect Pest Control

Ruud A. de Maagd

Abstract *Bacillus thuringiensis* (or *Bt*, as it has become generally known) is one of the oldest and widely used biological control agents and has a long history of use. *Bt* and a number of related bacteria produce a variety of toxins, mostly—but not exclusively- localized in the parasporal crystals, which are, together with the spores themselves, the components of the typical spore/crystal mixtures. These are used to control insect pests in agricultural crops. While *Bt* products quietly kept holding the first place in biological pesticide sales, interest in *Bt* was increased by the production and commercialization of transgenic crop plants expressing one or more *Bt* toxins since 1996. Here I will present a brief overview of the history, biology, and practical uses of *Bt* and its toxins.

20.1 History of Bacillus thuringiensis Research

Bt was first isolated in Japan in 1901 by Ishiwata, but first fully described by the German Berliner and named by him after the province in Germany (Beegle and Yamamoto 1992). Because of its origin (silkworm), *Bt* was long considered a risk for the silk industry, but a first product was launched in France in 1938 under the name Sporeine. In the 1950s, several studies focussed on the parasporal body or crystal, which had been noted before but was now characterized in terms of solubility, content, and insecticidal activity. Several European countries, the USSR, as well as the USA started commercial production again in the same period. Research and development received a further boost from formation of large collections of natural *Bt* isolates. The common belief that *Bt* was only active against Lepidoptera (the larvae of butterflies and moths) was refuted by the discovery of *Bt israelensis*, with activity against Diptera (larvae of blackflies and mosquitos) in 1977. Helped by the growth of large strain collections, held all over the world and thought to run in the tens of thousands of strains, *Bt*'s known activity range was further increased with

R. A. de Maagd (✉)
Plant Research International, Wageningen UR, P.O. Box 619,
6700 AP Wageningen, The Netherlands
Tel.: +31 317 480548
e-mail: ruud.demaagd@wur.nl

© Springer International Publishing Switzerland 2015
B. Lugtenberg (ed.), *Principles of Plant-Microbe Interactions*,
DOI 10.1007/978-3-319-08575-3_20

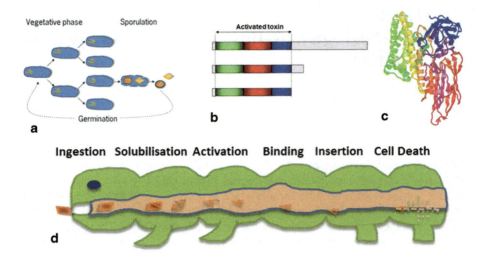

Fig. 20.1 a The life cycle of *Bacillus thuringiensis*. **b** Primary structure of three-domain protoxin proteins. The white parts are removed upon activation. Colors of the three domains correspond to those in the next panel. **c** Cartoon representation of the three-dimensional structure of the 3-domain toxins. **d** Schematic representation of the mode of action of *Bt* toxins in the insect gut

the discovery of *Bt tenebrionis*, with activity against Coleoptera (beetles) in 1983. Strains with further new (claimed) activities followed, such as against Hymenoptera (bees and wasps), Hemiptera, Mallophaga, Nematoda and Protozoa, although these have not all been confirmed. With the discovery of methods for genetic modification of plants in 1982, the idea of using genes encoding *Bt* toxins in transgenic plants to render them resistant to insect pests was rapidly put into practice, with the first report appearing in 1987, and the first transgenic crops in the field by 1996 (de Maagd et al. 1999).

20.2   Phylogeny and Ecology of *Bacillus thuringiensis*

Bt is a member of the *Bacillus cereus* group of gram-positive bacteria, and typically it will form so-called endospores when it senses adverse environmental conditions. Such spores form a though protective capsule and are dormant, non-reproductive and very hardy structures, that may "germinate", i.e. resume vegetative growth and division, when conditions are favourable again (Fig. 20.1a). Genetically, *Bt* is very similar to *B. cereus*, a causative agent of food poisoning, and to *B. anthracis*, the causative agent of anthrax in mammals. The distinction is in the details: only *Bt* forms parasporal crystals during sporulation, which make it a (facultative) insect pathogen, while it lacks the distinctive mammalian-active toxins of *B. anthracis*. Toxins are encoded (mostly) by genes on mobile genetic elements, plasmids, so that *Bt* strains that have lost those plasmids are indistinguishable from *B. cereus*.

Although originally associated with insect populations, both *Bt* as well as *B. cereus* are found in a large diversity of habitats, such as the phylloplane, soil, and stored grain products. Many isolated *Bt* strains have no demonstrable toxicity for any tested insect species, and with *Bt* occurring together with *B. cereus*, the role of parasporal crystal production in the bacterium's ecological niche, at a considerable cost of resources, remains somewhat a mystery. The consensus is to call *Bt* a facultative entomopathogen (Glare and O'Callaghan 2000).

There are several other bacterial species that produce insecticidal, and even parasporal crystal proteins related to those of *Bt*, such as *Lysibacillus sphaericus*, which is used against mosquito larvae, several *Paenibacillus* species, which cause diseases in honeybee larvae and in the white grubs of Japanese beetles, *Brevibacillus laterosporus* (toxic to Diptera), and *Clostridium bifermentans* (toxic to mosquitos) (de Maagd et al. 2003).

20.3   *Bt* Toxin Diversity, Mode of Action, and Specificity

Bacillus thuringiensis, as a species, has a vast armoury of different toxins, both parasporal crystal proteins as well as insecticidal and more general toxins produced during vegetative growth (de Maagd et al. 2003). The parasporal crystal proteins (and some vegetative, soluble proteins) make *Bt* an insect pathogen. In a natural setting the toxin's action on the insect gut may help the spore to get access to the body cavity, where it can germinate and cause septicaemia. In general, toxicity of a strain can be explained by the activity of the particular toxins that the strain produces. Each strain can produce one or more toxins, together determining the host specificity of the strain. By far the most common insecticidal proteins of *Bt* are the delta-endotoxins, which together form the crystal. Two major groups, Cry (crystal) and Cyt (cytolyic) delta-endotoxins, can be distinguished. The largest group are the Cry proteins, with over 120 different main types known today, and many more minor variants. While initially Cry proteins were classified according to their general insect order activity (CryI against Lepidoptera, CryII against Leps and Diptera, CryIII against Coleoptera, etc.), this classification was found to be untenable in the face of the growing number of holotypes. A new classification based on amino acid sequence homology was later accepted (see: http://www.btnomenclature.info/). Although all Cry1 proteins are still generally active against some Lepidoptera species, and Cry3 proteins against Coleoptera etc., the more recent higher rank number toxins have more varying activity. In general, each toxin has activity against only one or a few species within the same order, some toxins have a wider activity spectrum within an order (particularly so for Lepidoptera) and some have been found to be active against species from different orders. All this also depends on the number of target species that have actually been tested (see also: The *Bacillus thuringiensis* toxin specificity database at http://www.glfc.cfs.nrcan.gc.ca/bacillus/). Thus, it is important to keep in mind that to refer to "*Bt* toxin" is meaningless without knowing the specific type of toxin and its activity spectrum.

By far the most common type of Cry proteins appears to have a conserved three-dimensional structure, consisting of three structural domains, and are hence called the 3-domain toxins. These generally occur in crystals as protoxins, with N-terminal and C-terminal parts that after solubilisation from the crystal are removed by proteolytic action (Fig. 20.1b). In the remaining activated toxin, the first N-terminal domain consists of six amphipathic alpha-helices surrounding a central hydrophobic helix, which are thought to become part of the eventual membrane pore (see below). The second and third domain contain mostly beta-sheets, and are involved in binding to specific insect gut receptors (Fig. 20.1c). To add to the complexity of the *Bt* toxin arsenal, not all Cry proteins have this conserved structure, and other structures have been found (de Maagd et al. 2003).

The mode of action and its role in host specificity of the three-domain toxins has been extensively studied and is schematically shown in Fig. 20.1). Crystal proteins act on the gut of the insect host, so crystals need to be ingested to become active. Upon ingestion, the protoxins are solubilized from the crystal. This solubilization is dependent on pH. Subsequently protoxins are processed by proteolytic enzymes of the host's gut, by which the active toxin is produced. The activated toxin subsequently binds to one or several receptors on the surface of the epithelial cells lining the insect gut. The specific toxin/receptor-interaction is a major determinant of host-range, from the toxin side by domains II and III of the activated toxin. In a further not completely characterized, and somewhat contested sequence of events, possibly involving further processing, structural conformation changes and oligomerization of the toxin upon receptor binding, the toxin inserts into the epithelial cell membrane, forming pores. Both the extensive leakage of electrolytes as well as activation of host signal transduction pathways may contribute, but do eventually result in the death of the gut epithelial cells (Vachon et al. 2012). The insect stops feeding almost immediately, is paralyzed, and dies. From the above description it should be clear that each step of the mode of action contributes to the unique host specificities of the different toxins.

The other relatively common components of many parasporal crystals are the Cyt toxins, so called because *in vitro* they show general cytolytic activity against many cells, although *in vivo* they are generally restricted to Diptera, with some exceptions for coleopteran species reported. The protein structure is a three-layered α-β-α fold and it requires proteolytic activation. It acts on target gut epithelial cells by membrane pore formation, although unlike Cry proteins they have no specific receptors but more generally interact with membrane phospholipids (Soberón et al. 2013).

Other, less common *Bt* Cry proteins have very different structures from the 3-domain toxins, Cyt, and from each other. Some of these show more similarity to toxins from other pathogens, which may give clues to their mode of action. Except for Cry34/35 in transgenic plants (see below) these do not appear in products. Non-crystal toxins include the Vegetative Insecticidal Proteins (VIPs). Of these, VIP3 is toxic to a range of lepidopterans through pore formation in the insect gut and has been applied in transgenic plants (see below).

The evolution of the great diversity of strains, toxins and specificities, particularly for the 3-domain toxins, may be explained by the organization of the toxin-encoding genes in the bacterium. Most toxin genes are located on large plasmids, which can be mobilized and transferred between bacteria, thus creating new toxin gene combinations. On the plasmids, *cry* genes are often clustered in groups, and this may lead to intergenic recombination between homologous genes, which leads to reshuffling of gene fragments. This mechanism can be used for creating new domain combinations in the laboratory, and this mechanism probably played an important role in the generation of structural variation (de Maagd et al. 2001). Furthermore, most toxin genes are located close to sequences related to DNA transposition. Transposition may mobilize individual genes between plasmids and assemble new combinations of genes on plasmids.

20.4   Applications of *Bt* in Agriculture and Forestry

The first commercial use of microbial *Bt*-based insecticides occurred as early as 1938 in France, and in the 1950s in the United States. The history of *Bt* discovery and use have been extensively reviewed elsewhere (Beegle and Yamamoto 1992; Glare and O'Callaghan 2000). Commercial viability as an alternative for chemical insecticides arrived in the 1970s with more potent strains. *Bt* products are particularly used in high-margin crops, such as in vegetables and fruits. On the other hand, *Bt* has been used against forestry pests such as those caused by gypsy moth and spruce budworm, by means of wide-scale aerial applications. After its discovery, *Bt israelensis* was developed for use in the US and in Germany's Rhine valley for control of biting flies and mosquitos, and in Africa for disease vector control. *Bt tenebrionis* has been used to some extent for Colorado potato beetle control and for control of chrysomelid beetles in Eucalyptus plantations (Entwistle et al. 1993). *Bt* is the single most successful biopesticide product type on the market, in the 1990s taking up over 90 % of the global biopesticide market (which by itself is a small part of the total pesticide market). This number has decreased somewhat to an estimated 55 % in 2012, due both to decreased *Bt* spray use as well as to an increase in the number and volume of other biopesticide products.

Microbial *Bt* products usually consist of dried spore/crystal-mixtures that can be produced by fermentation on a large variety of media, together with other formulation ingredients. Most information about *Bt* products comes from the US market, where 13 different active ingredients in 123 products where registered (Walker et al. 2003). However, many more less well characterized products have been produced, particularly in the Soviet Union, in China or other countries (Glare and O'Callaghan 2000). Strain improvement for higher insecticidal protein content or wider target spectrum has been undertaken by conjugation of toxin encoding plasmids and site-specific recombination, while improvement of individual toxins has been studied using mutagenesis or domain swapping (Baum et al. 1999).

Despite its prominent position as a biopesticide, and advantages for environmental impact and non-target effects, *Bt* products have never taken a large part of the overall insecticide market. This is partially due to the lack of *Bt* products with activity against major insect pest orders such as aphids and white flies. Additionally, *Bt* is a not a systemic protectant and in most cases can only be used aboveground and on the outside of the plant, leaving the plant unprotected for pests attacking roots (such as Corn rootworm) or burrowing into plant tissues (such as European Corn borer). Expressing *Bt* toxins in transgenic crops, where they are expressed in these otherwise unprotected tissues and are continuously present without the need for repeated application, has proven to be an effective alternative.

Improvements in transgenic plants focussed on the differences between microbial and plant gene function, removing cryptic RNA splice sites and other elements that negatively affect RNA stability, and finally optimization of codon use for the target plants through the construction of synthetic genes. This results in *Bt* toxin levels of 0.2–1 % of total soluble protein and proper resistance to target insect species (de Maagd et al. 1999). Many different crop/gene-combinations for various pests have been developed. However, only a handful of these have reached the commercialization stage and are still being cultivated somewhere in the world today. The two most substantial crops containing *Bt* genes for insect control today are cotton and corn. Potato, tomato, eggplant, rice, and soybean varieties with *Bt* genes have been approved for cultivation but their application so far has been negligible. Cotton has been commercialized from 1996 with a varying array of *Bt* genes, mostly for resistance to cotton bollworm (Lepidoptera) in a large number of countries, including USA, Australia, India and China, and for Pink bollworm in the USA. This resulted in a vast majority of cotton in some countries being of a transgenic *Bt* variety. *Bt* maize has also been cultivated since 1996, first for resistance to European corn borer and Mediterranean corn borer (Lepidoptera), later also for resistance to cutworms (Lepidoptera) and for Corn rootworm (Coleoptera) (see also: www.isaaa.org). Recent developments concentrate on the so-called "stacking" of genes, combining several *Bt* genes in one variety by crossing, both for wider target spectrum as well as for resistance management (see below). For an extensive up-to-date overview of approved or pending *Bt* crops world-wide, the reader is referred to one of several online databases (http://www.cera-gmc.org/?action=gm_crop_database).

20.5 Safety and Resistance

Safety issues in relation with the use of *Bt* sprays and, particularly, with the use of *Bt* toxins in transgenic crops are a constant source of study and discussion. With regard to mammalian and human toxicity or pathogenicity, these issues may be partially overlapping as both applications involve similar toxins, but other issues are distinct: sprays may lead to ingestion or inhalation of live bacteria, albeit in usually small amounts, consumption of transgenic crops mostly leads to ingestion of individual Cry proteins. These issues are discussed in great detail elsewhere (Glare and O'Callaghan 2000).

Few reports exist of adverse effects of *Bt* sprays, despite its long history of use. Its close relation to the recognized food-pathogen *B. cereus* has raised some concerns.

Some *Bt* strains produce heat-stable β-exotoxins with demonstrated activity against many insects and even fish and mammals, and are banned from use as pesticides by regulatory authorities. The safety of *Bt* toxins in transgenic crops is part of a larger discussion of transgenic crop safety. Regulatory authorities such as the US EPA (Environmental Protection Agency) and EFSA (European Food Safety Agency) deem the currently approved *Bt* crops safe for human consumption and see no proof of adverse effects on the environment. Study of environmental and non-target effects of *Bt* toxins has been an integral part of the safety assessment of transgenic crops for environmental release (EFSA Panel on Genetically Modified Organisms 2010).

As for any insecticide, extensive and uninterrupted use of *Bt* sprays or toxins as in transgenic crops exerts strong selective pressure on insect pest populations leading to the increased presence of resistant individuals that, given time, may replace the sensitive population. This potential problem was recognized early on and observed in the field for some *Bt* sprays, and received considerable attention upon the introduction of *Bt* crops. Resistance alleles, for example genetic variants of genes for *Bt* receptors, are present in all populations at a (very) low frequency and can be readily selected in controlled laboratory conditions. Emergence of resistance in transgenic fields seems rare, although some has been reported. This delay may be partially due to resistance management strategies aimed at reducing the speed of resistance developing in a population (Tabashnik et al. 2013).

20.6   Concluding Remarks

Over a hundred years of discovery and use of *Bt* and its toxins has given us a vast domain of knowledge on toxin structure and action, as well as a valuable tool for insect pest control in agriculture. Further commercial development of new strains and crop varieties has slowed down since their first introduction, due to more stringent registration demands and high cost of the approval process for crops, in combination with increased weariness from the consumer side. Still, *Bacillus thuringiensis* will continue to fascinate researchers for a long time to come.

References

Baum JA, Johnson TB, Carlton BC (1999) *Bacillus thuringiensis*—natural and recombinant bioinsecticide products. In: Hall FR, Mean JJ (eds) Methods in biotechnology, vol 5. Humana, Totowa, pp 189–209

Beegle CC, Yamamoto T (1992) History of *Bacillus thuringiensis* berliner research and development. Can Ent 124:587–616

de Maagd RA, Bosch D, Stiekema WJ (1999) *Bacillus thuringiensis* toxin mediated insect resistance in plants. Trends Plant Sci 4:9–13

de Maagd RA, Bravo A, Crickmore N (2001) How *Bacillus thuringiensis* has evolved specific toxins to colonize the insect world. Trends Genet 17:193–199 de Maagd RA, Bravo A, Berry C et al (2003) Structure, diversity and evolution of protein toxins from spore-forming entomopathogenic bacteria. Annu Rev Genet 37:409–433

EFSA Panel on Genetically Modified Organisms (2010) Guidance on the environmental risk assessment of genetially modified plants. EFSA J 8:1879–1990

Entwistle PF, Cory JS, Bailey MJ et al (1993) *Bacillus thuringiensis*, an environmental pesticide: theory and practice. Wiley, Chichester

Glare T, O'Callaghan M (2000) *Bacillus thuringiensis*: biology, ecology and safety. Wiley, Chichester

Soberón M, López-Díaz JA, Bravo A (2013) Cyt toxins produced by *Bacillus thuringiensis*: a protein fold conserved in several pathogenic microorganisms. Peptides 41:87–93

Tabashnik BE, Brévault T, Carrière Y (2013) Insect resistance to *Bt* crops: lessons from the first billion acres. Nat Biotechnol 31:510–521

Vachon V, Laprade R, Schwartz JL (2012) Current models of the mode of action of *Bacillus thuringiensis* insecticidal crystal proteins: a critical review. J Invertebr Pathol 111:1–12

Walker K, Mendelsohn M, Matten S et al (2003) The role of microbial *Bt* products in U.S. crop protection. J New Seeds 5:31–51

Chapter 21
Post Harvest Control

Emilio Montesinos, Jesús Francés, Esther Badosa and Anna Bonaterra

Abstract Harvested fruits, vegetables, nuts and grains harbour a very reach microbiota that influence their shelf-life, quality and safety for human consumption. Spoilage due to fungal or bacterial rot, mycotoxin production and contamination by food-borne human bacterial pathogens are within the main problems of harvest products that are consumed fresh. Control of these problems is currently done by conventional methods of sanitation, disinfection or treatment with chemical fungicides. Biological control of postharvest problems can be achieved with certain strains of antagonistic viruses, bacteria, yeast and fungi. The mechanisms of action are very diverse, and several mechanisms may act simultaneously. Mechanisms include competition for nutrients and niches, antibiosis by means of antimicrobials and lytic enzymes, inhibitory volatile metabolites, pH decrease, parasitism, and induction of defence responses in the harvested plant product. Several commercial products containing strains of biological control agents are available as an alternative or complement to chemicals for postharvest rot control.

E. Montesinos (✉) · J. Francés · E. Badosa · A. Bonaterra
Laboratory of Plant Pathology, Institute of Food and Agricultural Technology,
University of Girona, Girona, Spain
Tel.: +34639763764
e-mail: emilio.montesinos@udg.edu

J. Francés
Tel.: +34 972419735
e-mail: jesus.frances@udg.edu

E. Badosa
Tel.: +34972418877
e-mail: esther.badosa@udg.edu

A. Bonaterra
Tel.: +34 972419734
e-mail: ana.bonaterra@udg.edu

© Springer International Publishing Switzerland 2015
B. Lugtenberg (ed.), *Principles of Plant-Microbe Interactions,*
DOI 10.1007/978-3-319-08575-3_21

21.1　Introduction

Plants provide a physical-chemical environment to complex microbial communities, comprising of viruses, bacteria and fungi. In addition, the different properties between aerial and soil plant parts determine that specific microbial communities are established. Microbiota composition also changes with the soil type, plant species, cultivar, season of the year, farm practices, and storage conditions (Setati et al. 2012). The diversity of the plant microbiota in fresh fruits and vegetables is very high as revealed by metagenomics studies (e.g. 100–1000 operational taxonomic units, OTUs) with only a few dominant groups (20–50 OTUs) (Leff and Fierer 2013) .

Since plants are important sources of fresh products, their in-field microbiota has a great influence in the harvested produce (see also Chap. 44). Plant products, that are consumed raw are rich in nutrients and can be contaminated at the field level or during the processing conditions at postharvest. The harvested produce maintains several functional and physiological activities (e.g. respiration, ripening in climacteric fruits), that can influence its microbiota during postharvest. These microorganisms can affect the quality and quantity (spoilage) (Snowdon 1990) as well as safety (food-borne human bacterial pathogens, mycotoxins) (Magan and Aldred 2007). For more details see Table 21.1.

Fresh fruit rot caused by fungi is a typical source of losses that affect all kinds of pome, stone, and citrus fruits, as well as berries, cucurbits and tropical fruits (Snowdon 1990; Prusky and Gullino 2010) . Fungal pathogens commonly found are *Penicillium* species in apple, pear and citrus fruits, *Rhizopus* and *Monilinia* in stone fruits, and *Botrytis* and *Colletotrichum* in berries. Bacterial rot is more common in vegetables. Dried grains and nuts are affected by fungi and bacteria that due to starch degradation can cause a decrease in weight, quality and nutritional properties.

Mycotoxigenic moulds, apart from deterioration, can produce toxic secondary metabolites that are very important in dried cereal grains and nuts stored under improper humidity conditions (Magan and Aldred 2007). The most common toxinogenic fungi are *Penicillium verrucosum* (ochratoxin), *Aspergillus flavus* (aflatoxins), *A. ochraceous* (ochratoxin) and several species of *Fusarium* (fumonisins, trichotecenes). Mycotoxinogenic fungi are also a source of mycotoxin contamination in rotten fresh fruits (Barkai-Colan and Paster 2008). Certain strains of *Bacillus cereus* produce heat stable toxins in cereals.

Food borne human bacterial pathogens can be found in postharvest produce and are an increasing problem of concern, especially in ready-to-eat fresh products, like minimally processed vegetables and sliced fruit, because they provide ideal conditions for bacterial growth (Badosa et al. 2008). These products have been related to outbreaks of foodborne diseases caused by the human pathogens *Escherichia coli* O157:H7, *Listeria monocytogenes*, *Salmonella enterica*, *Staphylococcus aureus* and *Pseudomonas aeruginosa* (Lin et al. 1996).

To prevent postharvest losses, the shelf-life of plant produce is increased by means of cold-storage or controlled atmosphere methodologies (low oxygen and increased CO_2, inert gases), or combinations of both systems. In spite of these limiting conditions for microbial activity, either the length of storage period or the environment

Table 21.1 Spoilage and safety problems of postharvest products caused by microorganisms

Product	Problem caused	Causal microorganism
Citrus fruits	Blue and green mold	*Penicillium italicum, P. digitatum*
	Black spot	*Alternaria* spp.
	Brown rot	*Phytophthora* sp.
Pomefruits	Blue mold	*Penicillium expansum*
	Neofabraea rot	*Neofabraea*
	Bitter rot	*Colletotrichum gloeosporioides*
	Mucor rot	*Mucor piriformis*
Stonefruits	Brown rot	*Monilinia* spp.
	Rhizopus rot	*Rhizopus stolonifer*
	Gray mold rot	*Botrytis cinerea*
Berries	Antracnose	*Colletotrichum acutatum*
	Gray mold	*Botrytis cinerea*
	Leather rot	*Phytophthora* sp.
Tropical/subtropical fruits	Banana fruit rot	*Gloeosporium, Colletotrichum, Fusarium*
	Mango fruit rot	*Colletotrichum, Phytophthora, Alternaria*
	Avocado fruit rot	*Colletotrichum, Phytophthora*
	Pineapple rot	*Fusarium, Phytophthora, Penicillium,* yeasts
Vegetables (leafy, roots and tubers, germinated seeds)	Soft rot	*Pseudomonas, Pectobacterium,* yeasts
	Food-borne human pathogens	*E. coli, Salmonella enterica, Listeria monocytogenes*
Grains and nuts	Spoilage and mycotoxin production	*Aspergillus, Penicillium, Erotium, Fusarium, Bacillus* and other G+ bacteria

may result in produce spoilage, in multiplication of pre-existing human pathogenic bacteria or in preservation of toxins.

Thus, storage conditions should be preceded by good production practices and preharvest treatments, either to decrease populations of undesirable microorganisms or to sanitize latent infections. Treatments of disinfection by chemical/physical means or application of authorized fungicides are a conventional practice. However, consumer concerns and strong restrictions about pesticide residues in plant products have stimulated the development of less aggressive and toxic means of preservation of postharvest products. Biological control, using beneficial microorganisms, has emerged as an alternative or complement to conventional methods of control of biotic postharvest losses.

Fig. 21.1 Examples of biological control of postharvest fungal rot. Non-treated control (*right panels*) and treated with the biocontrol agent (*left panels*). Pears and apples were treated with the biocontrol agent *Pseudomonas fluorescens* EPS288. Peach and strawberries were treated with *Pantoea agglomerans* EPS125

21.2   Biological Control Agents for Postharvest Control

Biological control agents (BCAs) of postharvest diseases have evolved rapidly since the first report on control of stone fruit rot with *Bacillus subtilis* (Pusey and Wilson 1984) and pomefruit by a non-pathogenic strain of *Pseudomonas syringae* (Janisiewicz and Korsten 2002). The mechanisms of action among BCAs are very diverse, and several mechanisms may act simultaneously. These mechanisms include competition for nutrients and niches (CNN; competitive exclusion), antibiosis by means of antimicrobials and lytic enzymes, inhibitory volatile metabolites, pH decrease, parasitism, and induction of defence responses in the harvested plant product (Fig. 21.1).

Biological control of fungal rot has been extensively studied using strains obtained from the microbiota of wild plants or postharvest produce (Janisiewicz and Korsten 2002; Bonaterra et al. 2003; Prusky and Gullino 2010) . A list of relevant BCAs is described in Table 21.2. Several bacterial strains pertaining to *Pseudomonas* (*P. syringae, P. fluorescens, P. graminis*), *Pantoea* (*P. agglomerans, P. ananatis*), *Bacillus* (*B. subtilis, B. amyloliquefaciens*), and *Rahnella* (*R. aquatilis*), yeast strains, mostly of *Candida* (*C. famata, C. oleophila, C. saitoiana, C. sake*), *Kloeckera apiculata*, *Metschnickowia pulcherrima*, *Cryptococcus laurentii*, and *Rhodotorula glutinis*, have been reported as biological control agents of fungal rot. Fungal strains like *Aureobasidum pullulans* and *Muscodor albus* were also described as effective.

Biocontrol to prevent mycotoxin production in postharvest products has been also the object of development (Magan and Adler 2007). Generally exclusion of colonization and growth of the toxinogenic fungus and degradation of the mycotoxins are the main strategies. Most efforts have focused on control of toxinogenic species of *Aspergillus* (ochratoxins) in nuts by using atoxigenic strains, lactic acid bacteria and *Flavobacterium aurantiacum*. Similarly, *Fusarium* (fumonisins) in grains is controlled by *Bacillus amyloliquefaciens, P. fluorescens* and several yeasts. Biological control of mycotoxigenic fungi in fresh fruit is based in the same BCAs as used for fruit rot control .

Biological control of food-borne human pathogens in fruit and vegetables has also been reported (Janisiewicz et al. 1999). Several beneficial bacteria are effective in preventing or decreasing population levels of *E. coli, Salmonella enterica* and *Listeria monocytogenes* in ready-to-eat vegetables and sliced fruits, like strains of *P. syringae* (Leverentz et al. 2006) and of lactic acid bacteria (Trias et al. 2008). Also lytic bacteriophages have been reported as being effective (Sulackvelidze 2013).

The success of biological control of postharvest losses, diseases and mycotoxins has stimulated commercial activities to bring products to the market. Thus, several BCA strains are currently, or have been in the past (some of them are no longer manufactured), the active ingredients of commercial biofungicide products registered for postharvest control in various countries. Some examples are *Cryptococcus albidus* (Yieldplus), *B. subtilis* B426 (Avogreen), *B. subtilis* QST713 (Serenade), *Metschnickowia fructicola* 277 (Shemer), *P. syringae* ESC10 (Biosave), *C. oleophila* I-182 (Aspire), *C. oleophila* O (Nexy) or *Aureobasidium pullulans* DSM14941 (BoniProtect).

Table 21.2 Microorganism strains reported to control rotting, toxin production or food-borne human pathogenic bacteria in harvested products. (For more details the reader is referred to the books from Barkai-Golan and Paster 2008; Janisiewicz and Korsten 2002; Magan and Aldred 2007; Prusky and Gullino 2010)

Microbial group	Species	Strain	Pathogen controlled	Harvested product
Pseudomonas	*P. syringae*	ESC10, ESC11	Pi, Pd	Citrus
	P. fluorescens	EPS288	Pe	Pome
	P. graminis	CPA-7	Ec, Se, Lm	Sliced fruits
Pantoea	*P. agglomerans*	EPS125, CPA-2	Pe, Pd, Bc, Mf, Rs	Pome, citrus, stone fruits
	P. ananatis	BLBT1-08	Bc	Grapes
Bacillus	*B. subtilis*	B-3, CPA-8, B426	Mf, Pi, Pd	Citrus, stone, avocado
	B. amyloliquefaciens	QST713	Mu, Bc, Co,	Peach, strawberry
Rahnella	*R. aquatilis*	BNM523	Pe, Bc	Pome
Candida	*C. oleophila*	O, I-182	P, Bc	Pome, citrus, stone
	C. saitoana	–	Pd, Bc	Pome, citrus
	C. sake	CPA-2	Pi, Pd, Pe	Pome, citrus
	C. famata	–	Pd	Citrus
Pichia	*P. guilliermondii*	M8	Bc	Apple
	P. anomala	K	Pe	Pomefruits
Cryptococcus	*C. albidus*	–	Bc, Pe	Pomefruits
	C. laurentii	YY6	Bc	Raddish
	C. infirmo-miniatus	HRA5	Mo, Pe	Sweet cherry, pomefruits
Metsnikowia	*M. fructicola*	277	Pe, Pd, Bc, Rs, Mo, Aa, Fu	Pome, citrus, grapes
Kloeckera	*K. apiculata*	34-9	Bc	Citrusfruit
Rhodotorula	*R. glutinis*	–	Pe	Pear, cherry
Muscodor	*M albus*	–	Co, Bc	Pome, stone, grapes
Aureobasidium	*A. pullulans*	DSM14941, L1, L8	Al, Gl, Pn, Bc, Mo	Pome, stone, strawberry
Trichoderma	*T. harzianum, T. viride*	T32	Bc, Co, Pd	Strawberry, tomato, apple, citrus
Lactobacillus	*L. plantarum*	CM160	Ec, Se, Lm	Lettuce, apple
Leuconostoc	*L. mesenteroides*	CM135	Ec, Se, Lm	Lettuce, apple

Table 21.2 (continued)

Microbial group	Species	Strain	Pathogen controlled	Harvested product
Weissella	*W. cibaria*	TM128	Ec, Se, Lm	Lettuce, apple
Bacteriophages	–	Mixtures	Ec, Lm	Vegetables, melon

Pi P. italicum, Pd P. digitatum, Pe P. expansum, Ec E. coli, Se S. enterica, Lm L. monocytogenes, Bc B. cinerea, Mf M. fructicola, Rs R. stolonifer, Mo Monilinia, Aa A. nigricans, Fu Fusarium, Co Colletotrichum, Al Alternaria, Gl Gloeosporium, Pn Penicillium

21.3   Factors Affecting Efficacy of Postharvest Biocontrol

The efficacy and success of biological control of postharvest diseases is greatly dependent on intrinsic properties of the BCA, but also on pathogen and host material (fruit, nut, grain, vegetable), and on environmental factors.

In BCA systems, there is a dose-effect relationship, doses of 10^6–10^7 for yeast and of 10^7–10^8 for bacteria are efficient for control of postharvest fruit rot (Montesinos and Bonaterra 1996). The relative dose of pathogen and biocontrol agent is an important factor determining the efficacy and consistency of biological control of postharvest pathogens. Infectivity titration of the pathogen in the presence of varying concentrations of the biocontrol agent provide data that can be fitted to dose–response models to derive efficacy parameters like the median effective dose (ED_{50}) of pathogen and biocontrol agent. These parameters can be useful to perform comparisons between strains, pathogens and hosts. Using this approach pathogen aggressiveness on the host was reported as an influencing factor in the efficiency of postharvest biocontrol on several fresh fruits (Francés et al. 2006).

The efficacy of control of postharvest problems depends on the time at which the treatment is applied relative to that of the pathogen arrival, and therefore depends on the strategy used for treatment, which can be preventive (before the infection of the pathogen), or curative (after the infection of the pathogen). Major progress in preventive treatments has been made using microorganisms effective against wound infecting pathogens, but considerably less advance has been reported in control of already developed, or in latent, infections (e.g. infections caused by *Monilia* or *Colletotrichum*).

Storage conditions can have a triple effect on the plant produce ecosystem. For example, reduced oxygen and increased CO_2 atmospheres greatly slow down fruit and seed respiration and pathogen activity, but may have a negative effect on the BCA in strict aerobes like *Bacillus* and *Pseudomonas* or do not affect facultative aerobes like yeasts. Low storage temperatures, close to 0°C, sufficiently slow down the activity of the BCA, pathogen and plant produce. However, at certain moments both the pathogen and the BCA may be active due to favourable environmental conditions. For example, during transition from field harvest to industrial cold-storage or when cold storage ends for commercial delivery (Fig. 21.2).

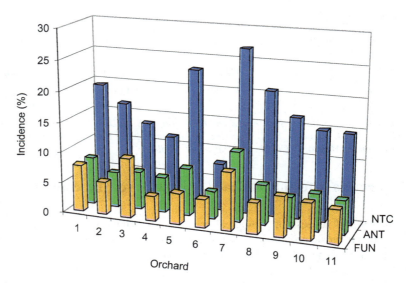

Fig. 21.2 Incidence of blue mold rot on *Golden* apple from eleven commercial orchards upon wounding, fungicide or biological control treatment with *Pseudomonas fluorescens* EPS288, *Penicillium expansum* inoculation and subsequent storage under Ultra Low Oxygen-cold storage (0.5–1.0°, 0–1.5 % CO_2, 1.25 % O_2) during 5 months, and a 7-day ripening period at 20°. Treatments consisted of either the chemical fungicide imazalil (*FUN*), the biocontrol agent *Pseudomonas fluorescens* EPS288 (*ANT*) and non-treated controls (*NTC*)

21.4 Production, Formulation and Application of Postharvest Microbial Pesticides

For the commercial development of a postharvest microbial pesticide, cells have to be grown, preserved, and formulated for storage and delivery (Montesinos and Bonaterra 2009). Methods used for industrial scale-up are solid or liquid-phase fermentations, but depend on the nature of the microorganism (bacteria, fungi, yeast, or bacteriophages) (Boytsensko et al. 1998).

Bacteria and yeast are generally grown by liquid fermentation in bioreactors, but some fungi are fermented by solid-state procedures. Subsequently, cells are harvested by centrifugation, either from liquid cultures or solid phase, homogenized and cleaned to obtain concentrated cell/spore suspensions that often contain some supernatant metabolites of interest (e.g. antimicrobials, lytic enzymes). Bacteriophages are produced in a double stage process in which the first step consists of preparing the bacterial target (e.g. *E. coli*), that is later used as the host to multiply the lytic bacteriophage.

In order to increase the shelf-life and ecological fitness of the BCA, several procedures have to be carried out for stabilizing the viability of the cells. Generally, formulations of commercial microbial pesticides consist of liquid-phase suspensions that are maintained under refrigeration, in a frozen state, or as a dried product.

Dehydration permits optimum storage conditions, handling, and distribution, but the associated processes are costly, especially lyophilisation, and the prefered method is spray drying or fluidized bed drying. However, spray drying generally results in a high loss of viability due to the thermal treatment. The final formulations are composed of an active ingredient (cells or spores and sometimes culture components), carriers or inert materials used to support cells, and adjuvants. Products can be stable for several months or even years.

Cell death of the biocontrol agent can occur after dehydration and delivery due to the sharp change from the optimal laboratory culture conditions to the stressing dehydration process and the growth-limiting fruit surface. However, stress tolerance can be induced by cultivation under suboptimal conditions by means of osmo-adaptation. This procedure has been used to improve drought stress tolerance, epiphytic survival and biocontrol efficacy of the apple blue mold biocontrol agent *Pantoea agglomerans* EPS125 (Bonaterra et al. 2005). Another strategy is the amendment of the formulation with specific nutrients that cannot be used by, or are toxic for, the pathogen and can be used or do not affect the biocontrol agent (Janisiewicz 1994).

The formulated product can be applied during preharvest (field spray) or before/after storage (spraying or drenching).

21.5   Limitations and Future Trends

The management of postharvest losses tends to use low impact (soft) strategies that often are less efficient than synthetic antimicrobial products (e.g. fungicides). Therefore, a multiple barrier strategy is necessary to optimize levels of control. Soft chemicals acting as barriers or affecting directly the spoilage microorganism (e.g. bicarbonates, silicates), surface disinfection compounds (e.g. ozone, chlorine, electrolyzed water), physical methods (e.g. hot water, microwaves, UV pulsed light), and response defence inducers on the host (chitosans, acibenzolar, salicylic acid) are among the systems used.

Biological control forms part of the list of these technologies. However, many of these systems are not compatible with the simultaneous use of biocontrol agents because they can inhibit its colonization, growth and metabolism. Fortunately, others are compatible, either as previous (e.g. hot water, disinfection), or as simultaneous treatments (defence inducers).

A limitation of certain BCAs of postharvest diseases is related to the biosafety of the antagonistic microorganism (see also Chap. 32 and Chap. 33). This aspect has greatly limited authorizations for commercial use in certain cases (e.g. *Burkholderia cepacia*) due to reports on clinical outbreaks associated to this species. Another issue is the acceptance and safety of improved biocontrol strains using recombinant DNA technology.

References

Badosa E, Trias R, Parés D et al (2008) Microbiological quality of fresh fruit and vegetable products in Catalonia (Spain) using plate counting normalized methods and QPCR. J Sci Food Agric 88:605–611

Barkai-Golan R, Paster N (2008) Mycotoxins in fruits and vegetables. Elsevier Inc, London

Bonaterra A, Mari M, Casalini L et al (2003) Biological control of *Monilinia laxa* and *Rhizopus stolonifer* in postharvest of stone fruit by *Pantoea agglomerans* EPS125 and putative mechanisms of antagonism. Int J Food Microbiol 84:93–104

Bonaterra A, Camps J, Montesinos E (2005) Osmotically induced trehalose and glycine betaine accumulation improves tolerance to desiccation, survival and efficacy of the postharvest biocontrol agent *Pantoea agglomerans* EPS125. FEMS Microbiol Lett 250:7–15

Boyetchko S, Pedersen E, Punja Z et al (1998) Formulations of biopesticides. In: Hall FR, Barry JW (eds) Methods in biotechnology. Humana Press, Totowa, pp 487–508

Francés J, Bonaterra A, Moreno MC et al (2006) Pathogen aggressiveness and postharvest biocontrol efficiency in *Pantoea agglomerans*. Postharvest Biol Technol 39:299–307

Janisiewicz WJ (1994) Enhancement of biocontrol of blue mold with the nutrient analog 2-deoxy-D-glucose on apples and pears. Appl Environ Microbiol 60:2671–2676

Janisiewicz WJ, Korsten L (2002) Biological control of postharvest diseases of fruits. Annu Rev Phytopathol 40:411–441

Janisiewicz WJ, Conway WS, Leverentz B (1999) Biological control of postharvest decays of apple can prevent growth of *Escherichia coli* O157:H7 in apple wounds. J Food Prot 12:1372–1375

Leff JW, Fierer N (2013) Bacterial communities associated with the surfaces of fresh fruits and vegetables. PLoS One 8(3):e59310

Leverentz B, Conway WS, Janisiewicz W et al. (2006) Biocontrol of the foodborne pathogens *Listeria monocytogenes* and *Salmonella enterica* serovar Poona on fresh-cut apples with naturally occurring bacteria and yeast antagonists. Appl Environ Microbiol 72:1135–1140

Lin C, Fernando SY, Wei C (1996) Occurrence of *Listeria monocytogenes*, *Salmonella* spp., *Escherichia coli* and *E. coli* O157:H7 in vegetable salads. Food Control 7(3):135–140

Magan N, Aldred D (2007) Post-harvest control strategies: minimizing mycotoxins in the food chain. Int J Food Microbiol 119:131–139

Montesinos E, Bonaterra A (1996) Dose–response models in biological control of plant pathogens. An empirical verification. Phytopathology 86:464–472

Montesinos E, Bonaterra A (2009) Microbial pesticides. In: Schaechter M (ed) Encyclopedia of microbiology, Elsevier Inc, Oxford pp 110–120

Prusky D, Gullino ML (eds) (2010) Postharvest pathology. Plant pathology in the 21st century, Vol 2, Springer, New York

Pusey PL, Wilson CL (1984) Postharvest biological control of stone fruit brown rot by *Bacillus subtilis*. Plant Dis 68:753–756

Setati ME, Jacobson D, Andong UC et al (2012) The vineyard yeast microbiome, a mixed model microbial map. PLoS One 7(12):e52609

Snowdon AL (1990) A colour atlas of post-harvest diseases and disorders of fruits and vegetables. General introduction and fruits. Wolfe Scientific, London

Sulakvelidze A (2013) Using lytic bacteriophages to eliminate or significantly reduce contamination of food by foodborne bacterial pathogens. J Sci Food Agric 93:3137–3146

Trias R, Baeras L, Badosa E et al (2008) Bioprotection of golden delicious apple and Iceberg lettuce against foodborne bacterial pathogens by lactic acid bacteria. Int J Food Microbiol 123:50–60

Part IV
Plant Growth Promotion by Microbes

Chapter 22
The Nitrogen Cycle

Martine A. R. Kox and Mike S. M. Jetten

Abstract This chapter focuses on the nitrogen (N) cycle, a complex network of mainly microbial transformations in which various nitrogen compounds are inter-converted. Both microorganisms and plants absorb N from and excrete N into the environment. First, N assimilation is addressed (22.1), after which N transformations by microorganisms are described (22.2). In paragraph 22.3 both plant and microbial N cycling are discussed at the ecosystem level, followed by paragraph 22.4, where the use of N by humans and the consequences for the N cycle are reviewed. Finally, in 22.5 the conclusions and outlook are presented.

22.1 N Metabolism of Plants

After the discovery in the early 1900s that N compounds could increase crop pro-ductivity, this topic was intensively studied. Generally N is a limiting nutrient for plant production and mineralization, hence N availability is an important controlling factor for ecosystem processes. The N cycle is also tightly coupled to the carbon (C) cycle. Access to N dictates both the photosynthetic activity, which is the main C input in plants, and the production of protein (Larcher 2001). N compounds are incorporated into plant material when (1) the N compound is available, when (2) the plant has adequate uptake systems and (3) when all assimilatory complexes are present and active.

N Sources In terrestrial ecosystems, the soil acts as a nutrient reserve for plants, where 98 % of the mineral nutrient supply is bound in humus, organic matter and insoluble compounds and only less than 0.2 % is dissolved in water. The soil's N

M. A. R. Kox (✉) · M. S. M. Jetten
Department of Microbiology, Institute of Water and Wetland Research,
Faculty of Science, Radboud University Nijmegen, Heijendaalseweg 135,
6525 AJ Nijmegen, The Netherlands
Tel.: +31 24 365 2569
e-mail: m.kox@science.ru.nl

M. S. M. Jetten
Tel.: +31 24 365 2941
e-mail: m.jetten@science.ru.nl

© Springer International Publishing Switzerland 2015
B. Lugtenberg (ed.), *Principles of Plant-Microbe Interactions*,
DOI 10.1007/978-3-319-08575-3_22

content depends mostly on microbial mineralization of organic matter into inorganic N such as ammonium (NH_4^+) or nitrate (NO_3^-). N availability for plants is affected by the root structure and density, Radial Oxygen Loss (ROL), root exudates and symbioses with microorganisms (Jackson et al. 2008). In anoxic environments such as wetlands, ROL is of special interest as it supplies oxygen (O_2) to microorganisms (Lamers et al. 2012).

Uptake of N Terrestrial plants control mineral uptake via their roots. Furthermore N uptake and assimilation are always dependent on the availability of other nutrients, especially phosphorus. Although both inorganic and organic N (i.e. amino acids) can be taken up by plants, most studies have focused on inorganic N sources. Energetic costs of N uptake and assimilation are highest for NO_3^-, followed by NH_4^+ and lowest for amino acids (Lambers et al. 1998). NO_3^- first has to be reduced to NH_4^+ before it can be assimilated. Terrestrial plants adapted to low pH and redox potential preferentially take up NH_4^+. Subsequently, the release of H^+ acidifies the rhizosphere, affecting the rhizosphere microbiome and causing reduced nitrification rates. Plants adapted to soils with a higher pH use preferentially NO_3^- as it causes less soil acidification compared to NH_4^+ (Lamers et al. 2012; Larcher 2001; Britto and Kronzucker 2002).

Terrestrial plants have both high and low affinity transport systems (HATS and LATS) for the active uptake of NO_3^- and NH_4^+. HATS take up inorganic N between 1 μM and 1 mM. LATS are only expressed when inorganic N is above 0.5 mM. Only for NO_3^- uptake, inducible transporters are known. Uptake of NO_3^- induces HATS, followed by positive feedback on the putative transporters, causing an increased uptake of NO_3^-. Subsequently, in the cytoplasm NO_3^- is first reduced to nitrite (NO_2^-) by nitrate reductase (Nas), followed by nitrite reductase (Nir) into NH_4^+. Uptake of NH_4^+ directly from the environment is mediated by ammonium transporters (AmtB) that are located in the cell membrane. Regulation of NO_3^- absorption via HATS depends on the N status of the whole plant, whereas NH_4^+ uptake via HATS is locally regulated in the roots (Jackson et al 2008). Passive uptake of inorganic N is mediated by LATS, which are energetically costly due to poor regulation.

The uptake and assimilation of N is not very different for (heterotrophic) microorganisms. AmtB's are responsible for the uptake of ammonium, NO_3^- uptake is mediated by ABC transport proteins or exchanged by NarK antiporters. NO_2^- uptake is catalyzed by FocA.

Assimilation and Incorporation of N The different steps in N assimilation in plant cells are comparable to assimilation in microorganisms. When NO_3^- is transported to the cytoplasm, it is reduced to NO_2^- by an assimilatory Nas. Subsequently Nir at the expense of ferrodoxin or NAD(P)H reduces NO_2^- to NH_4^+. NH_4^+ is used for the production of glutamate for which several systems exist. Glutamine synthetase forms glutamine from glutamate and NH_4^+. Next, glutamine 2-oxoglutarate aminotransferase (GOGAT; glutamate synthase), catalyzes the synthesis of 2 glutamate from glutamine and 2-oxoglutarate. Furthermore, glutamate dehydrogenase, can convert glutamate back into 2-oxoglutarate and NH_4^+ during N remobilization. Subsequently, glutamate and glutamine are used to produce various amino acids (mostly aspartate and aspargine) via transamination. Via negative feedback glutamine concentrations regulate the N uptake by HATS. Internal transport of N occurs primarily as amino acids via xylem and phloem, and can be stored in young shoots, leaves, buds, seeds and storage organs of plants, from where it can be remobilized if needed (Jackson et al. 2008).

N Requirements N requirements differ greatly between plant species. Severe insufficient N supply often results in dwarf growth. Ultimately this leads to sclerosis, where plants have small cells and thickened cell walls. Excessive amounts of N, especially of NH_4^+, can become toxic for plants. Most plausible processes involved in NH_4^+ toxicity are energetically costly membrane effluxes, reduced photosynthesis, displacement of cations and interference of hormone metabolism. Long term excessive N uptake will eventually lead to lowered resistance against abiotic and biotic stressors and delayed reproduction of the plant (Britto and Kronzucker 2002).

22.2 Microbial N Cycle

After reviewing the N processes of plants, the N transformations of microorganisms will be discussed in the next paragraphs.

N₂ Fixation Nitrogen fixation is the only known biological process where dinitrogen gas (N_2) is transformed into NH_4^+. This energetically costly process (16 ATP per molecule of N_2 fixed) is exclusively performed by N fixing microorganisms (diazotrophs). The enzyme nitrogenase (nif) catalyzes this reaction and becomes inactive when exposed to O_2. Diazotrophs occur free living and in symbiosis. The best studied diazotrophic symbiosis is that between legumes and *Rhizobia* sp., but it is also known for other plants, mosses and between microorganisms (Chap. 21). Even though direct evidence is missing, it seems likely that bryophytes were the first land plants that started to live in symbiosis with diazotrophs.

Nitrification The subsequent oxidation of ammonium via NO_2^- to NO_3^- is being studied since the nineteenth century, as this process causes significant losses of fixed N that are no longer available for crops. Early studies focused on the metabolism of *Nitrosomonas* spp. that thrive at high NH_4^+ concentrations. The key enzyme ammonium monooxygenase (AmoA) produces hydroxylamine (NH_2OH) and is present in large amounts in their membrane systems. NH_2OH is converted into NO_2^- by an hydroxylamine oxidoreductase (Hao). Further studies indicated that *Nitrosospira* spp. might be more relevant at lower NH_4^+ concentrations. NH_4^+ oxidizers mostly live in close proximity to NO_2^- oxidizing bacteria (NOB) that convert the toxic NO_2^- rapidly into NO_3^- by a nitrite oxidoreductase (NxrAB) system. *Nitrospira* and *Nitrospina* spp. thrive at low NO_2^- environments like freshwater and marine systems, respectively. In contrast to the expectation, only low numbers of bacterial nitrifiers were observed in marine surveys. About 20 % of the cells were of Archaeal origin, but their metabolism remained unknown until metagenomic inventories showed the presence of *amoA* genes on fosmids that also contained 16S rRNA gene copies of

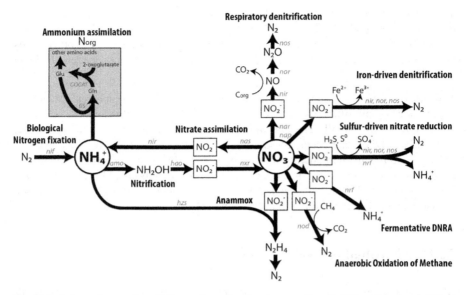

Fig. 22.1 Schematic overview of N transformation from ammonium (NH_4^+) and/or nitrate (NO_3^-) perspective. Corresponding genes are printed in italic (*grey*), where *nif* nitrogenase, *amo* ammonium monooxygenase, *hao* hydroxylamine oxidoreductase, *nxr* nitrite oxidoreductase, *nar* respiratory nitrate reductase, *nap* periplasmic nitrate reductase, *nir* nitrite reductase, *nor* nitric oxide reductase, *nos* nitrous oxide reductase, *nrf* multiheme nitrite reductase, *nod* putative NO dismutase and *hzs* hydrazine synthase. Enzymes *nar* and *nap* are involved in all conversions of NO_3^- to NO_2^-. Glutamine synthetase (*GS*) and glutamine 2-oxoglutarate aminotransferase (*GOGAT*) genes, both involved in assimilation of NH_4^+. Modified and extended with permission after Burgin and Hamilton (2007)

these Archaea. The isolation of the first archaeal NH_4^+ oxidizer *Nitrosopumilis maritimus* from a marine aquarium showed that Archaeal Ammonium-Oxidizers (AOA) have a very high affinity for NH_4^+ probably reflecting their low NH_4^+ habitats.

Denitrification Denitrification is the oldest known process of the N cycle. In this process NO and N_2O are produced from NO_3^- and NO_2^- and N_2 is the final product. Most denitrifiers are facultative anaerobes, but some species may continue to denitrify in the presence of O_2, a process known as aerobic denitrification. The denitrification trait is widespread among Bacteria, Archaea and can even be found in some Eukaryotes. The reduction of oxidized N species (NO_x) is catalyzed by metalloenzymes that contain molybdenum, iron or copper. The electrons needed for the reduction are derived from oxidation of inorganic or organic sources Fig. 22.1).

Denitrification is probably the most intensively studied process of the N cycle, because it may be responsible for more than 20 % of fixed N losses in agriculture. Initial studies focused on the identification of intermediates and later emphasis was on the regulation and enzymology of the process. Many organisms possess a nitrate reductase either located at the cytoplasmic membrane (NarGH) or in the periplasm (NapAB), that produces the important intermediate NO_2^-, which is subsequently converted into NO by either a nirK or a nirS nitrite reductase. Various membrane-bound nitric oxide reductases exist. Finally N_2O is converted by a nitrous oxide reductase (NosZ).

In laboratory cultures, one single species may convert NO_3^- all the way to N_2, in nature the reactions are probably divided between different species, and co-cultures are rather rule than exception. In natural habitats denitrifiers will have to compete with dissimilatory nitrate reduction to ammonia (DNRA) and anammox for NO_3^- and NO_2^-, and the outcome of the competition may be dependent on the C/N ratio (Fig. 22.2). For a longtime it was believed that methane (CH_4) could not be used by denitrifiers, because its activation would require O_2. However, in 2006 a co-culture of Bacteria and Archaea was enriched that could perform CH_4 dependent denitrification (Raghoebarsing et al. 2006). By increasing the NO_2^- concentration, the Archaea disappeared from the community and Bacteria named *Methylomirabilis oxyfera* became dominant. *M. oxyfera* was shown to have a peculiar denitrification pathway involving the dismutation of NO into O_2 and N_2 by a putative NO dismutase (Ettwig et al. 2010). Recently it has been established that Archaea can convert NO_3^- to NO_2^- and maybe even further to NH_4^+ at the expense of CH_4 (Haroon et al. 2013).

DNRA DNRA is one of the least studies aspects of the N cycle. Many microorganisms are able to perform the DNRA reaction especially at low NO_3^- and high C concentrations. First NO_3^- is converted to NO_2^- by a nitrate reductase. In the second step a multiheme nitrite reductase (nrfA) converts the NO_2^- directly to NH_4^+. Electrons needed for reduction are derived by fermentation of organic compounds or by sulfide oxidation. DNRA is a difficult pathway to detect and needs sophisticated stable isotope experiments. An elegant example is the study of Lam et al. (2009) that investigated the N cycle pathways in the Chilean OMZ (Oxygen Minimum Zone). By applying a complementary array of methods, they were able to show that DNRA may contribute up to 40 % of the N flux in this OMZ.

Anaerobic Ammonium Oxidation Only in 1995 the first publication on the disappearance of NH_4^+ from an anoxic denitrifying pilot plant was reported. After complaints by the citizens of Delft that the Gist Brocades pilot plants produced too much hydrogen sulfide, the waste water engineers added copious amounts of calcium nitrate to prevent sulfate reduction. Inadvertently they created favourable conditions for anaerobic ammonium oxidizing (anammox) bacteria to proliferate. Biomass of the pilot plant was subsequently used to start new more defined enrichment cultures, first as fluidized bed reactors, later as sequencing batch reactors, yielding enough anammox biomass to perform the necessary experiments. Inhibitors studies with antibiotics showed that the process was bacterial, while[15]N stable isotopes studies indicated the production of the rocket fuel hydrazine (N_2H_4). As the enrichments yielded 70–90 % anammox dominance, physical purification methods based on gradient centrifugation had to be applied. This gave sufficient purified cells to do crucial[15]N and ^{14}C experiments showing the autotrophic nature of the anammox bacteria. From the purified cells, the 16S rRNA gene could be amplified, and the anammox bacteria were shown to belong to the phylum of the Planctomycetes. Electron microscopic analysis showed that anammox bacteria have a unique cell plan

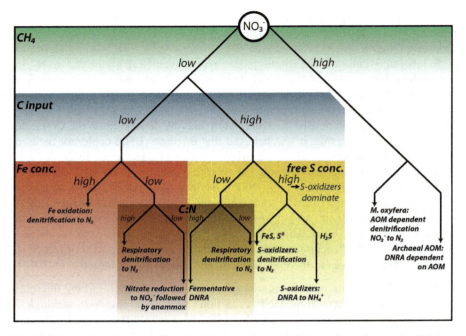

Fig. 22.2 Branching diagram with a simplified overview of NO_3^- transformations under different conditions, indicated by the different colors. Depicted in *green* is CH_4 availability, *blue* is the carbon input, *red* represents iron (*Fe*) concentrations, in *yellow* the free sulfide concentrations (H_2S, S^0, *FeS*), finally in different *brown* shades are the C/N ratio under the different Fe or free sulfide concentrations. (*Conc.* the concentration; *DNRA* Dissimilatory Nitrate Reduction to Ammonium; *AOM* Anaerobic Oxidation of Methane). Adapted and extended with permission from Burgin and Hamilton (2007)

with a specialized compartment harboring the enzymes responsible for the anammox reactions (Van Niftrik and Jetten 2012). Analysis of the fatty acids of the cells and organelle indicated that anammox bacteria have both ether and ester lipids of concatenated cyclobutane rings that from a kind of staircase structure, hence their name ladderane lipids. These are unique and can be used as specific anammox biomarkers.

After the availability of suitable diagnostic tools, several expeditions to OMZs were organized. Indeed it could be documented that in those OMZs, anammox bacteria were present and active. Taken together they could account for half of the loss of fixed N from the systems, making them important players in the global N cycle. Recently it was also shown that anammox can contribute significantly to the N-loss in terrestrial ecosystems such as wetlands and river sediments.

After the genome of the first anammox bacterium was resolved, the molecular mechanisms were elucidated (Kartal et al. 2011). The crucial intermediates were NO and N_2H_4 and a unique enzyme complex hydrazine synthase was identified. Application of anammox bacteria together with partial nitrification may result in more sustainable waste water treatment systems saving on O_2 and electricity usage, methanol consumption, and ecological footprint (Kartal et al. 2010). Based on these advantages, already more than 20 full scale anammox plants have been build worldwide, and many more are commissioned.

22.3 N Cycle in Ecosystems

After the introduction of the N cycle processes, the following section will focus on the cooperation and competition for N compounds between plants and microorganisms. The competition is controlled by metabolic limitations and environmental conditions.

Plant vs Plant N Competition Plants have evolved diverse adaptations to cope with nutrient limitations. Resource depletion has been hypothesized as the strategy in plant-plant competition for N. By taking up more N compounds than directly necessary, N can become rapidly depleted in the environment and thus limiting for competitors. Competition for N resources between plants also occurs indirectly. Microorganisms flourish in the rhizosphere due to high litter production by roots. Plants modulate their rhizospheric microbiome (see Chap. 43) by attracting certain species, which have the potential to enhance N uptake for the plant.

Plant vs Microorganisms The trade-off between plants and microorganisms in competition for N, is a much debated topic. In terrestrial ecosystems, the classical paradigm stated that microorganisms are stronger competitors for N than plants, hence plants would only use the microbial N left-overs. In the late 1990s, a new hypothesis was developed that put less emphasis on mineralization and underlined depolymerization of the complex N compounds present in soil by microorganisms, as the key process and bottleneck in N cycling (Jackson et al. 2008).

Short term experiments with ^{15}N additions showed that microorganisms take up organic and inorganic N faster than roots. Microorganisms have high substrate affinities, low volume to surface ratio's and fast turnover rates compared to plants and therefore are stronger competitors. In the long run plants assimilated most of the ^{15}N due to the gradual release of ^{15}N that was first mineralized by the microorganisms. It is this temporal difference that determines the competition for N between plants and microbes in the end (Jackson et al. 2008; Kuzyakov and Xu 2013). The most direct competition between plants and microbes occurs at the level of the available inorganic N. Nitrifiers have to compete for available NH_4^+ with all NH_4^+ assimilating plants and microorganisms, whereas the processes of denitrification and DNRA compete for NO_3^- with plant and microbial NO_3^- assimilation.

Competition for NH_4^+ Fertilization experiments with inorganic N have either shown about equal N uptake rates for both plants and microorganisms, hence both were simultaneously limited in N. In a NH_4^+ fertilization experiment plant removal resulted in increased nitrification rates. This indicated that plants and autotrophic nitrifiers compete for NH_4^+, with plants being the stronger competitors (Kaye and Hart 1997).

Competition for NO_3^- In rice soils, a study of NO_3^- assimilation by plant and microorganisms showed that fertilization with labeled NO_3^- always resulted in more $^{15}NO_3^-$ ending up in the microbial than in plant N pools. As rice grows in flooded soils under anaerobic conditions, ROL will release oxygen into the rhizosphere. The supplied O_2 stimulates nitrification, and subsequently NO_3^- can enter the anoxic zone where denitrification and anammox cause fixed N to be lost to the atmosphere. By blocking transport of O_2 and N_2 via the aerenchym through clipping of rice plants below the water surface, nitrification rates rapidly decreased and in the longer term also denitrification rates decreased (Arth et al. 1998; Matheson et al. 2002).

O_2 is an important regulator of NO_3^- partitioning between denitrification and DNRA. It was shown that microcosms planted with *Glyceria declinata* experienced more O_2 intrusion and higher redox potentials compared to unplanted microcosms, leading to higher denitrification than DNRA rates (Matheson et al. 2002). In a study with young-barley, short term rewetting increased nitrification rates, whereas long term rewetting also stimulated denitrification (Højberg et al. 1996). Since the role of denitrification in N loss might be overestimated (Burgin and Hamilton 2007), future studies in anoxic systems should also analyze the role of anammox and DNRA.

Microorganism vs Microorganisms Also microorganisms among each other have to interact and/or compete for NH_4^+ and NO_x. Nitrifiers (AOA, AOB, NOB) live in close proximity to ensure rapid conversion of NH_4^+ to NO_3^- without intermediate NO_2^- accumulation. Availability of O_2 is an important factor determining whether anammox or nitrifiers can convert NH_4^+ first, and whether NOB or anammox can convert NO_2^- subsequently. In systems with fluctuating O_2 and NH_4^+ concentrations like wastewater systems, sediments or OMZs, consortia of AOA, AOB, NOB and anammox have been found to be active (Lam et al. 2009). Low ammonium and O_2 concentrations seem to favor AOA. Laboratory studies under O_2 limitation showed that AOA can thrive very well in co-cultures with anammox bacteria. Several guilds may have to compete for NO_3^- and the outcome of the competition is mostly determined by the quality and quantity of the electron donor. Burgin and Hamilton (2007) made an elegant scheme to predict the occurrence of the various NO_3^- reducing processes (Fig. 22.2). At low organic carbon, denitrification and anammox will compete, and above a C/N ratio of 2, denitrification may be favored. Under high energy or carbon input DNRA may be the most important process, while denitrification is dominant at low C/N ratios in the absence of NH_4^+. At high CH_4 concentration AOM Archaea may play a significant role in nitrate reduction and DNRA, and *M. oxyfera* in NO_2^- conversion to N_2.

22.4 Anthropogenic N Use and Changes in N Cycle

The global cycling of N has doubled over the last century, starting with the application of the Haber-Bosch process ($N_2 + 3H_2 \rightarrow NH_3$) in 1913. At the expense of fossil fuel, artificial fertilizer could be produced, and thus increased crop productivity and harvest. Though, what was not realized at the time is that the use of (excess) N fertilizer has severe ecological impacts.

Studies on N fertilizer use have primarily focused on loss of N fertilizers into other (pristine) ecosystems. Only part of the N that is applied as fertilizer is taken up by microorganisms and plants and later on removed via harvest of the crops. The remainder will enter the N cycle of the ecosystem. According to Burgin and Hamilton (2007) the most desirable way to reduce high N levels in ecosystems is via permanent removal by denitrification (or anammox), because other N transformations may result in even more harmful N-compounds.

The main processes studied with respect to N loss from ecosystems are NO_3^- leaching, ammonia volatilization and loss as NO_x or N_2 gas (Cameron et al. 2013). NO_3^- leaching depends on nitrate loading of the soil and the levels of drainage that occur. NO_3^- is a large problem for water quality and can affect human health. Via groundwater the leached NO_3^- enters rivers and lakes, where it might stimulate algal blooms and cause biodiversity loss. NO_3^- losses from fertilizer-use can be reduced by using adequate and efficient fertilizer levels to prevent N excess, by optimizing plant N uptake to avoid N losses and if all else fails nitrification inhibitors can be applied. Fertilizer-use is nowadays highly restricted and managed so that fertilizers are used in an efficient manner, with amounts that are matched to the rate of plant growth.

Ammonia (NH_3) volatilization is especially a problem in areas surrounding intensive animal farms. Most important sources are animal urine and feces, but also N fertilizers contribute to NH_3 volatilization. Once volatilized, NH_3 deposits cause acidification and eutrophication. NH_3 volatilization can be best reduced by applying fertilizers beneath the soil surface or just before rain, and by reducing intensive animal farming.

Ultimately, N can be lost to the atmosphere via nitrification or denitrification in the form of NO, N_2O or N_2. In particular, NO and N_2O (NO_x) form a serious problem since they deplete the ozone layer and contribute substantially to climate change. The global warming potential of N_2O is 298 times that of CO_2. To reduce NO_x formation, nitrification inhibitors combined with optimized fertilizer application may diminish nitrification. Methods to reduce denitrification include changing the soil physiochemical parameters (i.e. increasing pH by applying lime, or increase aeration of the soil, as described in Cameron et al. 2013).

22.5 Conclusions and Outlook

N cycling has been studied intensively for over decades. Although the knowledge on the N cycle has increased, the role of anammox, the contribution of AOM dependent conversions of nitrogen and the interactions in the N cycle deserve more attention. Ultimately, improving our understanding of the N cycle will help to retain and restore balances in ecosystem N cycles which have been affected by anthropogenic activities.

Acknowledgements   We would like to thank our co-workers and collaborators and granting agencies for their continuous support (ERC 232937, ERC 339880, Spinozapremie 2012 and OCW-NWO Gravitation Grant SIAM 024.002.002).

References

Arth I, Frenzel P, Conrad R (1998) Denitrification coupleD to nitrification in the rhizosphere of rice. Soil Biol Biochem 30:509–515

Britto DT, Kronzucker HJ (2002) Ammonium toxicity in higher plants: a critical review. J Plant Physiol 159:567–584

Burgin AJ, Hamilton SK (2007) Have we overemphasized the role of denitrification in aquatic ecosystems? A review of nitrate removal pathways. Front Ecol Environ 5:89–96

Cameron KC, Di HJ, Moir JL (2013) Nitrogen losses from the soil/plant system: a review. Ann Appl Biol 162:145–173

Ettwig KF, Butler MK, Le Paslier D et al (2010) Nitrite-driven anaerobic methane oxidation by oxygenic bacteria. Nature 464:543–548

Haroon MF, Hu S, Shi Y et al (2013) Anaerobic oxidation of methane coupled to nitrate reduction in a novel archaeal lineage. Nature 500:567–570

Højberg O, Binnerup S, Sørensen J (1996) Potential rates of ammonium oxidation, nitrite oxidation, nitrate reduction and denitrification in the young barley rhizosphere. Soil Biol Biochem 28:47–54

Jackson LE, Burger M, Cavagnaro TR (2008) Roots, nitrogen transformations, and ecosystem services. Annu Rev Plant Biol 59:341–363

Kartal B, Kuenen JG, van Loosdrecht MCM (2010) Engineering. Sewage treatment with anammox. Science 328:702–703

Kartal B, Maalcke WJ, De Almeida NM et al (2011) Molecular mechanism of anaerobic ammonium oxidation. Nature 479:127–130

Kaye J, Hart S (1997) Competition for nitrogen between plants and soil microorganisms. Trends Ecol Evol 5347:139–141

Kuzyakov Y, Xu X (2013) Competition between roots and microorganisms for nitrogen: mechanisms and ecological relevance. New Phytol 198:656–669

Lam P, Lavik G, Jensen MM et al (2009) Revising the nitrogen cycle in the Peruvian oxygen minimum zone. Proc Natl Acad Sci U S A 106:4752–4757

Lambers H, Chapin III FS, Pons TL (1998) Plant physiological ecology, 1st edn. Larcher publisher Springer Verlag, Berlin

Lamers LPM, Van Diggelen JMH, Op den Camp HJM et al (2012) Microbial transformations of nitrogen, sulfur, and iron dictate vegetation composition in wetlands: a review. Front Microbiol 3:156

Larcher W (2001) Physiological plant ecology, 4th edn. Lambers publisher Springer Science + Business media, New York

Matheson F, Nguyen M, Cooper A (2002) Fate of 15 N-nitrate in unplanted, planted and harvested riparian wetland soil microcosms. Elsevier Ecol Eng 19:249–264

Van Niftrik L, Jetten MSM (2012) Anaerobic ammonium-oxidizing bacteria: unique microorganisms with exceptional properties. Microbiol Mol Biol Rev 76:585–596

Raghoebarsing AA, Pol A, Van de Pas-Schoonen KT et al (2006) A microbial consortium couples anaerobic methane oxidation to denitrification. Nature 440:918–921

Chapter 23
Biological Nitrogen Fixation

Frans J. de Bruijn

Abstract Biological nitrogen fixation (BNF) is the process of the reduction of dinitrogen from the air to ammonia carried out by a large number of species of free-living and symbiotic microbes called diazotrophs. BNF presents an inexpensive and environmentally sound, sustainable approach to crop production and constitutes one of the most important Plant Growth Promotion (PGP) scenarios. Here I will summarize various aspects of BNF, including the dinitrogen reduction catalysed reaction carried out by "nitrogenase" and the enzymes/genes involved and their regulation, the inherent "oxygen paradox", the identification of diazotrophs, sustainable agricultural uses of BNF, symbiotic plant-diazotroph interactions and endophytic diazotrophs, data from the field, and future prospects in BNF.

23.1 Introduction

Fixed nitrogen is a limiting nutrient in most environments, with the main reserve of nitrogen in the biosphere being molecular di-nitrogen from the atmosphere, which is an inert gas with a triple bond, that is energetically unfavourable to break. Nitrogen availability is limiting for plant growth and has long been overcome through the application of synthetic nitrogen-rich fertilizer. Using increasing amounts of fertilizers the yield of crop plants such as cereals has been greatly augmented, but this has been at a high economic and environmental cost (Ferguson et al. 2010). The industrial production of nitrogen fertilizer costs more than US$ 100 billion because the energetically difficult reduction of the triple bond carried out at high temperature and pressure requires the use of large amounts of fossil fuel, a limited resource. Thus, fertilizer costs are high and this affects especially resource-poor farmers worldwide. In addition, the use of fertilizer has a severe environmental impact, due to run-off of excess non-assimilated nitrate, and concomitant eutrophication of rivers, lakes and oceans, as well as contamination of the drinking water. Moreover, carbon dioxide is

F. J. de Bruijn (✉)
INRA/CNRS Laboratory of Plant-Microbe Interactions, 24 Chemin de Borde Rouge,
Auzeville CS 52627, 31326 Castanet-Tolosan Cedex, France
Tel.: +33561285320
e-mail: debruijn@toulouse.inra.fr

© Springer International Publishing Switzerland 2015
B. Lugtenberg (ed.), *Principles of Plant-Microbe Interactions,*
DOI 10.1007/978-3-319-08575-3_23

215

released during fossil fuel combustion which occurs during production of chemical fertiliser and contributes to the greenhouse effect, as does the decomposition of nitrogen fertilizer, which releases nitrous oxide. The latter gas is about 292 times more active as a greenhouse gas than carbon dioxide (Ferguson et al. 2010).

Biological Nitrogen Fixation (BNF), the reduction of atmospheric dinitrogen to ammonia, carried out by a large and diverse group of free-living and symbiotic microorganisms, presents an inexpensive and environmentally sound, sustainable approach to crop production and constitutes one of the most important Plant Growth Promotion (PGP) scenarios (see de Bruijn 2015a for a comprehensive coverage of BNF).

The increased need for fixed nitrogen, be it industrially- or biologically fixed nitrogen, is exemplified in the case of rice. Rice is the most important staple food for over 2 billion people in Asia and for hundreds of millions in Africa and Latin America. To feed the ever-increasing populations of these regions, the world's annual rice production must increase from 560 million tons in the year 2000 to 760 million tons by 2020 (IRRI 1993). If future increases in rice production have to come from the same or even reduced land area, rice productivity (yield ha-1) must be greatly increased to meet these goals (Ladha et al.1997). Nitrogen is the major nutrient limiting rice production. One kg of nitrogen is required to produce 15–20 kg of grain (Ladha and Reddy 2003). Enhancing rice production from the present 8–12 t per hectare by 2020 would require an increased application of 400 kg per hectare, doubling the amount of N fertilizer presently applied (Ladha and Reddy 2003). This obviates the need for alternative approaches, namely BNF in cereals (see de Bruijn 2015b; Chap. 42).

BNF occurs when atmospheric di-nitrogen is converted to ammonia by an enzyme called nitrogenase (Postgate 1998). The reaction for BNF is:

$$N_2 + 8\,H^+ + 8\,e^- + 16\,MgATP \rightarrow 2\,NH_3 + H_2 + 16\,MgADP + 16\,Pi$$

The process is coupled to the hydrolysis of 16 equivalents of ATP and is accompanied by the co-formation of one molecule of H_2. In free-living diazotrophs, the nitrogenase-generated ammonium is assimilated into glutamate through the glutamine synthetase/glutamate synthase pathway (Postgate 1998). In the case of associative or symbiotic nitrogen fixing diazotrophs (see below), the ammonia produced by the nitrogen fixating bacteria is excreted and assimilated by plant enzymes.

23.2   The Oxygen Paradox

Enzymes responsible for nitrogenase action are very susceptible to destruction by oxygen leading to the "Oxygen Paradox": while oxygen and respiration are generally involved in generating the large amount of ATP required for nitrogenase function, the enzyme is terminally inactivated by oxygen. How, one can ask, do microbes fix nitrogen on a planet covered by 20 % oxygen (Postgate 1998)? The problem is even acerbated in the case of oxygenic photosynthetic diazotrophs, including cyanobacteria (Flores et al. 2015; see below). The following strategies are used to deal with the Oxygen Paradox:

1. The simplest strategy is that of avoiding oxygen. For example, some diazotrophs are obligate anaerobes. Many bacteria are facultative anaerobes: capable of either aerobic or anaerobic growth. Diazotrophic members of this group usually fix nitrogen only anaerobically (Postgate 1998).
2. Some strains of rhizobia (see below) show microaerobic diazotrophy. Several other types of nitrogen-fixing aerobes show oxygen sensitivity amounting to microaerophily (Postgate 1998).
3. Respiration is another mechanism to allow (micro)aerobic nitrogen-fixation. In aerobes a high levels of respiration is a highly efficient means of generating ATP (necessary for nitrogenase action), while protecting nitrogenase from oxygen damage (respiratory protection). There exist also a "conformational protection" of nitrogenase (Postgate 1998).
4. The cyanobacteria are oxygenic phototrophic prokaryotes in which the capability to express the BNF machinery is widespread. Two main strategies have evolved to allow the two incompatible processes of oxygenic photosynthesis and BNF to be performed in a given organism: temporal expression in diel cycles and spatial separation through the formation of specialized cellular structures (heterocysts) in which BNF takes place (Flores et al. 2015).
5. The root- and stem nodules on leguminous plants are highly specialized structures in which bacteroids, differentiated forms of the inducing rhizobia fix nitrogen for the host plant (see below). In these nitrogen fixing nodules leghemoglobin is produced by the plant cells, which binds and transfers oxygen. Its affinity for oxygen is so high that it delivers it to the bacteroids at a free concentration harmless to their nitrogenase (Postgate 1998).

23.3   The Nitrogenase Mediated BNF reaction

Enzymatic conversion of dinitrogen to ammonia is catalysed by nitrogenase, an enzyme complex which is highly conserved in free-living and symbiotic diazotrophs. The convential nitrogenase or Mo-nitrogenase contains a prosthetic group with molybdenum (Iron-Molybdenum Cofactor; FeMoCo; Newton 2015).

The nitrogenase enzyme is composed of two metalloproteins. Component 1, also designated MoFe protein, is a tetramer, composed of two non-identical subunits α and β, while component 2, also designated as the Fe protein, is a dimer consisting of identical subunits (Franche et al. 2009; Newton 2015). Two FeMoCo's are bound to the α subunits of the MoFe protein. Nitrogen fixation is a highly complex process, which is not yet fully elucidated. For a detailed description of nitrogenases, their biosynthesis and their mode of action, see Newton (2015).

The Nitrogen Fixation Genes The genetics of nitrogen fixation was initially elucidated in *Klebsiella pneumoniae* where the *nif* genes required for the synthesis of nitrogenase are clustered in a 24 kb region of the chromosome. This entire region was sequenced early on by Arnold et al. (1988). The three structural genes encoding Mo-nitrogenase proteins are *nifD* and *nifK* for the Mo protein subunits and *nifH* for the Fe protein (Franche et al. 2009). The complete assembly of nitrogenase requires other *nif* genes involved in the synthesis of FeMoCo, including *nifB*, *nifQ*, *nifE*, *nifN*, *nifX*, *nifU*, *nifS*, *nifV*, *nifY* and *nifH*. In addition *nifS* and *nifU* are involved in the assembly of iron-sulfur clusters and *nifW* and *nifZ* in the maturation of the nitrogenase components (Franche et al. 2009). In addition, *Klebsiella* contains genes required for electron transport to nitrogenase (*nifF* and *nifJ*) as well as the regulatory *nifLA* genes involved in the regulation of *nif* gene expression in response to the oxygen and nitrogen status of the cell (Franche et al. 2009; Dixon and Kahn 2004; see also below). The *nif* gene cluster is not always this complex. Recently, a minimal nitrogen fixation gene cluster from *Paenibacillus* containing only 9 *nif* genes has been identified and shown to enable expression of active nitrogenase in *Escherichia coli* (Wang et al. 2013). This would greatly facilitate the engineering of nitrogen fixation in non-nitrogen fixing organisms such as plants (see Sect. 23.11).

Regulation of *nif* (*fix*) Gene Expression Nitrogen-fixing bacteria have evolved several mechanisms to sense multiple environmental signals in order to adapt the nitrogen fixation process to their physiological constraints. Availability of a nitrogen source is a key regulatory signal repressing the nitrogen fixation process. In several nitrogen-fixing γ-*Proteobacteria* (e.g.: *Azotobacter vinelandii, Pseudomonas stutzeri, K. pneumoniae*) the NifA activator and the anti-activator NifL proteins, encoded by the *nifLA* operon, control the expression of all other *nif* genes, in concert with the alternative sigma factor RpoN and the Integration Host Factor (IHF). The *nifLA* operon is in turn controlled by the general nitrogen regulatory protein NtrC, in concert with RpoN, and by the PII protein (GlnB or GlnK) in response to the fixed nitrogen status (Dixon and Kahn 2004).

In free-living diazotrophs oxygen regulation occurs via the *nifL* gene product, which serves as a repressor in the presence of oxygen. In symbiotic nitrogen-fixing rhizobia, transcription of nitrogen fixation genes (*nif* and *fix* genes) is induced primarily by low-oxygen conditions. Low-oxygen sensing and transmission of this signal to the level of *nif* and *fix* gene expression involve at least five regulatory proteins, FixL, FixJ, FixK, NifA, and RpoN (sigma 54) (Dixon and Kahn 2004).

23.4   Major Diazotropic Players and their Phylogeny

By 1960 the nitrogen-fixation capacities of free-living soil bacteria had been established for only a dozen genera. This is a long way from our present knowledge of the distribution of nitrogen fixation ability in most phyla of the *Bacteria* and *Archaea* domains (Franche et al 2009; see Fig. 23.1).

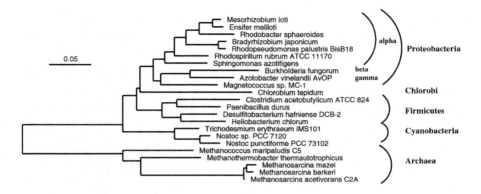

Fig. 23.1 Simplified phylogenetic 16S tree with prokaryotes carrying *nif* genes. Reprinted with permission from Springer from Franche et al. (2009)

The high degree of conservation of certain *nif* genes and the recent and rapid increase in the availability of microbial sequences affords novel opportunities to re-examine the occurrence and distribution of nitrogen fixation genes. The current practice for computational prediction of nitrogen fixation is to use the presence of the highly conserved *nifH* and/or *nifD* genes (Dos Santos et al. 2012 and references therein). Dos Santos et al. (2012) searched the fully sequenced genomes of 1002 bacterial and archaeal species for coding sequences for NifD and NifH, and identified 174 species which contain homologous sequences, suggesting that the phylogenetic distribution of diazotrophs is much broader than previously known. Nitrogen fixation activity has not been experimentally shown in 92 of these species (Dos Santos et al. 2012). The authors went on to look at the occurrence of nine additional *nif* genes and concluded that there existed a minimum gene set for nitrogen fixation, consisting of *nifHDK* (catalytic) and *nifENB* (biosynthetic), which they used to identify 92 species containing coding sequences similar to NifD and NifH, of which 67 met the minimum set requirement (Dos Santos et al. 2012). Based on gene content, these 67 species were proposed to have the capacity for nitrogen fixation (Dos Santos et al. 2012).

23.5 Sustainable Agricultural Uses of Biological Nitrogen Fixation; the Legume-*Rhizobium* Symbiosis

Biological nitrogen fixation is a highly valuable alternative to nitrogen fertilizer. It is most effective in the interaction between members of the legumes (*Leguminosae*) and the well known α-proteobacteria called rhizobia (including the genera *Azorhizobium, Allorhizobium, Bradyrhizobium, Mesorhizobium, Rhizobium* and *Sinorhizobium),* as well as more recently recognized rhizobia belonging to the β-proteobacteria (Moulin et al. 2015). Once the symbiosis is established, rhizobia fix atmospheric nitrogen

Fig. 23.2 Lentil field in Western Manitoba in which plants on the right received a commercial rhizobial inoculants; and in the middle rows are the control plants. (Reprinted with permission from the Crop Science Society of America from "Benefits of inoculating legume crops in the Northern Great Plains" (Vessey 2004. Crop Management Vol. 3)

and provide it to their host plant (symbiotic nitrogen fixation), in return for carbon (energy) provided by the plant. Because nitrogen is a key limiting factor for plant growth and development, the ability of legumes to enter into a symbiosis with nitrogen-fixing rhizobia provides them with a distinct advantage over other plant species (Ferguson et al. 2010) and constitutes highly proficient sustainable agriculture systems .

Legumes include major food and feed crop species, such as soybean, pea, clover, alfalfa, lentils and mungbean. They represent the third largest group of angiosperms and are the second largest group of food and feed crops grown globally (Ferguson et al. 2010; for the case of soybean, see Chap. 41). They are cultivated on 12–15 % of all available arable land and are responsible for more than 25 % of the world's primary crop production with 247 million tons of grain legumes produced annually (Ferguson et al. 2010). In addition to food and feed crops, nodulated legumes such as soybean and *Pongamia pinnata* have garnered a great deal of attention as future sustainable biofuel sources because of their high seed oil content (Ferguson et al. 2010; see also Chap. 41). The legume-Rhizobium symbiosis is the most important symbiotic association in terms of biological nitrogen fixation, producing roughly 200 million tons of fixed nitrogen annually (Ferguson et al. 2010; see Fig. 23.2).

23.6 The Actinorhizal-*Frankia* Symbiosis

Besides in legumes, symbiotic nitrogen fixation is also found in the Actinorhizal symbiosis. Actinorhizal plants represent about 200 species distributed among 24 genera in eight angiosperm families (Franche et al. 2009). All actinorhizal plants are woody trees or shrubs except for *Datisca*, a genus of flowering plants. Often they are considered "pioneer plants" in regions with poor soil and examples of well known genera include *Alnus* (alder), *Eleagnus* (autumn olive), *Hippophae* (sea buckthorn) and *Casuarina* (beef wood) (Franche et al. 2009).

The genus *Frankia* comprises high GC percentage Gram-positive bacteria belonging to the family *Frankiaceae* in the order *Actinomycetales*. One striking feature of *Frankia* is its ability to differentiate into two unique developmental structures that are critical to its survival: vesicles and spores. Vesicles are the site for actinorhizal nitrogen fixation. For an excellent review of this symbiotic system see Franche et al. (2009).

23.7   Cyanobacteria and Symbiosis

Cyanobacteria are widely distributed in aquatic and terrestrial environments. While nitrogen fixation occurs both in unicellular and filamentous species, associations with plants are essentially limited to heterocystous cyanobacteria, primarily of the genus *Nostoc* and *Anabaena*. Besides vascular plants, there exist a wide variety of non-vascular lower plants belonging to the bryophytes, including liverworts and hornworts, algae and fungi, that develop associations with cyanobacteria, as well as many marine eukaryotes (Franche et al. 2009). In terms of symbiotic associations with vascular plants, *Gunnera*, a genus of about 40 species, is the only angiosperm with which the cyanobacterium *Nostoc punctiforma* is associated and fixes nitrogen. The mechanism of infection and *Nostoc* differentiation is described in Franche et al. (2009) but once intracellular, a high frequency of differentiation of vegetative cells into heterocysts occurs and nitrogen is fixed at a high rate.

23.8   Nitrogen Fixation in the Ocean

BNF is an integral part of the marine nitrogen cycle and together with N losses through denitrification and anaerobic ammonia oxidation determines the size of the oceanic nitrogen pool (Zehr and Bombar 2015). BNF in the oceans is of a similar magnitude to anthropogenic BNF, and yet many questions remain on what the major N_2-fixing organisms are, and what controls their distributions and BNF rates. Using molecular and metagenomic approaches, surprising discoveries have been made, since many environmental microorganisms have yet to be obtained in pure culture. Cyanobacteria appear to be the main oceanic N_2-fixers, with several key species that include the filamentous, colonial, nonheterocyst-forming *Trichodesmium*, heterocyst-forming strains that are symbiotic with diatoms. The N_2-fixing microbial taxa generally differ among the different marine habitats, and include Archaea and diverse (photo)heterotrophic and chemolithotrophic bacteria and photoautrophic cyanobacteria. For example, archaeal nitrogenase (*nifH*) genes have been found in deep water and near hydrothermal vents (Zehr and Bombar 2015). It is accepted that BNF is a key component of the marine nitrogen cycle, but rather than being an easily quantifiable process carried out by few species in well constrained areas, the accumulating knowledge shows that we are just beginning to understand the impacts of diazotrophs in the pelagic ocean (and other) ecosystems (Zehr and Bombar 2015).

23.9 Associative and Endophytic Nitrogen Fixation

In addition to strictly associative diazotrophs, there has been an increasing interest in endogenous BNF systems, particularly nitrogen-fixing endophytic bacteria. Endophytic bacteria are defined as bacteria detected inside surface-sterilized plants or extracted from inside plants and having no visibly harmful effects on the host plant (see Chap. 5; see de Bruijn 2013).

The extensive Brazilian experience with associative and endophytic diazotrophs and Plant Growth Promoting Rhizobacteria (PGPR) on sugarcane, rice and other grasses is highly relevant. This experience is highlighted in a special issue of Plant and Soil (James and Baldani 2012), based on the BNF with non-Legumes International Symposium of 2010 in Brazil. In the case of sugarcane and other biofuel crops, the location of the meeting in Brazil was particularly pertinent since the highly advanced Brazilian bioethanol program, which produces over 27 billion liters of ethanol per year, is based on the cultivation of sugarcane, deriving much of its N-requirements via BNF (James and Baldani 2012). Field-based BNF quantitative studies revealed very substantial inputs into sugarcane of at least 40 kg N per ha per year (Urquiaga et al. 2012).

23.10 Field Data

Soybean and BNF have been the focus of many studies worldwide and field data are presented in Chap. 41. An evaluation of BNF in food grain legumes grown in experimental plots in Africa revealed high levels of symbiotic dependency on N_2 fixation for their N nutrition (Dakora et al. 2015). Cowpea could, for example, derive 30–96 % of its N nutrition from symbiosis, soybean 39–87 %, pigeon pea 27–92 %, groundnut 24–67 %, mungbean 66–86 %, chickpea (kabuli) 3–92 % and chickpea (desi) 21–82 % (Dakora et al. 2015).

23.11 Future Prospects in Biological Nitrogen Fixation

Several factors, such as efficient strain selection, inoculum production and quality, plant breeding for nitrogen fixation etc. can be improved upon, and associative (endophytic) nitrogen fixation clearly is of importance. However, the "holy grail" of nitrogen fixation research is the quest for nitrogen fixation in cereals, such as rice (de Bruijn 2015b). Two ways have been envisioned: (i) the transfer to and expression of the *nif* genes in transgenic cereal plants and (ii) the transfer of the ability to fix nitrogen symbiotically (de Bruijn 2015b). Although tremendous progress has been made in the characterization of *nif* genes for transfer into plants (de Bruijn 2015b) and the elucidation of the Common Symbiotic Signalling Pathway (CSSP or SYM; de Bruijn 2015b; see Chap. 42) in legumes and cereals, still a considerable amount of new information will be needed to achieve either goal (de Bruijn 2015b; Chap. 42). However, experimentation towards elucidating the essential *nif* genes and custom tailoring them for expression in plants, as well as studying the SYM pathway genes and identifying the "missing components" (de Bruijn 2015b; see Chap. 42), are now supported by large grants, for example from the Bill and Melinda Gates Foundation the BBSRC (UK) and NSF (USA), raising the hope for a bright future in this field.

Acknowledgements The writing of this review Chapter was supported by the Laboratory of Plant-Microbe Interactions (LIPM), INRA, CNRS and the Labex Tulip. Springer Verlag is gratefully acknowledged for their Permission to reprint Fig. 1 and quote and cite excerpts of the text of Franche et al. (2009). Claude Bruand is thanked for his critical review of the manuscript.

References

Arnold W, Rump A, Klipp W et al (1988) Nucleotide sequence of a 24,206 base-pair DNA fragment carrying the entire nitrogen fixation gene cluster of *Klebsiella pneumonia*. J Mol Biol 203: 715–738

Dakora F et al (2015) Food grain legumes: their contribution to soil fertility, food security and human nutrition/health in Africa. In: de Bruijn FJ (ed) Biological nitrogen fixation. Wiley, Hoboken (in press)

de Bruijn FJ (ed) (2013) Molecular microbial ecology of the rhizosphere. Wiley, Hoboken, pp 1–1269

de Bruijn FJ (ed) (2015a) Biological nitrogen fixation. Wiley, Hoboken (in press)

de Bruijn FJ (2015b) The quest for biological nitrogen fixation in cereals: a perspective and prospective. In: de Bruijn FJ (ed) Biological nitrogen fixation. Wiley, Hoboken (in press)

Dixon R, Kahn D (2004) Genetic regulation of biological nitrogen fixation. Nat Rev Microbiol 2:621–631

Dos Santos PC, Fang Z, Mason SW et al (2012) Distribution of nitrogen fixation and nitrogenase-like sequences amongst microbial genomes. BMC Genomics 13:162–174

Ferguson BJ, Indrasumunar A, Hayashi S et al (2010) Molecular analysis of legume nodule development and autoregulation. J Int Plant Biol 52:61–76

Flores E, Lopez-Lozano A, Herrero A (2015) Nitrogen fixation in the oxygenic phototrophic prokaryotes (cyanobacteria): the fight against oxygen. In: de Bruijn FJ (ed) Biological nitrogen fixation. Wiley, Hoboken (in press)

Franche C, Lindstrom K, Elmerich C (2009) Nitrogen-fixing bacteria associated with leguminous and non-leguminous plants. Plant Soil 321:35–59

IRRI (1993) Rice research in a time of change. International Rice Research Institute's medium plan for 1994–1998

James EK, Baldani JI (2012) The role of biological nitrogen fixation by non-legumes in the sustainable production of food and biofuels. Plant Soil 356:1–3

Ladha JK, Reddy PM (2003) Nitrogen fixation in rice systems: state of knowledge and future prospects. Plant Soil 252:151–167

Ladha JK, de Bruijn FJ, Malik KA (1997) Introduction: assessing opportunities for nitrogen fixation in rice- a frontier project. Plant Soil 194:1–10

Moulin L, James EK, Klonowska A et al (2015) Phylogeny, diversity, geographical distribution and host range of legume-nodulating Betaproteobacteria: what is the role of plant taxonomy? In: de Bruijn FJ (ed) Biological nitrogen fixation. Wiley, Hoboken (in press)

Newton WE (2015) Recent advances in nitrogenases and how they work. In: de Bruijn FJ (ed) Biological nitrogen fixation. Wiley, Hoboken (in press)

Postgate J (1998) Nitrogen fixation. Cambridge University Press, Cambridge, pp 1–109

Urquiaga S et al (2012) Evidence from field nitrogen balance and ^{15}N natural abundance data for the contribution of biological N_2 fixation to Brazilian sugarcane varieties. Plant Soil 356:5–21

Vessey K (2004) Benefits of inoculating legume crops with rhizobia in the northern great plains, Crop Management. Vol. 3, doi: 10.1094/CM-2004-0301-04-RV

Wang L et al (2013) A minimal nitrogen fixation gene cluster from *Paenibacillus* sp. WLY78 enables expression of active nitrogenase in *Escherichia coli*. PLOS Genet 9:e1003865

Zehr J, Bombar D (2015) Marine nitrogen fixation: organisms, significance, enigmas and future directions. In: de Bruijn FJ (ed) Biological nitrogen fixation. Wiley, Hoboken (in press)

Chapter 24
Phosphate Mobilisation by Soil Microorganisms

José-Miguel Barea and Alan E. Richardson

Abstract Microorganisms are fundamental to the cycling of phosphorus (P) in soil-plant systems as they are involved in a range of processes that govern P transformations and availability. Soil microorganisms in particular are able to release plant available P from otherwise sparingly available forms of soil P, through solubilisation and mineralisation reactions of inorganic and organic P, respectively. The potential of phosphate solubilising microorganisms (PSM) to improve plant P nutrition is widely recognised, and the mechanisms involved are being investigated. The feasibility of developing efficient management systems based on PSM as biofertilisers is of current interest in rhizosphere biotechnology. Mycorrhizosphere interactions involving PSM and their interaction with AM fungi is of further relevance for the acquisition, transport and supply of P to plant roots, and therefore to soil P cycling and plant P nutrition. Managing these interactions (mycorrhizosphere tailoring) provides an environmentally-acceptable agro-technological practice to improve agricultural sustainability.

24.1 Phosphorus in the Soil-Plant System

Phosphorus (P) is a vital element for life on earth. In particular, P is essential for plant growth and development, as it is a component of fundamental macromolecules involved in genetic, regulatory, structural, signal transduction and other metabolic processes. In addition to the orthophosphate anion, other plant P-integrating molecules include nucleic acids and ADP/ATP, indispensable for photosynthesis,

J.-M. Barea (✉)
Soil Microbiology and Symbiotic Systems Department, Estación Experimental del Zaidín, CSIC, Prof. Albareda 1, 18008 Granada, Spain
Tel.: + 34686404966
e-mail: josemiguel.barea@eez.csic.es

A. E. Richardson
CSIRO Plant Industry, PO Box 1600, Canberra 2601, Australia
Tel.: + 61 2 6246 5189
e-mail: alan.richardson@csiro.au

© Springer International Publishing Switzerland 2015
B. Lugtenberg (ed.), *Principles of Plant-Microbe Interactions*,
DOI 10.1007/978-3-319-08575-3_24

225

respiration and other biochemical processes involved in energy storage and transfer reactions. Plant P also occurs in storage compounds such as phytate and related compounds, pyrophosphate and as a component of membrane phospholipids and phosphoproteins (White and Hammond 2008).

Forms of Phosphorus in Soil Soil P occurs as either inorganic phosphates or organic phosphate derivatives. The primary mineral form of P in soil is apatite. The weathering of apatite results in the release of orthophosphate anions, primarily as HPO_4^{2-} and $H_2PO_4^{1-}$ to soil solution, but only in small quantities. Soil solution orthophosphate content typically ranges from 0.1 to 1 mg P kg^{-1} which represents about 1 % of the total soil P. Most orthophosphate in soil undergoes reactions which makes it only sparingly available to plants. Orthophosphate is rapidly adsorbed on clay mineral surfaces and other soil particles and colloids or precipitated as inorganic salts (e.g., with calcium in alkaline soils or with aluminum and iron in acidic soils), which are of low solubility. A significant amount of orthophosphate is also integrated in complex organic molecules (soil organic P), which can account for 30–60 % of the total soil P. Major identifiable fractions of organic P in soil include inositol phosphates, such as phytate (salts of *myo*-inositol hexakisphosphate), nucleic acids, phospholipids and phosphonates. Inositol phosphates are considered to be the dominant form of organic P in many soils. Phosphorus immobilised within the living soil microbial biomass is also significant, and typically represents about 5 % of the total soil P (Plante 2007; White and Hammond 2008).

The Soil Phosphorus Cycle From a functional point of view the various forms of P in soil are interconnected and integrated through the so called soil P cycle. As outlined by Plante (2007) the soil *solution P* pool is central to the P cycle and is the primary source of labile orthophosphate for biological uptake by microorganisms and plants. Soil *solution P* also provides the interconnection between the *biological subsystem* (including plant residues, soil microbial P, labile and stable organic P) and the *geochemical subsystem* (i.e., primary minerals, secondary minerals and adsorbed P, that includes P occluded with soil constituents). Whilst the availability of orthophosphate in the *geochemical subsystem* is mediated largely by physical-chemical reactions such as dissolution, precipitation, sorption-desorption and oxidation-reduction, these processes are also influenced strongly by biological activities. Soil microorganisms are able to interact across both *subsystems* through either solubilisation of inorganic P or mineralisation of organic P and thus play a key role in the cycling of soil P. Furthermore, soil microorganisms interact directly with soil *solution P* and may thus directly influence the availability of orthophosphate to plants through mobilisation or, conversely, in the short term, by competition with plants for available nutrient through P immobilisation.

Availability of Phosphorus for Plant Nutrition Plant roots acquire orthophosphate from soil solution via their associated volume of soil through either the rhizosphere (Chap. 3) or mycorrhizosphere (Chap. 25) (Fig. 24.1) However, because of the high reactivity of P in soil and the rapid uptake of orthophosphate by roots, the concentration of orthophosphate around roots is often low. This low concentration is further

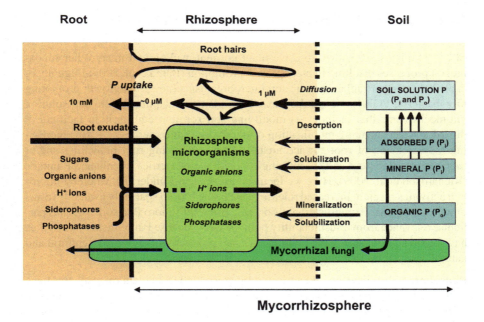

Fig. 24.1 Root-soil microbiome processes governing transformation and availability of phosphorus in soil-plant systems highlighting the importance of the rhizosphere and mycorrhizosphere. (Reproduced from Richardson et al. 2009 by permission of the publisher)

compounded by slow diffusion of orthophosphate anions in solution which results in a distinct zone of depletion in soil immediately surrounding the root system. The low rate of replenishment of orthophosphate in the rhizosphere/mycorrhizosphere soil solution from the bulk soil is therefore a major factor that regulates P availability to plants (White and Hammond 2008). Consequently, P-based fertilisers are routinely used in agricultural production systems to either maintain the P status of fertile soils or to increase P availability in deficient soils. The efficiency of P fertiliser use in most systems, however, is low, whereby only 10–50 % of applied P is recovered by crops in the year of application. The remainder of the P accumulates in soil in either inorganic or organic P fractions and, subject to efficient mobilisation, can provide a P benefit in subsequent years. Rock phosphates (RP), which are used for the production of water soluble P-fertilisers, are also widely used as a direct source of P albeit with relatively low agronomic efficiency. Development of strategies to increase the mobilisation of P from accumulated forms in soil or to enhance the utilisation of RP are thus promoted as being important to increase the efficiency of P fertiliser use. As microorganisms are a key component of the soil P cycle they are widely considered as the basis of some of these alternative strategies to improve the sustainability of P use in agriculture systems (Zapata and Roy 2004).

24.2   Microbial Mobilisation of Phosphorus in Soil

Microorganisms are known to drive plant nutrient cycling and many other fundamental processes resulting in plant growth promotion (Barea et al. 2007; Lugtenberg
et al. 2013). In particular, specific soil microorganisms (i.e., plant growth promoting
rhizobacteria; PGPR) change the capacity of plants to acquire P from soil solution
via mechanisms that include; (i) modifying soil sorption equilibria to facilitate P
diffusion, (ii) enhancing mobilisation of poorly available sources of P, (iii) increasing the extension of root surface area, (iv) by stimulating root branching and/or
root hair development and (v) altering root surface properties to enhance P uptake
(Richardson et al. 2009). Here we focus on mechanisms under (ii), whereby microbial activities result in increased release of available P from sparingly available forms
of either inorganic (solubilisation) or organic (mineralisation) P in soil. This has particular relevance from a sustainability point of view because P mobilisation activities
have broad significance in the maintenance and productivity of both agricultural and
natural ecosystems (Richardson 2007).

Phosphate Solubilisation   Bacteria and fungi isolated from plant rhizospheres have
been shown to solubilise *in vitro* various inorganic phosphates, such as calcium,
aluminum or iron salts. These microorganisms are collectively termed "phosphate
solubilising microorganisms" (PSM). They include *Bacillus, Enterobacter, Rhizobium, Bradyrhizobium, Enterobacter, Panthoea, Erwinia,* and *Pseudomonas* as
common bacterial genera, and *Aspergillus, Trichoderma* and *Penicillium* as fungal
representatives (Marschner 2008). In the case with sparingly soluble forms of calcium phosphates, the mechanism of solublisation is most commonly associated with
proton release and media acidification. For iron or aluminum phosphates, solubilisation due to acidification appears to be less effective and production of organic acids is
of greater importance. Organic anions are effective in chelation processes that result
in the sequestration of calcium, iron or aluminum which is associated with a release
of orthophosphate to solution. Commonly reported organic anions include citrate,
oxalate, lactate, succinate, gluconate and 2-ketogluconic acid. Siderophore production likewise has been reported to be effective for solubilisation of Fe phosphates
(Marschner 2008). The amount of orthophosphate released from sparingly soluble
forms is dependent on the microorganisms involved, culture conditions and the degree of solubility of the P substrate (Whitelaw 2000). Solubilisation of P is further
dependent on the presence of readily metabolisable carbon sources. As such, isolates
selected as being effective for P solubilistion under laboratory conditions may not
be effective in soil due to either carbon limitation or other unfavorable microhabitat
conditions (Richardson 2007).

Molecular-based approaches have recently been used to investigate the mechanisms involved in P solubilisation by specific microorganisms. For example, one
mechanism is based on the ability of *Pseudomonas* spp. to produce gluconic
acid from glucose by the oxidation reaction catalysed by glucose dehydrogenase which uses pyrroloquinoline quinone (PQQ) as a redox cofactor. Finally,
2-ketogluconate is produced which facilitates both the chelation of calcium and

release of protons. A genomic library of *Pseudomonas* spp. has recently been analysed for PQQ biosynthetic genes to determine their involvement in P solubilisation (Browne et al. 2013).

Phosphate Mineralisation The mineralisation of organic P in soil and release of orthophosphate to soil solution is largely mediated by microbial activities (Richardson et al. 2009). Bacteria and fungi isolated from plant rhizospheres have been shown to have capacity to hydrolyse organic P substrates either *in vitro* or when added to soil. Common microorganisms include *Bacillus* and *Pseudomonas* as bacteria and *Aspergillus* and *Penicillium* as fungi (Marschner 2008). Mineralisation of organic P often first requires solublisation of substrates with subsequent hydrolysis by phosphatase enzymes, which in many cases is synonymous with the activities of PSM. Microorganisms produce diverse types of enzymes which include non-specific acid and alkaline phosphates, and specific enzymes, such as phytases which release orthophosphate from phytate and other inositol phosphates. The importance of microorganisms for phytate mineralisation has been demonstrated in various studies, whereby the availability and plant uptake of orthophosphate can be improved by inoculation with PSM with P mineralisation capability. Nevertheless, the effectiveness of phytases in many soil environments remains less clear since enzymes may also readily be absorbed to soil particles or degraded, and inositol phosphates adsorb strongly or precipitate readily with iron or aluminum oxides and other soils constituents (Marschner 2008). Nonetheless, microbial utilisation of organic P substrates in soil and its turnover has potential to supply a significant amount of P to meet plant requirements. This is of particular importance in the rhizosphere and mycorrhizosphere where metabolisable carbon is more available and there is greater capacity to capture mobilised P. However, further experimental evidence to quantify microbial mineralisation of P and the direct value of immobilised P in the microbial biomass to plant nutrition is required (Richardson et al. 2009).

It is important to note that to date much of the work on P solubilisation and mineralisation has involved soil microorganisms that have been isolated and grown in culture media. More recent culture-independent molecular-based studies have shown that a high percentage (i.e., greater that 90 %, and possibly as high as 99 %) of soil microorganisms are unculturable, and that this includes therefore likely microorganisms that are involved in phosphate-solubilisation and P cycling (Barret et al. 2013). As culture-independent approaches are being used to further dissect plant-microbial interactions it is evident that plants play a significant role in shaping microbial communities in the rhizosphere and mycorrhizosphere. As such there is new opportunity for linking the structure and function of the root-soil microbiome to orthophosphate availability and P-solubilising capacity (Browne et al. 2013).

Significance of PSM in Improving Plant Nutrition While it is clear that soil microorganisms are integral to the operation of the soil P cycle, the extent to which P released by soil microorganisms actually benefits plant P acquisition remains to be more fully elucidated. Indeed because orthophosphate release from sparingly available soil P sources by microbiological-driven activities may be highly transient in nature, this has implications for its efficacy in promoting plant growth.

In addition, mobilised orthophosphates are also subject to further reaction in soil, for example through either 're-fixation' reactions or immobilistion into microbial biomass, and may thus not be considered to be immediately available to plants. Spatial interactions may likewise impede the availability of orthophosphate within the plant rhizosphere and/or mycorrhizosphere and thereby limit any potential agricultural benefit (Richardson 2007). However, if orthophosphate made available by PSB can be taken up more efficiently by plant roots or through an effective mycorrhizal mycelium (Chap. 25), then the resultant microbial interaction could synergistically act to improve P supply to the host plant (Richardson et al. 2009).

Managing PSM to Improve Plant Phosphorus Nutrition Given the potential of PSM to contribute to the development of more sustainable agricultural systems, a number of approaches have been proposed for their management and opportunity for capture of benefits. These are based on either the manipulation of naturally existing microbial populations, or by the development of microbial inoculants that contain specific microorganisms with recognised potential for P mobilisation. Management of naturally existing populations, whilst attractive, is a relatively untargeted approach, where it is often difficult to predict the response of microbial populations as a consequence of different agricultural practices (Richardson 2007). For example, crop rotation and amendment of soils with organic wastes (e.g., manure crops) is known to enhance P cycling or to increase the biological activity in soils, although specific effects on either solubilisation or mineralisation processes remain to be more fully investigated. Likewise, because of its distinct physical, chemical and biological properties, biochar (produced by pyrolitic transformations of organic materials) is also being used widely as a soil amendment that may also facilitate the efficacy of phosphate solubilising or mineralisation activities in soil (Lehmann et al. 2011).

A range of biofertiliser-based products using selected PSM (both bacteria and fungi) have been developed as commercial products for use in various agricultural systems across the world. These products, using specific microorganisms either individually or as part of a microbial consortium, have in many cases shown positive effects in various field trials (Antoun 2012). However, inconsistent results are commonly observed indicating the complexity of interactions in the soil-plant systems, where diverse ecological variables need to be considered. In some instances it seems that benefits of plant growth promotion derived from microbial inoculants occurs by stimulation of root growth, allowing a greater exploration of soil P, rather than by direct increase in P-mobilisation (Richardson et al. 2009).

Successful development of PSM requires that appropriate inoculant formulations and delivery systems be developed for specific microorganisms (Antoun 2012). Furthermore, selected microorganisms must be able to maintain their ability to solubilise P after repeated sub-culturing under laboratory conditions or following re-isolation from soil, and exhibit sufficient saphrophytic competence for persistence in soil environments. To be effective in soil it is critical that selected PSM are able to establish themselves either on the root surface or within in the root-soil habitat. *Rhizosphere competence* is a key trait involved in microbial establishment in the rhizosphere (Chap. 3) and there is need to develop non-disruptive visualisation techniques for assessing microbial colonisation of the rhizosphere and/or mycorrhizosphere (Barea et al. 2013).

PSM Interactions to Improve Key Rhizosphere Processes Interaction of PSM with other PGPR has been shown in several cases to be beneficial for plant growth. For example, PSM have been used in conjunction with N_2-fixing bacteria and mycorrhizal fungi to improve N_2-fixation by legumes though a greater P supply (Zaidi et al. 2010). Inoculation of lucerne (alfalfa) with phosphate-solubilising bacteria (PSB) enhanced both nodulation and N_2-fixation, as estimated by using ^{15}N and ^{32}P isotopic methods. Using the ^{32}P dilution approach, Barea et al. (2013) showed that inoculation with PSB increased the mobilisation of sparingly available P from either endogenous soil P or from P supplied as rock phosphate.

24.3 Mycorrhizosphere Interactions to Improve Plant Phosphorus Nutrition

The colonisation of roots by mycorrhiza affects diverse aspects of plant physiology resulting in quantitative and qualitative changes in root structure and composition of root exudates. This has a significant effect on the compositional structure and function of microbial communities in the rhizosphere. In addition, mycorrhizal mycelium directly modifies the physical characteristics of surrounding soil and interacts further with soil microorganisms. These mycorrhiza-induced interactions, has led to the development of the so-called mycorrhizosphere (Barea et al. 2013). Other rhizobacteria that favor the formation of mycorrhiza on roots have also been recognised (Frey-Klett et al. 2007), and are commonly referred to as "mycorrhiza-helper-bacteria" (MHB).

Mycorrhizosphere interactions involving PGPR can be managed (mycorrhizosphere tailoring) to benefit plant growth and health, and soil quality (Barea et al. 2013). Interactions between mycorrhizas and PSB in particular are relevant to P cycling and plant P nutrition (Richardson et al. 2009). PSB interactions with arbuscular mycorrhizal (AM) associations are considered in more detail below, as association with AM fungi is common with some 80 % of terrestrial plant species being able to be colonized, including those of agronomic interest (Chap. 25).

Biological and Ecological Basis of Mycorrhizosphere Interactions The extensive and highly branched external mycelium of AM fungi characteristically increases the zone of soil exploration and potential for plant nutrient uptake beyond the rhizosphere (Chap. 25). External AM mycelium are able to extend several cm (up to 25 cm) from the root and thus provide for the uptake and transport of P in soil that is independent of the diffusive rate of orthophosphate in soil solution. In addition, 1 cm of root length can harbour as much as 1 m of fungal hyphae, with densities of up to 40 m of mycelia per gram of mycorrhizosphere soil (Smith and Smith 2012). These properties of AM mycelium are especially relevant for acquisition of P from soil as has been widely demonstrated using compartmented devices and ^{32}P isotopic labelling studies. Importantly, orthophosphate made available by PSM has potential to be taken up and transported more effectively by roots that are colonised with AM fungi.

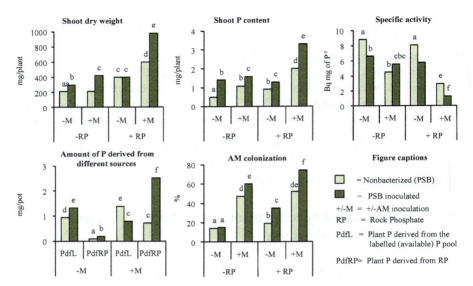

Fig. 24.2 Interactive effects of PSB and AM fungi in enhancing plant growth and phosphorus (P) uptake from endogenous soil P or P added as rock phosphate. Plants (*onions*) were grown in a soil microcosm system which integrated ^{32}P isotopic dilution approaches in an agricultural soil with indigenous microbiota either with or without inoculation. Plants inoculated with both PSB and AM fungi produced greater biomass, accumulated more shoot P and had a lower ^{32}P specific activity than non-inoculated or single-inoculated plants, thus indicating greater access to poorly-available sources of P. Up to 75 % of the P in dual-inoculated plants was derived from added RP and it was evident that inoculated PSB behaved as a mycorrhiza helper bacteria by promoting establishment of both indigenous and inoculated AM fungi. (Reproduced from Toro et al. 1997 by permission of the publisher)

Agronomical Application of PSB x AM Fungal Interactions The feasibility of capturing greater benefit for plant P nutrition through synergistic interactions between PSB and AM fungi has received wide interest since the pioneering work of Azcón et al. (1976). For example, various co-inoculation studies have shown significant increase in the mobilisation and plant uptake of P from sparingly available sources either directly from soil or when added as RP (Barea et al. 2013). In some studies the inoculated PSB behaved as MHB by promoting establishment of both indigenous and inoculated AM fungi (Fig. 24.2). Whilst consistent results have been reported for glasshouse studies, more variable response occurs under field conditions. Interestingly, in the study by Barea et al. (2007) the agronomic efficiency of P uptake by plants from RP was increased with dual inoculation, and was greatest when an organic matter amendment was applied. Whilst organic matter appeared to enhance the solubilisation activities of PSM, the underlying mechanisms involved and relative contribution of the various microbial partners in such interactions remains to be more fully investigated.

The Use of Isotopic Techniques to Assess Mycorrhizosphere Interactions Isotopic (^{32}P and ^{33}P) dilution approaches have commonly been used to investigate exchange rates in orthophosphate equilibrium between solution and solid phases of soil and to measure the availability of P from fertiliser sources (including RP) as influenced by management practices (Zapata and Roy 2004). Accordingly, ^{32}P-tracer methodologies have been used to determine the contribution of AM fungi and PSB to plant P uptake from different source of P (Toro et al. 1997). These techniques typically involve labelling of the exchangeable pool of soil P with ^{32}P-orthophosphate. Plants subject to different inoculation treatments are then grown in the soil amended either with or without RP. Difference in isotopic composition, or "specific activity" (SA $= ^{32}P/^{31}P$ quotient) in plant tissues can then be used to assess P uptake, whereby lower SA of the plant (relative to control plants) is indicative of greater access to otherwise sparingly available forms of P (Toro et al. 1997).

Results from several isotopic-labeling experiments using different plant species (mainly legumes) have shown increased biomass and P content of plants co-inoculated with AM and PSB, along with lower SA of shoot tissues compared with non-inoculated or singularly-inoculated plants (Azcón and Barea 2010). This suggests that PSB are effective in releasing orthophosphate anions from soil or RP sources and, that in the presence of AM fungi, the P is more available for plant uptake (Fig. 24.2). By using isotope dilution concepts (Zapata and Roy 2004) the relative contribution of the different P sources to plant P content (i.e., from added RP) can also be determined. As such, it is evident that co-inoculated plants generally have significantly greater access to P from RP, while plants inoculated with PSB only showed greater reliance on the exchangeable soil P pool (Barea et al. 2007). Collectively, plants inoculated with both AM fungi and PSM therefore appear to be more P efficient compared to those that were non-inoculated or singly-inoculated. In conclusion this demonstrates that opportunity exists to gain benefit in plant P nutrition from interactive effects of PSB and AM-fungi through tailored management of the mycorrhizosphere (Azcón and Barea 2010).

References

Antoun H (2012) Beneficial microorganisms for the sustainable use of phosphates in agriculture. Procedia Eng 46:62–67

Azcón R, Barea JM (2010) Mycorrhizosphere interactions for legume improvement. In: Khan MS, Zaidi A, Musarrat J (eds) Microbes for legume improvement. Springer, Vienna, pp 237–271

Azcón R, Barea JM, Hayman D (1976) Utilization of rock phosphate in alkaline soils by plant inoculated with mycorrhizal fungi and phosphate-solubilizing bacteria. Soil Biol Biochem 8:135–138

Barea JM, Toro M, Azcón R (2007) The use of ^{32}P isotopic dilution techniques to evaluate the interactive effects of phosphate-solubilizing bacteria and mycorrhizal fungi at increasing plant P availability. In: Velázquez E, Rodríguez-Barrueco C (eds) First international meeting on microbial phosphate solubilization. Series: developments in plant and soil sciences. Springer, Dordrecht, pp 223–227

Barea JM, Pozo MJ, Azcón R et al (2013) Microbial interactions in the rhizosphere. In: de Bruijn F (ed) Molecular microbial ecology of the rhizosphere. Wiley, Hoboken, pp 29–44

Barret M, Tan H, Egan F et al (2013) Exploiting new systems-based strategies to elucidate plant-bacterial interactions in the rhizosphere. In: de Bruijn F (ed) Molecular microbial ecology of the rhizosphere. Wiley, Hoboken, pp 57–68

Browne P, Barret M, Morrissey JP et al (2013) Molecular-based strategies to exploit the inorganic phosphate-solubilization ability of *Pseudomonas* in sustainable agriculture. In: de Bruijn F (ed) Molecular microbial ecology of the rhizosphere. Wiley, Hoboken, pp 615–628

Frey-Klett P, Garbaye J, Tarkka M (2007) The mycorrhiza helper bacteria revisited. New Phytol 176:22–36

Lehmann J, Rillig MC, Thies J et al (2011) Biochar effects on soil biota—a review. Soil Biol Biochem 43:1812–1836

Lugtenberg BJJ, Malfanova N, Kamilova F et al (2013) Plant growth promotion by microbes. In: de Bruijn FJ (ed) Molecular microbial ecology of the rhizosphere. Wiley, Hoboken, pp 561–573

Marschner P (2008) The role of rhizosphere microorganisms in relation to P uptake by plants. In: White PJ, Hammond J (eds) The ecophysiology of plant-phosphorus interactions series: plant ecophysiology, vol 7. Springer, Dordrecht, pp 165–176

Plante AF (2007) Soil biogeochemical cycling of inorganic nutrients and metals. In: Paul EA (ed) Soil microbiology, ecology, and biochemistry. Elsevier, Oxford, pp 389–432

Richardson AE (2007) Making microorganisms mobilize soil phosphorus. In: Velázquez E, Rodríguez-Barrueco C (eds) First international meeting on microbial phosphate solubilization. Developments in plant and soil sciences, vol 102. Springer, Netherlands, pp 85–90

Richardson AE, Barea JM, McNeill AM et al (2009) Acquisition of phosphorus and nitrogen in the rhizosphere and plant growth promotion by microorganisms. Plant Soil 321:305–339

Smith SE, Smith FA (2012) Fresh perspectives on the roles of arbuscular mycorrhizal fungi in plant nutrition and growth. Mycologia 104:1–13

Toro M, Azcón R, Barea JM (1997) Improvement of arbuscular mycorrhizal development by inoculation with phosphate-solubilizing rhizobacteria to improve rock phosphate bioavailability (^{32}P) and nutrient cycling. Appl Environ Microbiol 63:4408–4412

White PJ, Hammond JP (2008) Phosphorus nutrition of terrestrial plants. In: White PJ, Hammond JP (eds) The ecophysiology of plant-phosphorus interactions. Plant ecophysiology, vol 7. Springer, Netherlands, pp 51–81

Whitelaw M (2000) Growth promotion of plant inoculated with phosphate-solubilizing fungi. Adv Agron 69:99–151

Zaidi A, Ahemad M, Oves M et al (2010) Role of phosphate-solubilizing bacteria in legume improvement. In: Khan M, Zaidi A, Musarrat J (eds) Microbes for legume improvement. Springer, Vienna, pp 273–292

Zapata F, Roy R (2004) Use of phosphate rocks for sustainable agriculture. Fertilizers and Plant Nutrition, Bulletin 32. Food and Agriculture Organization of the United Nations and International Atomic Energy Agency, Rome, pp 148

Chapter 25
Arbuscular Mycorrhizas: The Lives of Beneficial Fungi and Their Plant Hosts

Paola Bonfante and Alessandro Desirò

Abstract When plants colonized the land 450 million years ago, they were already associated to soil fungi, which assisted them in facilitating the uptake of mineral nutrients. This symbiotic association is known as *mycorrhiza*, a word that covers all the symbioses established between plants and beneficial fungi. Their presence in most environments suggests that *evolution* has promoted mycorrhizas because of the benefits gained by both partners. The improved nutrient status has a positive impact on the overall plant physiology, as it influences growth, water absorption and protection from root diseases. Mycorrhizal fungi instead acquire organic carbon directly from their green hosts, and accomplish their life cycle. These features are considered as landmarks of mutualistic symbioses. Owing to the huge diversity of plant and fungal taxa involved, and the multiplicity of the resulting interactions, mycorrhizas are usually classified in two broad categories, known as *ecto- and endomycorrhizas*, depending on whether the fungus colonizes the root's intercellular spaces or develops inside the plant cells. The aim of this chapter is to focus on arbuscular endomycorrhizal symbiosis, which is the most ancient and common plant-fungal association, and to provide an overview that updates traditional knowledge with recent data.

25.1 Arbuscular Mycorrhiza: At the Root of Endosymbioses

Of all the amazing diversity hidden inside the mycorrhizal world, Arbuscular Mycorrhizal (AM) symbiosis is the most widespread type, as it occurs in more than 80 % of land plants, and involves, as symbiotic fungi, the Glomeromycota, an ancient

P. Bonfante (✉) · A. Desirò
Department of Life Sciences and Systems Biology, University of Torino,
Viale Mattioli 25, 10125 Torino, Italy
Tel.: + 39 011 6705965
e-mail: paola.bonfante@unito.it

A. Desirò
Tel.: + 39 011 6705775
e-mail: alessandro.desiro@unito.it

© Springer International Publishing Switzerland 2015
B. Lugtenberg (ed.), *Principles of Plant-Microbe Interactions,*
DOI 10.1007/978-3-319-08575-3_25

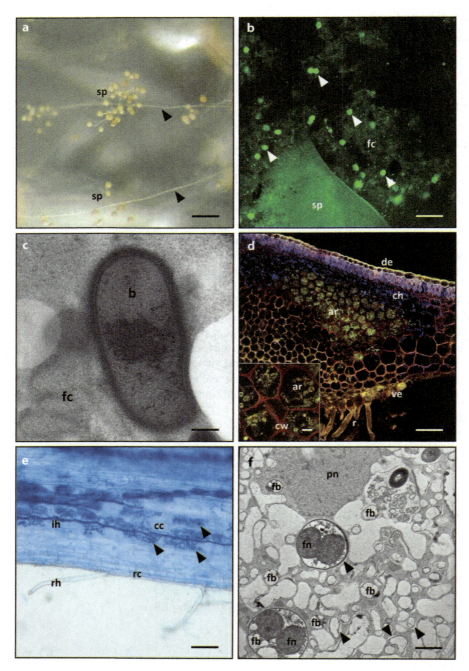

Fig. 25.1 AM symbiosis main features. **a** Cluster of asexual spores (*sp*) of *Rhizophagus intraradices* surrounded by extraradical hyphae (*arrow head*) which represent the extraradical AMF structures. **b** A squashed spore (*sp*) of *Gigaspora margarita* reveals its multinucleated nature. The nuclei, green

phylum that has coevolved with plants for at least 450 million years. *Arbuscular mycorrhizas* (AMs) contribute to the uptake of soil nutrients in plants, thus increasing their productivity and conferring resistance to stress. At the same time, as obligate biotrophs, *AM fungi* (AMF) cannot grow in pure culture, as they are unculturable in the absence of their host: they depend on the plant for their viability and in particular for carbohydrates. Their uniqueness is also due to other biological traits: the concept of species is poorly defined in this fungal group and reflects a high degree of genetic and functional variability; this situation has also led to difficulties in the assignment of a phylogenetic position. On the basis of the genome of *Rhizophagus irregularis*, a ubiquitous fungus which was the first AMF to be sequenced (Tisserant et al. 2013; Lin et al. 2014), we can state that Glomeromycota are phylogenetically closer to Mucoromycotina, a basal group of fungi, than to Asco-and Basidiomycota (Schüßler et al. 2001). A particular feature is their multinucleated status: spores and syncytial hyphae contain thousands of nuclei (Fig. 25.1b, f), and this makes classical genetic approaches unsuitable and opens questions on whether nuclei are genetically divergent. The genome sequencing of individual nuclei has revealed a consistent uniformity, thus pointing to the homokaryotic nature of *R. irregularis* (Lin et al. 2014). AMF are considered to be asexual, although genetically distinct strains anastomose with each other exchanging genetic material, and many mating-type related genes have been described. Finally, many AMF contain endobacteria in their cytoplasm, and this leads to an unexpected increase in their genetic complexity (Fig. 25.1c), even if the role of such novel fungal microbiota requires to be better investigated.

On the other hand, the availability of genetic tools and genomic information for several host plants, has shed light on the multiple aspects of plant-fungal interactions, including the process of root colonization, communication between the symbionts and the contribution of each partner to the functioning of the association (Gutjahr and Parniske 2013) .

Fig. 25.1 (continued) spots (*arrow head*) within the fungal cytoplasm (*fc*), were stained using SYTO 9 fluorescent dye. **c** AMF spores and mycelia may often contain endobacteria. The transmission electron micrograph shows a rod-shaped bacterium (*b*) embedded in the fungal cytoplasm (*fc*) and named *Candidatus* Glomeribacter gigasporarum. **d** AMF colonize not only roots, but also thalli of basal plants, like the liverwort *Conocephalum conicum* here illustrated. Fungal structures are labelled (*in green*) with an AMF-specific probe. Ventral (*ve*) and dorsal (*de*) epidermis, and chlorenchyma (*ch*) are never colonized. In the inset, a detail of arbuscules (*ar*) in the liverwort parenchyma cells. Rhizoids (*r*), liverwort cell wall (*cw*). **e** A root of grapevine colonized by *Funneliformis mosseae* which produces branched arbuscules (*arrow head*) in the cortical cells (*cc*). Root hairs (*rh*), rhizodermal cell (*rc*), intracellular hypha (*ih*). **f** Transmission electron micrograph of an arbusculated cell of *Lotus japonicus*. The details of the plant-fungal interface are shown: a membrane of plant origin (*arrow head*) surrounds the fungal branches (*fb*). This area is considered the main site for nutrient exchange between the two partners. Plant nucleus (*pn*), fungal nucleus (*fn*). Scale bars: **a** 400 μm; **b** 18 μm; **c** 0,2 μm; **d** 90 μm (10 μm in the inset); **e** 150 μm; **f** 1,5 μm

25.2   Arbuscular Mycorrhizal Fungi are an Important Component of the Plant Microbiota

Thousands of microbes are associated with plant roots, forming the so-called root microbiota (Chap. 3; Chap. 43). Among these, AMF assume a prominent position, as they play an essential role in ecosystem functioning: they influence organism interactions and provide a range of benefits to the host plant which, in return, supplies carbon to the fungus. AMF are important determinants of plant biodiversity, ecosystem variability and productivity: it has been demonstrated that an increase in the number of AMF species corresponds to an increase in aboveground biodiversity and productivity (van der Heijden et al. 2008). Hence, given their potential beneficial effects, it is essential to understand the factors that control the assembly, the distribution and the dynamics of AMF in order to identify the main drivers of the microbial communities in natural and agricultural ecosystems, and to monitor and maximize their ecosystem functions.

The occurrence of AMF in plants from most parts of the world has been studied since 1844 when the Tulasne brothers described the first AMF species, *Glomus macrocarpum* and *G. microcarpum*. After about 150 years, AMF were grouped within a monophyletic phylum, the Glomeromycota (Schüßler et al. 2001), which is distinct from the Zygomycota where they had previously been placed. The phylum Glomeromycota is currently represented by about 250 described species. Before the advent of molecular techniques, AMF identification was based solely on the microscopic examinations of their spores. However, spores are simple structures that offer a rather limited number of features to help discriminate among taxa and, in some cases, one species might easily be mistaken for another. In the nineties, the development of PCR-based approaches led to novel identification rules for AMF: since then, their taxonomy has been the subject of several changes and now, more than ever, it is extensively debated and controversial. New AMF species are frequently described, and their taxonomy is reorganized at different hierarchical levels (Redecker et al. 2013): until 2001, AMF were distributed in 1 class, 1 order, 3 families, and 6 genera; through the use of molecular phylogeny, they were later assigned to 1–3 classes, 4 to 5 orders, 11–14 families, and to 25–29 genera, depending on the scheme of the adopted classification. To make matters more complex, the use of DNA sequencing methods has allowed Glomeromycota to be directly detected from environmental samples (*i.e.* soil and plant root), thus the number of retrieved fungal phylotypes has increased. Many sequences assigned to the "uncultured" or "environmental" category have been uploaded in gene banks; however, the AMF phylotypes retrieved from soil and plant roots exceed the described AMF species, suggesting that the diversity of these fungi is still underestimated (Öpik et al 2013). Metagenetic approaches based on new technologies (see Chap. 30), such as high-throughput sequencing of DNA amplicons (*i.e.* a fragment of 18S rRNA gene) are excellent tools to increase our knowledge about diversity and distribution of AMF species. Studies on environmental samples have been carried out in temperate, boreal, tropical African and American woodlands and in grasslands as well as in natural ecosystems in Europe and North America. However, Asia, South America, Africa and Oceania are still poorly studied: a higher molecular diversity of AMF from these climatic and geographic zones can be expected (Öpik et al. 2013).

25.3 The Colonization Process

AMF establish symbiotic associations with almost 80 % of plants, including basal plants, such as liverworts and hornworts, ancient trees like *Ginkgo biloba*, most herbaceous plants and the cultivated crops that feed the world. Owing to this diversity, several plant tissue colonization patterns have been described that mirror the root anatomy of the involved host, but in spite of this, many features are shared (Fig. 25.1d). The propagules of AMF are asexual spores which germinate independently of the host plant to develop an *asymbiotic mycelium* (Fig. 25.1a), which cannot grow for more than a few days without the host root, due to the obligate biotrophic nature of Glomeromycota. When AMF perceive the host exudates, extraradical hyphae start branching, probably to improve their chances of contacting the root surface. This hypha-epidermis contact leads to the differentiation of hyphopodia, the swollen fungal structures that adhere to the plant cell wall and produce the first intraradical structures, *i.e.* penetration hyphae. Host cell integrity is maintained by the invagination of its plasma membrane, which proliferates and engulfs the developing hypha, physically separating the fungus from the plant cytoplasm. Colonization is often more massive in the cortical tissues, where hyphae spread the infection. Their final target is the inner cortical cells, where hyphae penetrate and branch repeatedly in order to differentiate into arbuscules, the small fungal trees (Fig. 25.1e) that give their name to this form of mycorrhizas and which represent the main site of nutrient exchange (Fig. 25.2).

In conclusion, the colonization process involves a host-independent and time-limited presymbiotic phase, and a symbiotic phase where extra- and intra-root fungal structures are developed. Root colonization is in fact accompanied by the proliferation of the extraradical mycelium, which extends beyond the nutrient depletion zone that surrounds the roots, and provides the structural basis for improving the ecological fitness of mycorrhizal plants. The description of such a colonization process has led to questions concerning the way in which the partners communicate in the rhizosphere and establish a physical contact at the surface of the roots, and how the plant accommodates the fungus inside its cells, making the symbiosis functionally active. Replying to these questions has required complementation between molecular, genetic, biochemical and cell biology approaches.

The Molecular Dialogue The release of soluble signals in the rhizophere is an easy way of allowing the mycorrhizal partners to be informed of their reciprocal presence, before physical contact. Plants are known to release a cocktail of molecules that attract AMF, in combination with chemiotropic responses. Among them, carotenoid-derived strigolactones (SL) have been well characterized. These molecules which

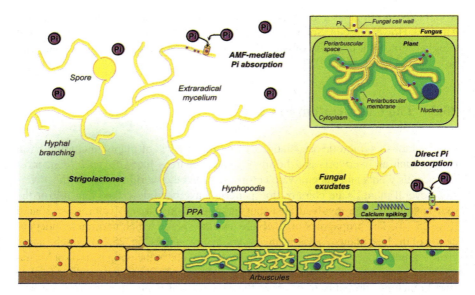

Fig. 25.2 The scheme illustrates the different steps of the root colonization process and the main pathways of the phosphate transfer from the soil to the plant cells. After germination of the resting spore, the mycelium starts to explore the surrounding soil. The perception of strigolactones, present in root exudates, induces hyphal branching. In the meantime, plant exudates induce the production of signalling molecules by the fungus, which activate the symbiosis signalling pathway and trigger calcium spiking in rhizodermal cells. The contact between the root and the fungus occurs after the formation of the hyphopodium on the root surface. This induces the aggregation of cytoplasm and the formation of the prepenetration apparatus (*PPA*) in the contacted rhizodermal cell and the underlying outer cortical cell. Thus, the fungus starts the intracellular invasion of the root, following the route chartered by the PPA from the surface to the inner cortical cells, where hyphae branch to form arbuscules. P_i is present in the soil at a low concentration. The plant itself directly takes up the phosphate from the soil, even if this event is limited to the proximity of the root. Differently, the hyphae of the extraradical mycelium grow over the depletion zone and actively absorb phosphate from the soil. The phosphate is condensed as polyphosphate granules which move along the hyphae from the outside to the inside of the root, till the arbuscules. Here (see the arbuscule details) the phosphate is released from the fungus towards the interface area and actively transferred to the plant by the AM-inducible plant phosphate transporters which are expressed on the periarbuscular membrane.

were first described as stimulators of seed germination of parasitic plants (Ruyter-Spira et al. 2013), elicit a branching in AMF and stimulate their metabolism at low concentrations. In addition, SLs have been shown to be powerful plant hormones, which are involved in shoot growth regulation by inhibiting the development of lateral buds. The multiple roles of SLs suggest that they play a basic role in plant development and that their leakage from the roots into the soil may have become useful to different root-interacting organisms (symbiotic fungi and parasitic plants) as they co-evolved with their hosts. Other plant molecules, which have been identified as cutin monomers, are considered important cues to allow hyphopodium formation (Oldroyd 2013).

Upon germination, AMF release diffusible signals, which are perceived by host plants, even in the absence of physical contact. These factors, often referred to as 'Myc factors' in analogy with the nodulation (NOD) factor of nitrogen fixing rhizobia, activate a number of plant responses: upregulation of genes involved in signal transduction, stimulation of lateral root development, starch accumulation, and repeated calcium oscillations in epidermal cells, in analogy with rhizobium-legume symbiosis. Bioactive exuded molecules have so far been identified as lipochitoligosaccarides (Myc-LCOs), which are very similar in structure to the NOD-factor (Maillet et al. 2011), and tetra- and penta-chitooligosaccharides (Genre et al. 2013). These chitin-related molecules activate a signalling pathway, which depends on the genes that are required for the establishment of both AM and nodule symbiosis, and for this reason are known as Common Symbiosis (SYM) genes. Such genes were first discovered in legumes but were then found to be present in many AM host plants. The proteins encoded by SYM genes activate a pathway that starts with the perception of microbial signals at the plant plasma membrane, thanks to lysine-motif (LysM) receptor kinases (Oldroyd 2013). Together with other co-receptors, the LysM proteins transduce the microbial signal to the cytoplasm and then to the nucleus, as suggested by the nuclear localization of all the downstream elements of the SYM pathway. Here, the generation of a nuclear calcium spiking is registered, and this offers an easily detectable marker for early symbiotic events. A calcium calmodulin-dependent protein kinase plays a central role in this cascade, probably decoding calcium oscillations and regulating a number of downstream genes. The proteins encoded by SYM genes are also involved in the intracellular fungal accommodation: mutants for SYM genes are not only defective in the signaling pathway (as testified by the lack of nuclear calcium spiking), but also in the assembly of the pre-penetration apparatus (PPA) and the subsequent fungal colonization (Bonfante and Genre 2010) (Fig. 25.2). It has been demonstrated that when the fungus produces the hyphopodium, the contacted epidermal cells start to assemble the machinery that is used to build the interface compartment where the fungus will be hosted. Cytoplasm aggregates at the contact site, concentrating endoplasmic reticulum membranes, Golgi bodies and secretory vesicles. Again, intense oscillations of calcium concentrations take place, and require the presence of SYM genes. Only when the PPA is completed, does the fungus start growing inside the roots. On this basis we can conclude that the molecular dialogue between AM partners acts at multiple levels: driving the fungus towards the plant, regulating its recognition upon physical contact and controlling early colonization events.

The Genetics and Molecular Basis of Plant-Fungal Interaction The current knowledge on AMs shows that plant cells actively control the colonization process. The development of mycorrhiza-defective mutants has resulted to be a powerful tool in deciphering the genetics basis of such a control. Mutants in SYM genes were the first to be investigated at the beginning of the twenty-first century: the genetic dissection of the colonization process became possible through a comparison of the impact of a specific mutation on nodule and AM formation in legumes (Oldroyd 2013). In recent years, other common genes have been found to be involved in the accommodation of both root microbes. These include Vapyrin, which is essential not only for arbuscule progression but also for rhizobial infection (Gutjahr and Parniske 2013) and a group of vesicle-associated membrane proteins (VAMPs). They are the molecular determinants for the biogenesis of the novel membrane which surrounds both rhizobia and AMF, when they have an intracellular location (Ivanov et al. 2012). The development of the perifungal membrane, which surrounds intracellular hyphae and defines the symbiotic interface, is in fact a common feature of all biotrophic interactions (Balestrini and Bonfante 2014). Other genes are, instead, AM specific and have an impact on arbuscule morphogenesis and functioning. Among them, the best characterized gene is the phosphate transporter that is induced by mycorrhization, is expressed at the periarbuscular membrane and is considered a functional marker for AMs. When silenced there is a block in the branching of the arbuscule which assumes a stunted morphology (Harrison 2012). Similarly, mutation in two ABC transporters results in arbuscular defects. Taken together, these findings suggest that the plant proteins which are essential for arbuscule development are mostly located in the perifungal membrane, regulate the nutritional exchanges and seem to be independent of the SYM signalling pathway.

Studies on mutants have been successfully integrated with molecular approaches based on the description of plant transcriptomic profiles upon AM colonization. The significant cell reorganization during root colonization is associated with changes in the transcriptome of AM roots. The pattern of gene expression of different root cell types during colonization has been investigated through genome-wide transcriptome profiling, combined with quantitative realtime-PCR on several model plants, including rice and tomato (Salvioli and Bonfante 2013). Thanks to the use of laser microdissection, the most prominent changes have been identified- as expected- in the arbusculated cells where hundreds of genes have resulted to be differentially regulated. Regardless of host and AM fungal identity, a core set of genes are consistently differentially regulated, leading to the identification of a "mycorrhizal signature". This set usually contains some gene classes, such as the nutrient transporters, including mycorrhiza-inducible phosphate, nitrogen and sulphate transporters, which are considered functional markers of an active AM. Another group of plant genes that are characteristic of mycorrhizal roots is related to membrane and cell wall synthesis: according to the notion that the fungus is limited by the interface compartment where cell-wall material is laid down, many cell wall related genes are upregulated, as also confirmed by immunocitochemical and *in situ* hybridization observations (Balestrini and Bonfante 2014) .

Transcriptome analyses have detected the effects of mycorrhization on plant defense mechanisms. Pathogenesis related proteins are reported to be activated in AMs thus confirming that mycorrhizal plants seem to react more promptly to pathogen attack, eventually showing enhanced resistance. The underlying mechanisms are not fully understood, since many events probably overlap: the fungus is perceived as a foreign microbe, since it releases elicitor molecules. Chitin is one of the best known fungal elicitors, and many chitin receptors have been identified as regulators of plant responses during their interaction with pathogens (Chap. 10). AMF are known to release a cocktail of chitin-related molecules which might be perceived as detrimental or beneficial signals. In order to respond to the former, the host develops a higher defense level, which is overcome by the fungal effectors. It is reasonable to presume that, at the end, positive signals are dominant, since the plant usually allows AMF colonization. A current hypothesis is therefore that mycorrhizal establishment stimulates the priming of a plant's innate immune system rather than the induction of specific defense mechanisms.

Taken as a whole, a combination of genetics, molecular and morhological approaches has revealed how the process of fungal colonization and arbuscule development can be dissected into phases that require a finely tuned activation of specific plant genes. Although fungal biology is now starting to be unravelled thanks to the genome sequencing of *R. irregularis* (Tisserant et al. 2013), a straightforward way of transforming AMF is still not available, and—as a result—the functional meaning of a fungal gene/protein cannot be validated easily. All the current knowledge therefore indicates a dominant role of the plant on fungal development.

25.4 The Functioning Mycorrhiza

Although AM host plants can survive without their symbiont, this condition is virtually unknown in natural ecosystems, where AMF function as helper microorganisms, and improve the overall plant fitness. AMF have so far proved to be unculturable in the absence of a host. Being unable to absorb carbohydrates, except from inside a plant cell, they depend totally on their green hosts for the organic carbon metabolism, which gives them the status of obligate biotrophs. Although carbon transfer from plants to AMF was demonstrated already in the 1960s, its molecular mechanisms are still unclear. The genome sequencing of *R. irregularis* has provided clear evidence that this AMF at least does not possess genes coding for polysaccharide and plant-cell wall degradation (Tisserant et al. 2013), while a high-affinity monosaccharide transporter MST2 has been shown to play a major role in the uptake of glucose and xylose during the symbiotic and extraradical phase (Helber et al. 2011). However, AMF have an excellent capacity to take up all the required minerals from the soils as well as organic nitrogen products. Their biotrophy therefore seems to be related to organic carbon, while they may successfully exploit the other soil nutrients, unlike pathogenic biotrophs, that have no access to soil nutrients at all. AMF were first shown to possess high-affinity inorganic phosphate (P_i) transporters many years ago and this provided a breakthrough in the understanding of fungal functions (Fig. 25.2). Accumulated as polyphosphate, P_i is then rapidly translocated along the aseptate mycelium to the host plant.

Nitrogen is the other important element taken up by AMF, and the genes that are involved in the transport of ammonium and amino acids have been identified, whereas arginine is probably the preferred molecule for long-distance transport to the host plant. In conclusion, AMF seem to act as an active bridge between the soil and the plant. However, not all AMF perform equally well in releasing minerals to the plant: it has been demonstrated using an *in vitro* system that a reward process exists thanks to which the fungal partners enforce cooperation by increasing nutrient transfer to only those roots that provide more carbohydrates. Vice versa, plants can detect, discriminate, and reward the best fungal partners that provide them more minerals (Kiers et al. 2011). This description, which is based on the use of stable isotope probing to track and quantify both plant and fungal resource allocation to their reciprocal partners, offers a physiological view of the so-called biological market, and describes the behavior of mycorrhizal roots at an organismic level. Molecular approaches have identified the arbusculated cell and, at a sub-cellular level, the interface area as the compartment in which the nutritional exchanges take place.

25.5   From Root to Food: Conclusions and Perspectives

In addition to the expected effects on roots, mycorrhization also affects gene expression in shoots, leaves and fruits, and modulates the genes involved in diverse metabolic processes, such as photosynthesis, defence, transport and hormonal metabolism. These systemic effects open new questions on whether mycorrhiza establishment can improve the nutritional value of specific crops, by enhancing the content of nutritionally valuable compounds in their edible parts (Salvioli and Bonfante 2013). Interestingly, AMF biodiversity itself has been shown to promote plant productivity, and also to buffer, in some way, the productivity fluctuations that occur under differing environmental conditions. These latter findings clearly suggest that the multifunctional ability of AMF to improve plant performance is still far from being fully understood, and this makes their use in agriculture much more valuable than as a simple substitute of chemical fertilizers/pesticides (Salvioli and Bonfante 2013).

Maximizing the beneficial effects of AMF requires a better understanding of the molecular mechanisms of symbiosis, by on one hand studying the different responses of plant species and genotypes and on the other hand by developing a deeper knowledge on the still enigmatic biology of AMF. This could lead to the formulation of more efficient fungal inoculants that are suitable for specific in field applications: their availability in the market is thus a urgent need upon scientific validation. As a prerequisite, a closer link between field research and laboratory studies must be established, in order to fill the gap that exists between basic knowledge, availability of novel molecular tools and the systemic effects of AMF in plants of agronomic relevance.

Acknowledgements   Contributions to this review were partly funded by the projects: MY-COPLANT (Compagnia di San Paolo and UNITO), MIUR PRIN 2012 and UNITO (60 % 2014).

References

Balestrini R, Bonfante P (2014) Cell wall remodeling in mycorrhizal symbiosis: a way towards biotrophism. Front Plant Sci 5:237
Bonfante P, Genre A (2010) Mechanisms underlying beneficial plant-fungus interactions in mycorrhizal symbiosis. Nat Commun 1:48

Genre A, Chabaud M, Balzergue C et al (2013) Short-chain chitin oligomers from arbuscular myc-
    orrhizal fungi trigger nuclear Ca2 + spiking in *Medicago truncatula* roots and their production
    is enhanced by strigolactone. New Phytol 198:190–202
Gutjahr C, Parniske M (2013) Cell and developmental biology of arbuscular mycorrhiza symbiosis.
    Annu Rev Cell Dev Biol 29:593–617
Harrison MJ (2012) Cellular programs for arbuscular mycorrhizal symbiosis. Curr Opin Plant Biol
    15:691–698
Helber N, Wippel K, Sauer N et al (2011) A versatile monosaccharide transporter that operates
    in the arbuscular mycorrhizal fungus *Glomus* sp is crucial for the symbiotic relationship with
    plants. Plant Cell 23:3812–3823
Ivanov S, Fedorova EE, Limpens E et al (2012) *Rhizobium*–legume symbiosis shares an exocytotic
    pathway required for arbuscule formation. Proc Natl Acad Sci USA 109:8316–8321
Kiers ET, Duhamel M, Beesetty Y et al (2011) Reciprocal rewards stabilize cooperation in the
    mycorrhizal symbiosis. Science 333:880–882
Lin K, Limpens E, Zhang Z et al (2014) Single nucleus genome sequencing reveals high similarity
    among nuclei of an endomycorrhizal fungus. PLoS Genet 10:e1004078
Maillet F, Poinsot V, André O et al (2011) Fungal lipochitooligosaccharide symbiotic signals in
    arbuscular mycorrhiza. Nature 469:58–64
Oldroyd GED (2013) Speak, friend, and enter: signalling systems that promote beneficial symbiotic
    associations in plants. Nat Rev Microbiol 11:252–263
Öpik M, Zobel M, Cantero JJ et al (2013) Global sampling of plant roots expands the described
    molecular diversity of arbuscular mycorrhizal fungi. Mycorrhiza 23:411–430
Redecker D, Schüßler A, Stockinger H et al (2013) An evidence-based consensus for the
    classification of arbuscular mycorrhizal fungi (Glomeromycota). Mycorrhiza 23:515–531
Ruyter-Spira C, Al-Babili S, van der Krol S et al (2013) The biology of strigolactones. Trends Plant
    Sci 18:72–83
Salvioli A, Bonfante P (2013) Systems biology and "omics" tools: a cooperation for next-generation
    mycorrhizal studies. Plant Sci 203–204:107–114
Schüßler A, Schwarzott D, Walker C (2001) A new fungal phylum, the Glomeromycota: phylogeny
    and evolution. Mycol Res 105:1413–1421
Tisserant E, Malbreil M, Kuo A et al (2013) Genome of an arbuscular mycorrhizal fungus provides
    insight into the oldest plant symbiosis. Proc Natl Acad Sci U S A 110:20117–20122
van der Heijden MGA, Bardgett RD, van Straalen NM (2008) The unseen majority: soil microbes
    as drivers of plant diversity and productivity in terrestrial ecosystems. Ecol Lett 11:296–310

Chapter 26
Plant Hormones Produced by Microbes

Stijn Spaepen

Abstract Plant hormones or phytohormones are historically classified into five major classes: auxins, cytokinins, gibberellins, abscisic acid and ethylene. Nowadays, many other phytohormones have been identified. Diverse microbial species possess the ability to produce phytohormones, with most data accumulated for the production and role of auxin. In this chapter the microbial biosynthesis, its regulation and the role of the different phytohormones in the interaction with the plant are discussed. Microbial phytohormonal production is a potent mechanism to alter plant physiology, leading to diverse outcomes from pathogenesis to promotion of plant growth. However, genetic evidence for the role of many phytohormones in microbe-plant interactions is still lacking, thus questioning the importance of the microbial production. Targeted approaches focusing on genetic evidence for the role of phytohormones together with plant experiments in an agronomic setting will allow unraveling the importance and potential of this fascinating microbial trait.

26.1 Phytohormones in Plants

Hormones are defined as chemical compounds that are produced in small amounts in a certain tissue controlling and regulating various functions related to growth, metabolism and reproduction in the receptive tissue. Plants produce different hormones, also called phytohormones, but the structures of these hormones are—in contrast to those of animals -rather simple and the molecules are not produced and stored in specific glands.

The five classical phytohormone classes are auxins, cytokinins, gibberellins, abscisic acid and ethylene. More recently discovered phytohormones include strigolactones, brassinosteroids, jasmonate, salicylic acid, polyamines and nitric oxide.

S. Spaepen (✉)
Centre of Microbial and Plant Genetics, KU Leuven, Kasteelpark Arenberg 20,
Box 2460, 3001 Heverlee, Belgium
Tel.: + 32 16321631
e-mail: stijn.spaepen@biw.kuleuven.be; spaepen@mpipz.mpg.de

Department of Plant Microbe Interactions, Max Planck Institute for Plant Breeding Research,
Carl-von-Linné-Weg 10, 50829 Köln, Germany
Tel.: + 32 16321631

© Springer International Publishing Switzerland 2015
B. Lugtenberg (ed.), *Principles of Plant-Microbe Interactions,*
DOI 10.1007/978-3-319-08575-3_26

247

The physiological functions of these hormones have been studied in detail in the past decades but only recently the molecular mechanisms of how these hormones exert their effects have been unraveled. (Davies 2010; Santner et al. 2009).

Most phytohormones exert their activity at very low concentrations and changes in hormone concentrations can drastically alter plant growth and development both in a positive and negative way as illustrated by the dose-response curve for auxins. Additionally, most phytohormones do not act alone on a certain growth or developmental aspect and therefore balances between different hormonal classes can be important. Another overlooked aspect of phytohormonal action is the conjugation of the active hormone to other molecules such as sugars and amino acids, (reversibly) inactivating the hormone.

26.2 Phytohormones in Microbes

Phytohormones have also been detected and identified in the supernatant of culture medium of many soil and plant-associated bacteria and fungi. In these organisms, the phytohormones do not induce typical hormonal or major physiological changes. Microbial phytohormone production has been linked to changes in root architecture and plant growth promotion. However, the degree of proof for their involvement can vary a lot depending on the phytohormone and the studied microbial strain. The presence of a certain phytohormone in the supernatant of a microbial culture is not enough to prove a functional role of this molecule in its interaction with the plant. Further evidence can be the correlation of growth responses of plants with the hormone levels measured in the culture medium or on/in the colonized plant tissues. The ultimate proof is the inoculation with a bacterial mutant strain, impaired in the phytohormonal biosynthesis, to directly demonstrate the involvement of the phytohormone. The five main classes of phytohormones can be identified in the culture medium of many microbes (see Table 26.1 for an overview of the discussed phytohormones) but the subsequent analyses necessary for proving their role in plant growth promotion are lacking for many described cases.

Auxins

This group of phytohormones has the ability to induce cell elongation in the subapical region of the stem. Besides this ability, auxins are involved in almost all aspects of plant growth and development such as stem and root elongation, stimulation of cell division, lateral and adventitious root initiation, apical dominance, vascular tissue differentiation, gravitropism and phototropism (Davies 2010). The most important naturally occurring auxin is indole-3-acetic acid (IAA). Other molecules such as indole-3-butyric acid and phenylacetic acid are also considered active auxins but their biosynthetic pathways and roles have not been extensively studied.

Table 26.1 Most important phytohormones, produced by microbes

Class	Structure [1]	Effect on plant
Auxin		Cell elongation and division Root initiation Apical dominance
Cytokinin		Inhibition of root elongation Stimulation of cell division Leaf expansion by cell enlargement Delay of senescence
Gibberellin		Seed germination Stem elongation Floral induction and fruit growth
Abscisic acid		Stomatal closure Inhibition of shoot growth Bud dormancy Abiotic and biotic stresses
Ethylene	$H_2C=CH_2$	Stress and ripening hormone Senescence and abscission Abiotic and biotic stresses

[1] One representative is shown if multiple compounds belong to one class. Auxin, indole-3-acetic acid; cytokinin, zeatin; gibberellin, GA_3 or gibberellic acid.

IAA Biosynthetic Pathways in Microbes Until now, six biosynthetic pathways have been described in microbes, with most pathways being suggested based on the presence of metabolic intermediates in the culture medium (see Spaepen et al. 2007 for details). For many pathways no genetic evidence is available and therefore the presence and importance of these pathways needs to be uncovered. Most pathways use the aromatic amino acid tryptophan as precursor. Despite the multitude of pathways, there are apparently two dominant microbial pathways based on both the abundance and genetic evidence for these pathways: one via the intermediate indole-3-acetamide (IAM) and one via the intermediate indole-3-pyruvate (IPyA) (see Fig. 26.1). In the IAM pathway, tryptophan is first converted by a tryptophan monooxygenase to IAM, which is then catalyzed to IAA by an IAM hydrolase. The pathway has been well characterized in many phytopathogenic bacteria and also in some rhizobia. In the IPyA pathway abundant in beneficial plant-associated bacteria, tryptophan is in a first step transaminated to IPyA by an aromatic aminotransferase. In the second, rate-limiting step, IPyA is converted to indole-3-acetaldehyde (IAAld) by a decarboxylation reaction catalyzed by an IPyA decarboxylase (IPDC, encoded by the *ipdC* gene). Finally, IAAld is converted into IAA. In this pathway, the regulation (see below) and biochemical characterization of the second step (IpdC protein/*ipdC* gene) has been intensively studied in multiple bacterial species.

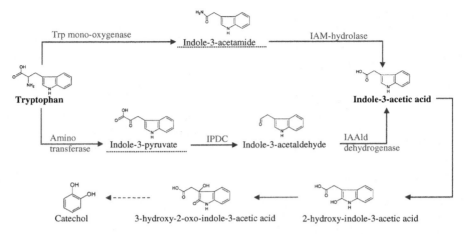

Fig. 26.1 Most important microbial IAA biosynthesis and degradation pathways. The *dashed arrow* refers to multiple steps. Intermediates underlined with a *dashed line* refer to the name of the pathway. IAAld, indole-3-acetaldehyde; IAM, indole-3-acetamide; IPDC, indole-3-pyruvate decarboxylase; Trp, tryptophan

Besides the two pathways described above, other microbial pathways for IAA biosynthesis have been proposed, but for most of these pathways genetic evidence is lacking. In the tryptamine pathway, tryptophan is first decarboxylated by a tryptophan decarboxylase and subsequently catalyzed to IAAld by an amine oxidase. Similar to plants, a pathway via indole-3-acetonitrile has been suggested, based on the conversion of indole-3-acetonitrile by nitrilases or nitrile hydratase to IAA directly or via IAM respectively.

Important to note is that in one single organism multiple IAA biosynthetic pathways may be present and active as demonstrated for *Pantoea agglomerans* which genome encodes both the IAM as well as the IPyA pathway (see below for details). As for plants, some storage products and conjugates also exist in microbes but their role is still unexplored.

Regulation of IAA Biosynthesis Since IAA is costly to produce, mainly due to the high cost to synthesize tryptophan, IAA biosynthesis in bacteria is tightly regulated. One exception is found for pathogenic *Agrobacterium* strains. The IAA biosynthetic genes are under control of strong (plant-specific) constitutive promoters. Since this DNA region is transferred to the plant upon infection, the tryptophan pool of the plant is used to produce massive amounts of IAA causing gall formation in combination with a high level of cytokinins (Jameson 2000). In other phytopathogens (not transferring DNA to the plant), expression is mostly linked to other virulence factors such as the type III secretion system (TTSS). In *Ralstonia solanacearum*, auxin biosynthesis is reduced in a TTSS mutant, while in *Erwinia chrysanthemi* (reclassified as *Dickeya dadantii*) the expression of TTSS genes is reduced in an IAA biosynthesis mutant. In *P. agglomerans*, IAA is a master regulator not only for TTSS genes, but also for quorum sensing related genes (Spaepen et al. 2007).

The regulation of IAA biosynthesis has also been studied in detail for some beneficial plant-associated strains (see for review Spaepen et al. 2007 and Patten et al. 2013). However, it is difficult to draw a general consensus and framework based on the available information. In this section, I will focus on some well-documented cases to highlight important regulatory factors and mechanisms. In general, the addition of the precursor tryptophan to the culture medium will increase IAA production. However, this increase is not always correlated with the induction of expression of IAA biosynthetic genes by tryptophan as illustrated for the effects of tryptophan in *Pseudomonas putida* and *Azospirillum brasilense* (expression induced and not-induced by tryptophan respectively). A very particular case of regulation is observed for *A. brasilense*: IAA itself is a main regulator. The key gene in IAA biosynthesis is regulated by a positive feed-back regulation. In this way, the expression increases with the cell density and reaches its maximum at the stationary phase coinciding with the highest IAA levels. The autoinduction is specific for auxin molecules and is not observed for auxin conjugates and tryptophan (Vande Broek et al. 2005).

Many environmental factors are involved in the regulation of IAA biosynthesis and these factors comprise pH, osmotic and matrix stress and nutrient limitations. Again, the effect of a certain environmental factor can vastly differ from strain to strain, but most of the observed effects reflect the environmental cues important for IAA production in that particular strain. Since IAA is important in the interaction with the plant, certain plant-derived molecules and cues will induce IAA biosynthesis and gene expression as demonstrated for flavonoids in *Rhizobium*, leavy gall extract for *Rhodococcus fascians* and the plant surface itself for *P. agglomerans*. In many gamma-Proteobacteria, IAA biosynthesis is regulated by the alternative sigma factor RpoS, a master regulator for the response upon stress conditions and starvation as demonstrated for different *Pseudomonas* and *Pantoea* strains. Other regulators comprise the GacS/GacA two-component system in *Pseudomonas chlororophis* and TyrR (regulatory protein for transport and metabolism or aromatic amino acids) in *Enterobacter cloacae*.

Effect of Microbial IAA on Plants and Microbes The effects of auxin production (mostly in combination with aberrant cytokinin production) by phytopathogens on plants are very pronounced (gall and tumor formation). Inactivation of auxin biosynthesis leads to a reduced or no gall formation directly demonstrating the link between bacterial IAA and plant disease. In the gall-inducing bacterium *P. agglomerans* pv. *gypsophilae* both the IAM and IPyA pathway are present, allowing to study the role of both pathways. Inactivation of the IAM pathway leads to a significant reduction in gall size without compromising colonization capacity while inactivation of the IPyA pathway leads to no significant decrease in gall size but a reduced epiphytic fitness as measured by colonization capacity (Manulis et al. 1998).

In beneficial bacteria, IAA production has always been linked to plant growth promotion since inoculation experiments with these strains result in increased root and shoot biomass especially under sub-optimal nitrogen levels. For *P. putida* and *A. brasilense*, mutant strains impaired in IAA production have been used to analyze the importance of bacterial IAA biosynthesis. For *A. brasilense*, inoculation of wheat roots results in typical auxin-like responses (shortening of primary root and induced number of lateral roots and root hairs) ultimately leading to an increased root surface allowing better uptake of nutrients. The changes in root architecture are also translated into higher biomass of the shoot. From an agronomic point of view, inoculation of these strains would allow reducing the N fertilizer input without compromising plant yield. However, from these inoculation studies it is also apparent that IAA biosynthesis is not the only mechanism responsible for plant growth promotion since the IAA-impaired mutants still exhibit residual promoting capacity. Also some fungi such as *Trichoderma* and *Fusarium* strains promote plant growth by the microbial auxin production, although for *Piriformospora indica* auxin biosynthesis is only necessary for root colonization.

Auxins not only affect plants, but also induce physiological changes along with altered gene expression in microbes. For *Agrobacterium tumefaciens*, it was shown that IAA shuts down *vir* gene expression, possibly being a cue for the bacterium for a successful plant transformation. In yeast, IAA induces adhesion and filamentation mediated by the surface protein FLO11 in a YAP-1 dependent manner. Here it was hypothesized that IAA can occur at plant wounding sites and serves as attractive signal for yeast. In *Escherichia coli*, IAA protects cells against adverse stress conditions. In addition, genes encoding cell envelope components and proteins overcoming stress conditions are upregulated upon IAA treatment. In beneficial bacteria, IAA seems to operate as a signal molecule to alter gene expression for presence of the plant environment. In *Rhizobium etli*, IAA-regulated genes are involved in flavonoid signal processing, attachment to the roots and switching off motility. Also in *A. brasilense*, IAA is a signal molecule as assessed by whole genome transcriptional analysis. In presence of IAA, it adapts itself to the interaction with the plant by altering the expression of genes coding for transport and cell surface proteins, transcription factors and the type VI secretion machinery (Spaepen and Vanderleyden 2011; Patten et al. 2013).

Auxin Degradation In *P. putida*, the *iac* locus is encoding for the enzymes responsible for IAA degradation. In a three-step reaction, IAA is degraded to catechol which can be further degraded to β-ketoadipate. In addition, a MarR-type repressor of the *iac* gene expression was identified which is probably released in the presence of IAA. Despite this genetic knowledge on IAA degradation, the ecological function of auxin degradation in plant interactions is unknown, although it has been suggested that this activity could alter the plant auxin homeostasis for the benefit of the bacterium. Alternatively, bacterial auxin degradation may provide a nutritional benefit (Scott et al. 2013).

Cytokinins, Gibberellins, Abscisic Acid and Ethylene

Biosynthesis and Role of Cytokinins Naturally occurring cytokinins (CK) are mostly derived from adenine and modified by substitutions at the N^6, including the respective ribotides, ribosides and glycosides. CKs promote cell division and differentiation in meristematic tissues both in plant roots and shoots. They are also involved in processes such as senescence delay, organ formation, root and root hair development and leaf expansion (Davies 2010).

In plants, CK biosynthesis starts with the transfer of an isoprenoid moiety (mostly dimethylallyl pyrophosphate) to adenosine phosphate catalyzed by adenosine phosphate-isopentenyltransferase to generate isopentenyl adenosine-5'-phosphate. The isoprenoid-derived side-chain can further be modified by hydroxylation. The cytokinin compound is then hydrolyzed to a free base by a (phospho-)ribohydrolase (Frébort et al. 2011).

The massive production of auxins and cytokinins by phytopathogenic bacteria or by the transfer of bacterial oncogenes into the infected plant exemplified by *Pseudomonas savastanoi* and *A. tumefaciens* are a strategy to induce tumor/gall formation (Jameson 2000). For the pathogen *R. fascians*, the *fas* operon, comprising six genes, is responsible for the biosynthesis of CKs necessary for leafy gall formation (Pertry et al. 2010).

For these pathogenic interactions, the role of CKs in their interaction with plants is clearly demonstrated. Also for many non-pathogenic bacteria, CK production has been demonstrated. Although plant growth-promoting action is claimed for CKs, firm evidence is lacking due to the absence of mutant strains defective in CK biosynthesis. However, it is proven that bacterial CKs are perceived by plant CK receptors, pointing towards the potential of bacterial CKs to influence CK signaling in plants.

Biosynthesis and Role of Gibberellins The class of gibberellins (GAs) is a broad group of more than 100 compounds that can be classified as tetracyclic diterpenoid acids, with *ent*-gibberellane as backbone. GAs are involved in developmental processes such as cell division and elongation during almost all stages of plant growth (from seed germination to fruit growth). In addition, the balance with other phytohormones is important in determining the role of GAs (Davies 2010).

Both in some fungi as bacteria, GA production has been detected and the biosynthetic pathways have been proposed and/or unraveled. In the fungal rice pathogen *Gibberella fujikuroi*, which is used to commercially produce GA_3, GA biosynthesis (starting from the precursor GA_{12}-aldehyde) has independently evolved from the plant pathways and differs especially at the stage in which the 3β- and 13-hydroxylation occurs. In *G. fujikuroi*, the biosynthetic genes are clustered together in the genome (Bömke and Tudzynski 2009). In bacteria, genetic evidence for GA biosynthesis is minor. Operons containing genes encoding for putative enzymes involved in GA biosynthesis were identified in some *Rhizobium* and *Bradyrhizobium* strains. Only a biochemical analysis of diterpene synthases of *Bradyrhizobium japonicum* provides some evidence for the role of this operon in GA biosynthesis. In *Azospirillum lipoferum*, a detailed GC-MS analysis could pinpoint the different synthesized GAs, allowing to suggest a putative pathway for biosynthesis (Cassan et al. 2014).

Some plant-associated microbes produce GAs as assessed by measuring the GA content of the culture medium. The best documented case for the role in plant growth promotion is the reversion of the dwarf phenotype of plants induced by GA inhibitors by inoculation with GA-producing *Azospirillum* strains (Lucangeli and Bottini 1997). Since most of these strains can also release GAs from conjugated inactive forms, it is unclear whether bacterial GA production directly or plant endogenously released GAs are responsible for this reversion. In root nodules of legumes infected with rhizobial strains, a higher amount of GAs is measured in comparison with non-infected roots. Since the identified blend of GAs in nodules resembles the GAs found in the culture medium of the rhizobial strains, it was suggested that nodule-derived GAs originated from the rhizobial strains.

Abscisic Acid Abscisic Acid (ABA) induces stomatal closure and fruit ripening and inhibits seed germination. In addition it is involved in bud dormance and protective responses against abiotic stresses such as drought, salt stress and metal toxicity (Davies 2010).

In plants, the "indirect pathway" is the route for ABA biosynthesis. In short, the carotenoid lycopene is modified in several enzymatic steps to violaxanthin, which is cleaved by a dioxygenase to xanthoxin. The latter compound is further converted to ABA in two enzymatic steps (dehydrogenase/reductase and aldehyde oxidase) (Nambara and Marion-Poll 2005).

Bacterial production of ABA has been reported for *A. brasilense* and *B. japonicum* strains, although the biosynthetic pathways are unknown. Since ABA inhibits cytokinin biosynthesis, bacterial ABA production can interfere with the cytokinin levels in plants. In addition under stress conditions, the bacterial ABA production might sustain the internal ABA pool in plants, alleviating the negative effects of the imposed stress.

Altered Ethylene Levels by Microbes The gaseous phytohormone ethylene is involved in physiological and developmental processes such as seed germination, cell expansion, senescence and abscission. It is sometimes called the ripening hormone since it induces fruit ripening (Davies 2010). Ethylene is also involved in the plant defense responses against pathogens. It can affect the outcome of the jasmonate-dependent defense responses by acting synergistically with jasmonate on one branch of the pathway leading to resistance to necrotrophic pathogens. However, ethylene antagonizes the MYC branch of the jasmonate pathway, leading to higher susceptibility to insect attacks (Pieterse et al. 2012). Ethylene also has a role in abiotic stress conditions. Elevated ethylene levels upon stress conditions may lead to inhibitory effects on plant growth.

Microbial ethylene production has been reported but mainly for bacterial pathogens such as *Pseudomonas*, *Xanthomonas* and *Erwinia*. The ethylene production is contributing to the bacterial virulence by inducing hormonal imbalances in the plant. The suggested bacterial biosynthetic pathways are distinct from the plant biosynthetic pathway. In the first bacterial pathway, methionine is the precursor, while in the second pathway ethylene is produced from 2-oxoglutarate (Tsavkelova et al. 2006). The full understanding of the putative ethylene biosynthetic pathways and the role of ethylene production in disease are still unclear.

Since ethylene accumulates in plants under stress conditions (e.g. drought and wounding), strategies to lower ethylene levels might alleviate stress-induced growth retardation. One such strategy is employed by some beneficial bacteria. In plants, ethylene is produced from methionine. Methionine is converted to 1-aminocyclopropane-1-carboxylate (ACC) and 5'-deoxy-5'methylthioadenosine via *S*-adenosylmethionine (SAM) by the consecutive action of a SAM synthase and ACC synthase. Finally ACC oxidase converts ACC to ethylene, CO_2 and cyanide. Bacteria that can lower ethylene levels encode for an enzyme ACC deaminase (AcdS) which can degrade the direct ethylene precursor ACC to α-ketobutyrate and ammonia, decreasing plant ethylene biosynthesis. This strategy has intensively been studied and is discussed in detail in Chap. 27 by BR Glick.

26.3   Concluding Remarks

The microbial production of phytohormones is a potent mechanism allowing microbes to alter plant physiology. For some phytohormones, the biosynthesis and role in interaction with the plant are well studied and sustained by genetic evidence as discussed above. However, for many examples this is not the case questioning the importance of this phytohormone in microbe-plant interactions. Further experiments are necessary to prove whether the microbial production is a real effector in the interaction or whether this is a by-product of the microbial metabolism without any substantial role. Futhermore, plant experiments in an agronomic setting are necessary. Under these conditions, the introduced organisms will need to compete with the indigenous microflora, a factor that can explain the low reproducibility and high variability.

Acknowledgements   Stijn Spaepen is a recipient of a postdoctoral fellowship grant from Research Foundation—Flanders (FWO-Vlaanderen). I wish to thank Jos Vanderleyden for fruitful discussions on this topic.

References

Bömke C, Tudzynski B (2009) Diversity, regulation and evolution of the gibberellin biosynthetic pathway in fungi compared to plants and bacteria. Phytochemistry 70:1876–1893

Cassan F, Vanderleyden J, Spaepen S (2014) Physiological and agronomical aspects of phytohormone production by model plant-growth promoting rhizobacteria (PGPR) belonging to the genus Azospirillum. J Plant Growth Regul 33:440–459

Davies PJ (2010) Plant hormones: Biosynthesis, signal transduction, action! Springer, Dordrecht

Frébort I, Kowalska M, Hluska T et al (2011) Evolution of cytokinin biosynthesis and degradation. J Exp Bot 62:2431–2452

Jameson P (2000) Cytokinins and auxins in plant-pathogens interactions—an overview. Plant Growth Regul 32:369–380

Lucangeli C, Bottini R (1997) Effects of Azospirillum spp. on endogenous gibberellin content and growth of maize (Zea mays L.) treated with uniconazole. Symbiosis 23:63–71

Manulis S, Haviv-Chesner A, Brandl MT et al (1998) Differential involvement of indole −3-acetic acid biosynthetic pathways in pathogenicity and epiphytic fitness of Erwinia herbicola pv. gypsophilae. Mol Plant Microbe Interact 11:634–642

Nambara E, Marion-Poll A (2005) Abscisic acid biosynthesis and catabolism. Annu Rev Plant Biol 56:165–185

Patten CJ, Blakney AJC, Coulson TJD (2013) Activity, distribution and function of indole −3-acetic acid biosynthetic pathways in bacteria. Crit Rev Microbiol 39:395–415

Pertry I, Vaclavikova K, Gemrotova M et al (2010) Rhodococcus fascians impacts plant development through the dynamic Fas-mediated production of a cytokinin mix. Mol Plant Microbe Interac 23:1164–1174

Pieterse CMJ, Van der Does D, Zamioudis C et al (2012) Hormonal modulation of plant immunity. Annu Rev Cell Develop Biol 28:489–521

Santner A, Calderon-Villalobos LIA, Estelle M (2009) Plant hormones are versatile chemical regulators of plant growth. Nat Chem Biol 5:301–307

Scott JC, Greenhut IV, Leveau JHJ (2013) Functional characterization of the bacterial iac genes for degradation of the plant hormone indole-3-acetic acid. J Chem Ecol 39:942–951

Spaepen S, Vanderleyden J (2011) Auxin and plant-microbe interactions. Cold Spring Harb Perspect Biol 3 pii: a001438

Spaepen S, Vanderleyden J, Remans R (2007) Indole-3-acetic acid in microbial and microorganism-plant signaling. FEMS Microbiol Rev 31:425–448

Tsavkelova EA, Klimova SY, Cherdyntseva TA et al (2006) Hormones and hormone-like substances of microorganisms: a review. Appl Biochem Microbiol 42:229–235

Vande Broek A, Gysegom P, Ona O et al (2005) Transcriptional analysis of the Azospirillum brasilense indole-3-pyruvate decarboxylase gene and identification of a cis-acting sequence involved in auxin responsive expression. Mol Plant Microbe Interact 18:311–323

Chapter 27
Stress Control and ACC Deaminase

Bernard R. Glick

Abstract During its lifetime, a plant is subject to a wide range of both biotic and abiotic stresses that can limit the growth and development of the plant. The one thing that all of these stresses have in common is that they induce the plant to synthesize growth-inhibiting stress ethylene. Plants that are treated with plant growth-promoting bacteria that synthesize the enzyme 1-aminocyclopropane-1-carboxylate (ACC) deaminase produce lower levels of stress ethylene as a consequence of the consumption of the ethylene precursor ACC by the enzyme. These treated plants are damaged/inhibited to a significantly lesser extent following a biotic or abiotic stress than are plants that are not treated with ACC deaminase-containing plant growth-promoting bacteria.

27.1 Plant Stress

Plants grown in the field are subject to a large number of both biotic and abiotic stresses. The biotic stresses include infection by fungal and bacterial pathogens as well as plant viruses, insect predation and nematode infection. Abiotic stresses include high and low temperature, drought, flooding, high salt concentrations, high metal concentrations, organic contaminants, mechanical wounding and excessive levels of radiation (Abeles et al. 1992). Plants respond to various stresses by synthesizing defensive/stress proteins and by modifying their physiology and biochemistry. The response of most plants to stress includes the synthesis of increased amounts of the low molecular weight gas ethylene, a plant hormone that acts to turn on the synthesis of other plant genes. Depending on the conditions, ethylene can both alleviate and exacerbate many of the effects of plant stress. Within a few hours following a biotic or abiotic stress, there is typically a small peak of ethylene synthesized by the affected plant (Glick et al. 2007). This ethylene peak consumes the small amount of 1-aminocyclopropane-1-carboxylate (ACC) that is generally present in non-stressed

B. R. Glick (✉)
Department of Biology, University of Waterloo, 200 University Avenue West,
Waterloo, ON N2L 3G1, ON, Canada
Tel.: 519-888-4567 x 32058
e-mail: glick@uwaterloo.ca

© Springer International Publishing Switzerland 2015   257
B. Lugtenberg (ed.), *Principles of Plant-Microbe Interactions*,
DOI 10.1007/978-3-319-08575-3_27

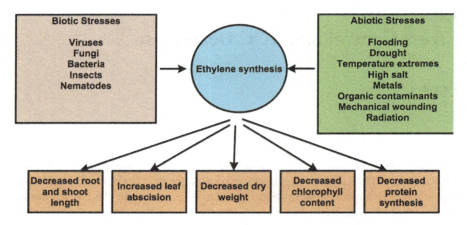

Fig. 27.1 Schematic view of how biotic and abiotic stresses cause plants to synthesize ethylene that increases plant senescence, chlorosis and abscission

plants and is believed to activate the synthesis of plant defensive proteins. The first ethylene peak is followed (several hours to a few days later) by a much larger peak of ethylene that results from conversion of newly synthesized ACC, and initiates processes such as senescence, chlorosis and abscission, all of which are inhibitory to plant growth and development (Fig. 27.1).

To avoid some of the deleterious effects of various stresses, plants may be treated with plant growth-promoting bacteria (PGPB) that provide the plant with various means of altering the plant's metabolism and thereby decreasing the severity of the stress. In this regard, biocontrol PGPB may act to inhibit the functioning of various plant pathogenic organisms by: (i) producing antibiotics or pathogen cell wall degrading enzymes, (ii) out-competing pathogens, (iii) stimulating induced systemic resistance, (iv) producing pathogen-lysing volatile organic compounds, (v) synthesizing iron-sequestering siderophores or (vi) by lowering plant ethylene levels through the action of the enzyme ACC deaminase (Glick 2010, 2014). On the other hand, to avoid the deleterious effects of abiotic stresses, plants may be treated with PGPB that: synthesize (i) siderophores and thereby provide the plant with iron, (ii) trehalose a non-reducing disaccharide that acts as an osmoprotectant, (iii) indole-3-acetic acid (IAA), a plant hormone that promotes plant growth, (iv) cytokinin, another plant hormone [Note that much of what is believed to be the role of bacterially-produced cytokinin is based on the knowledge of what the addition of this hormone itself does to plants.] or (v) ACC deaminase which lowers plants ACC concentrations thereby limiting the extent of plant ethylene synthesis (Fig. 27.2).

Fig. 27.2 Schematic view of the interrelatedness of plant stress, ethylene and IAA. SAM (S-adenosyl-methionine) is converted to ACC (1-aminocyclopropane-1-carboxylate) by the plant enzyme ACC synthase and ACC is converted to ethylene by the plant enzyme ACC oxidase. IAA (some of it produced by the plant and some by the bacteria) can both promote plant growth and turn on the transcription of ACC synthase. ACC may be converted to ammonia and α-ketobutyrate by the bacterial enzyme ACC deaminase. Accumulating ethylene both inhibits auxin signal transduction and increases the effects of the stress. Exogenous biotic or abiotic stress can increase the ethylene and/or IAA

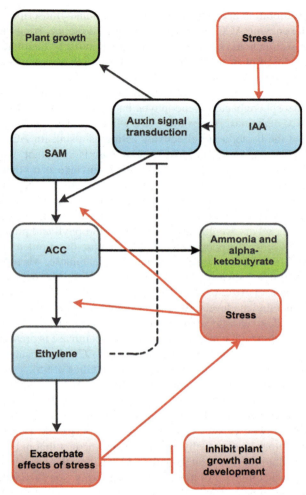

27.2 ACC Deaminase and IAA Working Together

The enzyme ACC deaminase, which catalyzes the cleavage of ACC, into ammonia and α-ketobutyrate (compounds readily metabolized and completely consumed by most bacteria and most plants), is commonly found in many soil bacteria as well as in some soil fungi. In a model that describes the role of ACC deaminase in promoting plant growth (Glick et al. 2007; Glick 2014), it is envisioned that PGPB bind either to plant seeds or roots, or colonize the interior surfaces of the plant (i.e. they are endophytes) and, in response to low levels of the amino acid tryptophan, that are exuded from the plant, the bacteria produce and secrete low levels of IAA (Fig. 27.2). The IAA produced by the bacteria and subsequently taken up by plant cells will, together with the endogenous plant-synthesized IAA, either promote plant growth (cell proliferation and cell elongation) or stimulate transcription of plant genes encoding the enzyme ACC synthase. In the latter instance, bacterial IAA will ultimately act to stimulate the synthesis of ethylene. In order to facilitate cell elongation, IAA acts to loosen plant cell walls with the result that root exudation is increased providing additional nutrients to rhizosphere bacteria. In addition to other small molecules such as sugars, amino acids and organic acids, plants exude some ACC that is taken up by rhizosphere bacteria (endophytic bacteria can also do this). If these bacteria contain ACC deaminase, the exuded ACC will be cleaved and metabolized with the net result that the bacteria are effectively acting as a sink for ACC. The consequence of the functioning of bacterial ACC deaminase is that plant ACC levels are decreased so that the plant cannot produce as much ethylene as it otherwise might (had the ACC not been cleaved). A plant's physiology changes as a consequence of a lower level of ethylene; thus, plants associated with ACC deaminase-containing PGPB typically have longer and more extensive roots and shoots, a lower amount of leaf abscission, increased dry weight, and increased chlorophyll and protein content.

Because the plant enzyme ACC oxidase has a much greater affinity (i.e. a lower K_m) for ACC than does bacterial ACC deaminase, when ACC deaminase-producing bacteria are present, plant ethylene levels will be dependent upon the ratio of the amount of ACC oxidase to the amount of ACC deaminase. To effectively reduce plant ethylene levels, ACC deaminase must function before any significant amount of ACC oxidase is induced (by either a particular stress or attainment of a specific developmental phase). Bacterial ACC deaminase is present at a low, basal, level and is induced by the increasing amounts of ACC that ensue from the induction of ACC synthase in the plant following a stress (Glick 1995).

Since IAA activates the transcription of ACC synthase and > 85 % of rhizosphere bacteria synthesize IAA, according to the model that has been described so far, these bacteria should all eventually produce relatively high concentrations of ACC within plant cells. The ACC should subsequently be utilized to synthesize plant inhibitory levels of ethylene. However, the fact that not all IAA-producing bacteria are inhibitory to plant growth indicates that the model that has been described so far is incomplete. In this regard, microarray experiments that examined the effects of wild-type and ACC deaminase minus mutants of PGPB on plants indicated that with the mutant but not the wild-type bacteria, as plant ethylene levels increased, the ethylene that is produced feedback inhibits IAA signal transduction thereby limiting the extent that IAA can activate ACC synthase transcription (Stearns et al. 2012). However, with plant growth-promoting bacteria that both secrete IAA and synthesize ACC deaminase, plant ethylene levels do not become elevated to the same extent as when plants interact with bacteria that secrete IAA but do not synthesize ACC deaminase. In the presence of ACC deaminase, ethylene levels in the plant are lower as is the subsequent ethylene feedback inhibition of IAA signal transduction. Thus, in the presence of ACC deaminase the bacterial IAA can continue to both promote plant growth and increase ACC synthase transcription with only a small amount of ethylene feedback inhibition of this pathway. The net result of this synergism between IAA and ACC deaminase is that by lowering plant ethylene levels, ACC deaminase permits IAA to (do its job to) stimulate plant growth.

27.3   Stress Control in Action

A number of different chemical compounds can either lower ethylene levels in plants or alter a plant's sensitivity to ethylene. However, most of these chemicals are either harmful to the environment, expensive or limited in their potential use. However, if the model shown schematically in Fig. 27.2 has any validity, ACC deaminase-containing bacteria should effectively decrease the deleterious effects that result as a consequence of different stresses, and the ability of these bacteria to synthesize IAA should, at the same time, stimulate plant growth. Of course, not all plants are equally sensitive to ethylene and different plant growth-promoting bacteria synthesize different amounts of IAA.

When plant roots are stressed because they are exposed to decreased oxygen concentrations as a result of flooding, ACC synthase is induced and relatively large amounts of ACC are synthesized. However, the conversion of ACC to ethylene cannot occur in oxygen-deprived roots so the accumulated ACC is transported through the plant to the aerobic environment of the shoots where ACC is converted to ethylene by ACC oxidase. The ethylene synthesis causes wilting, leaf chlorosis and necrosis, and a significant loss of biomass. However, plants that are first treated with ACC deaminase-containing PGPB suffer a significantly decreased level of damage from flooding (Grichko and Glick 2001) .

Various plant diseases may reduce plant biomass yields by $\sim 10\%$ per year in more developed countries and by $> 20\%$ per year in less developed countries of the world. A wide variety of infectious organisms can cause plant diseases including fungi, oomycetes, bacteria, viruses, phytoplasmas, protozoa and nematodes although fungi and bacteria cause the majority of common plant diseases. Many PGPB can act as biocontrol agents, utilizing a wide range of strategies to limit the growth and pathogenicity of fungal and bacterial phytopathogens. Many disease symptoms of pathogen-infected plants arise as a consequence of the stress imposed by the infection; a significant part of the damage to pathogen-infected plants is a result of the response of the plant to the increased stress ethylene that forms in response to the infection rather than from the direct action of the pathogen. When PGPB contain the enzyme ACC deaminase, they can modulate the level of ethylene in pathogen-infected plants thereby significantly limiting the damage caused by the pathogen. This is the case for both fungal and bacterial pathogens (Wang et al. 2000; Toklikishvili et al. 2010) as well as for nematodes (Nascimento et al. 2013).

Some scientists have argued that drought, or soil drying, limits crop yield more than any other abiotic stress. Moreover, this problem is exacerbated by climate change that may result in decreased annual rainfall in many agricultural regions that provide the world's major staple crops. Thus, as drought is an ethylene-inducing stress, it is expected that ACC deaminase-containing plant growth-promoting bacteria will facilitate plant growth under drought conditions, and this is precisely what has been observed (Mayak et al. 2004a; Belimov et al. 2008). In addition, Timmusk et al. (2011) examined the prevalence and distribution of ACC deaminase-containing *Paenibacillus* spp. among bacterial strains isolated from the rhizospheres of wild barley (*Hordeum spontaneum*) growing in Northern Israel. The strains were all isolated from a region termed 'Evolution Canyon' in which the South Facing Slope was sparsely vegetated as a consequence of the harsh conditions on this side of the canyon including excessive sunlight and frequent drought while the North Facing Slope featured conditions much more conducive to plant growth including lower levels of sunlight and the apparent absence of drought. These workers observed that $\sim 50\,\%$ of the bacteria isolated from around the roots of plants from the South Facing Slope contained ACC deaminase while only approximately 4 % of the bacteria isolated from the North Facing Slope contained this enzyme. These researchers suggested that bacterial ACC deaminase was selected for by the plants growing under the more perennially stressful conditions on the South Facing Slope, thereby protecting plants and facilitating their survival; without plant growth and root exudation the bacteria themselves would likely not proliferate. On the other hand, with the much better growth conditions on the North Facing Slope, there is apparently much less selective pressure for bacteria to retain genes for ACC deaminase.

Salinity is an enormous worldwide problem for agriculture, especially for crops grown under irrigation. This is because salt is inhibitory to the growth of a large number of plants. In fact, it is estimated that around half of the land worldwide devoted to the growth of irrigated crops is adversely affected by salt. While a number of strategies have been developed to facilitate the growth of plants in the presence of salt, arguably one of the most successful approaches has been the use of ACC deaminase-containing PGPB (Mayak et al. 2004b; Gamalero et al. 2009a). Thus, researchers have found this strategy to be effective with a wide variety of crops both in the lab and in the field.

Ethylene is a key signal in the initiation of plant flower senescence although different flowers may be more or less sensitive to ethylene. Chemical ethylene inhibitors are currently used to prolong the shelf-life of cut flowers, however, many of these chemicals are either expensive or environmentally hazardous or both. As an alternative to the use of chemicals, it has been shown that cut flowers may be treated with endophytic (but not with rhizospheric) ACC deaminase-containing PGPB and these bacteria can significantly delay the senescence of cut flowers (Ali et al. 2012).

The problem of toxic waste disposal is enormous, with billions of tons of hazardous waste produced worldwide every year. One strategy that has been developed relatively recently, phytoremediation, includes the use of plants to either remove or detoxify many environmental contaminants, both metals and organics (Gamalero et al. 2009b; Glick 2010). Unfortunately, the presence of many of these contaminants is inhibitory to plant growth. However, a significant portion of the inhibition caused by the contaminants may be overcome by the addition of PGPB. Where it has been examined in some detail, the bacterial traits that are most effective in stimulating plant growth in the presence of environmental contaminants are siderophore synthesis, IAA production and ACC deaminase activity.

27.4 Future Prospects

Notwithstanding the fact that there are still only a very limited number of commercialized PGPB, the widespread use of these bacteria is a technology whose time has come. The current understanding of the mechanisms employed by these bacteria should hasten the commercial development of this technology, both in agriculture and in environmental cleanup protocols (phytoremediation). In the short term, it is likely that ACC deaminase-containing PGPB will be used in conjunction with existing strains of Rhizobia to inoculate and promote the growth of legumes. Secondly, the use of ACC deaminase-containing PGPB as an adjunct to the phytoremediation of organic contaminants such as polycyclic aromatic hydrocarbons already works well enough for this technology to be commercialized. Third, the use of ACC deaminase-containing PGPB should be able to effectively complement and augment the job done by many of the existing commercialized biocontrol strains. Once farmers, growers and environmental engineers are convinced that this approach works effectively in the field, the use of these organisms, selected for the presence of the enzyme ACC deaminase and highly effective at lowering plant stress levels, should continue to grow.

References

Abeles FB, Morgan PW, Saltveit ME Jr (1992) Ethylene in plant biology, 2nd edn. Academic Press, New York

Ali S, Charles TC, Glick BR (2012) Delay of carnation flower senescence by bacterial endophytes expressing ACC deaminase. J Appl Microbiol 113:1139–1144

Belimov AA, Dodd IC, Hontzeas N et al (2008) Rhizosphere bacteria containing 1-aminocyclopropane-1-carboxylate deaminase increase yield of plants grown in drying soil via both local and systemic hormone signaling. New Phytol 181:413–423

Gamalero E, Berta G, Glick BR (2009a) The use of microorganisms to facilitate the growth of plants in saline soils. In: Khan MS, Zaidi A, Musarrat J (eds) Microbial strategies for crop improvement. Springer-Verlag, Berlin, pp 1–22

Gamalero E, Lingua G, Berta G et al (2009b) Beneficial role of plant growth promoting bacteria and arbuscular mycorrhizal fungi on plant responses to heavy metal stress. Can J Microbiol 55:501–514

Glick BR (1995) The enhancement of plant growth by free-living bacteria. Can J Microbiol 41: 109–117

Glick BR (2010) Using soil bacteria to facilitate phytoremediation. Biotechnol Adv 28:367–374

Glick BR (2014) Bacteria with ACC deaminase can promote plant growth and help to feed the world. Microbiol Res 169:30–39

Glick BR, Cheng Z, Czarny J et al (2007) Promotion of plant growth by ACC deaminase containing soil bacteria. Eur J Plant Pathol 119:329–339

Grichko VP, Glick BR (2001) Amelioration of flooding stress by ACC deaminase-containing plant growth-promoting bacteria. Plant Physiol Biochem 39:11–17

Mayak S, Tirosh T, Glick BR (2004a) Plant growth-promoting bacteria that confer resistance to water stress in tomato and pepper. Plant Sci 166:525–530

Mayak S, Tirosh T, Glick BR (2004b) Plant growth-promoting bacteria that confer resistance in tomato to salt stress. Plant Physiol Biochem 42:565–572

Nascimento FX, Vicente CSL, Barbosa P et al (2013) The use of the ACC deaminase producing bacterium *Pseudomonas putida* UW4 as a biocontrol agent for pine wilt disease. Biocontrol 58:427–433

Stearns JC, Woody OZ, McConkey BJ et al (2012) Effects of bacterial ACC deaminase on *Brassica napus* gene expression. Mol Plant-Microbe Interact 25:668–676

Timmusk S, Paalme V, Pavlicek T et al (2011) Bacterial distribution in the rhizosphere of wild barley under contrasting microclimates. PLoS One 6. doi:e17968

Toklikishvili N, Dandurishvili N, Tediashvili M et al (2010) Inhibitory effect of ACC deaminase-producing bacteria on crown gall formation in tomato plants infected by *Agrobacterium tumefaciens* or *A. vitis*. Plant Pathol 59:1023–1030

Wang C, Knill E, Glick BR (2000) Effect of transferring 1-aminocyclopropane-1-carboxylic acid (ACC) deaminase genes into *Pseudomonas fluorescens* strain CHA0 and its *gacA* derivative CHA96 on their growth-promoting and disease-suppressive capacities. Can J Microbiol 46: 898–907

Chapter 28
Plant-Microbe Interactions and Water Management in Arid and Saline Soils

Daniele Daffonchio, Heribert Hirt and Gabriele Berg

Abstract Drought and salinity are major factors limiting agriculture in many regions in the world, and their importance is predicted to even increase in the near future in parallel with the ongoing global warming and climate changes. Soil and rhizosphere microbes are potential resources for counteracting such abiotic stresses in plants. The knowledge on the roles of root microorganisms in retaining soil humidity and promoting plant growth under such abiotic stresses is analyzed in this chapter. The importance of microbial diversity in the rhizosphere for alleviating drought and salinity effects on the plant physiology is discussed in the light of "Desert Farming", the general crop management practice that is frequently used in arid regions. The plant growth promoting functional services exerted by microorganisms within the rhizosphere in arid soils are presented in relation to the plant response under water stress.

28.1 The Effects of Drought and Salinity on Plant Physiology

A major challenge for future agriculture is to cope with the increasing demand for food production, facing a constantly increasing world population. This growing demand for agricultural production is paralleled by dramatic losses of arable land due to enhanced soil destruction and erosion. Drought and soil salinity are the two major

D. Daffonchio (✉)
DeFENS, Department of Food, Environmental and Nutritional Sciences,
University of Milan, 20133 Milan, Italy
Tel.: + 393339742943
e-mail: daniele.daffonchio@unimi.it

H. Hirt · D. Daffonchio
BESE Division, King Abdullah University of Science and Technology,
Thuwal, 23955-6900, Kingdom of Saudi Arabia
Tel.: + 00966 (012) 808 2959
e-mail: heribert.hirt@kaust.edu.sa

G. Berg
Institute of Environmental Biotechnology, Graz University of Technology,
8010 Graz, Austria
Tel.: + 43664608738310
e-mail: Gabriele.berg@tugraz.at

© Springer International Publishing Switzerland 2015
B. Lugtenberg (ed.), *Principles of Plant-Microbe Interactions*,
DOI 10.1007/978-3-319-08575-3_28

environmental factors that limit plant growth and development, thereby negatively affecting agricultural yield by more than 60 %. Water shortage is critical in many areas of the world and is usually countered by extensive irrigation. Although the planet earth is rich in water, not only the water of the oceans, but also most inland water resources are highly saline. Moreover, irrigation usually results in soil salinization making drought and soil salinity increasing agricultural problems (Bartels and Sunkar 2005).

To convert deserts into arable, green landscape is a global vision as well as competent answer to world hunger and climate change (Clery 2011). Desert farming, which generally relies on irrigation, is part of this vision. Agriculture systems were already developed in arid landscapes by ancient cultures, yet nowadays, there is a dramatically increasing need for large-scale desert farming to feed the population. Desert farming is not only a challenge for irrigation systems with impact on the global water balance, it can also have an impact on soil microbial diversity. Changes of soil bacterial diversity, especially reduced beta diversity, was shown to occur in a semi-arid ecosystem as a consequence of land use for agriculture with potential irreversible consequence on the natural soil ecosystem (Ding et al. 2013). To avoid this risk and to establish sustainable desert farming systems it is important to understand diversity in arid and saline environments.

Salt and drought stress share some properties and generally result in impaired key physiological functions. One component of salinity is hyperosmotic stress, resulting in a water deficit that is comparable to a drought-induced water deficit. The other component of salt stress is ion toxicity, resulting in metabolic imbalance. Membranes may become disorganized, proteins may become inactive and excess levels of reactive oxygen species (ROS) can be produced leading to oxidative damage. As a consequence, inhibition of photosynthesis, metabolic dysfunction, and damage of cellular structures contribute to plant growth inhibition, reduced fertility and premature senescence.

Plants basically counteract the negative effects of salinity and drought by activation of genetic and biochemical responses. These responses include the synthesis and accumulation of osmolytes, such as proline or raffinose, that are able to stabilize proteins and cellular structures but also to maintain cell turgor pressure by osmotic adjustment. Moreover, plants also enhance scavenging of ROS, which are generated as a secondary effect of salt and drought stress. Several plant stress signaling pathways have been dissected in detail in the model organisms *Arabidopsis thaliana* and rice and a number of central transcription factors have emerged that regulate cohorts of downstream genes. Despite this tremendous progress, our knowledge on the coordination and integration of these regulatory pathways into the complex matrix of plant stress physiology is still limited. Moreover, different plant species developed different strategies to cope with stress. For example, some halophytic Brassicacae, such as the *Arabidopsis*-related halophyte *Thellungiella halophyla*, avoid stress by salt exclusion, whereas *Lobularia maritima* accumulates and detoxifies salt by compartmentalisation. Moreover, harsh environmental conditions, which are harmful for one plant species, might not be stressful for another species. These differences correlate with different stress-response mechanisms and two main strategies of stress avoidance and stress tolerance can be found in the plant kingdom. Stress avoidance in some species is a genetically inherited trait that delays or prevents the negative impacts of a stress. For example, cacti show a permanently adapted morphology and physiology to hot and arid climates. Stress tolerance is an adaptive strategy to counterbalance stress conditions and most plants can adapt to drought conditions by closing their stomates to reduce transpirational water loss. Most plants can also acclimate to stress conditions upon a gradual and repeated increase of a stress factor. Acclimation induces various physiological changes that are reversed when the adverse environmental conditions disappear. Overall, whereas the avoidance mechanisms are usually constitutive features that are genetically inherited, acclimation mechanisms are plastic and reversible. At the molecular level, acclimation involves the modification of gene expression but also epigenetic mechanisms. Stress-inducible genes comprise genes involved in direct protection from stress, including the synthesis of osmoprotectants, detoxifying enzymes, and transporters, as well as genes that encode regulatory proteins such as transcription factors, protein kinases, and phosphatases.

An issue that is worth to be explored for the future of agriculture is the effects that plant-associated microorganisms may exert on all these metabolic processes for improving plant resistance and resilience to drought and salinity stresses.

28.2   Rhizosphere Bacterial Diversity in Arid and Saline Soils and under "Desert Farming"

Arid and Saline Soil Habitats Harbour a Unique Bacterial Diversity Deserts represent in general extreme environments for microorganisms. Although the conditions varied strongly in the different regions of the world, all of them are characterized by a combination of extreme temperatures and desiccation, resulting in arid and saline soils. These extreme abiotic factors, as well as the absence of plants at many sites, contribute to the visual appearance of a sterile environment. While early studies supported this "sterility" by very low levels of viable/cultivable microorganisms, applications of DNA/RNA-based molecular methods led to interesting new insights and showed a contrasting picture. For example, in their global-scale study, Fierer and Jackson (2006) found that the acidic soils of tropical forests harbour fewer bacterial taxa than the neutral pH soils of deserts. In different sites in the Negev Desert, archaeal and bacterial diversity was not constrained by precipitation, although the taxonomic composition differed (Angel et al. 2010). In soil of the Atacama Desert, a high diversity of microorganisms known to live in hypersaline environments was found by analysis of Denaturing Gradient Gel Electrophoresis profiles of amplified 16S rRNA gene (de Los Ríos et al. 2010). Most of the desert microbial communities seem to be structured solely by abiotic processes. However, if desert plants are present, they strongly shape soil microbial diversity. Also desert animals play an important role as ecosystem engineers and affect the microbial complex of diazotrophs in soils. All these investigations showed a unique and extraordinary microbial diversity in belowground desert environments, strongly adapted to their unique conditions.

The SEKEM Farms as Example for Sustainable Desert Farming On the SEKEM farms in Egypt, desert land was converted into arable land, and biodynamic agriculture is operated for over 30 years now (www.sekem.com). Today SEKEM is carrying out organic agriculture on more than 4100 ha and has the largest market for organic products outside Europe and North America. They produce organic foods, spices, tea, cotton textiles and natural remedies. However, the cultivation especially of medical plants is more and more affected by soil-borne phytopathogens, which lead to significant yield losses. Therefore, this is an excellent study example (i) to understand the impact of desert farming on microbial diversity and, (ii) to develop a specific biocontrol strategy for desert agriculture.

To examine the impact of organic agriculture on bacterial diversity and community compositions in desert soil, soil from a SEKEM farm in comparison to the surrounding desert soil from different sites were assessed by pyrosequencing-based analyses of 16S rRNA and *nifH* gene sequences (Köberl et al. 2011). In addition, fingerprinting and cultivation-dependent methods were included in this study. Altogether, a strong impact on the structure and function of the bacterial and fungal communities was found. The computed Shannon indices of diversity (H') calculated for bacterial communities on the basis of amplicon libraries were much higher for agricultural soil than for desert soil (H' at a dissimilarity level of 20 %: SEKEM soil: 4.29; desert soil: 3.54). This indicates a higher bacterial diversity in soil due the agricultural use of the desert. For both soil types together 18 different phyla were identified; dominant groups present in both soils were especially Proteobacteria (30.2 %), Firmicutes (27.3 %) and Actinobacteria (10.5 %). In detail, Firmicutes were highly enriched in agricultural soil (from 11.3 % in desert soil to 36.6 % in SEKEM soil), while Proteobacteria (46.0 % in desert soil and 21.0 % in SEKEM soil) and Actinobacteria (20.7 % in desert soil and 4.6 % in SEKEM soil) occurred in SEKEM in lower abundances than in the surrounding desert. In addition, in both soils Bacteroidetes (4.6 and 5.3 %) and Gemmatimonadetes (1.4 and 1.9 %) were present. Whereas Acidobacteria (7.9 %) and Planctomycetes (1.1 %) were only present in the agricultural soil, *Deinococcus-Thermus* (1.1 %) was only detectable in the desert sand. In contrast to other studies, the most important difference was the high abundance of Firmicutes. Most of the Firmicutes were classified as *Bacillus*; in the agricultural soil also the phylogenically related genus *Paenibacillus* was found (5 % of classified Firmicutes). These differences between the two soil types found for the bacterial community could be confirmed for the fungal community analyzed by fingerprinting and for the functional studies regarding antagonistic microorganisms and diazotrophs (Köberl et al. 2013a; Köberl et al. 2013b). After long-term farming, a drastic shift in the microbiomes in desert soil was observed. Bacterial communities in agricultural soil showed a higher diversity and a better ecosystem function for plant health but a loss of extremophilic bacteria. Interestingly, we detected that indigenous desert microbes promoted plant health in desert agro-ecosystems.

In desert soils, plant roots provide important nutrient sources for bacteria and act as microbial hot spot. The composition of the microbial community in rhizosphere and endorhiza of three different medical plants (*Matricaria chamomilla* L., *Calendula officinalis* L. and *Solanum distichum* Schumach. & Thonn.) grown under organic conditions on SEKEM farms confirmed this hypothesis. These results show that dominant bacteria, e.g. *Ochrobactrum* and *Rhodococcus*, are taken up by the plants from the soil and that soil is the main reservoir for plant-associated bacteria. Further, nearly in all samples *Bacillus* sp. was found. The fungal community fingerprints included a quite high diversity in all microenvironments. *Verticillium dahliae*, a prominent phytopathogenic fungus, was found nearly in all samples on the SEKEM farms. In general, mainly potential plant pathogens were found within the fungal communities. The obligate root-infecting pathogen *Olpidium*, (fungal phylum Chytridiomycota), was found especially in the rhizosphere and endorhiza of *M. chamomilla*. *Alternaria* and *Acremonium* were primarily present in the rhizosphere samples. There were significant differences between the rhizosphere and the endorhiza of the medical plants. In general, samples from the rhizosphere generated more strains than samples from the endorhiza of the medical plants, which indicate that a sub-set of rhizobacteria was able to invade the root. Diazotrophs are key organisms for providing nitrogen in natural or organically managed agricultural ecosystems, especially under desert conditions. By combining *nifH*-specific quantitative PCR, fingerprints, amplicon pyrosequencing and fluorescent *in situ* hybridization-confocal laser scanning microscopy, a generally high *nifH* abundance and diversity in native and agricultural desert soil and in the rhizosphere of medicinal plants was detected, the highest reported until now compared with other ecosystems. Statistically significant differences were found between both soil types (native and agriculturally used), between the different microhabitats (bulk soil, rhizosphere, endorhiza), and between the three investigated medicinal plants. Again, a considerable community shift from desert to agriculturally used soil was observed with higher abundance and diversity in the agro-ecosystem. In comparison to the rhizosphere, the endorhiza was characterized by lower abundances and a subset of species. Comparing root-associated communities, remarkable differences were found. While the microbiomes of *M. chamomilla* and *C. officinalis* were similar and dominated by potential root-nodulating rhizobia mainly acquired from soil, the perennial *S. distichum* formed primarily associations with free-living nitrogen fixers most likely transmitted between plants, possibly by mean of the seeds, as they are undetectable in soils. The results underline the importance of diazotrophs in desert ecosystems and moreover identified plants as important drivers for functional diversity.

The major problems in the cultivation of plants on SEKEM farms are caused by the soil-borne pathogenic fungi *Verticillium dahliae* Kleb., *Rhizoctonia solani* J.G. Kühn and *Fusarium culmorum* (Wm.G. Sm.) Sacc. as well as by the soil-borne pathogenic bacterium *Ralstonia solanacearum*. Bacterial isolates obtained from the soil of the SEKEM farm exhibited a higher *in vitro* antagonistic potential towards soil-borne phytopathogenic fungi in comparison to the bacteria isolated from the desert soil (SEKEM $21.2 \pm 1.2\%$; desert $12.6 \pm 0.8\%$). From the agricultural soil 17.4% (27 isolates) demonstrated antagonism towards all three fungal pathogens, while only 10.6% (21 isolates) from the desert soil. Already the desert soil harbors a high proportion of antagonists, which were augmented by organic agriculture in SEKEM soil. The soil from the farm seems to be supplied with antagonists in such an optimal way, that there was no detectable enrichment of antagonists in the rhizosphere and endorhiza of the investigated medical plants. In general, *M. chamomilla* and *S. distichum* showed a better antagonistic potential than *C. officinalis*. Especially the endorhiza from *M. chamomilla* harbors a high proportion of antagonists. Whereas most antagonistic bacteria against *Fusarium culmorum* were found in the soil and in the rhizosphere of the medical plants, those against *Verticillium dahliae* were found in the endorhiza. A representative selection of promising biological control agents was identified by partial 16S rRNA gene sequencing as members of the *Bacillus subtilis* and *Bacillus cereus* groups, and of the genera *Paenibacillus*, *Streptomyces* and *Lysobacter*. Except for one isolate of *Lysobacter*, only gram-positive antagonists were found. All microenvironments were dominated by antagonists from Firmicutes with *Bacillus* and *Paenibacillus* isolated from all habitats. Antagonistic *Streptomyces* were found exclusively in desert soil. Promising strains of *Streptomyces subrutilus*, *Bacillus subtilis* and *Paenibacillus polymyxa* were tested for their activity under field conditions on chamomile seedlings. The *Bacillus* and *Paenibacillus* strains enhanced the production of the bioactive secondary plant metabolite apigenin-7-O-glucoside (Schmidt et al. 2014). Other *Bacillus subtilis* isolates were able to control root-knot nematodes by inducing systemic resistance of tomato plants (Adam et al. 2014).

28.3   Functional Services of Rhizosphere Bacteria under "Desert Farming"

The wide diversity of bacteria in the rhizosphere is exploited in a series of services that help plants in counteracting drought and salinity stresses. Such services, ranging from general ecological services to contributions to the physical protection of the root from the mechanical stress or in the plant hormone homeostasis, are important features under Desert Farming.

Microbial Ecological Services   The ecology of the arid systems modulates the type of microorganisms and the extent of the services provided to plants. For instance the potential plant growth promoting (PGP) services provided by the root associated bacteria appear to be invariant respect to macroecological factors such as latitude, soil type and plant cultivar, despite provided by different bacterial communities, according to observations across an aridity macrotransect from North Italy to North Tunisia and Egypt (Marasco et al. 2013a). Rhizosphere and endophytic bacteria are capable of promoting the growth of plants challenged by drought. It has been shown that such a trait is not a per se capability of the selected microorganisms but it is associated to the water stress (Rolli et al. 2014). So, root-associated bacteria that normally do not present *in-vivo* promotion activities can determine plant resistance to drought, suggesting that the services to plant are stress-dependent.

Protection from the Soil Mechanical Stress   Water evaporation and drought determine important changes in the soil aggregates with modifications in the soil particle architecture. Such changes influence the mechanical interactions of the soil with the root surface and the overall physical condition of the soil including water circulation, air exchanges and temperature in the soil. Rhizodeposition determines favorable conditions for bacterial growth. Bacterial biofilms produce exopolymers (EPS) that affect the binding of soil particles to the rhizoplane, improve the water retention in the rhizosphere and protect the root surface from mechanical damages determined by the soil hardness. Plants treated with EPS-producing strains and maintained under drought conditions have increased biomass and cause increased macroporosity in the root adhering soil compared to the non-treated controls.

Protection from Osmotic and Oxidative Stresses Many bacteria are capable of mitigating water stress in plants by stimulating the production of osmoprotectants in the associated plants. The capacity of both Gram negative (*Azospirillum* and *Pseudomonas*) and Gram positive (*Bacillus*) strains to promote resistance of basil plants to water stress has been associated to increased concentrations of proline and soluble carbohydrates in root and leaf tissues. In the treated plants chlorophyll content increased significantly, confirming the overall beneficial effects of bacteria against drought (Heidari et al. 2011). Bacteria-mediated protection against water stress was also associated to the accumulation of anthocyanin in the vacuoles of mountain laurel (*Kalmia latifolia* L.), when the stressed plant was colonized by an endophytic *Streptomyces padanus* actinomycete (Hasegawa et al. 2004). Bacteria engineered for overexpressing genes for threhalose synthesis increased their osmotic stress tolerance as well as resistance to drought of their host plants to which they were associated to (Rodriguez-Salazar et al. 2009).

Cell turgor is an essential process for maintaining normal physiological activities in plant tissues. During drought the aqueous vacuoles in the plant cells regulate the concentrations of osmolytes to decrease the vacuolar osmotic potential and improve water uptake (Park et al. 2005). Different compatible solutes are accumulated in the vacuoles, including sugars, glycine betaine, amino acids, organic acids and pigments. These molecules contribute to protection of the plant from stress through osmotic adjustment, stabilization of the native structure of enzymes and cell membranes and detoxification of ROS induced by the stressful conditions. The increased concentration of such solutes raises the cell's osmotic pressure, contributing to maintain turgor by preventing water loss and promoting water uptake. The overexpression in *Arabidopsis* as well as in crop plants (such as tomato) of the vacuolar H^+-pyrophosphatase, an enzyme implicated in the vacuolar ion turnover, determines improved resistance to drought and enhanced growth compared to the non-engineered controls. The engineered plants have enhanced pyrophosphate-driven cation transport into root vacuoles and increased root biomass and are capable of recovering from exposure to water stress.

The concentration of ROS, such as superoxide, hydrogen peroxide, hydroxyl radical and singlet oxygen, increases in plant tissues under water stress (Apel and Hirt 2004). ROS react with proteins, lipids and nucleic acids, impairing cell physiological functions. In parallel to the promotion of the synthesis of osmolytes in plants exposed to water and salinity stresses, PGP bacteria (PGPB) can enhance tolerance to oxidative stress through the synthesis of stress-related enzymes. For instance, inoculation of drought stressed *Lactuca sativa* with PGP microbes resulted in increased activity of catalase enzymes (Kohlera et al. 2009).

Effects on Hormone Homeostasis PGPB are capable of producing plant hormones including auxins, gibberellins (GAs), cytokinins, abscisic acid (ABA) and ethylene (see Chap. 26). Despite the difficulty of separating the effects of bacteria vs. plant-produced hormones, experiments using hormones synthesis inhibitors as well as *Arabidopsis* mutants impaired in specific hormone synthetic pathways have shown some potential effects of different bacteria-produced hormones on stimulating resistance to drought and salinity. By using chemical inhibitors of GAs and ABA synthesis in planta, *Azospirillum lipoferum* treated maize seedlings showed drought stress alleviation, suggesting that bacterial ABA and GAs may play roles in water stress mitigation. Bacterial production of volatile organic compounds (VOCs) (see Chap. 8), such as 2R,3R-butanediol, have been proposed to influence the overall plant hormone balance affecting transpiration and water loss under drought by influencing stomata closure or contributing to shape root architecture potentiating the uptake of water, minerals and microelements in water stressed plants (Cho et al. 2008).

Bacteria can also produce enzymes capable of disrupting plant hormone synthetic pathways, such as the enzyme 1-aminocyclopropane-1-carboxylic acid (ACC) deaminase, which hydrolyzes the plant ethylene precursor ACC to ammonia and α-ketobutyrate. Such traits may have roles in promoting resistance of plants to drought and salinity. ACC deaminase activity has been associated with stimulation of resistance to salinity stress. For instance, an ACC deaminase-producing *Achromobacter piechaudii* strain decreased the ethylene production by tomato seedlings grown in the presence of up to 172 mM NaCl, and increased the plant fresh and dry weights (Mayak et al. 2004).

28.4 The Plant Response to Plant Growth Promoting Bacteria Under Stress

Plants can generate a large portion of their energy by photosynthesis, but plant growth requires significant quantities of nitrate (see Chap. 23), phosphate (see Chap. 24), and other minerals, which are often not freely available in the soil. Root-associated beneficial microbes are important partners of plants and in exchange to carbohydrates, provide many of the limiting minerals. The best-known beneficial microbes are mycorrhizal fungi (see Chap. 25) and rhizobia (see Chap. 23). Mycorrhiza interact with about 80 % of all terrestrial plant species and provide phosphate and nitrate to plants. Free-living or endophytic rhizobia can fix atmospheric nitrogen, but only the family of leguminosae profits from such an interaction through their ability to house rhizobia in root nodules. Although the interaction of plants with mycorrhizal fungi and rhizobial bacteria is well understood, other rhizosphere microbes have received much less attention.

Soil-grown plants are immersed in a sea of microbes and diverse beneficial microorganisms such as PGPB as well as plant-growth promoting fungi (PGPF) can stimulate plant growth and/or confer enhanced resistance to biotic and abiotic stresses (de Zelicourt et al. 2013). The establishment of beneficial plant-microbial interactions requires the mutual recognition and a considerable orchestration of the responses at both the plant and the microbial side. Rhizobial and mycorrhizal symbioses share a common plant signaling pathway that is activated by rhizobial and mycorrhizal factors (Corradi and Bonfante 2012; Geurts et al. 2012) and this signaling pathway also seems to be activated by certain beneficial bacteria, suggesting that different beneficial and pathogenic microbes initiate common plant signaling pathways.

The Role of Plant Growth Promoting Bacteria PGPB belong to a number of different bacterial genera, including *Rhizobium, Bacillus, Pseudomonas* and *Burkholderia*. PGPB can improve plant growth under abiotic stress conditions. Enhanced salt tolerance of *Zea mays* upon co-inoculation with *Rhizobium* and *Pseudomonas* is correlated with decreased electrolyte leakage and maintenance of leaf water contents. Some microbes produce plant hormones, such as indole acetic acid and gibberellic acid, which induce increased root growth and thereby lead to enhanced uptake of nutrients.

Plants have the ability to acquire a state of induced systemic resistance (see Chap. 14) to pathogens after inoculation with PGPB. In association with plant roots, PGPB can prime the plant's innate immune system and confer resistance to a broad spectrum of pathogens with a minimal impact on yield and growth. Several PGPB colonize roots and protect a large variety of plant species, including vegetables, crops and even trees, against foliar diseases in greenhouse and field trials.

Plant-Associated Fungi Confer Stress Tolerance to Plants Mycorrhizal and/or endophytic fungi interact with many plant species and significantly contribute to the adaptation of these plants to a number of environmental stresses including drought, heat, pathogens, herbivores or limiting nutrients. Some plants are unable to withstand stress conditions in the absence of their associated microbes. It appears that stress tolerance of the host plant can be a habitat-specific feature of the interaction. For example, *Curvularia protuberata* confers heat tolerance to its geothermal host plant *Dichanthelium lanuginosum*. However, neither the fungus nor the plant can survive alone at temperatures above 38 °C (Redman et al. 2002). Moreover, only *C. protuberata* isolates from geothermal plants can confer heat tolerance (Rodriguez et al. 2008). A comparison of different fungal endophytes unravels a further layer of specificity: *C. protuberata* confers heat but neither disease nor salt tolerance. In contrast, *Fusarium culmorum* only confers salt tolerance and *Curvularia magna* only disease tolerance. It appears that these specific features contribute to the ability of some plants to establish and survive in extreme habitats.

Symbiotically conferred disease tolerance appears to involve different mechanisms, depending on the endophyte. For example, a non-pathogenic *Colletotrichum* strain that confers disease resistance does not activate host defense in the absence of pathogen challenge (Redman et al. 1999). Moreover, disease resistance is localized to tissues that the fungus has colonized, but is not systemic.

In contrast, *Piriformospora indica* confers disease resistance systemically. *P. indica* colonizes the roots of many plant species and stimulates growth, biomass and seed production. *P. indica* promotes nitrate and phosphate uptake and confers resistance against abiotic and biotic stress (Waller et al. 2005). The fungal colonization stimulates the host to synthesize phosphatidic acid, which triggers the OXI1 pathway (Camehl et al. 2011). This pathway is usually activated only in response to pathogen attack to activate host defense. A defect in the OXI1 pathway negatively affects plant growth by the fungus, resembling a pathogenic interaction. Overall, the differences between *Colletotrichum* spp.- and *P. indica*-conferred disease resistance indicate a number of different mechanisms yet to be elucidated.

Further evidence indicates that our present concepts of categorizing microbes as pathogenic or beneficial are inadequate. For example, *Fusarium culmorum* can cause disease on a variety of crop plants. However, the *F. culmorum* isolate FcRed1 is beneficial and confers salt tolerance to its host dunegrass *Leymus mollis*, but isolates from non-coastal dunegrass do not. *C. protuberata* is a plant pathogen for several monocots, but isolate Cp4666D confers heat and drought tolerance to its host plant *Dichanthelium lanuginosum*. While *Curvularia* species are not known to have broad disease-host ranges, *C. protuberata* from the monocot *D. lanuginosum* also confers heat tolerance on tomato (Márquez et al. 2007; Rodriguez et al. 2008). Some microbes can also be present in plants without showing disease symptoms. For example, *Colletotrichum acutatum* can colonize pepper, eggplant, bean and tomato without causing disease, but with other plants, such as strawberry, disease symptoms become evident (Freeman et al. 2001). So it appears that a number of microbes have a host-dependent lifestyle as pathogenic or beneficiary partner of plants, but the molecular basis of the plant-microbe interactions remains to be unraveled.

28.5 Desert Farming Exploits Plant-Microbe Interaction to Improve Water Management

Beneficial plant-associated microbes can help plants suppress diseases, stimulate growth, occupy space that would otherwise be taken up by pathogens, promote biotic stress resistance, and increase crop yield and quality by nutrient mobilization and transport (Berg et al. 2013). Therefore, the plant microbiome is a key determinant of plant health and productivity. While the possibility to control biotic stress by plant-associated microorganisms is known since more than 100 years, less is known about controlling abiotic stress. There are many biocontrol products on the market but our understanding how plant-associated microbes can compensate abiotic stress is only at the beginning. However, several promising examples are already reported in the literature. Armada et al. (2014) showed that strains of *Bacillus megaterium*, *Enterobacter* sp., *Bacillus thuringiensis*, and *Bacillus* sp. have the potential to alleviate drought stress in *Lavandula* and *Salvia* by increasing K content, by depressing stomatal conductance, or by controlled shoot proline accumulation. The production of osmoprotectans was also identified as a key mechanism of the Stress Protecting

Agent *Stenotrophomonas rhizophila* DSM14405T (Alavi et al. 2013). In addition, this strain produced spermidine, which is a general, highly efficient stress protectant. In another example, pepper plants exposed to bacterial isolates from plants cultivated under desert farming exhibited a higher tolerance to water shortage, compared with untreated control (Marasco et al. 2012; Marasco et al. 2013b). This promotion was mediated by a larger root system (up to 40 %), stimulated by the bacteria that enhanced the plant's ability to take up water from dry soil. Altogether, to exploit plant-microbe interaction to improve water management in desert farming is a challenging but also promising task for the future.

References

Adam M, Heuer H, Hallmann J (2014) Bacterial antagonists of fungal pathogens also control root-knot nematodes by induced systemic resistance of tomato plants. PLoS One 9:e90402

Alavi P, Starcher MR, Zachow C et al (2013) Root-microbe systems: the effect and mode of interaction of Stress Protecting Agent (SPA) *Stenotrophomonas rhizophila* DSM14405 T. Front Plant Sci 4:141

Angel R, Soares MI, Ungar ED et al (2010) Biogeography of soil archaea and bacteria along a steep precipitation gradient. ISME J 4:553–563

Apel K, Hirt H (2004) Reactive oxygen species: metabolism, oxidative stress, and signal transduction. Ann Rev Plant Biol 55:373–399

Armada E, Roldán A, Azcon R (2014) Differential activity of autochthonous bacteria in controlling drought stress in native lavandula and salvia plants species under drought conditions in natural arid soil. Microb Ecol 67:410–420

Bartels D, Sunkar R (2005) Drought and salt tolerance in plants. Critical Rev Plant Sci 24:23–25

Berg G, Zachow C, Müller H et al (2013) Next-generation bio-products sowing the seeds of success for sustainable agriculture. Agronomy 3:648–656

Camehl I, Drzewiecki C, Vadassery Y et al (2011) The OXI1 kinase pathway mediates *Piriformospora indica*-induced growth promotion in *Arabidopsis*. PloS Pathog 7:e1002051

Cho SM, Kang BR, Han SH et al (2008) 2R,3R-butanediol, a bacterial volatile produced by *Pseudomonas chlororaphis* o6, is involved in induction of systemic tolerance to drought in *Arabidopsis thaliana*. Mol Plant-Microbe Interact 21:1067–1075

Clery D (2011) Environmental technology. Greenhouse-power plant hybrid set to make Jordan's desert bloom. Science 331:136

Corradi N, Bonfante P (2012) The arbuscular mycorrhizal symbiosis: origin and evolution of a beneficial plant infection. PLoS Pathog 8:e1002600

de Los Ríos A, Valea S, Ascaso C et al (2010) Comparative analysis of the microbial communities inhabiting halite evaporites of the Atacama Desert. Int Microbiol 13:79–89

de Zelicourt A, Al-Yousif M, Hirt H (2013) Rhizosphere microbes as essential partners for plant stress tolerance. Mol Plant 6:242–245

Ding GC, Piceno YM, Heuer H et al (2013) Changes of soil bacterial diversity as a consequence of agricultural land use in a semi-arid ecosystem. PLoS One 8:e59497

Fierer N, Jackson RB (2006) The diversity and biogeography of soil bacterial communities. Proc Natl Acad Sci U S A 103:626–631

Freeman S, Horowitz S, Sharon A (2001) Pathogenic and non-pathogenic lifestyles in *Colletotrichum acutatum* from strawberry and other plants. Phytopathology 91:986–99.

Geurts R, Lillo A, Bisseling T (2012) Exploiting an ancient signalling machinery to enjoy a nitrogen fixing symbiosis. Curr Opin Plant Biol 15:438–443

Hasegawa S, Meguro A, Nishimura T, Kunoh H (2004). Drought tolerance of tissue-cultured seedlings of mountain laurel (*Kalmia latifolia* L.) induced by an endophytic actinomycete I. Enhancement of osmotic pressure in leaf cells. Actinomycetology 18:43–47

Heidari M, Mousavinik SM, Golpayegani A (2011) Plant growth promoting rhizobacteria (PGPR) effect on physiological paramters and mineral uptake in basil (*Ociumum basilicum* L.) under water stress. ARPN J Agric Biol Sci 6:6–11

Köberl M, Müller H, Ramadan EM et al (2011) Desert farming benefits from microbial potential in arid soils and promotes diversity and plant health. PloS ONE. 6:e24452

Köberl M, Ramadan EM, Adam M et al (2013a) *Bacillus* and *Streptomyces* were selected as broad-spectrum antagonists against soilborne pathogens from arid areas in Egypt. FEMS Microbiol Lett 342:168–178

Köberl M, Schmidt R, Ramadan EM et al (2013b) Biocontrol strategies and next generation sequencing: organic desert agriculture in Egypt. iConcept online

Kohlera J, Hernández JA, Caravacaa A et al (2009) Induction of antioxidant enzymes is involved in the greater effectiveness of a PGPR versus AM fungi with respect to increasing the tolerance of lettuce to severe salt stress. J Exp Bot 65:245–252

Marasco R, Rolli E, Ettoumi B et al (2012) A drought resistance-promoting microbiome is selected by root system under desert farming. PLoS One 7:e48479

Marasco R, Rolli E, Fusi M et al (2013a) Plant growth gromotion potential is equally represented in diverse grapevine root-associated bacterial communities from different biopedoclimatic environments. BioMed Res Int 2013:491091

Marasco R, Rolli E, Vigani G et al (2013b) Are drought-resistance promoting bacteria cross-compatible with different plant models? Plant Signal Behav 10:e26741

Márquez LM, Redman RS, Rodriguez RJ et al (2007) A virus in a fungus in a plant-three way symbiosis required for thermal tolerance. Science 315:513–515

Mayak S, Tirosh T, Glick B (2004) Plant growth-promoting bacteria that confer resistance to water stress in tomatoes and peppers. Plant Sci 166:525–530

Park S, Li J, Pittman JK et al (2005) Up-regulation of a H^+-pyrophosphatase (H^+-PPase) as a strategy to engineer drought-resistant crop plants. Proc Natl Acad Sci U S A 102:18830–18835

Redman RS, Freeman S, Clifton DR et al (1999) Biochemical analysis of plant protection afforded by a nonpathogenic endophytic mutant of *Colletotrichum magna*. Plant Physiol 119:795–804

Redman RS, Sheehan KB, Stout RG et al (2002) Thermotolerance conferred to plant host and fungal endophyte during mutualistic symbiosis. Science 298:1581

Rodriguez RJ, Henson J, Van Volkenburgh E et al (2008) Stress tolerance in plants via habitat-adapted symbiosis. ISME J 2:404–416

Rodríguez-Salazar R, Suárez R, Caballero-Mellado J et al (2009) Trehalose accumulation in *Azospirillum brasilense* improves drought tolerance and biomass in maize plants. FEMS Microbiol Lett 296:52–59

Rolli E, Marasco M, Vigani G et al (2014) Improved plant resistance to drought is promoted by the root-associated microbiome as a water stress-dependent trait. Environ Microbiol. doi:10.1111/1462-2920.12439

Schmidt R, Köberl M, Mostafa A et al (2014) Effects of bacterial inoculants on the indigenous microbiome and secondary metabolites of chamomile plants. Front Microbiol 5:64

Waller F, Achatz B, Baltruschat H et al (2005) The endophytic fungus *Piriformospora indica* reprograms barley to salt-stress tolerance, disease resistance, and higher yield. Proc Natl Acad Sci USA 102:13386–13391

Chapter 29
Rhizoremediation

Sofie Thijs and Jaco Vangronsveld

Abstract Over the past centuries, technological revolutions have brought about new sources of soil and (ground)water pollution. The clean-up costs by conventional remediation methods are often exorbitantly high, retarding soil remediation if performed at all. This is a severe problem as the health consequences of soil pollution can be considerable. Against these drawbacks, rhizoremediation which is an inexpensive and sustainable technology, based on the actions of biodegradative microorganisms in the rhizosphere and the plant phytoremediation capacity, has gained increased attention. During the symbiotic interactions in the rhizosphere, ectomycorrhizal fungi extend the belowground surface area of plants where billions of root-associated bacteria help to take-up minerals and pollutants, produce vitamins and plant hormones and degrade organic compounds or sequestrate metals. Genomics technologies and systems-based approaches have tremendously advanced the way we investigate the plant "black box" and lead to new insights about how we can exploit the beneficial plant-microbe association in terms of soil remediation, the topic of this chapter.

29.1 Introduction

The rapid pace of technological advances since the nineteenth century, the Green Revolution, Second Industrial Revolution, transportation, electronics and medical developments have brought distinctively benefits in life quality, crop yields and human health but also have its inextricably drawbacks for the environment: the over-exploitation of natural resources and the enormous increases in wastes, releases of pollutants and soil/water quality degradation that occur at high rates. According to the European Environment Agency (EEA), industrial activities account for 60 %

J. Vangronsveld (✉) · S. Thijs
Centre for Environmental Sciences, Hasselt University, Agoralaan, building D,
3590 Diepenbeek, Belgium
Tel.: +32 11 268 331
e-mail: Jaco.vangronsveld@uhasselt.be

S. Thijs
Tel.: +32 11 268 331
e-mail: sofie.thijs@uhasselt.be

© Springer International Publishing Switzerland 2015
B. Lugtenberg (ed.), *Principles of Plant-Microbe Interactions*,
DOI 10.1007/978-3-319-08575-3_29

of Europe's soil contamination from which the oil sector takes a one fourth part. According to UNESCO, in the developing world, 70 % of the industrial waste is dumped untreated in the waterways. In the next decades, it is estimated that each oil and polycyclic aromatic hydrocarbons (PAHs)-contaminated site will cost hundreds of thousands of euros to treat by conventional techniques and costs can be upwards to billions when groundwater is also polluted. Besides the high costs, dump and treat techniques are not always sufficient, they are invasive, time-consuming, detrimental to soil and biological life and can lead to pollution of the biosphere and release of greenhouse gasses. With the impetus demand for new agricultural and industrial areas, the development of cheap, eco-friendly remediation strategies such as rhizoremediation are important.

29.2 Rhizoremediation

Rhizoremediation is a specific subset of phytoremediation and involves the biodegradation of organic pollutants in the root zone ascribed to the microbes in the rhizosphere of plants used for phytoremediation or of plants which naturally occur on contaminated soils (Fig. 29.1).

Plant roots can be considered as a substitute for tilling of soil, to spread the root-associated microorganisms through the soil and to penetrate layers normally not accessible to bacteria and to incorporate nutrients, bring oxygen and provide better redox conditions which helps to stimulate and activate the rhizosphere microorganisms. In addition, the majority of plants, in particular trees, live in symbiosis with ectomycorrhizal (ECM) and/or arbuscular mycorrhizal (AM) fungi (Chap. 25). ECM fungi act as a web-like extension of the root system enhancing the absorptive surface area of plants. ECM play a crucial role in nutrient recycling and they are often more resistant against (a)biotic stresses. In this mycorrhiza-root continuum, billions of bacteria help to take-up minerals and pollutants, produce vitamins and plant hormones and degrade organic compounds or sequestrate metals. The enhanced degradation of pollutants in the rhizosphere is beneficial to the plant and results in improved plant growth on contaminated soils.

Compared to conventional techniques, plants are very attractive clean-up technology tools, (1) they provide their own solar energy by photosynthesis and pump–up gratuity pollutants with the water-stream making it much cheaper (it depends on the site and pollutant concentration but in general it falls within 10–50 % of the costs for conventional techniques, (2) remediation is in situ, (3) it generates much less wastes and instead even creates economic benefits such as wood, feed stock or bioenergy production. Rhizoremediation is suitable for vast sites with low to moderate pollutant concentrations. Sites are often co-contaminated with organic and heavy metal pollutants and are characterized by poor soil conditions (pH, texture, low nutrient availability, salinity) which all have retarded the implementation. However, new knowledge led to designing novel strategies that can overcome some of the inherent limitations and commercialization of bio-inoculants is taking place, showing the promising growing interest in rhizoremediation.

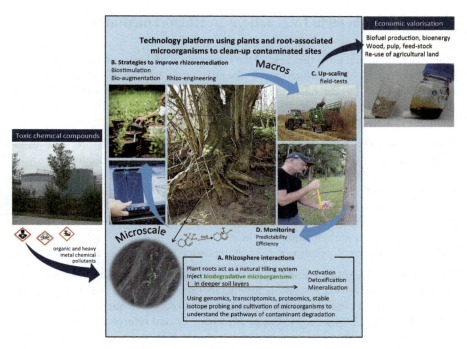

Fig. 29.1 Schematic overview of the processes involved in rhizoremediation. **a** Summary of the rhizosphere interactions at the micro-scale. On the bottom, scanning electron microscopy picture of bacteria (colored in green) on the root surface of the grass *Agrostis capillaris*, from Thijs, *unpublished*. Picture of the split-root vertical agar plate system to analyse root-growth responses of *Arabidopsis thaliana*, from dr Remans, *unpublished*. **b** Engineering techniques to improve rhizoremediation. Picture of the inoculation of *Salix sp.* cuttings with PGP and heavy-metal resistant bacteria in cadmium contaminated soil, from Janssen *unpublished* (Hasselt, Belgium). **c** The need for up-scaling laboratory and microcosm tests to the field; picture of the harvest of short-rotation coppice of willow trees on cadmium polluted soils in Belgium, from Janssen, *unpublished*. **d** The development of effective monitoring methods. (Picture: prof dr Joel Burken (Missouri S&T, Missouri, US) taking a sample from a tree at the campus of Hasselt University, Belgium, picture from Vangronsveld *unpublished*. In the right panel, the economic benefits and value of rhizoremediation are described. (Picture of bio-oil, Hasselt University, Belgium; Picture of the oil tanks is from Weyens, *unpublished*; Picture of the roots is from Thijs, *unpublished*)

In this chapter, we explore the plant-bacteria interactions in rhizoremediation with a focus on organic contaminants degradation. Next, we comment on strategies how to successfully improve rhizoremediation and highlight key case studies of bacterial-ectomycorrhizal interactions in rhizoremediation. We conclude with the perspectives and needs brought by new analytical techniques.

29.3   Plant-Bacteria Interactions in Rhizoremediation

Plant-bacteria interactions that lead to the degradation of contaminants in the rhizo-sphere occur at the soil-root interface. At the microscale, the crucial processes involve root-colonization by the bacteria, selection and maintenance of the degradation genes, the role of root exudates in the activation of catalytic genes and inter-kingdom communication which shape the community.

Colonization The biodegradative bacteria in the rhizosphere can originate from (bacterized) seeds or they can be recruited from the main soil reservoir during growth of the plants on the contaminated soil (Fig. 29.1). These bacteria can subsequently be spread through the soil during root emergence and growth. In addition, bacteria can also actively colonize the roots by chemotactic movement. Using in vivo expression technology (IVET), transcriptomics and mutants defective in motility, the mechanisms of root-colonization and the identification of genes which are activated during rhizosphere colonization are being unraveled (Ramos-Gonzalez et al. 2005).

Root Exudates Regulate Gene Expression Root exudate compounds such as sugars, organic acids, fatty acids, secondary metabolites, nucleotides and also inorganic compounds play a crucial role in structuring the rhizosphere microbial population (Chap. 17, Chap. 23, Chap. 24, Chap. 26, Chap. 27 and references therein). Exudates select those microbial populations which are able to respond to the exudate buffet with rapid growth responses so that eventually only a small subset of the whole soil microbial diversity is colonizing the roots and to a much higher density than the surrounding bulk soil (Bais et al. 2008). A significant role in plant selection processes is attributed to aromatic compounds that are exuded (i.e. terpenoids, phenols, flavonoids and other lignin-derived components), the structures are similar to those of contaminants and can act as co-substrates for difficult to degrade compounds such as high-chlorinated biphenyls (PCBs) and PAHs or as inducers of degradation pathways. Recently, -omics technologies and systems-based approaches are applied to study the gene-regulation network in contaminant degradation and for sure they will provide promising new insights about how we can exploit the beneficial plant-microbe association in terms of soil remediation (Matilla et al. 2007).

Communication and Dynamics Multiple signals are sent and received by plants and microorganisms. These signals are responsible for recognition of microorganisms, recruitment of catalytic potential, mycorrhization, resistance to stresses and quorum sensing, i.e. the gene-expression regulation on the population level and includes bacterial traits such as motility, biofilm formation, symbiosis and conjugation (Chap. 3). Some of the communication systems, e.g. the interaction between arbuscular mycorrhiza (AM) and trees, are symbiosis of several hundreds of millions years old and probably evolved during co-evolution. Changing a(biotic) factors, e.g. growth stage of the plant, soil type, contamination type, season, and level and density of other microorganisms are all thought to cause significant shifts in the rhizosphere and (root) endophytic community structure. For example, Siciliano et al. (2001) demonstrated that the catabolic alkane monooxygenase genes were more prevalent in root endophytic and rhizospheric microbial communities than in bacteria present in bulk soil contaminated with hydrocarbons.

Bio-degradation Potential of Bacteria Many bacteria possess the catabolic/catalytic tools to mineralise/transform recalcitrant pollutants including petroleum hydrocarbons, PAHS, chlorinated aliphatic compounds, solvents, PCBs, nitro-aromatics and also for the removal of nitrate and ammonia in waste-streams and metal immobilization or exclusion. The online Minnesota Biocatalysis/Biodegradation Database supports almost 1200 compounds, over 800 enzymes, 1300 reactions and almost 500 microorganisms with bio-degradative characteristics and is increasing to grow (Gao et al. 2010).

29.4   Strategies to Improve Rhizoremediation

The presence of the biodegradative strains, the expression of the catalytic genes and maintenance are all crucial factors which determine the success of rhizoremediation. However under natural conditions, rhizoremediation can be slow, indicating that some of these factors or others are limiting the removal of pollutants. Therefore, scientists and biotechnology companies developed strategies to improve the rhizoremediation efficiency (Fig. 29.1b).

Bio-stimulation Fertilisation of the plants by addition of compost, minerals (nitrogen, phosphorous, potassium) or fertilisers on a base of organic carboxylic acids which can enhance root exudation, has been reported to enhance the degradation of pollutants in the rhizosphere by stimulating growth and activity of microorganisms. In addition, additives are used to improve the physico-chemical properties of the soil and to increase the bio-availability of pollutants e.g. biosurfactants for oil degradation and siderophores for metal-bioavailability.

Bio-augmentation The introduction of a microbial strain or consortium is another way to enhance degradation of contaminants in the rhizosphere. Delivery methods for introducing these beneficial degrading microorganisms in soil include seed coating or inoculation of plant roots with a bacterial suspension by root-dipping or soil drench, in analogy to probiotics (Chap. 33). Growth of the inoculated root system acts as a 'bio-injector' of the degrading strains in the soil.

For 'newer' pollutants, such as high-level chlorinated PCB's and chloro-ethylenes, sometimes no microbial catabolic pathways have evolved. In this situation, construction of genetically engineered (GE) and genetically modified (GM) strains might be interesting (Brazil et al. (1995). In addition, in case of co-contamination, construction of microbial resistance for multiple organics and/or heavy metals is required. However, genetic modification has also limitations, the costs, the amount of work involved and the often limited success with the new synthetic vectors make this area extremely challenging (Hernández-Sánchez and Wittich 2012).

With the introduction of bacteria in the rhizosphere it is important to consider their rhizosphere competence. Firstly, Kuiper et al. (2002) developed novel enrichment techniques whereby the strains are selected based on the combination of effective root colonization, assessed by sequential bacteria inoculation/re-isolation from the roots of model plants and the abilities to degrade the pollutant. Secondly, bacteria which show in vitro PGP-properties and abilities to increase plant biomass *in planta* are receiving more attention. We hypothesize that the beneficial PGP-bacteria settle in the root-zone and increase plant biomass in contaminated soil, alleviate plant stress by ACC-deaminase and thereby enhance rhizoremediation (Thijs et al. 2014). Thirdly, it is important to select the appropriate plant-bacterium combinations for efficient degradation of the contaminants (Kuiper et al. 2004). There exists a plant-cultivar specificity of microbial communities and furthermore variations in exudates, contaminant levels and soil conditions influence the microbial community. Fourthly, inoculating a consortium of bacteria whereby the partners can deliver nutrients and other growth factors or use the various intermediates in the organic degradation pathway more efficiently often out-performs the action of single (GM) strains. Fifthly, because of the complex rhizosphere interactions, also the inoculation with 'competent endophytes' (see Chap. 5) is considered, i.e. bacteria that successfully colonize the internal plant tissues and which possess the capacity to incite plant physiology can be selectively favored.

Transgenic Plants Engineering of plants with genes that allow superior degradation abilities was endeavored, by e.g. overexpression of genes involved in metabolism, uptake and transport of pollutants. Very often bacterial genes are used whereby complete degradation pathways are introduced in plants. This was already proven to be successful for enhanced degradation of highly recalcitrant compounds such as explosives, PCBs and PAHs (Mackova et al. 2006). Another approach is rhizo-engineering and is the use of transgenic plants which secrete more root exudates or other rhizodeposits, which all indirectly have a significant effect on the rhizosphere microbial community (Van Aken et al. 2010).

29.5   Realisations in the Field

During recent years, numerous studies have been published that describe the combined use of plants and bacteria in rhizo-and phytoremediation through successes and failures. Studies have focused on phreatophytic trees that are deep-rooting, fast growing and are 'pumping' huge amounts of water such as *Populus* and *Salix* or grasses and small shrubs with an extensive fine root system enhancing the bacterial transformation of pollutants in the rhizosphere. Good reviews on phyto- and rhizoremediation studies exist (Gerhardt et al. 2009; Kanaly and Harayama 2010; Kavamura and Esposito 2010; Megharaj et al. 2011; Mackova et al. 2006). Bio-augmented rhizoremediation was reported to not always lead to enhanced successes compared with natural rhizoremediation. Often, this results from the poor root colonization ability of the introduced strains. Therefore, as mentioned previously, this trait, together with other rhizosphere competence abilities, is crucial for successes in the field.

The first studies dealing with ECM-fungi in remediation, mainly focused on the re-vegetation and stabilization of polluted soils. Most of the results from the involvement of ECM in contaminant degradation originate from laboratory experiments and microcosm studies but field studies remain scarce. Therefore, up-scaling of ECM-bacteria rhizoremediation is needed (Bücking 2011). Mycorrhizal fungi can importantly contribute to rhizoremediation: ECM fungi can directly enable ligninolytic and cell-wall degrading enzymes to degrade various recalcitrant pollutants such as TPH, PAHs, pesticides, explosives and PCBs, increase the plant tolerance to withstand toxicity, improve nutrition and protection from pathogens, enhance the bio-availability of organics and heavy metals or by affecting the composition and activity of the bacterial contaminant degrading population (Chaudhry et al. 2005). Under some circumstances, ECM can also inhibit bacterial growth or reduce bacterial activity, so not only positive and synergistic effects must be considered.

Remediation of total petroleum hydrocarbons (TPH) is often hindered by low plant growth, particularly when TPH-concentrations cause stress to plants. Application of PGPR-enhanced phytoremediation (PEP) was demonstrated to be successful in remediating a highly contaminated petroleum hydrocarbon site in Southern Ontario, California (Gurska et al. 2009). It was hypothesized that the PGPR-bacteria reduce the plant stress level and improve plant growth, thereby stimulating rhizoremediation. When *Pinus sylvestris* seedlings were inoculated with the ECM *Suillus bovinus* or *Paxillus involutus* and grown in soil contaminated with petroleum hydrocarbons, a bacterial biofilm of strains was formed on the exterior of the hyphae (Sarand et al. 1998). Moreover, these strains harbored plasmids involved in the degradation of mono-aromatics.

Because of their hydrophobicity and chemical stability, PCB's are only slowly taken up by plants and degraded by associated microbes, resulting in incomplete degradation or accumulation of toxic intermediates. Narasimhan et al. (2003) demonstrated that *A. thaliana* flavonoid-overexpressing mutants were colonized at a much higher level than non-mutant lines by the PCB-degrading bacterial strain *P. putida* PML2, suggesting that rhizo-engineering of plants producing altered exudates is a valuable tool to enhance PCB-degradation.

Forty percent of the waste sites in the United States are co-contaminated with organics and trace elements (arsenic, barium, cadmium, chromium, lead, copper, nickel and zinc). Todd and Reina (2003) described several approaches to increase organic biodegradation in the presence of metals and include the application of metal immobilizing additives or clay minerals and the reduction of metal bio-availability by use of metal-resistant bacteria or mycorrhizae. Dual inoculation with ectomycorrhizal fungi and bacteria has been shown to improve growth and metal accumulation of mycorrhized trees and it is suggested that the bacteria facilitate the ECM colonization and increase the bio-availability of metals to the mycorrhized plant (Zimmer et al. 2009). In return, the bacteria receive nutrients from the plant.

29.6   Perspectives and Research Needs

A large number of man-made chemicals still lack good biological catalysts. To date, for most of the 10 million organic compounds described, biological degradation has not been investigated. With the intense research that is now occurring in the nanotechnology, material sciences and electronics sectors, new challenges are arising in e.g. the biodegradability of fullerenes and carbon nanotubes, considered as the ultimate high molecular weight PAHs. More than 6000 ECM fungal species are likely to exist worldwide and only a fraction of the potential of ECM fungi to degrade pollutants has so far been determined. In addition, with the characterization of cultivable biodegradative isolates we are only scratching the top of the iceberg of the whole microbial diversity. The detection and capture of novel broad-host range plasmids from soil bacteria and subsequent sequencing can provide a wealth of new information and insights into the role of plasmids in bioremediation and ecology (Heuer and Smalla 2012).

Technological advances and reduced costs of next generation sequencing technologies, whole-genome approaches, metatranscriptomics, metaproteomics, metabolomics, advances in micro-arrays (PhyloChip, GeoChip) and ecological tools such as stable isotope probing (SIP), mark the start of a more wide use of these NGS-techniques in rhizoremediation, providing unprecedented insights in the complex interactions in the rhizosphere (Ramos et al. 2011). Using gene-expression technologies, Govantes et al. (2009) elucidated a complex regulator circuit in bacteria that involves nitrogen control of the herbicide atrazine utilization. Based on these findings, they developed valuable strategies to improve atrazine degradation in fertilized agricultural soil, e.g. they propose the application of inhibitors of nitrogen assimilation or the use of mutant bacterial strains that are impaired in the nitrogen control.

In many cases, the most effective remediation solution is a combination of several techniques depending on the soil pollution type/degree, its environmental and health risks, the economic value of the polluted site and juridical and legislative policies (Segura and Ramos 2013). Because rhizoremediation needs time and is under influence of environmental factors e.g. temperature, pH, new pollution over time, it is important to establish reliable monitoring methods to estimate the efficacy of rhizoremediation and to predict the clean-up time. Innovative tools such as the use of indicator species, phytophorensics i.e. the chemical analyses of xenobiotics in the sap-stream of plants to trace contaminants in soil and groundwater and plant nanobionics, which all take much less time, efforts and costs than traditional detection methods are being developed (Fig. 29.1d).

A last important consideration is the **preservation and maintenance** of degradation genes for rhizoremediation. Noor et al. (2014) pointed out that ongoing selection is taking place in newly isolated strains from French agricultural soils recurrently exposed to atrazine and simazine herbicides in 2000, proven by 6 amino-acid differences in the *atzA* dechlorinase gene in the majority of the new strains compared with the original isolates from 1990. This suggests that the environment remains the major reservoir for the discovery of novel strains for rhizoremediation better adapted to the current conditions. As such, rhizoremediation still is a promising and fertile research area.

Conclusion

At this point, our understanding of the plant system and its microorganisms in the rhizosphere is fairly good. Studies involving newly isolated organisms have enhanced our knowledge of degrading activities and regulation in the rhizosphere over an expanding range of contaminants and environmental conditions. At the same time, the challenge of future investigations is to further unravel how the microorganisms function together in their natural environment, the mechanisms of ECM in biodegradation of recalcitrant xenobiotics and fortuitous metabolism, the efficacy of consortia in rhizoremediation and the dynamics in rhizosphere communities. This information will allow the manipulation of rhizoremediation systems to accomplish soil remediation to the greatest extent, predictability and over the shortest time periods.

References

Bais H, Broeckling C, Vivanco J (2008) Root exudates modulate plant—microbe interactions in the rhizosphere. In: Karlovsky P (ed) Secondary metabolites in soil ecology, vol 14. Soil biology. Springer, Berlin, pp 241–252

Brazil G, Kenefick L, Callanan M et al (1995) Construction of a rhizosphere pseudomonad with potential to degrade polychlorinated biphenyls and detection of bph gene expression in the rhizosphere. Appl Environ Microbiol 61:1946–1952

Bücking H (2011) Ectomycoremediation: an eco-friendly technique for the remediation of polluted sites. In: Rai M, Varma A (eds) Diversity and biotechnology of ectomycorrhizae. Springer, Berlin, pp 209–231

Chaudhry Q, Blom-Zandstra M, Gupta S, Joner EJ (2005) Utilising the synergy between plants and rhizosphere microorganisms to enhance breakdown of organic pollutants in the environment. Environ Sci Pollut Res Int 12 (1):34-48. doi:10.1065/espr2004.08.213

Gao J, Ellis L, Wackett L (2010) The University of Minnesota Biocatalysis/ Biodegradation Database: improving public access. Nucleic Acids Res 38 (Database issue):91. doi:10.1093/nar/gkp771

Gerhardt KE, Huang X-D, Glick BR et al (2009) Phytoremediation and rhizoremediation of organic soil contaminants: potential and challenges. Plant Sci 176:20–30

Govantes F, Porrúa O, García-González V et al (2009) Atrazine biodegradation in the lab and in the field: enzymatic activities and gene regulation. Microb Biotechnol 2:178–185

Gurska J, Wang W, Gerhardt KE et al (2009) Three year field test of a plant growth promoting rhizobacteria enhanced phytoremediation system at a land farm for treatment of hydrocarbon waste. Environ Sci Technol 43:4472–4479

Hernández-Sánchez V, Wittich R-M (2012) Possible reasons for past failures of genetic engineering techniques for creating novel, xenobiotics-degrading bacteria. Bioengineered 3:260–261

Heuer H, Smalla K (2012) Plasmids foster diversification and adaptation of bacterial populations in soil. FEMS Microbiol Rev 36:1083–1104

Kanaly R, Harayama S (2010) Advances in the field of high-molecular-weight polycyclic aromatic hydrocarbon biodegradation by bacteria. Microb Biotechnol 3:136–164

Kavamura V, Esposito E (2010) Biotechnological strategies applied to the decontamination of soils polluted with heavy metals. Biotechnol Adv 28:61–69

Kuiper I, Kravchenko LV, Bloemberg GV et al (2002) Pseudomonas putida strain PCL1444, selected for efficient root colonization and naphtalene degradation, effectively utilizes root exudate components. Mol Plant-Microbe Interact 15:734–741

Kuiper I, Lagendijk EL, Bloemberg GV et al (2004) Rhizoremediation: a beneficial plant-microbe interaction. Mol Plant Microbe Interact 17:6–15

Mackova M, Barriault D, Francova K et al (2006) Phytoremediation of polychlorinated biphenyls. In: Mackova M, Dowling DN, Macek T (eds) Phytoremediation and rhizoremediation: theoretical background. Focus on biotechnology, vol. 9 A. Springer, Dordrecht, pp 143–167

Matilla MA, Espinosa-Urgel M, Rodriguez-Herva JJ et al (2007) Genomic analysis reveals the major driving forces of bacterial life in the rhizosphere. Genome Biol 8:R179

Megharaj M, Ramakrishnan B, Venkateswarlu K et al (2011) Bioremediation approaches for organic pollutants: a critical perspective. Environ Int 37:1362–1375

Narasimhan K, Basheer C, Bajic VB et al (2003) Enhancement of plant-microbe interactions using a rhizosphere metabolomics-driven approach and its application in the removal of polychlorinated biphenyls. Plant Physiol 132:146–153

Noor S, Changey F, Oakeshott J et al (2014) Ongoing functional evolution of the bacterial atrazine chlorohydrolase AtzA. Biodegradation 25:21–30

Ramos-Gonzalez MI, Campos MJ, Ramos JL (2005) Analysis of Pseudomonas putida KT2440 gene expression in the maize rhizosphere: in vitro expression technology capture and identification of root-activated promoters. J Bacteriol 187:4033–4041

Ramos J-L, Marqués S, van Dillewijn P et al (2011) Laboratory research aimed at closing the gaps in microbial bioremediation. Trends Biotechnol 29:641–647

Sarand I, Timonen S, Nurmiaho-Lassila E-L et al (1998) Microbial biofilms and catabolic plasmid harbouring degradative fluorescent pseudomonads in Scots pine mycorrhizospheres developed on petroleum contaminated soil. FEMS Microbiol Ecol 27:115–126

Segura A, Ramos J (2013) Plant-bacteria interactions in the removal of pollutants. Curr Opin Biotechnol 24:467–473

Siciliano SD, Fortin N, Mihoc A et al (2001) Selection of specific endophytic bacterial genotypes by plants in response to soil contamination. Appl Environ Microbiol 67:2469–2475

Thijs S, Weyens N, Sillen W et al (2014) Potential for plant growth promotion by a consortium of stress-tolerant 2,4-dinitrotoluene-degrading bacteria: isolation and characterization of a military soil. Microb Biotechnol 7(4):294–306. doi:10.1111/1751-7915.12111

Todd RS, Raina MM (2003) Impact of metals on the biodegradation of organic pollutants. Environ Health Perspect 111:1093–1101

Van Aken B, Correa P, Schnoor J (2010) Phytoremediation of polychlorinated biphenyls: new trends and promises. Environ Sci Technol 44:2767–2776

Zimmer D, Baum C, Leinweber P et al (2009) Associated bacteria increase the phytoextraction of cadmium and zinc from a metal-contaminated soil by mycorrhizal willows. Int J Phytoremediation 11:200–213

Part V
Important Technologies

Chapter 30
Microbial Communities in the Rhizosphere Analyzed by Cultivation-Independent DNA-Based Methods

Susanne Schreiter, Namis Eltlbany and Kornelia Smalla

Abstract The development of methods to extract nucleic acids directly from the rhizosphere or from microbial cells detached by a mechanical treatment from roots opened new dimensions to study the rhizosphere microbiome and to overcome limitations of cultivation-dependent methods. This chapter summarizes the potentials and limitations of cultivation-independent methods used by our group in the last 15 years to investigate microbial communities in the rhizosphere and their response to changing environmental conditions. We showed that rhizosphere microbial communities are highly dynamic, and that their composition is mainly shaped by the plant and the soil type and factors influencing these drivers of microbial diversity in the rhizosphere.

30.1 Introduction

The importance of the plant microbiome for plant growth and health is increasingly recognized. The fraction of soil influenced by the plant, termed rhizosphere, is an interface that connects the soil with the plant. Understanding the complex interactions in the rhizosphere remained a challenge until tools allowing cultivation-independent analysis of DNA or RNA extracted directly from the rhizosphere became available. Here we provide a short overview of some of these tools which were used to study the influence of different factors on the microbial community compositions in the rhizosphere. The chapter is biased towards our own work and for more comprehensive compilation the reader is referred to recent reviews

K. Smalla (✉) · S. Schreiter · N. Eltlbany
Julius Kühn-Institut, Federal Research Centre for Cultivated Plants (JKI),
Messeweg 11–12, 38104 Braunschweig, Germany
Tel.: + 49 531 299 3814
e-mail: kornelia.smalla@jki.bund.de

S. Schreiter
e-mail: susanne.schreiter@jki.bund.de

N. Eltlbany
e-mail: namis.eltlbany@jki.bund.de

© Springer International Publishing Switzerland 2015   289
B. Lugtenberg (ed.), *Principles of Plant-Microbe Interactions*,
DOI 10.1007/978-3-319-08575-3_30

(Berg and Smalla 2009; Berendsen et al. 2012; Bulgarelli et al. 2013). Before discussing the different nucleic acid-based methods and the major findings we would like to draw the reader's attention to critical prerequisites for obtaining meaningful data.

30.2 Experimental Design and Sampling

The adequate experimental design and sampling strategy depends on the hypotheses to be tested and often pre-experiments might assist in determining the numbers of samples to be analyzed. Furthermore, the strategy used to sample the rhizosphere influences the data obtained. Typically composite samples from the root system of several plants are analyzed for reporting in a representative manner on the structural and functional diversity in the rhizosphere and on the variation among replicates within the same treatment and between treatments. We have studied the rhizosphere microbial communities of different plant species from various sites and geographic regions. Usually the plants were destructively sampled by uprooting them and vigorously shaking the roots. Different protocols have been used also in our laboratory and the protocols had to be adapted for various reasons. Therefore it is highly recommended for comparison of data from different studies to carefully read the sampling protocols described, as the fractions of the rhizosphere microbial communities analyzed differed—depending on the protocols applied—in the proportion of rhizoplane and bulk soil microorganisms present. Rhizosphere total community (TC-) DNA was extracted from the soil brushed off from the root (Marques et al. 2014). This technique is typically used when long-distance transport of samples is needed and microbes residing on the rhizoplane or in soil particles glued to the roots or from fine roots were likely missing. However, in most studies performed by our group the complete root system, after vigorous shaking, was cut into pieces and mixed. Subsamples placed in plastic bags were treated in the Stomacher® Circulator after adding saline or water. Via paddle movement cells were detached from the root and soil particles and the Stomacher treatment step was repeated three times. To obtain the microbial pellets, the combined supernatants were centrifuged and the TC-DNA was extracted from the complete pellet with of commercial soil DNA extraction kits (Weinert et al. 2009; Schreiter et al. 2014a). When a combination with cultivation-dependent analysis was done, e.g. to determine the potentially antagonistic fraction (Berg et al. 2002) or to monitor inoculant strains (Adesina et al. 2009; Xue et al. 2013; Schreiter et al. 2014a) an aliquot from the combined supernatant was used for plating of serial dilutions. Recently, we had to modify the protocol in a project aiming to compare the effect of three soil types on the rhizosphere communities. An additional root washing was performed in order to remove big clumps of soil adhering to the roots of plants grown in clay rich soils before the Stomacher® protocol (Schreiter et al. 2014b). The TC-DNA obtained from the pellet gained with this protocol was assumed to represent the genetic information of microbes colonizing the rhizoplane and rhizosphere. Although a complete dislodgment of cells adhering to the roots and soil particles seems to be impossible, it is important that cells bound to soil particles with different degrees of strength are released with similar efficiency. Another crucial step for the recovery of representative DNA that mirrors the genomes of all microbes present in a rhizosphere sample is the efficient lysis of microbial cell walls. This can be achieved by mechanical cell disruption and by enzymatic or chemical disintegration of cell walls, or a combination of these methods. The efficiency of the different methods used might not only influence the yield but also the presence of genomic DNA in cells difficult to lyse. However, obviously the strength of lysis needs to be a trade-off as too rigorous lysing methods might shear DNA released from cells that are easy to lyse. The DNA yield might vary considerably for different DNA extraction kits used for the same rhizosphere soils. Commercial kits for extraction from soil after a harsh lysis with the FastPrep®-24 Instrument were major achievements and allowed a simplification and miniaturization of the method. Extraction kits are less time-consuming and efficiently remove co-extracted humic acids which would disturb PCR-amplification. Finally, it should be stressed that strict precautionary measures need to be taken to prevent contamination of the DNA during the extraction. In particular, when PCR is used to amplify a target gene that occurs less frequently, e.g. antibiotic resistance genes or transgenic DNA, the extraction of DNA, preparation of PCR reactions and analysis of PCR products need to be done in separate rooms.

30.3   Bacterial and Fungal Community Composition in the Rhizosphere

PCR-based amplification of 16S and 18S rRNA gene or ITS fragments from rhizosphere DNA and their subsequent analysis by fingerprinting, cloning and/or sequencing are most frequently used to study the composition of microbes in the rhizosphere and the effects of treatments. The rapidly growing database of ribosomal rRNA gene sequences contains presently more than a million good quality 16S rRNA gene sequence entries deposited in Ribosomal Database. A disadvantage of using ribosomal rRNA gene fragments is that bacteria possess different numbers of ribosomal RNA operons. The numbers of 16S rRNA operons are assumed to reflect different ecological strategies of bacteria (Klappenbach et al. 2000) and sequence heterogeneity of the different operons might occur (Nübel et al. 1996). Costa et al. (2007) proposed the *Pseudomonas*-specific *gacA* gene as an alternative marker for studying their community composition. However, no matter which gene is targeted, one major limitation that remains is that gene fragments of less common populations are often not represented in clone libraries or fingerprints, especially when primers targeting all bacteria, archeae or fungi are used for PCR amplification. Bent and Forney (2008) termed this problem "the tragedy of the uncommon". The application of group-specific primers targeting the 16S rRNA gene can assist in studying less common populations (Heuer et al. 1997; Heuer et al. 2002; Gomes et al. 2001; Costa et al. 2006a; Costa et al. 2006b; Weinert et al. 2009). The sequence diversity among 16S and 18S rRNA gene or ITS amplicons from TC- DNA can be analyzed by various techniques such as the terminal restriction fragment length polymorphism (T-RFLP), single strand conformation polymorphism (SSCP) or denaturing gradient gel electrophoresis (DGGE) that were developed in the end of the 1990's. A comparison of these fingerprinting techniques showed that they had similar resolution levels and provided similar results despite the different 16S rRNA gene regions used (Smalla et al. 2007). At that time the great advantage of the fingerprinting techniques was that a sufficient number of replicates could be analyzed in parallel, and when combined with statistical analysis, testing of different biotic and abiotic factors influencing the bacterial and fungal community composition became possible (Kropf et al. 2004). A clear drawback of the molecular fingerprinting techniques was that bands with treatment-dependent intensity had to be excised from the fingerprints, re-amplified, cloned and sequenced. Sequencing of dominant bands with identical electrophoretic mobility detected in the rhizosphere of strawberry and oilseed rape were shown to represent taxonomically different populations (Costa et al. 2006a). On the one hand, 16S rRNA gene fragments from taxonomically distinct populations might have the same electrophoretic mobility due to similar melting behavior while, on the other hand, one population might generate more than one band due to operon sequence heterogeneity. The DGGE fingerprints based on 16S and 18S rRNA gene and ITS fragments were used to study the influence of the following factors on the composition of the bacterial and fungal communities in the rhizosphere: (i) plant species (Smalla et al. 2001; Costa et al. 2006a), (ii) plant growth developmental stage (Smalla et al. 2001; Gomes et al. 2001; Gomes et al. 2003), (iii) the cultivar Weinert et al. 2009), (iv) the site (Costa et al. 2006a, 2006b) (v), the soil type (Schreiter et al. 2014a), and (vi) the effects of inoculants or pathogens (Adesina et al. 2009; Xue et al. 2013).

To obtain information on the taxonomic affiliation of the dominant bacteria in the rhizosphere cloning and sequencing of 16S rRNA gene amplified from TC- DNA of three potato genotypes grown at two sites were used. This approach was rather time and cost intensive and thus typically was not applied for replicates but for pooled samples. The TC-DNA from the potato rhizosphere of replicates of the same samples was also analyzed by PhyloChips (DeSantis et al. 2007). The PhyloChip was hybridized with Biotin-labelled 16S rRNA gene fragments. By means of the PhyloChip a total of 2432 operational taxonomic units (OTUs) were detected in the rhizosphere of potatoes and 864 were detected in all replicates. The major limitation of the PhyloChip approach is that the diversity detected depends on what is on the Chip, and that the hybridization signal intensity cannot be directly related to relative abundance. Nevertheless, PhyloChips are great tools for comparing the relative abundance of particular OTUs within and between treatments (Weinert et al. 2011). Thus OTUs differing in relative abundance in the rhizosphere of the same potato cultivars between sites and, more importantly, between cultivars could be identified. In recent years amplicon sequencing technology became an important tool in rhizosphere microbiology and revolutionized the field. With increasing read length and sequencing depth this technology now allows analyzing multiple replicates as previously done by DGGE to determine the effect of various biotic and abiotic factors on the microbial community composition in the rhizosphere (Marques et al. 2014; Schreiter et al. 2014a). The community composition analysis done by pyro- or illumina sequencing at a much higher resolution level largely confirmed data obtained by DGGE. The main advantage of amplicon sequencing is that at the same time insights into the taxonomic composition and identification of genera differing in relative abundance depending on the treatment becomes feasible. Although the assignment to species level is only achieved for a fraction of the sequence reads, the situation will improve with increasing read lengths. However, researchers should keep in mind that there is a large diversity beyond the 16S rRNA gene level (Eltlbany et al. 2012). Recent insights come from the determination of the plant microbiome by direct sequencing of DNA (metagenome) or cDNA (metatranscriptome). Presently, a major limitation of the direct sequencing approach is that typically no replicates were sequenced. The enormously large sequence data sets can provide insights into metabolic pathways, plant effectors, and mobile genetic elements (MGE) which can be the basis for generating new hypotheses.

The TC-DNA can be also used to quantify the abundance of beneficial or plant pathogenic bacteria by PCR-Southern blot hybridization (Eltlbany et al. 2012). The presence of antibiotic resistance genes and MGE in TC-DNA can be determined by quantitative real-time PCR (qPCR) and Southern blot hybridization. The latter approach was shown to be more sensitive and specific than qPCR but remained semi-quantitative. Quantitative real-time PCR should be done, if possible, with Taqman probes instead of Evagreen in order to achieve a high specificity.

30.4   Main Findings Obtained by Molecular Analysis of Rhizosphere Plant Species and Growth Stage-Dependent Diversity

DGGE fingerprints of bulk soil and rhizosphere samples from strawberry, oilseed rape and potato plants that were grown in a randomized plot design at the same field site revealed an enrichment of specific bacterial populations in the rhizosphere (rhizosphere effect) and plant species-dependent bacterial community composition (Smalla et al. 2001; see Fig. 30.1). Bulk soil fingerprints were characterized by many equally intense bands indicating a high evenness while in the rhizosphere fingerprints of the several stronger bands were detected, indicating an enrichment of some populations in response to root exudates and a reduced evenness. Some bands showed a plant species-dependent enrichment. Bands that were detected only in the rhizosphere fingerprints of strawberry plants were identified after cloning of the re-amplified PCR products and sequencing indicated an enrichment of *Actinobacteria* in response to the growing strawberry plants. Furthermore, the early studies by Smalla et al. (2001) and Gomes et al. (2001) already showed that different plant developmental stages were characterized by different bacterial community compositions. This finding was also observed for lettuce grown in three soils by means of amplicon sequencing (Schreiter et al. 2014a). When strawberry and oilseed rape plants were grown at different field sites, the rhizosphere fingerprints were influenced by both the site and the plant species. Interestingly, the actinobacterial DGGE fingerprints of

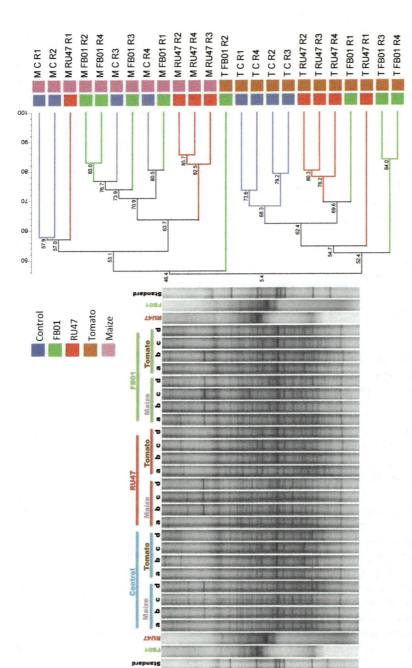

Fig. 30.1 The effect of *Pseudomonas jessenii* RU47 (RU47) and *Bacillus amyloliquefaciens* FZB42 (FB01) on total bacterial community in rhizosphere obtained by DGGE. The unweighted pair group method with arithmetic mean (UPGMA) analysis of this gel revealed a strong effect of the plant species on the composition of the rhizosphere bacterial community

strawberries grown at three sites displayed highly similar actinobacterial community compositions indicating that *Actinobacteria* did strongly respond to the strawberry exudates (Costa et al. 2006a). Similarly, *Pseudomonas* populations were enriched in the rhizosphere of strawberries as revealed by sequencing of bands from the *Pseudomonas* fingerprints from strawberry. The sequences of populations enriched in the strawberry rhizosphere grown at three sites were identical to those from isolates with in vitro antagonistic activity towards *Verticillium dahliae* (Costa et al. 2007).

Plant Genotype-Dependent Diversity In contrast to the effect of the plant species and the site, the influence of the plant genotype on the microbial community composition in the rhizosphere is much more subtle. In order to investigate the effect of transgenic potato plants, five different potato cultivars and two transgenic lines grown at two sites were investigated by DGGE fingerprints (Weinert et al. 2009). Transgenic potatoes were found to be in the normal range of variability among different cultivars. PhyloChip analysis revealed that OTUs differing between cultivars belonged to the *Pseudomonadales*, *Enterobacteriales* and *Actinomycetales* (Weinert et al. 2011). Moreover, the bacterial community compositions in the tuber rhizosphere of three sweet potato genotypes were recently compared by DGGE and amplicon sequencing of 16S rRNA gene fragments. While DGGE fingerprints showed only minor plant genotype-dependent differences at both sampling times, amplicon sequencing allowed identifying plant genotype-specific populations which were linked to low starch content. However, in the tuber rhizosphere of all plant genotypes, *Bacillus* and *Paenibacillus* were significantly enriched compared to bulk soil.

Site- and Soil Type-Dependent Microbial Diversity DGGE and amplicon sequencing analysis was used to analyze the effect of soil types on the microbial community composition in the rhizosphere under field conditions. Earlier studies from our group had already provided insights into the effects of different sites (Heuer et al. 2002; Costa et al. 2006a; Weinert et al. 2009) as it was assumed that the microbial community composition was not only influenced by the soil type but also by cropping history, agricultural management practices and climate. The study by Schreiter et al. (2014a) was the first to show under field conditions that three soils that had been kept in an experimental plot system under identical cropping history and weather conditions at the same field site for more than 10 years still displayed a distinct bacterial community composition in the rhizosphere of lettuce, indicating that the soil properties (mineral and organic composition, pH) are indeed a major factor shaping the microbial community composition in the rhizosphere.

Taxonomic Composition Cloning and sequencing of 16S rRNA gene fragments amplified from TC-DNA of the rhizosphere of three potato varieties grown at two sites showed a similar composition of major phyla and classes in the potato rhizosphere with the phylum *Proteobacteria* being the most abundant, followed by *Acidobacteria, Actinobacteria, Firmicutes, Bacteroidetes* and *Verrucomicrobia*. Interestingly, despite the low coverage (approx. 150 sequences per site) sequences affiliated to the *Acidobacteria, Verrucomicrobia* or phylum TM7 were detected in the rhizosphere of potatoes of all cultivars and from both sites (Weinert et al. 2011).

Cultivation-independent analysis clearly showed that these organisms which are difficult to culture are abundant in the rhizosphere but their role, e.g. in the dialogue with the plant, still needs to be revealed. Amplicon sequencing of 16S rRNA gene fragments from the rhizosphere of lettuce grown under field conditions in three soils revealed that *Proteobacteria* were strongly enriched in the rhizosphere of lettuce in all three soils compared to bulk soil while the relative abundance of *Actinobacteria* decreased. Several genera such as *Sphingomonas*, *Variovorax*, *Pseudomonas*, and *Rhizobium* were enriched in the rhizosphere of lettuce, independent of the soil type. Many dominant OTUs (defined at 97 % sequence identity) in the rhizosphere of lettuce were shared among the three soil types although some were soil type-specific (Schreiter et al. 2014a). Whereas in the tuber rhizosphere of sweet potato in particular the relative abundance of *Firmicutes* (*Bacillus* and *Paenibacillus*) was enriched compared to bulk soil (Marques et al. 2014).

Effects of Inoculants DGGE fingerprints and amplicon sequencing of 16S rRNA gene fragments were also used to investigate the effect of inoculants on the indigenous microbial communities in the rhizosphere (Götz et al. 2006; Adesina et al. 2009; Grosch et al. 2012; Xue et al. 2013). Compared to the effect of the plant species, the soil type or the year-to-year variation, inoculants influenced rhizosphere microbial communities to a lesser extent. Interestingly, the composition of the indigenous microbial community was most strikingly influenced by a mixture of *Trichoderma viridae* and *Serratia plymuthica*.

Detection of Antibiotic Resistance Genes and Mobile Genetic Elements The importance of horizontal gene exchange for short-term bacterial adaptability and successful colonization of new ecological niches has only recently been fully recognized (Heuer and Smalla 2012). The rhizosphere provides a natural hot spot of horizontal gene transfer as nutrient availability, bacterial cell numbers and activity are increased compared to the bulk soil. We used PCR-Southern blot hybridization and qPCR to detect resistance genes and MGE-specific sequences. We could show that the abundance of sulfonamide resistance genes (*sul1*, *sul2*) was unexpectedly lower in the rhizosphere of maize and grass grown in manure-treated soils compared to control. Another interesting observation was the enrichment of class 1 integrons and IncP-1 plasmids in the rhizosphere of lettuce grown in three soils (Jechalke et al. 2014). This increased abundance might be caused by aromatic compounds in the root exudates of lettuce (Neumann et al. 2014).

Conclusion

The analysis of TC-DNA from the rhizosphere became an important component of the tool set available in rhizosphere microbial ecology and provided important insights of practical relevance, e.g. for plant breeding or biocontrol. Conclusions from 16S rRNA gene based analysis should be drawn cautiously as the resolution level is limited and diversity beyond 16S rRNA gene sequences is high and not captured. Therefore, methods analyzing TC-DNA should be combined with microscopy and cultivation approaches. Likely, new image analysis tools and sensitive chemical detection methods will be more and more integrated to better understand the complex interactions in the rhizosphere.

References

Adesina MF, Grosch R, Lembke A et al (2009) In vitro antagonists of *Rhizoctonia solani* tested on lettuce: rhizosphere competence, biocontrol efficiency and rhizosphere microbial community response. FEMS Microbiol Ecol 69:62–74

Bent SJ, Forney LJ (2008) The tragedy of the uncommon: understanding limitations in the analysis of microbial diversity. ISME J 2:689–695

Berendsen RL, Pieterse CMJ, Bakker PAHM (2012) The rhizosphere microbiome and plant health. Trends Plant Sci 17:478–486

Berg G, Smalla K (2009) Plant species and soil type cooperatively shape the structure and function of microbial communities in the rhizosphere. FEMS Microbiol Ecol 68:1–13

Berg G, Roskot N, Steidle A et al (2002) Plant-dependent genotypic and phenotypic diversity of antagonistic rhizobacteria isolated from different *Verticillium* host plants. Appl Environ Microbiol 68:3328–3338

Bulgarelli D, Schlaeppi K, Spaepen S et al (2013) Structure and functions of the bacterial microbiota of plants. Annu Rev Plant Biol 64:807–838

Costa R, Götz M, Mrotzek N et al (2006a) Effects of site and plant species on rhizosphere community structure as revealed by molecular analysis of microbial guilds. FEMS Microbiol Ecol 56: 236–249

Costa R, Salles JF, Berg G et al (2006b) Cultivation-independent analysis of *Pseudomonas* species in soil and in the rhizosphere of field-grown *Verticillium dahliae* host plants. Environ Microbiol 8:2136–2149

Costa R, Gomes NCM, Krögerrecklenfort E et al (2007) *Pseudomonas* community structure and antagonistic potential in the rhizosphere: insights gained by combining phylogenetic and functional gene-based analyses. Environ Microbiol 9:2260–2273

DeSantis TZ, Brodie EL, Moberg JP, et al (2007) High-density universal 16 S rRNA microarray analysis reveals broader diversity than typical clone library when sampling the environment. Microb Ecol 53:371–383

Eltlbany N, Prokscha Z-Z, Castaneda-Ojeda MP et al (2012) A new bacterial disease on *Mandevilla sanderi*, caused by *Pseudomonas savastanoi*: Lessons learned for bacterial diversity studies. Appl Environ Microbiol 78:8492–8497

Gomes NCM, Heuer H, Schönfeld J et al (2001) Bacterial diversity of the rhizosphere of maize (*Zea mays*) grown in tropical soil studied by temperature gradient gel electrophoresis. Plant Soil 232:167–180

Gomes NCM, Fagbola O, Costa R et al (2003) Dynamics of fungal communities in bulk and maize rhizosphere soil in the tropics. Appl Environ Microbiol 69:3758–3766

Götz M, Gomes NCM, Dratwinski A et al (2006) Survival of *gfp*-tagged antagonistic bacteria in the rhizosphere of tomato plants and their effects on the indigenous bacterial community. FEMS Microbiol Ecol 56:207–218

Grosch R, Dealtry S, Schreiter S et al (2012) Biocontrol of *Rhizoctonia solani*: complex interaction of biocontrol strains, pathogen and indigenous microbial community in the rhizosphere of lettuce shown by molecular methods. Plant Soil 361:343–357

Heuer H, Smalla K (2012) Plasmids foster diversification and adaptation of bacterial populations in soil. FEMS Microbiol Rev 36:1083–1104

Heuer H, Krsek M, Baker P et al (1997) Analysis of actinomycete communities by specific amplification of genes encoding 16 S rRNA and gel-electrophoretic separation in denaturing gradients. Appl Environ Microbiol 63:3233–3241

Heuer H, Kroppenstedt RM, Lottmann J et al (2002) Effects of T4 lysozyme release from transgenic potato roots on bacterial rhizosphere relative to communities are negligible natural factors. Appl Environ Microbiol 68:1325–1335

Jechalke S, Schreiter S, Wolters B et al (2014) Widespread dissemination of class 1 integron components in soils and related ecosystems as revealed by cultivation-independent analysis. Front Microbiol 4:420

Klappenbach JA, Dunbar JM, Schmidt TM (2000) rRNA operon copy number reflects ecological strategies of bacteria. Appl Environ Microbiol 66:1328–1333

Kropf S, Heuer H, Gruening M et al (2004) Significance test for comparing complex microbial community fingerprints using pairwise similarity measures. J Microbiol Meth 57:187–195

Marques JM, Da Silva TF, Vollu RE et al (2014) Plant age and genotype affect the bacterial community composition in the tuber rhizosphere of field-grown sweet potato plants. FEMS Microbiol Ecol 88:424–435

Neumann G, Bott S, Ohler M et al (2014) Root exudation and root development of lettuce (*Lactuca sativa* L. cv. Tizian) as affected by different soils. Front Microbiol 5:2

Nübel U, Engelen B, Felske A et al (1996) Sequence heterogeneities of genes encoding 16 S rRNAs in *Paenibacillus polymyxa* detected by temperature gradient gel electrophoresis. J Bacteriol 178:5636–5643

Schreiter S, Ding G, Heuer H et al (2014a) Effect of the soil type on the microbiome in the rhizosphere of field-grown lettuce. Front Microbiol 5:144

Schreiter S, Sandmann M, Smalla K et al (2014b) Soil type dependent rhizosphere competence and biocontrol of two bacterial inoculant strains and their effects on the rhizosphere microbial community of field-grown lettuce. Plos ONE:9:e103726

Smalla K, Wieland G, Buchner A et al (2001) Bulk and rhizosphere soil bacterial communities studied by denaturing gradient gel electrophoresis: plant-dependent enrichment and seasonal shifts revealed. Appl Environ Microbiol 67:4742–4751

Smalla K, Oros-Sichler M, Milling A et al (2007) Bacterial diversity of soils assessed by DGGE, T-RFLP and SSCP fingerprints of PCR-amplified 16 S rRNA gene fragments: Do the different methods provide similar results? J Microbiol Meth 69:470–479

Weinert N, Meincke R, Gottwald C et al (2009) Rhizosphere communities of genetically modified zeaxanthin-accumulating potato plants and their parent cultivar differ less than those of different potato cultivars. Appl Environ Microbiol 75:3859–3865

Weinert N, Piceno Y, Ding G, et al (2011) PhyloChip hybridization uncovered an enormous bacterial diversity in the rhizosphere of different potato cultivars: many common and few cultivar-dependent taxa. FEMS Microbiol Ecol 75:497–506

Xue Q, Ding G, Li S et al (2013) Rhizocompetence and antagonistic activity towards genetically diverse *Ralstonia solanacearum* strains—an improved strategy for selecting biocontrol agents. Appl Microbiol Biotechnol 97:1361–1371

Chapter 31
Visualization of Plant-Microbe Interactions

Massimiliano Cardinale and Gabriele Berg

Abstract Plants form complex mosaics of different microhabitats, each of them colonized by adapted microorganisms ranging from beneficials to pathogens. Since only a minor fraction of environmental microbes can be cultivated in the laboratory, molecular methods are usually employed to characterize the whole microbiome. However, spatial information at micro-scale, which is fundamental to understand the dynamics of host-microbe interactions, is usually lost during the sample processing. Therefore, it is useful to complement the indirect molecular techniques with direct visualization methods. Confocal laser scanning microscopy (CLSM) allows the exploration of microbial habitats at a spatial resolution level unattainable with molecular methods. In this chapter, we will show how CLSM played a fundamental role in understanding plant-microbe interactions and their significance. Moreover, this chapter is expected to be an inspiration for integrating microscopy with molecular methods in research on plant microbiology.

31.1 Role of Microscopy in Modern Microbial Ecology

"One look is worth a thousand words", the famous phrase attributed to Fred R. Barnard (1921), expresses well the reasons why microscopy is a highly valuable tool in microbial ecology and plant microbiology. Spatial information, crucial for understanding microbial interactions, is irretrievably lost after sample processing for molecular analysis (for example PCR- or metagenomic-based approaches), because the spatial scale of microbes does not match that of the molecular methods.

M. Cardinale (✉)
Institute of Applied Microbiology, Justus-Liebig-University Giessen,
Heinrich-Buff-Ring 26–32, 35392 Giessen, Germany
Tel.: +49 641 99-37354
e-mail: Massimiliano.Cardinale@umwelt.uni-giessen.de

G. Berg
Institute of Environmental Biotechnology, Graz University of Technology,
Petersgasse 12, 8010 Graz, Austria
Tel.: +43664608738310
e-mail: Gabriele.berg@tugraz.at

© Springer International Publishing Switzerland 2015   299
B. Lugtenberg (ed.), *Principles of Plant-Microbe Interactions,*
DOI 10.1007/978-3-319-08575-3_31

Box 31.1 Basics of Confocal Microscopy and Image Analysis

Confocal laser scanning microscopes are hightech machines based on the acquisition of fluorescent light and composed of hundreds of mechanical, optical and electronic parts. Confocal microscopy allows the detection of (1) fluorescently stained molecules, cells and tissues, and (2) organisms able to synthesize fluorescent molecules, either naturally or after genetic engineering. Usually, CLSM systems give also the possibility to add a further (non-confocal) transmission light channel to the confocal stacks, such as a Differential Interference Contrast (DIC) image. The "heart" of a confocal microscope is the *pinhole*. Here, the photons of light emitted by the specimen are filtered, and only those coming from the focal plane can reach the detector (*photomultiplier*) that converts them into an electron flow, which is finally digitalized by the software. The result of this process is an extremely sharp image, representing an "optical slice" of the sample. The thickness of this optical slice depends on both the magnification and the pinhole diameter. Successive optical slices sequentially acquired along the Z-axis form a *confocal stack*, which can then be used to create bi-dimensional projections, or visualized in the three-dimensional space. Proprietary software, such as Imaris (Bitplane, Zuerich, Switzerland) and AMIRA (TGS Inc., US), as well as freeware, such as DAIME (www.microbial-ecology.net/daime/), Image Surfer (http://imagesurfer.cs.unc.edu/) and ImageJ implemented with appropriate plugins (http://rsbweb.nih.gov/ij/plugins/) allow imaging and quantitative analysis of confocal stacks. Dedicated software tools can create isosurfaces and spheres, three-dimensional objects that replace the original fluorescent signals. Such *three-dimensional models* facilitate the investigation of host-microbe and microbe-microbe relationships by dramatically increasing the resolution level of the image.

Moreover, microorganisms interact with the environment as whole cells and at a micro-scale. Advanced microscopy methods such as electron microscopy, CLSM, Raman spectroscopy, super resolution microscopy, atomic force microscope (AFM) show the world of microbes as observed by microbes or their hosts.

Molecular methods deliver significant results, as it is possible to process numerous representative (pooled) samples in parallel, which allow robust statistics. In the past few years, microbial ecology was revolutionized by affordable deep sequencing and -omics technology (see Chap. 30), which allowed the study of natural microbiomes at an unprecedented level of depth. Once more, the effect was that the scientists' attention was focused on the sequence-based information delivered by the new techniques, and microscopy was consequently overshadowed. The direct visualization at micrometric scale delivers qualitative and quantitative information on bacterial populations and their variation within the hosts' tissues. However, a statistical approach in microscopy analysis of host-associated microorganisms is challenging, and extremely time-consuming for two main reasons: (1) it is not possible to pool samples, and (2) different samples cannot be run in parallel by the same operator, in the same laboratory. Therefore, only the combination of indirect molecular characterization of microbiomes with direct observation in situ allows drawing conclusions that are more reliable. CLSM not only remains as a widely applicable methodology for studying plant-microbe interactions, but also can be extended and integrated by other microscopic techniques, and therefore serves as a central technology in such studies. In the following paragraphs, after a technical box explaining basic principles of the technique, we will discuss relevant topics and applications of confocal microscopy in plant microbiology, including analyses of host-colonization and interactions with pathogens, critical methodological aspects, and integration with complementary methods.

31.2 Application of CLSM to Plant-Microbe Interactions

Confocal microscopy is used in plant microbiology with different purposes. The extremely accurate localization of microorganisms within the host, along with targeted staining of specific structures or taxa, allow gaining insights into their ecology and interaction with the host (Cardinale 2014). Plant-associated microbes can exhibit very diverse behaviors, such as ectophytic vs. endophytic colonization. With molecular approaches, these two groups are often separated by surface sterilization of plant material, but with CLSM it is possible to discriminate between different colonization patters. This is relevant, because, in terms of interactions, it makes a huge difference whether an endophytic bacterium is able to enter the plant cells, or it grows in the apoplastic spaces only. In the first case, the closer intimacy is expected to confer to this microorganism the ability to interfere more with the host, at both physiological and genetic level. Such a case is represented by the colonization of root nodules, usually considered the unique niche of rhizobial symbiosomes (Chap. 23), by non-rhizobial symbionts, which were shown by CLSM to actually nodulate *Cyclopia* ssp., *Macroptilium atropurpureum* and *Mimosa pigra* (Elliott et al 2007), and *Paenibacillus polymyxa* (Annapurna et al. 2013). Such evidences force us to re-evaluate the extent of microbial diversity in the context of plant-microbe interactions.

The identification of colonization strategies sheds light on the microbial ecology of host-associated microbes, and this is important to understand the dynamics of microbial establishment in plants. This relationship between a microorganism and its host can indeed exhibit an unexpected level of specificity: *Bacillus amyloliquefaciens* FZB42 showed three different colonization patterns on three different host plants (Fan et al. 2012). This suggests that it is imprudent to draw general conclusions from single studies, and that each plant-microbe system may display unique features. The interaction with microbes is apparently required for good plant fitness, since it was recognized that probably all plants in nature host complex microbiomes (Vorholt 2012; Philippot et al. 2013). The nature of these associations and their importance for the plant just began to be elucidated. Like in animals, the microflora associated with the different plant organs seems to contribute to their correct functioning and to the stability of the host's immune system. Therefore, to investigate the niche specialization *vs.* ubiquity of a microorganism across the different plant microhabitats, can help us in understanding its role and possible effects on plant fitness, especially if coupled with appropriate molecular methods. Such investigations can be performed under controlled conditions, for evaluating the specific behavior of single strains (or known mixtures of organisms), or on plants grown in nature, in order to understand the natural associations with the native microbiota. Bragina et al. (2012) showed by fluorescence *in* situ hybridization (FISH)-CLSM that the hyalocytes (dead cells which contain water) of *Sphagnum* mosses are the preferred colonization sites of diverse bacteria. Complementary molecular studies demonstrated their potential involvement in nitrogen fixation and methane degradation. The integration of these results suggested that *Sphagnum* hyalocytes are not only water reservoirs, but instead could represent "micro-bioreactors" for nutrient production, which support the host growth. Comparative analysis of environmental samples can highlight the importance of certain groups within the plant-associated microbiome, as shown in an Egyptian desert farm for *Bacillus* and *Streptomyces*, which appeared to be recruited by plants as effective biocontrol agents (Köberl et al. 2013; Chap. 28).

Intrusion of transport vessels at root level was shown already by CLSM for several bacteria, such as *Enterobacter gergoviae* (An et al. 2006) and *Bacillus subtilis* (Ji et al 2008). The implication of such observations are particularly relevant in the case of potential human pathogens, such as *Enterobacteriaceae*, which were already proven to persist in the soil, colonize crop roots, and from there move to the edible parts of the plant (Cooley et al 2007). Thus, shedding light onto the mechanisms underlying bacterial intrusion and dynamics of their internal translocation can improve food safety by development of targeted measures for outbreak prevention. *Escherichia coli* was recently shown by CLSM to colonize *Lactuca sativa* (green lettuce) leaves preferentially beneath the epithelium (Chap. 44); similarly, *Salmonella enterica* intrudes the lettuce leaves via the open stomata (Kroupitski et al. 2009) and resides in the endosphere. This niche specialization may explain why conventional washing fails in removing pathogens from raw leaves, which may cause the recurrent bacterial outbreaks in lettuce.

Pathogens represent a special case of plant-microbe interactions; their virulence relies on a combination of genetic and ecological traits, since the infections usually target specific plant tissues, and occur only when the responsible microbial genes are expressed. Only when the density of the pathogen exceeds a certain critical level, the plant response will not succeed in containing the disease. CLSM was used as a tool to investigate such synchronized mechanisms, thus contributing to understand the dynamics of the infection process. The mechanism responsible for the *Xylella fastidiosa*-induced degenerative disease of *Vitis vinifera* was identified as the vessel-to-vessel movement of the bacterial cells (Newman et al. 2003), and the modulation of *vir* gene expression in *Agrobacterium tumefaciens* during the infection process (Chap. 37) was studied after downstream insertion of *gfp* genes (Li et al. 1999).

CLSM helps elucidating the mechanisms of action of beneficial microbes. Plant growth promoting bacteria (PGPB; Chap. 22–29) and biocontrol agents (BCAs;

Chap. 18) caught the attention of the scientific community due to their promising biotechnological potential for sustainable agriculture. They are regarded as environmental friendly measures to replace chemical fertilizers and pesticides, respectively, although unfortunately the positive effects observed in laboratory under gnotobiotic conditions often disappear in the field. Rhizosphere competence (the ability to colonize plant roots stably) and endophytism are regarded as indicators of potentially beneficial bacteria (Zachow et al. 2010). Interestingly, in several case studies, CLSM showed that neither endophytism nor direct contact with the pathogen was the discriminative feature of efficient biocontrol strains (Maldonado-González et al. 2013; Gasser et al. 2012). Notably, it was also shown that an in vitro inefficient *Collimonas fungivorans* strain, successfully reduced disease incidence in vivo (Kamilova et al. 2007). These results indicate that further research is needed to understand the mechanisms of actions underlying the interactions with beneficial microorganisms, in order to develop efficient bio-products with consistent results in the field (Chap. 32–34).

Lichens, traditionally considered as the typical example of mutualistic symbiosis between fungi and algae/cyanobacteria, were intensively studied by CLSM, which unexpectedly showed them to be microbial hot spots (Grube et al. 2009). A clear succession of bacterial communities from young to older thallus regions suggests that the lichens harbor a stable, active and functionally adapted microbiome, which may contribute to the growth and survival under extreme conditions. Such CLSM observations lead to the conclusion that the lichen paradigm should be reconsidered: from bipartite symbiosis to autonomous mini-ecosystems supported by high microbial diversity (Fig. 31.1).

31.3 Methodological Aspects of Confocal Microscopy in Plant Microbiology

Interpretation of the Images Confocal stacks are usually presented as either maximum projections or single optical slices. For correct interpretation of the images, the thickness of the confocal stack, as well as the Z-step dimension, must be given. The scale bar alone does not inform about the spatial volume included in a bi-dimensional projection of an image series. This is particularly important when analyzing endophytic microbes, cell-cell interactions and huge micro-colonies. In the case of single optical slices, the thickness of the section should also be mentioned. However, in certain cases, a proper interpretation of the confocal images is only possible when displaying the three-dimensional space, either as volume rendering of the signals or as three-dimensional model made of isosurfaces and spheres, which can drastically enhance the digital resolution of the image (Fig. 31.2).

Autofluorescence Biological tissues, including plant roots, are usually autofluorescent to some extent. Although in some cases it is useful to eliminate or reduce it by specific pretreatments, more often it is suitable to exploit the genuine autofluorescence, to achieve a spatial contextualization of the targeted microorganisms.

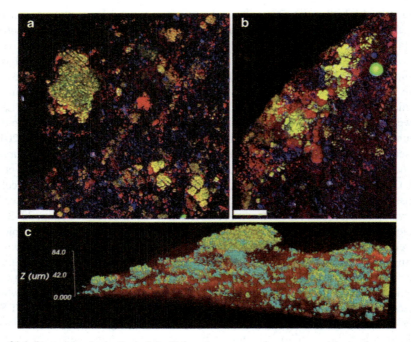

Fig. 31.1 Bacterial colonization of the lichen *Lecanora polytropa*. **a** and **b** maximum projections of confocal stacks showing bacteria stained by FISH with the universal bacterial probe EUB338MIX (*red*) and the Alphaproteobacteria-specific probe ALF968 (*yellow*). Fungal and algal autofluorescence appear as *blue/purple* and *green*, respectively; scale bars: 25 μm. **c** volume-rendering/isosurfaces three-dimensional model of panel **a**; EUB338MIX: *cyan*; ALF968: *yellow*; lichen autofluorescence: *red*

This applies in most studies of plant-microbe interactions (see Figs. 31.1 and 31.2). In multichannel confocal systems, one channel can be dedicated to the wavelength range of autofluorescence. This requires preliminary CLSM observations of unstained samples, to define the appropriate autofluorescence range and its intensity (Cardinale 2014). In case of relatively weak signals, signal accumulation during image acquisition can subsequently help to properly visualize an autofluorescent host matrix.

Combination of CLSM with Other Microscopy Techniques The optical resolution of the confocal microscopes is constrained by the physical properties of the light. Therefore, fluorescent microscopy was used in a correlative approach in combination with electron microscopy, in order to identify target objects first, and then imaging them at nanometric scale (Jahn et al. 2012). A further correlative approach already employed in medical sciences but not yet in plant microbiology, is the combination of CLSM with AFM (Haupt et al. 2006), which could potentially deliver structural information, such as interaction forces between beneficial microorganisms, pathogens and hosts. Correlative FISH-CLSM with nano-SIMS (secondary

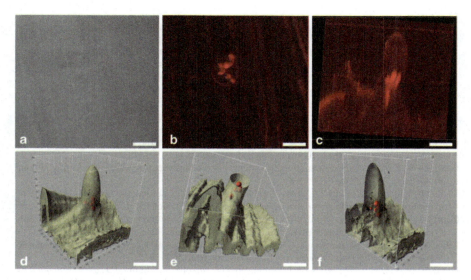

Fig. 31.2 Cells of *Stenotrophomonas rhizophila* P69 colonizing an emerging root hair in tomato endophytically. Bright field image (**a**), CLSM stack (**b**, maximum projection; **c**, volume rendering) and corresponding isosurfaces/spots three-dimensional model (**d**) with respective cutting planes (**e-f**), to reveal the endophytic colonization; scale bars: 10 μm

ion mass spectrometry) could be of special interest, as this will provide information on functional contributions of specific taxonomical groups of bacteria. Until now, nanoSIMS has been used in plant-microbe interactions only to visualize differential partitioning of $^{15}NH_4^+$ between roots and soil microbial communities at nanometric scale (Clode et al. 2009).

31.4 Conclusions

Investigation of host-microbe systems requires a polyphasic approach to untangle the relationships and their ecological significance. As a complementary approach to molecular methods, CLSM is best suited to both integrate and corroborate results of -omics methodologies. Localization at microscale, colonization pattern and cell-cell interactions are not detectable by cultivation, fingerprinting, or deep-sequencing analysis, yet they represent the basic information necessary to understand mechanisms and dynamics of plant-microbe interactions in microbial ecology studies.

References

An Q, Dong Y, Wang W et al (2006) Constitutive expression of the *nif*A gene activates associative nitrogen fixation of *Enterobacter gergoviae* 57-7 an opportunistic endophytic diazotroph. J Appl Microbiol 103:613–620

Annapurna K, Ramadoss D, Bose P et al (2013) *In situ* localization of *Paenibacillus polymyxa* HKA-15 in roots and root nodules of soybean (*Glycine max* L). Plant Soil 373:641–648

Barnard FR (1921) "One look picture is worth a thousand words" Advertisement appeared on Printers' Ink 08 December 1921

Bragina A, Berg C, Cardinale M et al (2012) *Sphagnum* mosses harbour highly specific bacterial diversity during their whole lifecycle. ISME J 6:802–813

Cardinale M (2014) Scanning a microhabitat: plant-microbe interactions revealed by confocal laser microscopy. Front Microbiol 5:94

Clode PL, Kilburn MR, Jones DL et al (2009) In situ mapping of nutrient uptake in the rhizosphere using nanoscale secondary ion mass spectrometry. Plant Physiol 151:1751–1757

Cooley M, Carychao D, Crawford-Miksza L et al (2007) Incidence and tracking of Escherichia coli O157:H7 in a major produce production region in California. PLoS One 2:e1159

Elliott GN, Chen WN, Bontemps C et al (2007) Nodulation of *Cyclopia* spp (Leguminosae Papilionoideae) by *Burkholderia tuberum*. Ann Bot 100:1403–1411

Fan B, Borriss R, Bleiss W, Wu X (2012) Gram-positive rhizobacterium *Bacillus amyloliquefaciens* FZB42 colonizes three types of plants in different patterns. J Microbiol 50:38–44

Gasser F, Cardinale M, Schildberger B et al (2012) Biocontrol of *Botrytis cinerea* by successful introduction of *Pantoea ananatis* in the grapevine phyllosphere. Int J Wine Res 4:53–63

Grube M, Cardinale M, De Castro JV Jr et al (2009) Species-specific structural and functional diversity of bacterial communities in lichen symbiosis. ISME J 3:1105–1115

Haupt BJ, Pelling AE, Horton MA (2006) Integrated confocal and scanning probe microscopy for biomedical research. Sci World J 15:1609–1618

Jahn KA, Bartonb DA, Kobayashia K et al (2012) Correlative microscopy: providing new understanding in the biomedical and plant sciences. Micron 43:565–582

Ji X, Lu G, Gai Y et al (2008) Biological control against bacterial wilt and colonization of mulberry by an endophytic *Bacillus subtilis* strain. FEMS Microbiol Ecol 65:565–573

Kamilova F, Leveau JHJ, Lugtenberg B (2007) *Collimonas fungivorans* an unpredicted in vitro but efficient in vivo biocontrol agent for the suppression of tomato foot and root rot. Environ Microbiol 9:1597–1603

Köberl M, Ramadan EM, Adam M et al (2013) *Bacillus* and *Streptomyces* were selected as broad-spectrum antagonists against soilborne pathogens from arid areas in Egypt. FEMS Microbiol Lett 342:168–178

Kroupitski Y, Golberg D, Belausov E et al (2009) Internalization of *Salmonella enterica* in leaves is induced by light and involves chemotaxis and penetration through open stomata. Appl Environ Microbiol 75:6076–6086

Li L, Li Y, Lim TM et al (1999) GFP-aided confocal laser scanning microscopy can monitor *Agrobacterium tumefaciens* cell morphology and gene expression associated with infection. FEMS Microbiol Lett 179:141–146

Maldonado-González MM, Prieto P, Ramos C et al (2013) From the root to the stem interaction between the biocontrol root endophyte *Pseudomonas fluorescens* PICF7 and the pathogen *Pseudomonas savastanoi* NCPPB 3335 in olive knots. Microb Biotechnol 6:275–287

Newman KL, Almeida RP, Purcell AH et al (2003) Use of a green fluorescent strain for analysis of *Xylella fastidiosa* colonization of *Vitis vinifera*. Appl Environ Microbiol 69:7319–7327

Philippot L, Raaijmakers JM, Lemanceau P et al (2013) Going back to the roots: the microbial ecology of the rhizosphere. Nat Rev Microbiol 11:789–799

Vorholt JA (2012) Microbial life in the phyllosphere. Nat Rev Microbiol 10:28–40

Zachow C, Fatehi J, Cardinale M et al (2010) Strain-specific colonization pattern of *Rhizoctonia* antagonists in the root system of sugar beet. FEMS Microbiol Ecol 74:124–35

Part VI
Products for Plant Growth-promotion and Disease Suppression

Chapter 32
Commercialisation of Microbes: Present Situation and Future Prospects

Willem J. Ravensberg

Abstract Microbes are used in biopesticides and biostimulants. The definition of biopesticides includes microorganisms, and beneficial arthropods; biopesticides are developed for the control of biotic stresses which are caused by pests and diseases. Biostimulants influence the plant's responses to abiotic stresses. The market for both categories of products is described. An overview of the biopesticides in Europe is given as well as their targets and crops. Critical failure and success factors in terms of commercialisation are discussed. The implementation of a product in an integrated crop management system needs to be investigated in order to develop guidance for proper use. Limiting factors and promoting trends determine the growth of the markets for these products. Continued growth is expected since alternatives to chemicals are demanded by legislation and consumer demand for residue-free food. Products based on bacteria are anticipated to increase the most for disease and insect control as well as for plant growth.

32.1 Definitions of Biopesticides and Biostimulants

The Use of Biopesticides The commercial use of products based on living organisms in agriculture started in 1938 in France with Sporeine. The bacterium *Bacillus thuringiensis* was used as a biopesticide for control of caterpillars. Gradually, from the 1960s on, products based on other bacteria, fungi and nematodes reached the market, and the range of products for pest and disease control has grown ever since. Today, the worldwide turnover of biopesticides is approximately 1.8 billion US$ (at grower level). Annual growth for the last decade has been around 15 % (CPL 2013). Precise figures on the global market are hard to find. This is partly due to the various definitions used for biopesticides. These could include living organisms, like beneficial arthropods and microorganisms, but also pheromones, natural and

W. J. Ravensberg (✉)
International Biocontrol Manufacturers Association (IBMA), Rue de Trèves 61,
1040 Brussels, Belgium

Koppert Biological Systems, Veilingweg 14, 2651 BE Berkel en Rodenrijs, The Netherlands
Tel.: + 31651410068
e-mail: willem.ravensberg@ibma-global.org; wravensberg@koppert.nl

© Springer International Publishing Switzerland 2015   309
B. Lugtenberg (ed.), *Principles of Plant-Microbe Interactions*,
DOI 10.1007/978-3-319-08575-3_32

biochemical products, and even plant-incorporated genes. They are applied for bio-control purposes in crops as well as beyond agriculture. Microbes used for plant growth promoting purposes are defined as biostimulants.

Biopesticides There is no widely accepted use of the term biopesticide. Various regulatory authorities have their own definition, some do not even use the word biopesticide, but refer to active substances based on microorganisms. In EU legislation the word biopesticide is not used. Regulation (EC)1107/2009 distinguishes chemical substances and microorganisms and defines the latter as *any microbiological entity, including lower fungi and viruses, cellular or non-cellular, capable of replication or of transferring genetic material.* The USA EPA definition is as follows: *biopesticides include naturally occurring substances that control pests (biochemical pesticides), microorganisms that control pests (microbial pesticides), and pesticidal substances produced by plants containing added genetic material (plant-incorporated protectants or PIPs).* A microbial pesticide is a *microbial agent . . . ., that: (1) Is a eucaryotic microorganism including, but not limited to, protozoa, algae, and fungi; (2) Is a procaryotic microorganism, including, but not limited to, Eubacteria and Archaebacteria; or (3) Is a parasitically replicating microscopic element, including, but not limited to, viruses.* The Canadian authority PMRA uses the following definition for a microbial pesticide: *a naturally occurring or genetically modified micro organism including but not limited to fungi, bacteria, and viruses.* The OECD has no formal definition for biopesticides despite having a BioPesticide Steering Group (BPSG). Working practice covers microbials, semiochemicals, botanicals and beneficial arthropods. The FAO describes a biological pesticide as – *A generic term, not specifically definable, but generally applied to a biological control agent , usually a pathogen, formulated and applied in a manner similar to a chemical pesticide, and normally used for the rapid reduction of a pest population for short-term pest control.*

In this chapter I will use the term *'biopesticide'* for products based on microorganisms, arthropods and beneficial nematodes. A *'microbial pesticide'* or simply a *'microbial'* is used when a microorganism is the active substance. Products based on arthropods (predatory mites, parasitoids, predatory bugs, lacewings, etc.) and entomopathogenic nematodes (EPNs) are usually regulated in Europe by national laws. They are often referred to as Invertebrate BioControl Agents (IBCAs) or *'macrobials'*.

Microorganims used to control pests and diseases are considered plant protection products (PPPs) and need to be registered as such, following similar procedures as chemical plant protection products. The EU definition for plant protection products is, however, wide and includes the use of a substance influencing the life processes of plants (Article 2.1b of Reg. (EC)1107/2009). That means that any microorganism used for promoting plant growth, yield enhancement, etc. is considered a PPP. This is not the case in the USA and Canada. The wider definition of the EU causes considerable confusion and many microorganism-based products are not being registered, but sold as biostimulants, biofertilizers, soil improvers, etc.

Biostimulants The EU as well as the biostimulant industry have taken the initiative to solve this issue, and a revision of the Fertilizer Regulation (EC)2003/2003 offers the opportunity to develop a regulatory framework that defines and covers bio-stimulants. The proposed definition by the European Biostimulant Industry Council (EBIC) is the following: *plant biostimulants contain substance(s) and/or micro-organisms whose function, when applied to plants or the rhizosphere, is to stimulate natural processes to enhance/benefit nutrient uptake, nutrient efficiency, tolerance to abiotic stress, and crop quality* (http://www.biostimulants.eu). Biostimulants differ from crop protection products because they act only on the plant's vigour and do not have any direct actions against pests or diseases. When direct action against biotic plant stress, like pests and diseases, is claimed the product is considered a PPP. In order to have clarity on these claim related categories both (EC)1107/2009 and (EC)2003/2003 need to be amended. This is foreseen to be enforced in 2016–2017.

32.2 Overview of Commercially Available Biopesticides

Several sources providing overviews of approved biopesticides are available on the market. The 5th edition of BCPC's Manual of Biocontrol Agents is an online database (www.bcpcdata.com) that lists all biocontrol agents, mainly from the Western world. In the EU Database (www.ec.europa.eu/sanco_pesticides/public/) all registered plant protection products can be found. A list of active ingredients and products approved in the USA can be found on the EPA website (www2.epa.gov/pesticide-registration). Many competent authorities in the EU member states have similar websites. In the EU there are currently 44 microorganisms approved (on strain level) as active substances, 11 are pending approval. In the USA approximately 110 strains are approved. CPL (2013) reports the number of registered microorganisms for plant protection world wide: 77 bacteria, 68 fungi and yeasts, 36 viruses and phages, and 2 protozoa, in more than 2300 products.

Overview of Microbial Pesticides in Europe The approved and pending microor-ganisms in the EU (in March 2014) are given below as well as the pest/disease targets and the main crops in which they are applied. The products based on these active ingredients are only approved in a given number of EU member states, depending on the company's market interests.

Fungi—for Insect and Nematode Control Three entomopathogenic fungi, *Beau-veria bassiana, Lecanicillium muscarium, Isaria* (= *Paecilomyces*) *fumosorosea*, have been approved for control of whitefly, thrips and some other soft-bodied insects in greenhouse crops mainly. *Metarhizium anisopliae* is approved for control of black vine weevil in soft fruit and tree nurseries. *Paecilomyces lilacinus* is approved for control of root knot nematodes.

Fungi—for Disease Control The majority of approved fungi are *Trichoderma* (12 strains from 5 species) for use as a fungicide. The target diseases are predominantly soil diseases caused by pathogens such as *Fusarium, Pythium* and *Rhizoctonia* spp., in greenhouse vegetables, herbs and ornamentals, turf, field vegetables, and some are applied for control of wood diseases in grapes. Other fungi used against soil diseases are *Gliocladium catenulatum* and *Coniothyrium minitans*. The latter is used to control *Sclerotinia sclerotiorum*. Fungi for control of foliar diseases are limited to *Ampelomyces quisqualis* for powdery mildews in vegetables and grapes. *Verticillium albo-atrum* is approved against Dutch elm disease, and *Phlebiopsis gigantea* against Heterobasidion root rot of pine.

Bacteria—for Insect and Nematode Control   Most bacteria are approved for control of caterpillars in greenhouse and field vegetables, in orchards, and in forestry. These are *Bacillus thuringiensis* strains (11). *Bacillus firmus* and *Pasteuria penetrans* are used to control several nematode species in a variety of plants such as soybean, cotton, vegetables and turf.

Bacteria—for Disease Control   *Bacillus amyloliquefaciens* and *B. subtilis* strains are approved for disease control. They target similar diseases and crops as the above-mentioned *Trichoderma*'s. The same holds for *Streptomyces* K61 and *Streptomyces lydicus*. *Pseudomonas chloroaphis* can be used for the control of diseases in cereals and *Pseudomonas* sp. 'proradix' for *Rhizoctonia* control in potatoes. *B. pumilus* has been developed for control of powdery mildew in greenhouse crops and grapes.

Viruses—for Insect Control   Four species of baculoviruses have been approved for control of caterpillars like the codling moth (*Cydia pomonella*) in apples and pears, the beet army worm (*Spodoptera exigua*), the cotton leaf worm (*S. littoralis*) and the cotton boll worm (*Helicoverpa armigera*) in greenhouse vegetables. Two plant viruses have been developed, as weak strains, for control of pepino mosaic virus and zucchini yellow mosaic virus, in tomato and zucchini respectively.

Yeasts—for Disease Control   Four yeast-based products have been developed for control of diseases. *Aureobasidium pullulans* is approved for fire blight control in pome fruit, *Candida oleophila* for post harvest diseases (*Botrytis cinerea* (grey mould), *Penicillium expansum* (blue mould)) in apple and pear. *Pseudomyza flocculosa* is targeted at powdery mildew in roses and greenhouse vegetables. Registration for baker's yeast *Saccharomyces cerevisiae* is pending as a systemic resistance inducer aimed at bacterial and fungal (downy mildew, botrytis) diseases in field and greenhouse vegetables.

Microbes used beyond Agriculture   Microbial pesticides are also applied in forestry, in amenity, in home and garden, for vector control (mosquitoes) as well for exoparasites on livestock. Uses in forestry are considerable for control of caterpillars like the gypsy moth and for sawflies. The use in turf for soil diseases is increasing.

Invertebrate Biocontrol Agents   Commercial augmentative biological control is conducted by releasing invertebrate biocontrol organisms. There are about 230 species of the orders of Nematoda and Arthropoda, one species of Mollusca and one Chilopoda (Van Lenteren 2012). All these IBCAs are used for control of insect and mite pests, none for plant parasitic nematodes or diseases. One beneficial nematode (*Phasmarhabditis hermaphrodita*) is applied for control of slugs. Furthermore, 9 species of beneficial nematodes (5 *Steinernema* and 4 *Heterorhabditis* spp.) are used for control of soil-dwelling stages of insects, mainly. The majority of IBCAs are Hymenopteran parasitoids (120), predatory mites (30), and species of Neuroptera, Heteroptera, Diptera and Coleoptera.

The first use of augmentative biocontrol ocurred in 1902. The expansion of the number of used species took place in the 1980 and 1990s. The world market of macrobials is estimated at 480 million US$ in 2011 (CPL 2013), most IBCAs are sold for greenhouse crops in Europe.

The Market for Biostimulants There are neither reliable sources on the number of biostimulants available nor is there a consistent and accepted definition for this type of products. Products are marketed as soil improvers, biofertilizers, organic soil amendments, biostimulants, etc. They can be substances, microorganisms or complex mixtures. According to EBIC, the European market for biostimulants reached 500 million € in 2013 of which a part consists of microbes. The area of use in Europe is estimated at 3 million ha with multiple applications. EBIC foresees a steady annual growth for biostimulants of 10 % or more. In the USA the BioStimulant Coalition has been founded with similar goals as EBIC. No data on the use and market of biostimulants in the USA are available.

Biostimulants based on microorganisms often contain multiple species. A general purpose is solubilizing nutrients like nitrogen and phosphorus for an improved nutrient uptake by the plant, production of hormones for plant growth promotion, yield enhancement, and making the plant more resilient towards abotic stresses caused by drought and salt. Here, similar bacteria and fungi are used as in plant protection products such as *Bacillus, Pseudomonas, Rhizobium* and *Trichoderma* spp. as well as mycorrhizae.

32.3   Critical Failure and Success Factors

The annual growth in the biopesticide market is currently around 15 % and the prospects are promising. Still it requires a great amount of effort to achieve a successful commercialisation. Decisive is customer satisfaction based on consistent performance of the product in relation to the costs and ease of use. Many factors determine success or failure of a product. The disadvantages of microbial pesticides are frequently reported as the cause of why biopesticides only reach a small percentage, today 3 %, of the total crop protection market which was approximately 60 billion US$ in 2013.

Strengths, Weaknesses, Opportunities and Threats The success or failure of a microbial is determined by the attributes of a biopesticide. The chance for success, however, is not only dependent on the product, but also on the macro-environment in which it is used. A SWOT (strengths, weaknesses, opportunities and threats) analysis can be made of these internal and external factors in general, and for each product (Table 32.1). It cannot be ignored that the list of weaknesses of biopesticides is long, which presently often restricts their use to niche markets. On the other hand, biopesticides have unique strengths, and the opportunities to exploit these will increase.

Table 32.1 SWOT analysis for biopesticides

Strengths	Weaknesses
Unique mode of action	Speed of kill slow and insect stage-dependent
Specific host range	Narrow target spectrum
No residue (exempt of MRL)	Efficacy moderate and variable
No or short pre-harvest interval	Short residual activity
No or short worker re-entry interval	Sensitive to abiotic factors
Compatible with natural enemies, pollinators	Limited storage stability (at low temp.)
Compatible with microbial pest control agents	More complicated application technology, spraying necessary, good coverage essential, often direct contact needed
Excellent tool in IPM systems	Incompatible with chemical pesticides
Excellent tool in resistance management programmes	Knowledge-intensive products
Probability of (cross-) resistance low	Often only work as part of IPM programme
Safe for humans and the environment	Relatively high end-user price
Safe for plants	
Approved for organic production	
Opportunities	Threats
BCAs are a natural, renewable resource	Novel safer chemical pesticides
Environmentally benign, no environmental pollution due to production and use	Growers' scepticism based on experience with chemicals
Increase or maintain biodiversity	Only for niche-markets
Developmental costs low to reasonable	Increasing regulatory burden
Demand for residue-free food	Transgenic crops
Withdrawal and reduction of chemicals	Biodiversity and access benefit sharing regulations
Combination with chemicals extend the product life of resistant-risk pesticides	Bioterrorism regulations
Growing of sustainable agriculture and IPM	Fear of microbes by the public
Organic food production	Limited public research funding
Increasing availability of high quality biopesticides	Lack of user education and extension services
Increasing users' confidence in biopesticides	Weak economic position of farmers
Increasing number of sector where use of chemicals is forbidden such as forestry and amenity areas	Unregistered ineffective "snake oils"
	Import of low quality products

Still, robust field performance is the key factor. Users may decide to use a biopesticide for other reasons such as a short re-entry period or pre-harvest interval, or for resistance management. Market demands to deliver residue-free produce becomes a strong incentive to use microbials. Still, costs and efficacy remain the most important decision factor for a grower.

The Development of a Microbial Pesticide The development of a biopesticide is an extremely complex process that includes many phases such as discovery, production, product development, efficacy testing, registration and finally commercialisation (see Chap. 33). Developmental time can be between 3–5 years and registration likewise. It will cost many millions of Euros and adoption in the market and the time to a considerable sale volume usually takes a few years. It is therefore imperative that enough resources are available as well as a thorough and long term company commitment to allow for such a long and expensive product development process. The complete process from ideation to successful commercialisation has been described by Ravensberg (2011) (http://link.springer.com/book/10.1007%2F978-94-007-0437-4) for insect-pathogens as microbial insecticides. Key factors of each phase in the process are described there in detail, the model can be used for any kind of biopesticide.

32.4 Implementation in an Integrated Crop Management Programme

Both biopesticides and biostimulants are used in agriculture where a multitude of practices is applied. Proper imbedding of these products is needed for an optimal use. Particularly the use of chemical pesticides may interfere with, or even kill, the beneficial organisms.

Application Strategy In order to achieve the most effective results with a biopesticide or biostimulant an optimal application strategy must be developed. For a good field performance efficient delivery is essential, both in terms of efficacy and costs. Furthermore, timing, frequency of the application and intervals between multiple applications need to be established by field trials and in commercial settings. The environmental conditions, host plant-mediated effects, crop and cultivation effects influence the field performance, and for every situation testing is required before proper advice on the use and expected results can be provided to the farmer. The most influential factors must be identified and investigated.

Integrated Use Products will be used in an integrated programme which means many variables and multi-trophic interactions need to be considered and studied. Compatibility with chemical pesticides need to be investigated as well as with natural enemies and pollinators. Where necessary a safe interval period has to be established. For biostimulants, optimal use within a regime of use of artificial fertilizers requires studies. Combined use of biopesticides, biostimulants, chemicals and IBCAs in an integrated crop management programme needs to be studied in the developmental phase of a product.

Costs of an Integrated Programme Integrated crop management programmes can become quite complex, and for adopting a new product into such a system, the monetary aspect is a significant factor for a grower's decision. This means not just the product's end-user price, but also the labour involved in the application and the potential advantage in the market. Demonstrating the costs and benefits to the grower is vital for the adoption of the product.

The Role of a Distributor The perception of biopesticides often is a barrier to the adoption of such products, and the role of distributors is pivotal for a successful biopesticide. Marrone (2007) recommended that biocontrol companies invest in the education of customers and distribution channel partners to improve adoption of biopesticides. Feedback on the customer's satisfaction or complaints allows improvements of the product and its use.

32.5 Conclusions

The driving forces in crop protection and crop health management are many and varied, they continue to develop and alter. The following macro-environmental factors play an important role: governmental policy and its influence on crop protection via legislation, the diminishing funding of research and extension, environmental programmes minimizing pollution, trading policies, biodiversity topics, and supermarket demands.

Limiting Factors and Threats Research funding by governments is generally getting reduced and without collaboration between academic institutes and companies new product development is under threat. The Access and Benefit Sharing of genetic resources and the "Nagoya Protocol" (Convention of Biological Diversity 1992) (Cock et al. 2009) makes exploration for new microorganism an administrative burden and threatens new product development. Biopesticides, integrated crop management and sustainable agriculture tend to be more expensive for farmers, while agricultural returns are declining. "Food-scare", i.e. the risk of a negative association with insects and microbes ("germs") on our food is a risk. The regulatory issue remains to be critical, there is no or a too slow innovation and – as a consequence – time to registration is too long and too costly.

Promotional Factors and Trends The demand for reduced levels of residue on food (legislation, and consumer-retailer demands), the Global-GAP rules for growers, the resulting supermarket competition on extra-legal residue requirements, resistance to chemicals and the increase of organic agriculture favour biocontrol. Scientific and technological trends create a flow of new ideas and results, e.g. insect pathogens used as antagonists of plant diseases, endophytes against insects (Vega et al. 2009). Other developments occur in strain improvement (hybridization, genetic modification), production technology, and formulation (encapsulation, polymers, adjuvants).

The regulatory climate challenges the registration and use of chemicals more and more. Directive 2009/128/EC on sustainable use of pesticides "encourages the use of alternative approaches......", and EU Member States have to make a National Action Plan showing how they will reach those political goals. As an example, France launched an ambitious plan -'Ecophyto'-that should lead to 50 % reduction of chemicals in 2018. Furthermore, the Convention of Biological Diversity requires governments to improve biodiversity and to mitigate negative impacts. Chemical pesticides have a negative influence on biodiversity (Geiger et al. 2009). Directive 2009/128/EC states that the use of pesticides should be minimized due to biodiversity concerns. National Action Plans should realize this goal.

New invasive pests continue to spread, and the lack of registered chemicals or resistance creates chances for biocontrol agents. Examples in EU are the red palm weevil, which can be controlled by nematodes, and the tomato leafminer which is now controlled by a predatory bug in Spanish greenhouse tomatoes. The biopesticide industry has made considerable progress in production, formulation, efficacy, quality, and application strategy. Biopesticides have become more reliable and efficacious, and markets are expanding, even in agricultural crops. Several large agro-chem multinational companies acquired biocontrol companies indicating that biocontrol is a serious trend in crop protection today.

Future Prospects The market for biopesticides is estimated to reach between 3–4 billion US$ in the next five years. The demand for sustainable agriculture and the requirement of supermarkets and consumers for residue-free food is the incentive for the use of biopesticides. I anticipate particularly growth in the use of bacteria for insect and disease control, and of antagonistic fungi for disease control. The biocontrol and biostimulant industry will continue to grow through new companies, agro-chem companies' activities, and expansion of the existing players. New opportunities, as described in this book, will lead to the development of new biocontrol and biostimulant products for sustainable agriculture.

References

Cock MJW, van Lenteren JC, Brodeur J et al (2009) Do new access and benefit sharing procedures under the convention on biological diversity threaten the future of biological control? BioControl 55:199–218

CPL (2013) Biopesticides worldwide market. Wallingford, UK

Geiger F, Bengtsson J, Berendse F et al (2009) Persistent negative effects of pesticides on biodiversity and biological control potential on European farmland. Basic Appl Ecol 11:95–105

Marrone PG (2007) Barriers to adoption of biological control agents and biological pesticides. CAB Reviews: perspectives in agriculture, veterinary science, nutrition and natural resources, vol. 2(51). CAB International, Wallingford, pp 1–12

Ravensberg WJ (2011) A roadmap to the successful development and commercialization of microbial pest control products for control of arthropods. Springer, Dordrecht

Van Lenteren JC (2012) The state of commercial augmentative biological control: plenty of natural enemies, but a frustrating lack of uptake. BioControl 57:1–20

Vega FE, Goettel MS, Blackwell M et al (2009) Fungal entomopathogens: new insights on their ecology. Fungal Ecol 2:149–159

Chapter 33
Commercialization of Microbes: Manufacturing, Inoculation, Best Practice for Objective Field Testing, and Registration

Faina Kamilova, Yaacov Okon, Sandra de Weert and Katja Hora

Abstract Commercialization of plant growth stimulating and biocontrol microbes plays a significant role in providing environmentally friendly and efficient alternatives for chemicals used in agriculture and horticulture. This chapter describes principles of the manufacturing process, including mass production and formulation, of microbes and specifies existing inoculation techniques. Moreover, since evaluation of the efficacy of products is essential for estimation of their commercial potential and because it is also required for their registration, rules and approaches used for the official efficacy tests based mostly on guidelines of the European Plant Protection Organization are included. In order to be placed on the market, microbiological plant protection products and biostimulants/biofertilizers have to be registered. Therefore, also requirements for registration of microbial products in different countries are described and discussed.

F. Kamilova (✉) · S. de Weert · K. Hora
Koppert Biological Systems, Veilingweg 14, 2651 BE Berkel en Rodenrijs, The Netherlands
Tel.: +31 10 5140546
e-mail: fkamilova@koppert.nl

S. de Weert
Tel.: +31 10 5140202
e-mail: sdweert@koppert.nl

K. Hora1
Tel.: +31 10 5140461
e-mail: khora@koppert.nl

Y. Okon
Department of Plant Pathology and Microbiology, Faculty of Agriculture,
Food and Environment, The Hebrew University of Jerusalem, 76100 Rehovot, Israel
Tel.: +972 8 9489216
e-mail: yaacov.okon@mail.huji.ac.il

© Springer International Publishing Switzerland 2015
B. Lugtenberg (ed.), *Principles of Plant-Microbe Interactions,*
DOI 10.1007/978-3-319-08575-3_33

33.1 Introduction

Many rhizobacteria (such as strains of *Azospirillum, Bacillus, Bradyrhizobium, Pseudomonas, Rhizobium and Streptomyces*) and fungi (such as *Clonostachys and Trichoderma*) have the potential to directly promote crop yield, help plants to withstand abiotic stress, and/or protect plants from pathogens (Harman 2006; Jensen et al. 2007; Lugtenberg and Kamilova 2009). Primarily studies on the identification of the strains and eliminating of putative pathogens, on the evaluation of the mode of action, and on laboratory and/or small scale efficacy testing can indicate the commercial potential of such strains. Development of the manufacturing process and efficacy testing of the product and its subsequent registration are critically important stages for the successful commercialization of beneficial microbes for application in agriculture and horticulture.

33.2 Manufacturing

The industrial manufacturing process includes several steps: the production of a starting inoculum, small scale production, big scale (up-scale) production, harvest, formulation and packaging.

Production After the selection of a microorganism with properties of interest and proven to be safe for humans, animals and the environment, the screening for a well defined production medium and for the establishment of the optimal process parameters starts. The goals of industrial or mass production are to minimize the fermentation costs and to produce the highest quantities of the microbe in the best physiological and metabolic state. The latter can be illustrated by two examples. Firstly, Bacilli have a relatively low yield and sporulation level which becomes a serious issue during mass production. Combination of academic research with modification of the cultivation process allowed reaching a maximum cell density of approximately 1.0×10^{10} CFU/g, with $> 95\%$ sporulation of the cells. This achievement significantly stimulated the manufacturing of products containing *B. amyloliquefaciens* (Junge et al. 2000). Secondly, for *Azospirillum* the capability to use intercellular storage material such as polyhydroxybutyrate (PHB), is an important trait for plant growth promotion (Dobbelaere and Okon 2007). Thus, industrial cultivation of bacteria under conditions providing intracellur accumulation of PHB, is a prerequisite for a consistantly efficient *Azospirillum* inoculant.

Production of the starting inoculum begins from a pure culture, usually stored in glycerol at $-80\,^\circ$C or as freeze-dried cells, and occurs in laboratory flasks. This so called pre-culture is used for inoculation of small size (15–50 L) fermenters. The resulting microbial culture is used for up-scale production. Industrial fermentation can be done as solid or semisolid state cultivation or as liquid (submerged) cultivation.

During solid state fermentation, with little or no "free" water, the microorganisms are grown on a solid carrier, either mixed with solid nutrient substrate or impregnated with liquid nutrient broth. Sometimes solid nutrient can also play the role of a carrier. The final product includes microbial biomass, exometabolites and solid carrier with or without remnants of nutrients. Advantages of solid state fermentation are the use of relatively inexpensive substrates, continuality of the process and low waste volumes. A disadvantage is the risk of microbiological contamination due to non-sterile conditions typical for this type of cultivation. The process often includes a lot of manual operations and thus high labour costs. Solid state fermentation is frequently used for the production of fungal products, mostly because of the inability of some fungi to produce robust spores in liquid culture.

Liquid or submerged fermentation allows a wide range of use of water-dissolved substrates including refined and unrefined carbon and nitrogen sources. This provides more freedom for optimization of technological process. Other advantages of liquid fermentation are (i) a high degree of process control due to the use of modern bioreactors, (ii) a substantially reduced probability of contamination due to sterile conditions and (iii) the microbial biomass can be separated from culture broth and concentrated. A disadvantage of submerged cultivation is the high volume of waste. The choice for either solid state or liquid fermentation is determined by the biology of the strain and by the economy of up-scaling: low initial investments but higher labor cost versus high investments and lower labor costs (Friedman 1990).

Formulation The goal of formulation is to preserve the microbial biomass obtained after fermentation and to deliver the microbes in a good condition to their targets and, after delivery, to enhance their activity (Burges 1998). The process of production of dry formulations, such as powders and granules, depends mostly on the biological properties of the microbial cells and on their ability to withstand the drying procedures (Kamilova and De Bruyne 2013). Formulations may contain various additives such as carriers, nutrients, stickers, protectants and emulsifiers. Among the most often used carriers are sterilized peat, conditioned cereal grains, talc, agricultural clays and diatomaceous earth. Nutrients provide a primary boost for the microorganism immediately after application of the product. Stickers allow microbes to attach better to—and stay on—plant surfaces. Protectants defend microbes from desiccation, UV light, and temperature changes. Microbial exometabolites that are directly or indirectly involved in beneficial effects of strains can also be included in the formulations. Microbial products can be formulated as (water or oil) suspensions, powder, and wettable or insoluble granules. Three important factors determine the choice of formulation type and composition: biological properties and characteristics of the microorganism, inoculation techniques and types of irrigation systems involved in the agricultural practice.

Packaging Only materials that secure intact biological, physical and chemical properties of the end product during storage and transportation should be used for packaging. The type of packaging is determined by the physical state of the formulated product. The size of packaging can vary and depends on logistics and on market demand in different countries.

33.3 Inoculation

The choice of the inoculation technique depends on the mode of action of the beneficial microbe, the plant growth stage at the time of application, and the type of formulation. Powder and liquid inoculants can be applied either to the seed or directly to the soil in the seeding furrow. Seed treatment can be performed by specialized companies and treated seeds of some crops can be stored for a considerable period of time. Certain crops can be treated by farmers at site, right before sowing. Granular formulations are most suitable if slow release of a microbe is needed e.g. in situations when multiple applications of the product is not feasible. Granular inoculants are applied in-furrow by using suitable granular applicators. If water dispersible formulations or liquid inoculants are applied directly to soil, they are first suspended in clean potable water so that they can be evenly distributed over the cropping area by hand or by mechanical spraying equipment. Suspended products can be delivered directly to the root zone of individual plants by drip irrigation or by drenching furrows. The latter techniques are particularly suitable for multiple applications during crop cultivation.

33.4 Best Practice for Objective Field Testing

The term "efficacy" was coined by European legislation authorities to evaluate the benefit of a plant protection product (PPP) in terms of quality and quantity of the crop of interest under naturally occurring or artificially introduced disease conditions. It also includes an estimation of the risk of a phytotoxic effect that a PPP might have on the treated crop (if the product is applied in excessive amounts) and on adjacent and successive crops as well as its effect on the quality of the products produced from the treated crops.

The European Plant Protection Organization (EPPO) has set up a number of guidelines or standards (http://pp1.eppo.int/). Products can only come on the market after their efficacy has been positively evaluated by government bodies. The guidelines describe in detail what information needs to be presented in order to enable evaluators to weigh the benefits and risks of PPPs. The European legislation uses these guidelines to prescribe the producers of PPPs how they should test their products. Two of these standards (PP1/181 and PP1/152) contain a summary of the basic principles for the design and performance of agronomically and statistically sound field trials (OEPP/EPPO 2012a, b). Governmental requirements for the demonstration of efficacy vary between continents and individual countries. Europe tends to be most strict, forcing the producer to demonstrate the efficacy level claimed on the label of a product with a high number of field trials, executed according to the EPPO guidelines by recognized research organizations (6–8 trials per EPPO climate zone, for each unique crop/target combination). Canada, Australia, California and Florida follow the European approach. Other Northern American states do not strictly require evidence of efficacy, but reserve the right to ask for it. Here, the aspects of safety of the product for health and environment are more important than efficacy. Countries in South America, Africa and Asia differ widely in their legislation on this point, requiring case-by-case studies and consultation with the authorities before starting field trials. Besides basic guidelines, a number of standards for the individual combination crop/disease (pathogen) were developed by EPPO. It is clear that these standards can not cover all possible combinations. Nevertheless, at least some of them can be adapted or modified for trials with other crops and similar pathogens. In case no specific crop/disease EPPO standard exists and modification of current standards is impossible, a scientifically sound experimental protocol for appropriate testing of the product against the disease of interest in the crop of interest should be developed in collaboration with a testing institute.

For a good estimation of the efficacy of a product, field trials need to conform to four basic principles. (i)Trials have to be carried out with the end product, in the formulation as will be used by the grower.(ii) A dose range around the recommended label dose should be included in the trial. Establishing the minimal effective dose provides information on the effectiveness and safety of the product. Moreover, it will also predict the approximate costs per hectare. This information will be very valuable for the acceptance of the product by the farmers. Inclusion of higher dosages is equally important to make sure that, in the event of accidental overdosing by the farmer, there still will be no detrimental effects on the crop. (iii) Trials should be performed under a wide range of environmental and agronomic conditions, preferably in the climate where the product will be used. For example, one cannot compare the effects on the crop of a large scale corn-grower in the United States with that of a small-holder with a patch of maize in Africa. EPPO provides a table on comparable climates on a global level in the standard PP1/269(1) (OEPP/EPPO 2010). To be able to compare reports or papers on efficacy of microbial products in different geographic localities, it is vital that the agronomic practice and climate conditions of the trial are described in detail. (iv) The trial should be carried out as closely as possible to the local agronomic practice.

The product should be evaluated in relation to farmer's practice: when the farmer is looking for a solution of his problem, his scope is wider than the application of just one product. Effectiveness of the product may be increased or decreased in combination with practices such tilling, the choice of fertilizers, pesticides/herbicides, soil amendments (compost), irrigation, plant density, hygienic measures, crop rotation, and the use of other products based on microorganisms. To be able to advise the grower on the correct use of the product, and to be able to predict the effect that a product will have in his situation, these aspects need to be addressed during product development. Therefore, a detailed inventory of the local agronomic practice is important, and a multidisciplinary overview of the agronomic setting needs to be kept.

What are the hurdles in efficacy trials of microbial products? Why are the full benefits of the product not always observed in a well designed fully randomized trial? This could be due to the generally small plot sizes (often practiced because of economic restrictions), and the tendency to secure the investment in a trial. Testing institutions generally try to increase this security by increasing the disease pressure in case of trials with products aimed at disease control, or growing the crop under too optimal conditions in case of trials aiming at yield assessments. The level of disease pressure can play a dual negative role in the evaluation of a product. For instance, the benefit of root colonizing microorganism for soil borne fungal root diseases might be hardly visible when the disease level is impractically high. This will lead to rejection of a product that may have worked very well for growers that maintain an economic threshold of 5 % disease incidence. On the other hand, a trial with a disease incidence lower than 10 % in the untreated control may not lead to statistically significant differences. Another example is that a microorganism meant to improve phosphate uptake will not demonstrate its beneficial properties under conditions which do not restrict the availability of phosphate for the plant. However, a phosphate deficit should not cause severely compromised plants in the untreated control since this would not reflect a realistic situation in the farms.

33.5   Registration

Plant growth promoting strains can posses properties that help plants (i) to fight bi-otic stress (biocontrol) (see Chap. 18), (ii) to tolerate abiotic stress (draught, elevated temperature, salinity) (see Chap. 27), (iii) to directly provide plants with essential nutrients e.g. nitrogen (see Chap. 23),or (iv) to stimulate natural process which benefit the efficiency of nutrient uptake e.g. of phosphorus (see Chap. 24). The first category is represented by microbial PPPs, often called biopesticides. The last three categories are represented mostly by plant strengtheners/ biostimulants and microbial fertilizers. It seems that only bacterial strains directly involved in nitrogen fixation e.g. *Rhizobium, Bradyrhizobium* and *Azospirillum* are easily fitting in the definition of microbial fertilizers. For other PGP microorganisms a definition of biostimulants is more suitable. Many biocontrol strains often combine antifungal activity with stimulation of plant growth under unfavorable abiotic conditions. Whereas for the researchers and end-users of the microbial products this combination of beneficial properties looks very interesting and attractive, it seems to be complicated for the regulatory authorities worldwide. Regulatory authorities apply to agricultural microbial products the same approach as to chemical products: PPPs and fertilizers are regulated differently. So, despite the fact that many beneficial microbes combine more than one beneficial trait, from the regulatory point of view currently microbial inoculants can be assigned only as either biopesticides or as biostimulants/plant strengtheners/biofertilizers. This requirement plays a critically important role in the registration of microbial products and in the way the products can be placed in the market. Companies should critically evaluate the properties and mode of action of the potential microorganism from the point of view of the most consistently reproducible effects and the marketability of the future product. More details can be found in Kamilova and De Bruyne, 2013.

Plant Protection Products   Registration of PPPs in the EU is regulated by EU Regulation 1107/2009 (2009). The procedure includes the authorization of the active microorganism(s) and formulation(s). The registration dossier consists of two parts:

Active substance (microorganism) dossier and Product dossier. Information on the identity/biology of microorganism, on the methods used for production of the microbes as well as of the final product, on analytical methods used for the active microorganism and formulation are parts of the registration dossier. Safety evaluation of both a microbe and the end product includes data on human toxicology and ecotoxicology, fate and behavior in the environment. Tests should be performed by laboratories according to guidelines and standards of international organizations such as Collaborative International Pesticides Analytical Council (CIPAC), Organization for Economic Co-operation and Development (OECD) or United States Environmental Protection Agency Office of Chemical Safety and Pollution Prevention (US EPA OCSPP), and International Organization for Standardization (ISO) according to Good Laboratory Practice (GLP). Outside Europe these tests can be asked to be performed according to specific national standards. Recognition of the microorganism as safe on the EU level involves communications between company applicant designated Rapporteur Member State, national agencies, and the European Food Safety Authority (EFSA). This leads to publication of the scientific opinion by EFSA. Based on this document, the European Commission makes a decision on inclusion of a microorganism into the list of safe active substances. Then a submission of a Product dossier should be done at the national level. In the EU, efficacy trials must be performed by authorized institutions according to Good Experimental Practice (GEP) in many respects based on EPPO standards in the intended geographical Zone as described earlier in this chapter.

In the USA, registration of the active microorganism and the product is regulated by the Federal Insecticide, Fungicide, and Rodenticide Act (FIFRA, http://www.epa.gov/agriculture/lfra.html) and implemented by US EPA. In most aspects related to data requirements for the registration dossier there is no difference between USA and EU. However, a major difference is that in the EU efficacy trials are an obligatory part of a Product dossier, while in the USA efficacy data are not required by EPA. However, they are required by the regulatory authorities of some states e.g. California and Florida.

African and South -American countries, China and Mexico do accepts data present in the European or the US dossiers. However, all of them require efficacy trials performed locally.

Biostimulants/Microbiogical Fertilizers Currently in Europe registration of microbial non-PPP inoculants is regulated at the national level. Details about the situation in some EU countries has been described in Kamilova and De Bruyne (2013). To harmonize the legislation basis for these products, the recently established European Biostimulants Industry Council (EBIC) underlines that plant biostimulants *"contain substance(s) and/or micro-organisms whose function when applied to plants or the rhizosphere is to stimulate natural processes to enhance/benefit nutrient uptake, nutrient efficiency, tolerance to abiotic stress, and crop quality. Biostimulants have no direct action against pests, and therefore do not fall within the regulatory framework of pesticides"* (http://www.biostimulants.eu). It is clear that inoculants based on Bradyrhizobium, Rhizobium, Azospillum and other PGP microbes fall under this definition. EBIC puts a lot of efforts to include biostimulants into Regulation

(EC) 2003/2003, which regulates fertilizers in the EU in order to provide among others safe and efficient microbial products. Nevertheless, it is still unclear how in practice registration of these products will be regulated in the EU.

The situation in South America with microbial inoculants is somewhat less obscure since all MERCOSUR countries (Argentina, Brazil, Paraguay, Uruguay and Venezuela) signed "Recommendation on use of inoculants in agriculture" (MERCOSUR/GMC/RES # 28/98 1998), which includes the obligation to register inoculants and to establish a "Reference Institution " for each country, as well as defining its responsibilities. Actually, each country has adapted its own regulations according to this recommendation. The quality of microbial inoculants, particularly of rhizobia, is a major concern of MERCOSUR countries. For example, the network REDCAI (Inoculants Quality Control Network) was established as a member of the Division for Agricultural and Environmental Microbiology of the Argentinian Association of Microbiology, with the aim of creating and validating a set of methodological tools for the evaluation of inoculants. The Normative Instructions to the Fertilizer's National Laws in the MERCOSUR countries sets the official methods for inoculant analysis. In these countries a the list of recommended and recognized strains does exist. MERCOSUR may allow registration of an inoculant with strains still not included in the said list, but the registrant company shall submit reports of efficacy under laboratory and greenhouse conditions, together with data from at least 3 years of field trials in three different agro-ecological zones. For other, non $-(Brady)rhizobium$ strains, companies must provide evidence on safety and efficacy. Efficacy trials have to be performed in institutions appointed by the national governments.

The Canadian Food Inspection Agency regulates biofertilizers (to improve plant growth) via Fertilizers ACT and Regulations C.R.C.666 (http://laws-lois.justice. gc.ca/eng/acts/F-10/) paying attention to the safety for human and animal health, and environmental as well as for the quality of the product. Under product quality the number of viable cells sufficient for providing a significant beneficial effect is meant. In Canada there are no specific requirements for efficacy data for biofertilizers.

In the USA there is no federal law regulating biofertilizers and microbiological soil amendments. The Departments of Food and Agriculture of the individual states regulate this type of products. Requirements may differ drastically: authorities in some states can ask to perform local efficacy trials, while in other states only notification of the fact that a product is on the market is sufficient.

In the countries of the Commonwealth of Independent States (CIS) the registration of microbiological fertilizers as well as of PPPs is performed by the department of Phytosanitary and Veterinary Surveillance at the Ministries of Agriculture. Efficacy tests should be performed in the country of application. In general, all tests are required to be performed by the national institutions or specialized laboratories. However toxicological tests performed in the EU or in the USA after evaluation by the designated national institutions can be accepted.

In the African countries and China, departments responsible for registration of biostimulants/ biofertilizers may be different from those that regulate PPPs. However, requirement to perform efficacy tests are common for these countries.

In conclusion, the increasing demand for safe and environmentally friendly microbial products as alternatives for chemicals stimulates the microbiological industry.

Improvement of the production process, development of the most suitable and stable formulation, and proof of efficacy of the products, will allow building up the market and successfully introducing microbials particularly in areas where chemicals are either ineffective or not available. Correctly performed registration will provide a legal basis for the products and will therefore help to fit in—or create—new commercial niches for beneficial microbes.

References

Burges HD (1998) Formulation of microbial biopesticides. Kluwer Academic, Dordrecht

Dobbelaere S, Okon Y (2007) The plant growth promoting effects and plant responses. In: Elmerich C, Newton WE (eds) Associative and endophytic nitrogen-fixing bacteria and cyanobacterial associations (Nitrogen fixation: origins, applications and research progress), vol V. Springer, Heidelberg, pp 145–170

European and Mediterranean Plant Protection Organization (2010) PP 1/269 (1). Efficacy evaluation of plant protection products. Comparable climates on a global level. Bulletin OEPP/EPPO Bulletin 40:266–269

European and Mediterranean Plant Protection Organization (2012a) PP 1/181(4). Efficacy evaluation of plant protection products. Conduct and reporting of efficacy evaluation trials, including good experimental practice Bulletin OEPP/EPPO Bulletin 42:382–393

European and Mediterranean Plant Protection Organization (2012b) PP 1/152(4). Efficacy evaluation of plant protection products. Design and analysis of efficacy evaluation trials. Bulletin OEPP/EPPO Bulletin 42:367–381

Federal Insecticide, Fungicide, and Rodenticide Act. (1947) Last amended in 2003, Recent update 27 June 2012. http://www.epa.gov/agriculture/lfra.html. Accessed 15 Sept 2014

Fertilizers ACT and Regulations C.R.C. 666. (1985) Last amended on 28 June 2006. http://laws-lois.justice.gc.ca/eng/acts/F-10. Accessed 1 Sept 2014

Friedman MJ (1990) Commercial production and development. In: Gaugler R, Kaya HK (eds) Entomopathogenic nematodes in biological control. CRC Press, Boca Raton, pp 153–172

Harman GE (2006) Overview of mechanisms and uses of *Trichoderma* spp. Phytopathology 96:190–194

Jensen DF, Knudsen BIM, Lübeck M et al (2007) Development of a biocontrol agent for plant disease control with special emphasis on the near commercial fungal antagonist *Clonostachys rosea* strain 'IK726'. Australas Plant Pathol 36:95–101

Junge H, Krebs B, Kilian M (2000) Strain selection, production and formulation of the biological plant vitality enhancing agent FZB24® *Bacillus subtilis*. Pflanzenschutz-Nachr. Bayer 1:94–104

Kamilova F, de Bruyne R (2013) Plant growth promoting microorganisms: the road from an academically promising result to a commercial product. In: de Bruijn FJ (ed) Molecular microbial ecology of the rhizosphere, vol 1 & 2. Wiley, Hoboken, pp 677–686

Lugtenberg B, Kamilova F (2009) Plant-growth promoting rhizobacteria. Annu Rev Microbiol 63:541–556

MERCOSUR/GMC/RES. N° 28/98 (1998) Disposiciones para el comercio de inoculantes visto: El Tratado de Asunción, el Protocolo de Ouro Preto y la Recomendación N° 9/97 del SGT N° 8 "Agricultura". http://www.mercosur.int/msweb/Normas/normas_web/Resoluciones/ES/Res_028_098_Disp-Rel_Comercio%20Inoculantes_Acta%202_98.PDF. Accessed 14 Sept 2014

Regulation (EC) No 1107/2009 of the european parliament and of the council of 21 October 2009 concerning the placing of plant protection products on the market and repealing Council Directives 79/117/EEC and 91/414/EEC Official Journal L 309, 24.11.2009 P. L 309/1-309/50

Chapter 34
Towards a New Generation of Commercial Microbial Disease Control and Plant Growth Promotion Products

Rainer Borriss

Abstract Biofertilizer and biocontrol formulations prepared from plant growth-promoting bacteria are increasingly applied in sustainable agriculture. However, these bioformulations of the first generation are sometimes hampered in their action and do not fulfill in each case the expectations of the appliers. Unraveling specific responses of selected model bacteria on the global level will greatly stimulate the rational design of a new generation of biological fertilizer and biocontrol agents.

34.1 Plant Growth-Promoting Bacteria Used as Bioinoculants

Selective pressure on the microbial population in the rhizosphere is posed by the plant, in the form of root exudates containing specific nutrients, secondary metabolites, oxygen radicals (ROS), and other stress signals (Ramos-Gonzalez et al. 2013). Many of the bacteria, which are able to overcome this selective pressure and to propagate within the plant rhizosphere, act beneficial on plant growth [this process is designated as *plant-growth-promotion* (PGP); Lugtenberg and Kamilova 2009), and, simultaneously, suppress growth of plant pathogens (this process is designated as *biocontrol*, (BC); Haas and Défago 2005). Microbial inoculants, prepared from such rhizobacteria, are presently known under different names, such as bioeffectors, biofertilizers, biostimulants, or biocontrol agents (biopesticides). In the following, I will use the general term *bioeffector*, which includes the biological fertilizer and biocontrol function, since both functions are difficult to separate in plant-growth-promoting bacteria (PGPB) (Kloepper et al. 1980).

Since the genomes of *Pseudomonas fluorescens* Pf5 and *Bacillus amyloliquefaciens* FZB42 have been sequenced as the first representatives of Gram-negative and Gram-positive bacteria with BC and PGP activity several years ago (Paulsen et al. 2005; Chen et al. 2007), our knowledge base about plant-bacteria interactions has been steadily increased. Numerous genes and gene clusters that may contribute to a

R. Borriss (✉)
ABiTEP GmbH, Glienicker Weg 185, Berlin, Germany
Tel.: + 49 30 67057-14
e-mail: rborriss@abitep.de

© Springer International Publishing Switzerland 2015   329
B. Lugtenberg (ed.), *Principles of Plant-Microbe Interactions*,
DOI 10.1007/978-3-319-08575-3_34

plant-associated life-style of beneficial rhizobacteria have been identified, and can be used in developing improved bioeffectors.

34.2   The 'Ideal' Bioinoculant (Bioeffector)

Example Bt   Despite increasing acceptance of bioinoculants, many features of this environmental-friendly agents need to improve before they are widely used as a promising alternative or substitute of chemical pesticides and fertilizers. Especially variable application results hinder their further distribution. However, there is one remarkable exception, which might serve as a guideline for further product development: *bioinsecticides*. These products prepared from endospores of *Bacillus thuringiensis* (Bt) and related species are by far the most successful biopesticides (market share 79 %). Bt insecticides are most commonly used against some leaf- and needle-feeding caterpillars. In recent years, there has been tremendous renewed interest in Bt. Several new products have been developed, largely because of the safety associated with Bt-based insecticides, their stability, and their narrow action spectrum which is specifically directed to the target organisms, but is not harmful to other insects. Moreover, the molecular features of their action are well investigated (Chap. 20).

Bioeffectors of the Next Generation has to be compatible with routine field practices under different field conditions and types of soil (Fig. 34.1), and should fulfill the following tasks: (i) the formulation should be easy to handle and to place at the target region (roots, leaves etc.); (ii) persistence at the plant; (iii) shelf-life should last for at least one or two seasons; (iv) high efficacy, thereby yielding constant reproducible results in the field based on known action principles; (v) safe: no undesired side effects and not causing health problems for human beings and animals (e.g. skin irritation); (vi) no negative impact on the environment and not affecting biodiversity of concomitant microbial communities; (vii) not allowing development of resistance of pathogens after application; (viii) economically feasible for the customer.

34.3   Manufacturing of Bioinoculants with Superior Stability

Candidate Microbes   At present, representatives of three different groups of microbes seem to be attractive candidates for developing improved bioinoculants. (i) Gram-negative *Pseudomonas* spp., such as *P. fluorescens*, are one of the most powerful colonizers of the plant rhizosphere and are able to enhance crop yield dramatically and to control many plant diseases (Haas and Defago 2005; Chap. 18; Chap. 38). (ii) Gram-positive *Bacillus* spp., such as *B. amyloliquefaciens plantarum*, are the main constituents of many biocontrol and biofertilizer agents presently on the market (Chap. 40). (iii) Last but not least, beneficial microfungi such as *Trichoderma* spp., are also applied in agriculture to support plant growth and to suppress plant pathogens.

Fig. 34.1 Bioeffectors can be applied by using a conventional seed drilling machinery, AMAZONEN-WERKE H. Dreyer GmbH & Co. KG, working width of 6 m, row distance: 75 cm. Under foot inoculation of the bioeffector is performed during maize seed drilling (by courtesy of Dr. Frank Eulenstein, ZALF e. V., Müncheberg, DE)

Other beneficial microbes, but without biocontrol action, such as rhizobia and *Azospirillum brasiliense* (Chap. 23), mycorrhiza fungi (Chap. 25), and Sebacinales, are not treated in this review.

Carriers Protect Bioinoculants However, in many instances, the numbers of cfu's in microbe suspensions without a proper carrier start to decline immediately after drenching into the soil. Consequently, a major task in formulating inoculants is to provide a more suitable microenvironment, combined with physical protection for a prolonged period of time to prevent a rapid decline of the introduced bacteria. Carriers used for inoculants should be preferentially sterile. They include many different types, from cheap liquid, organic, inorganic, up to highly sophisticated encapsulated formulations using several polymers. Very promising for durable and efficient immobilization of inoculant(s) are alginate beads. Unfortunately, due to relative high production costs, their use is restricted mainly to medical applications (Bashan et al. 2014) .

Spore-Forming Microbes There is a possibility to circumvent the relatively expensive use of carriers, when spore-forming microbes are used. Especially, endospore-forming *Bacillus* spp are suitable for preparing simple liquid or dry formulations in which it is not necessary to add stabilizing carriers (Borriss 2011; Chap. 40). Dormant spores are naturally resistant against extreme temperatures, desiccation, ultraviolet radiation, and chemical disinfectants. Formulations consisting of *Bacillus* endo-spores can be stored for nearly unlimited time, given that premature germination of spores is avoided, e.g. by adding alcohols, such as iso-propyl alcohol in concentrations of less than 1 %, or complete removal of germination stimulating nutrients. Despite that, *Trichoderma* conidiospores do not reach the longevity of *Bacillus* endo-spores although their shelf-life is much higher than that of vegetative cells and allows the preparation of formulations with sufficient stability.

34.4   Improved Inoculation Techniques

Appropriate Placement of a sufficient amount of the bioinoculant to the host plant is crucial for the success of its application. Independent of the technique used, a threshold concentration of the bioeffector is a precondition for its successful application. For bacteria, cell numbers of at least 10^6 per plant/seed is recommended. This figure corresponds roughly to 10^{13}–10^{14}cfu per ha. May be this figure could be reduced when more target-specific placement techniques are developed. At present, several techniques for inoculating bioeffectors are in use, but further development and refinement is necessary for obtaining reproducible results in their application. It ruled out that different bioeffectors have specific requirements. For example, experiments performed with *Bacillus amyloliquefaciens* FZB42 revealed that it is necessary to use endospores, not vegetative cells, when directly applied to plants. Furthermore, clearly better results in growth promotion of lettuce plants were obtained, when the bioinoculant was applied twice, before and after transplanting, than the corresponding amount only once.

Seed and Soil Inoculation Success in 'coating' of seeds with bioinoculants depends on proper adhesives and on the surface of the seeds. Therefore, this method works not always appropriate, especially in small seeds with very smooth surfaces, to which the bioeffector does not well attach. If the inoculation process damages seed coats (e.g. in the case of peanuts) this may prevent germination. Some species release anti-bacterial compounds from their seeds, which can inhibit the inoculant (Bashan et al. 2014). A common soil inoculation technique is band application in which the inoculant is placed directly in the seed furrow. Granulated or liquid bioinoculants can be used. Using an appropriate machine, which allows combining the drilling and inoculation processes, is desirable (Fig. 34.1). For more details about the inoculation techniques presently recommended, the reader is referred to Chap. 33.

34.5   How to Improve the Plant Growth-Promoting Function of Bioinoculants?

It is clear that the PGP effect exerted by every PGPB is due to many different features within a complex network of plant-bacteria interactions. However, there is common sense that the first precondition for beneficial actions on plant growth is the ability to colonize plant roots (rhizosphere competence) and/or other (aerial) parts of the plant.

Spore Germination and Root Colonization of Fungi, and Bacilli Chemotactic movement towards roots, and competitive root colonization is essential for root colonizing microbes (Chap. 3). In case of plant-associated bacilli it was shown that spores start to germinate when nutrients present in seed or root exudates become available. Transposon mutants impaired in chemotactic motility and biofilm formation were unable to support plant growth, suggesting that rhizosphere competence is a *'conditio sine qua non'* for PGP. Therefore, selection of mutant or novel wild type strains, especially aggressive in root colonization, is a promising tool for developing enhanced PGPB (Kamilova et al. 2005)

Careful Characterization of the Molecular Principles Underlying PGP Although our present knowledge about the different aspects of PGP is far from complete, intensive efforts should be made for each PGPB to elucidate the molecular principles determining its beneficial effect on plant growth. For example, we have to know whether the PGP effect is due to biofertilizer functions (mobilization of nutrients for plant nutrition, fixation of nitrogen), general plant strengthening, or biostimulant function (synthesis of plant growth hormones and/or volatiles as is the case in the *B. subtilis* and *P. fluorescens* groups and so on. Only, when we understand, at least in part, the mechanisms acting between a single PGPB and the host plant, the customer will be in a position to use such bioinoculants successfully under the appropriate conditions.

Mixtures of Different Bioinoculants There are many products on the biofertilizer market consisting of several microbes which are claimed to have a beneficial effect on plant growth. The problem with such products is their variable quality. It seems impossible to keep a constant product quality when vegetative cells of beneficial Gram-negative bacteria, such as *Pseudomonas*, *Rhizobium*, *Azospirillum*, and *Agrobacterium*, are mixed with dormant spores of Gram-positive bacilli and—possibly—of some fungi. In addition, PGP effects coming from these beneficial microbes can only be expected when threshold concentrations, e.g. at least 10^6cfu, become attached at the host plant, which is highly unlikely. There is a better application perspective for products consisting of only two plant-beneficial microbes with similar shelf life, which could act synergistically to each other. In the following a few examples for promising combinations are given.

PGPB and Phosphate Solubilizing Bacteria Bacteria making available fixed phosphates for plant nutrition (biofertilizer) are highly desirable (Chap. 24). Their combination with PGPB should enhance their beneficial effect. An example for combining two spore-forming inoculants is a bioformulation consisting of endospores of the silicate bacterium *Paenibacillus mucilaginosus*, which is able to solubilize phosphate and potassium from minerals, and the PGPB *Bacillus amyloliquefaciens plantarum*. The combination was successfully applied in tobacco cultures in Yunnan province, China.

Bacillus spp. and *Trichoderma* spp. Many PGP *Bacillus* spp., including *B. amyloliquefaciens* and *B. subtilis*, synthesize fungicidal lipopeptides under laboratory conditions (Chap. 40). It is therefore surprising that mixtures of *Bacillus* spores and *Trichoderma* conidiospores were proven to be extremely efficient bioagents under field conditions. Possible explanations for 'coexistence' of both microbes are that *Bacillus* is unable to produce reasonable amounts of fungicides *in planta*, or that both bioinoculants occupy different niches within the plant rhizosphere, where they act synergistically on the host plant.

Mycorrhiza and Mycorrhiza Helper Bacteria (MHB) The same seems to be true for mixtures of mycorrhiza fungi and Bacilli. It is known that mycorrhizal formation is enhanced by co-inoculation with MHB, e.g. *Bacillus* spp., which promote rapid root colonization by ectomycorrhizal fungi. In addition, some MHB promote the functioning of the mycorrhizal symbiosis. This was illustrated for three critical functions of practical significance: nutrient mobilization from soil minerals, fixation of atmospheric nitrogen, and protection of plants against root pathogens.

Combined Bioformulations Bioeffectors enhancing harvest yield of crops and vegetables are based on PGPB, but can also contain plant and seaweed extracts, humic acids, strigolactones and other organic substances stimulating plant growth. It is likely that synergistic effects occur when formulations consisting of living microbes and such organic materials in specific combinations are applied to plants. Exploiting these possibilities is still in an early stage of research.

34.6   How to Improve the Efficacy of Biopesticides?

B. amyloliquefaciens plantarum and *P. fluorescens* are prototypes for plant-associated bacteria with BC activity. It is generally assumed that suppression of plant pathogens is based on two features of these PGPR: (1) production of antibiotics, cyclic lipopeptides, and siderophores—all of which can be efficient against bacterial and fungal pathogens -, and (2) induced systemic resistance (ISR), which stimulates the plant's defense system against harmful microbes and viruses.

Antimicrobial Compounds Secondary metabolites toxic to phytopathogenic fungi, oomycetes and bacteria synthesized in *P. fluorescens* are phenazines (phenazine-1-carboxylic acid, PCA; and phenazine-1-carboxylic acid, PCN), hydrogen cyanide, the chlorinated tryptophan derivative pyrrolnitrin, and the polyketides 2,4-diacetylphloroglucinol (DAPG), rhizoxin and pyoluteorin (reviewed in Chap. 18). In addition, several cyclic lipopeptides (c-LPs) are non-ribosomally synthesized in some representatives of the *P. fluorescens* species complex. *B. amyloliquefaciens plantarum* devotes around 10 % of its whole genome capacity to the synthesis of antimicrobial compounds. Non-ribosomal synthesized cLPs, such as iturins and fengycin, are mainly directed against fungal pathogens, whilst the polyketides difficidin, macrolactin and bacillaene are mainly efficient against plant pathogenic bacteria. Bacilysin and some bacteriocins are efficient against bacterial antagonists (reviewed in Chap. 40).

Expression of Antimicrobial Compounds in the Plant Rhizosphere For a long time the plant protective activity of PGPR has been correlated with the potential to secrete a wide array of antibiotic compounds. However, in most cases this ability was only demonstrated upon growth as planktonic cells under artificial conditions. Notably, under environmental conditions, fluorescent pseudomonads are able to produce redox-active phenazines which contribute to the natural soil suppressiveness of Fusarium wilt disease and may act in synergy with carbon competition by resident non-pathogenic *F. oxysporum*. Notably, PCA was detected at up to nanomolar concentrations in the rhizosphere of wheat plants growing in the suppressive soils near Lind and Ritzville, WA, U.S.A., suggesting that a natural antibiotic can be transiently accumulated across a terrestrial ecosystem in amounts sufficient for the direct inhibition of sensitive organisms (Mavrodi et al. 2012). By contrast, except the c-LP surfactin, secondary metabolites synthesized by PGPR *B. amyloliquefaciens* are hardly to detect within the plant rhizosphere, suggesting that the BC effect exerted by the bacterium is not due to direct antibiosis (Debois et al. 2014). These findings are important for future strategies for screening of powerful PGPR and BC strains. It is known for a long time, that high efficiency in suppressing fungal or bacterial pathogens under laboratory conditions do not necessarily reflect the potential of these selected strains for their performance under field conditions. Also other mechanisms, such as siderophore-mediated competition for iron, production of lytic enzymes (e.g. chitinases), and induced systemic resistance play a role in disease suppression.

Induced Systemic Resistance (ISR) Representatives of PGPR, including *Pseudomonas* spp. and *Bacillus* spp, are able to trigger plant defense responses against pathogens. Independent of the location of the bacteria triggering that plant defense reaction, ISR is effective against pathogens infecting aerial or belowground parts of the plant. Determinants of *Pseudomonas* involved in ISR are c-LPs, the siderophore pseudobactin, salicylic acid, and 2,4 diacetyl phloroglucinol (Bakker et al. 2007). In case of Bacilli, it seems that ISR stimulation is a multifactorial process dependent on several compounds produced by the rhizobacteria. Candidate compounds are surfactin and volatiles, especially acetoin and 2,3 butanediol (Chap. 8). Taken together, at least for BC Bacilli it is likely that ISR is more important than direct antibiosis in suppressing plant pathogens. The recent findings about direct antibiosis and ISR have to be taken into account when developing improved biopesticides.

Improved Biocontrol Formulations Many of the biocontrol formulations which are currently in use are based on living microbes, such as Bacilli, *Pseudomonas*, *Trichoderma*, and *Paecilomyces*. Unfortunately, for many of such formulations the active principle(s) for their pathogen suppressing action it is not known. c-LPs, for example, have been claimed to be responsible for the antimicrobial effects exerted by *B. subtilis*. However, as shown above, it is very unlikely that the concentration of antifungal c-LPs (iturins and fengycins) within the plant rhizosphere reach levels sufficient for antibiosis. A possibility to circumvent this problem are bioformulations consisting of both, *Bacillus* spores and concentrated culture supernatants containing antimicrobial metabolites (Serenade, Double Nickel 55, see Chap. 42). Unfortunately, also in those products, only the number of spores is considered as active ingredient of the biofungicide by regulators. In contrast to chemical fungicides, there is no indication about metabolites and their concentration, except for a specific treatment of pathogen-infected plant parts. Improved biopesticides rely on both principles, direct antibiosis by antimicrobial metabolites, and ISR stimulated by the plant-associated bacterium. In case that the bacterium does not produce sufficient amounts of the antibiotic, when growing *in planta*, external metabolite 'cocktails' should be added. But in that case it is necessary, for having a standardized product quality, to guarantee a fixed concentration of the active principle for efficiently suppressing the target pathogen. This would enable comparison of chemical and biological pesticides.

34.7 Is There Any Perspective for GMO-Bioinoculants in Agriculture?

We have to acknowledge that at present the use of genetically engineered bioinoculants under field conditions is refused by the public, at least in Europe. It is therefore not surprising that, by contrast to GM-crops (Chap. 15), until now no GM-PGPB has been registered by the governmental authorities, neither in Europe nor in the U.S. However, in the light of a steadily increasing world population, growing from 7 billion now to 8.3 billion in 2025, innovative approaches for getting higher harvest yields without using increasing amounts of agrochemicals should not longer be excluded, given that their use is safe and without harmful consequences for human beings and nature. Like for other releases, careful environmental studies are a precondition before releasing genetically engineered bacteria into the environment. To illustrate the possibilities of using of GMOs in agriculture, a short example will be given here.

Expression of the Harpin Gene Enhances Biocontrol Activity of FZB42 The *hrp* ("harp") genes encode type III secretory proteins enabling many phytopathogenic bacteria to elicit a hypersensitive response (HR) on non-host or resistant host plants and induce pathogenesis on susceptible hosts. The HR is a rapid localized death of the host cells that occurs upon pathogen infection and, together with the expression of a complex array of defense-related genes, is a component of plant resistance. The plant genes create a cascade of effects, which promote a Systemic Acquired Resistance (SAR) throughout the plant. Beneficial effects on plant growth and health have been reported. In the laboratory of Xuewen Gao, Nanjing Agriculture University, China, the PGPR *B. amyloliquefaciens* FZB42 was engineered to express the harpin gene product by chromosomal integration of two *hpa1* genes cloned from *Xanthomonas oryzae*. Greenhouse experiments demonstrated efficacy of FZBHarpin in bio-controlling of the disease rice bacterial blight in tobacco plants (Qiao et al. 2013).

References

Bakker PAHM, Pieterse CMJ, van Loon LC (2007) Induced systemic resistance by fluorescent *Pseudomonas* spp. Phytopathology 97:239–243

Bashan Y, de-Bashan LE, Prabhu SR et al (2014) Advances in plant growth-promoting bacterial inoculant technology: formulations and practical perspectives (1998–2013). Plant Soil 378:1–33

Borriss R (2011) Use of plant-associated *Bacillus* strains as biofertilizers and biocontrol agents. In: Maheshwari DK (ed) Bacteria in agrobiology: plant growth responses. Springer, Germany, pp 41–76

Chen XH, Koumoutsi A, Scholz R et al (2007) Comparative analysis of the complete genome sequence of the plant growth–promoting bacterium *Bacillus amyloliquefaciens* FZB42. Nat Biotechnol 25:1007–1014

Debois D, Jourdan E, Smargiasso N et al (2014) Spatiotemporal monitoring of the antibiome secreted by *Bacillus* biofilms on plant roots using MALDI Mass Spectrometry imaging. Anal Chem doi:10.1021/ac500290s

Haas D, Défago G (2005) Biological control of soil-borne pathogens by fluorescent pseudomonads. Nat Rev Microbiol 3:307–319

Kamilova F, Validov S, Azarova T et al (2005) Enrichment for enhanced competitive plant root tip colonizers selects for a new class of biocontrol bacteria. Environ Microbiol 7:1809–1817

Kloepper JW, Leong J, Teintze M et al (1980) Enhancing plant growth by siderophores produces by plant-growth-promoting rhizobacteria. Nature 286:885–886

Lugtenberg B, Kamilova F (2009) Plant growth-promoting rhizobacteria. Annu Rev Microbiol 63:541–556

Mavrodi DV, Mavrodi O, Parejko JA et al (2012) Accumulation of the antibiotic phenazine-1-carboxylic acid in the rhizosphere of dryland cereals. Appl Environ Microbiol 78:804–812

Paulsen IT, Press CM, Ravel J et al (2005) Complete genome sequence of the plant commensal Pseudomonas fluorescens Pf-5. Nat Biotechnol 23:873–878

Qiao J, Wu HJ, Huo R et al (2013) Construction of Harpin expression engineering strain FZBHarpin and evaluation of its biocontrol activity. J Nanjing Agricult Univ 36:37–44

Ramos-Gonzalez M-I, Matilla MA, Quesada JM et al (2013) Using genomics to unveil bacterial determinants of rhizosphere life style. In: de Bruijn FJ (ed) Molecular microbial ecology of the rhizosphere. Wiley-Blackwell, Hoboken, New Jersey pp 7–16

Chapter 35
Important Organizations and Companies

Ben Lugtenberg

Abstract Many people, companies and organizations are professionally involved in - or are interested in - the field of plant-microbe interactions and in the roles microbes can play in making agriculture and horticulture more sustainable. These include academic scientists, industrial professionals working in agriculture, horticulture, biotech and food industry, as well as students, teachers, government officials, decision makers and consumers who want to make themselves quickly familiar with particular aspects of this broad field. In this chapter I have listed a rather randomly chosen number of important organizations and companies.

Novozymes, Monsanto and The BioAg Alliance
www.novozymes.com; www.monsanto.com

As the global population's rapid growth is set to continue, the need to significantly increase agricultural output without increasing pressure on the environment also grows. Microbial solutions enable farmers to drive yield and productivity in a sustainable way. Deriving from various naturally-occurring microorganisms such as bacteria and fungi, these solutions can protect crops from pests and diseases and enhance plant productivity and fertility.

Microbial solutions make up approximately two thirds of the agricultural biologicals industry. Representing roughly US$ 2.3 billion in annual sales, agricultural biologicals has posted double-digit sales growth each of the last several years. There are numerous biological products currently on the market that contain microorganisms as active ingredients, including seed treatment and foliar applied products. Microbial technologies can help improve nutrient acquisition, promote growth and yield, control insects and protect against disease. These emerging agricultural biological technologies complement the integrated systems approach that is necessary in modern agriculture, bringing together breeding, biotechnology and agronomic practices to improve and protect crop yields.

In December 2013, Novozymes and Monsanto established The BioAg Alliance, with a goal to discover, develop and sell microbial solutions that enable farmers

B. Lugtenberg (✉)
Institute of Biology, Sylvius Laboratory, Leiden University,
Sylviusweg 72, 2333 BE Leiden, The Netherlands
Tel.: + 31629021472
e-mail: Ben.Lugtenberg@gmail.com

© Springer International Publishing Switzerland 2015
B. Lugtenberg (ed.), *Principles of Plant-Microbe Interactions,*
DOI 10.1007/978-3-319-08575-3_35

Applied Microbiology
_____

Biostimulants for substrates
_____

Biological Control Agents
_____

Contract Fermentation

MISSION

The agronomic solutions developed at **Agrifutur** are essential for **organic farming** and integrated farming, to reduce or eliminate the use of synthetic pesticides and along with them residues. **Agrifutur** develops and produces beneficial microorganisms. The agronomic benefits are then reflected in better food safety, environment sustainability and farmers' health.
Agrifutur is proud to contribute to sustainable agriculture, with a corporate mission that has not changed since the company was first established.

VISION

Thanks to a brave and innovative vision going back over 30 years, **Agrifutur** is finally able to make concrete what we have been dreaming of for a long time - the implementation of a type of agriculture able to grow healthy food produce with profitable sustainability. A growing number of achievements increases the commitment and enthusiasm that enables **Agrifutur** to cultivate life year after year.

RESEARCH

Applied microbiology is as useful in agriculture as it is complex in all its multiple aspects: identification, genetic stability, optimisation of microbial growth, final product formulation and application in the field. From the very beginning we have always favoured the setting up of **research networks**, in which each collaborator investigates a key aspect.

EUROPEAN PROJECT

For the past 20 years, **Agrifutur** has regularly taken part in **European and national research** projects, thanks to these it has established enduring cooperation with universities and centres of excellence in Europe and other parts of the world. **Agrifutur** coordinated the European project **BCA GRAPE** (2008-2011) "New biocontrol agents for powdery mildew on grapevine" (EU-FP7-SME). **Agrifutur** is currently involved in the new European project **DROPSA** "Strategies to develop effective, innovative and practical approaches to protect major European fruit crops from pests and pathogens" (Seventh Framework Programme KBBE.2013.1.2-04).

AGRIFUTUR srl
Via Campagnole, 8 - 25020 ALFIANELLO (Brescia) - ITALIA
Tel. +39 030 9934776 - Fax +39 030 9934777
www.agrifutur.com

supported by

TiS
innovation park

Fig. 35.1 Example of a flyer of a company which sells microbial products for plant growth promotion

worldwide to increase crop yields with less input. Novozymes brought an established product portfolio and strengths within microbial discovery, application development and fermentation to this partnership. Combined with Monsanto's highly-developed seeds and traits discovery, field-testing and extensive commercial network, the aim is to deliver a comprehensive research, development and commercial collaboration that can benefit of agriculture, consumers, the environment and society at large.

Microbial solutions provide more choice for farmers and help meet the demand for more sustainable agricultural practices. Such solutions can increase crop yields and develop a more sustainable industry impact profile, ultimately resulting in more food to feed the growing world and new opportunities to protect the planet.

BISOLBI-INTER LLC

Bisolbi-inter@rambler.ru www.bisolbi.ru

The innovative company Bisolbi-Inter was established on the basis of the All Russia Research Institute for Agricultural Microbiology (ARRIAM) in Saint-Petersburg. The company is developing and producing microbial preparations and fertilizers for agriculture, horticulture and forestry. Besides in Russia, some products have been registered in Kazakhstan, Bulgaria, Serbia, and South Africa. Registration in Australia and Turkey is in progress.

Institute of Biology Leiden

The Institute of Biology is proud and happy with this book on Plant-Microbe Interactions, a topic of major interest in our past and future. We greatly acknowledge emeritus professor Ben Lugtenberg for all the work and time that he has put into this project. We wish Jos Raaijmakers a successful professorship in our Institute and we look forward to a bright future for Plant-Microbe Interaction research at Leiden University.

International Society for Molecular Plant-Microbe Interactions

www.ismpmi.org

The International Society for Molecular Plant-Microbe Interactions (IS-MPMI) is a globally diverse organization of scientists who research molecular aspects of microorganisms interacting with plants and the consequences of such interactions. IS-MPMI provides opportunities for building, extending, and nurturing collaborations and scientific community through its congress, society communications and its journal *Molecular Plant-Microbe Interactions* (*MPMI*).

ABiTEP GmbH

www.abitep.de

ABiTEP GmbH is a German biotech company founded in 2005. We produce and distribute natural microbial products for use in agriculture and gardening as well as biological cleaning agents. Other important activities are contract production and research. In addition we are involved in various research projects developing modern and ecologically beneficial methods of plant production.

INCOTEC Group BV

www.incotec.com

INCOTEC's Coating and Seed Technology Companies around the world provide products and services for seed coating, pelleting, seed enhancements like priming, disinfection, application of actives, additives and beneficial microbes, and analytical services for genetic analysis. By providing key solutions, INCOTEC contributes significantly to the development of sustainable agriculture worldwide.

Koppert Biological Systems

info@koppert.nl

Koppert provides biological crop protection and natural pollination for professional growers worldwide in agriculture and horticulture since 1967. We make use 1f natural enemies and micro-organisms to control and prevent infestations and diseases. In addition we also supply bumblebees for the natural pollination of plants. Research, production, distribution and advice are all major activities of our company.

The International Biocontrol Manufacturers' Association (IBMA)

http://www.ibma-global.org/

President: Willem Ravensberg; email: willem.ravensberg@ibma-global.org

International Organisation for Biological Control (IOBC)

http://www.iobc-wprs.org/

IOBC was established in 1955 to promote environmentally safe methods of pest and disease control in plant protection.

Part VII
Paradigms in Plant-Microbe Interactions

Chapter 36
Trichoderma: A Multi-Purpose Tool for Integrated Pest Management

Matteo Lorito and Sheridan L. Woo

Abstract *Trichoderma* spp. are mainly known as biocontrol and beneficial microbes useful for a range of applications, from seed coating to post-harvest, from soil to foliar, and able to provide a variety of benefits by using a plethora of mechanisms. No other beneficial fungus in the agriculture field has received so much combined attention from science and the commercial market. However, as indicated from the many hundreds of related publications normally produced each year, we are far from fully understanding the potential of these incredibly successful, from an ecological point of view, bionts. This chapter briefly summarizes the main knowledge of the interactions established by agriculturally useful Trichodermas, and discusses the next future scenario of the use of these natural, multi-purpose tools.

36.1 The Multiple Interactor and Integrated Pest Management Tool

The large body of literature concerning this ubiquitous fungal genus indicates that Trichodemas are among the most active microbes found in natural environments, as they manage to modify whatever substrate they colonize and establish functional interactions with a variety of other living entities. Stable root colonization, endophytism, mycoparasitism, competition for nutrients, symbiosis, pathogenicity, antibiosis, induced resistance, seed germination and growth promotion, increased nutritional value of produce; all of these activities or properties have been fully demonstrated as affecting plants, animals, invertebrates, fungi, bacteria and viruses.

M. Lorito (✉) · S. L. Woo
Department of Agriculture, University of Naples Federico II, and Institute of Plant Protection
IPP—CNR, Via Università, 100, 80055 Portici (NA), Italy
Tel.: + 390812539376
e-mail: lorito@unina.it

S. L. Woo
Tel.: + 390812539010
e-mail: woo@unina.it

© Springer International Publishing Switzerland 2015   345
B. Lugtenberg (ed.), *Principles of Plant-Microbe Interactions*,
DOI 10.1007/978-3-319-08575-3_36

Further, the capacity of these fungi to substantially modify the biological and chemical characteristics of the colonized substrates, which includes detoxification and enrichment in organic matter, is a common knowledge.

The many thousands of publications, both scientific and divulgative, made since Weindling's first description of the fungus in the mid 1930's, have produced a variety of patents filed in dozens of countries, and activated the interest of several hundreds of small-, middle- and large-size companies. This has created a market for *Trichoderma*-based products estimated to be over 1 billion USD worldwide. Virtually, anyone working in the agriculture sector (including gardening and ornamental crops), is aware of the usefulness of these fungi, which are by the way making a particularly notable impact on crop production in developing countries (i.e. see Cumagun 2012; Ha 2010).

Scientific work on *Trichoderma* is considered a successful model of "translational research", where data obtained by using genomics, proteomics and metabolomics techniques can be effectively implemented in agricultural practices, i.e. by allowing fast selection of "elite" strains from the immense natural *Trichoderma* germoplasm (Lorito et al. 2010; Mukherjee et al. 2013; Studholme et al. 2013).

36.2 The Pathogen Killer and Inhibitor

Trichoderma strains seem to have ancestrally evolved with the genome of predators (Druzhinina et al. 2011). They are able to directly kill other microbes, and notably other soil and plant-associated fungi, and, in terms of host range, to make interkingdom jumps as well as major shifts in ecology (Chaverri and Samuels 2013). Different mycoparasitic strategies may be used (Atanasova et al. 2013), with the role of many factors still to be clarified (Ramão-Dumaresque et al. 2012) since the first demonstrations based on gene knock-out (Woo et al. 1999). Aspects involved in the process include cell wall degrading enzymes acting together with powerful sets of secondary metabolites (Schirmböck et al. 1994), of which many have been found to affect the activity of all the "The Top 10 fungal pathogens in molecular plant pathology" (Dean et al. 2012). Direct killing may occur not only on plant pathogenic microbes, although some of these appear to be preferred targets, possibly because these microorganisms are more readily encountered around the plant hosts on which many Trichodermas act basically as symbionts. Predation is not limited to fungi, since numerous strains are also proposed on the agricultural market as nematode killers (Spiegel et al. 2007). Direct inhibition or parasitism of pathogens is not the only process; also the ability to sequester nutrients and to physically exclude the pathogen from the suitable site of infection are known to be used.

36.3 The Disease Suppressor and Resistance Stimulator

Mycoparasitism has been reported to be used by over one hundred species of fungi besides *Trichoderma*. Obviously, genome studies will reveal that also for these fungi a major component is related to living as predators. However, only in the case of

Trichoderma strains such, has a large variety of beneficial effects produced on the plant been reported. In fact, the latest research reveals more and more genetic characters related to a plant symbiont lifestyle, which, in the case of strains selected for agricultural applications, may represent a more significant behaviour over mycoparasitism (Harman et al. 2004; Lorito et al. 2010; Seidl et al. 2006; Shoresh et al. 2010; Studholme et al. 2013; Vinale et al. 2008). Effects on the plant are typically divided into those enhancing resistance to disease and those promoting plant growth (see next section), although they depend on numerous interconnected or interplaying mechanisms (Shoresh and Harman 2008). The *Trichoderma*-mediated enhanced resistance is today a paradigm in plant-microbe interaction (see Vos et al. 2014 for a recent review). Many studies have demonstrated its occurrence even at the level of open field cultivation, and that this is related to the activation of either the ISR (Induced Systemic Resistance) or the SAR (Systemic Acquired Resistance) pathways, or both, depending on the conditions and the strain used. In fact, the mechanism of activation is still far from being fully clarified, with the demonstrated involvement of MAMPs (Microbe Associated Molecular Patterns) (proteins, sugars and secondary metabolites) and effectors recognized by specific plant receptors (in the case of beneficial interaction effectors; they may be considered equivalent to elicitors) as well as defense related hormones (Martinez-Medina et al. 2013). The plant response to pathogen attack in the presence of *Trichoderma* may be substantially affected, with dozens of genes differentially expressed, resulting in enhanced PTI (PAMP Triggered Immunity) and ETI (Effector Triggered Immunity) (Lorito et al. 2010). Defense priming, possibly divided in a ISR-prime and ISR-boost phase, has been reported to occur at no metabolic cost (Perazzolli et al. 2011), and is able to suppress diseases caused also by foliar pathogens (Vos et al. 2014).

Finally, there is mounting evidence that the increased resistance to abiotic stresses (mainly drought and saline) of a variety of plant species treated with *Trichoderma*, that has also been observed in the field, is based on specific mechanisms of interaction (Brotman et al. 2013; Mastouri et al. 2012).

36.4 The Plant Growth Promoter

Trichoderma spp. may be PGPMs (Plant Growth Promoting Microbes). This property is widely diffuse among the genus, but not present in every strain. The effect can be so clear and diffuse, ranging from horticultural crops to trees, that this ability is now regularly tested in newly selected agricultural strains. The molecular mechanisms supporting the observed effect are complex and related to the plant genotype (Tucci et al. 2011). They include increased nutrient availability (Yedidia et al. 2001) and/or stimulation by fungal metabolites (Vinale et al. 2008), with consequent changes in the plant hormonal profile (Contreras-Cornejo et al. 2009; Hermosa et al. 2013, Roldán et al. 2011). The transcriptional changes in the plant resulting in the combined effect of increased resistance and PGP (Plant Growth Promotion) are extensive and may be induced by using either the living fungus or some of its secreted metabolites. The recent appearance on the market of new strains specifically selected for biofertilizer activity has further driven the widespread use of *Trichoderma*, although this has raised legal issues about registration for use that need to be resolved.

36.5 The Source of Useful Compounds

The secretome of many *Trichoderma* strains used in agriculture is a rich source of bioactive compounds. Over 250 different secondary metabolites and dozens of different enzymes are known to be produced, although only a limited number of them are normally associated to a single strain. Cell wall degrading enzymes, peptides like peptaibols, water soluble metabolites such as harzianic acid and its derivatives, volatile compounds such as 6-pentyl- α-pyrone and isocyanide derivates, proteins such as hydrophobins and swollenins etc., have all been demonstrated to strongly affect plant growth and/or resistance to biotic/abiotic stresses when applied in the absence of the living fungus (Keswani et al. 2014; Mukherjee et al. 2012; Vinale et al. 2012). Evidence is accumulating on the role of many of these compounds in the symbiotic association with plant roots, which further supports the concept that most Trichodermas selected for agricultural applications have evolved from a main ancestral mycoparasitic lifestyle to that of a plant beneficial colonizer. The direct use of *Trichoderma* metabolites, in alternative or combination with the living fungus or other beneficials, is a declared future target of several R&D programs in the public and private sectors worldwide. The usefulness of the rich *Trichoderma* genome was first demonstrated by the transgenic use of genes encoding cell wall degrading enzymes, with several papers published after the first report (Lorito et al. 1998), and this will be further exploited following the now common practice to sequence the genome of commercially important strains.

36.6 The Soil Cleaner and Enricher

A recently highlighted application of *Trichoderma* concerns the use of strains as a soil cleaner and/or soil enricher. Many reports propose selected Trichodermas as bioremediators of soils polluted with arsenic, cyanide, hydrocarbons—i.e. crude oil, naphthalene, phenanthrene, benzo[α]pyrene—pesticides, heavy metals—i.e. iron, lead, copper, manganese, zinc—phenols etc. (Tripathi et al. 2013). On the other hand, the ability of *Trichoderma* to simply enrich the quality of soils and composts, as a saprophyte, decomposer and a major component of a well balanced microflora, has been known for a long time (Bernard et al. 2012). This has made *Trichoderma* one of the key microbial ingredients of commercial composts or products that help composts to mature, based on the evidence that a high level of *Trichoderma* is also a positive indicator of both disease suppressiveness and soil fertility.

36.7 The Commercially Successful Microbe

Trichoderma spp. and *Bacillus* spp. could be considered today as the most used IPM biological tools available for the agriculture industry for infectious disease control, although other BCAs (Biological Control Agents) and PGPMs are also widely applied in the real world for growing food and non-food crops. However, if one considers the diversity of potential uses, *Trichoderma* appears as a multi-purpose "Swiss army knife" compared to other single-use microbial products. The number of commercial products, microbe combinations, formulations, claims and promises (some realistic, others not) based on *Trichoderma*, distributed in over 100 countries worldwide, are endless and ever increasing. India alone may have over 300 products on the market. Regardless, a significant positive impact has already been achieved, especially on food crop production in some developing countries. For instance, there is virtually no medium-to-large size farm south of California and Texas, all the way to Chile, that is not using or has not used *Trichoderma*; often combined with other beneficial biologicals as well as using low-impact agronomic practices, in an attempt to cut costs on pesticides and fertilizers. In countries such as Cuba and Venezuela, the use of *Trichoderma* and other "insumos biológicos" is a government promoted and supported agricultural practice (Harman et al. 2010). Actually, it is becoming a common procedure for larger farms to acquire the necessary technological know-how and to grow by themselves on-site *Trichoderma* and other BCAs in fermentors for direct application of freshly-made cultures to crops (Fig. 36.1). For instance, it has been calculated that one out of four melons or pineapples consumed in US or Europe has been treated with *Trichoderma* alone or in combination with other BCAs. The applications range includes use in the nursery, seed coating, greenhouse, open field, soil-less systems, post-harvest, as well as directly on tree trunk and wounds.

36.8 The IPM Tool of the Future

The road taken by many countries toward IPM implementation is accelerating the development and use of "non-chemical methods of plant protection and pest and crop management". For instance, European Directive 128/2009 and the Regulation 1107/2009 impose the use of IPM practices to all 28 EU Countries by 2014, and determined the loss of registration for an estimated 40 % of the available synthetic pesticides. This will permit *Trichoderma*—and other BCAs—based products to finally hatch from their niche market share, and boost related R&D activities, with the subsequent expected innovations appearing already in the next few years.

New "multi-action" strains will be selected by using genome era-generated information; they will be able to kill pathogens, produce large amount of propagules, act as PGPMs, increase resistance against a variety of pathogens—microbes, insects (Battaglia et al. 2013), viruses and abiotic stresses, enhance nutrient use efficiency and soil fertility, degrade pollutants in the soil, and act compatibly with other BCAs. These strains will have a limited efficacy determined by cultivar/crop-associated

Fig. 36.1 Successful performance of a *Trichoderma* plus *Bacillus* integrated application made on a large scale in the field. The picture on the left, shows a 5000 ha production of melons (cv. Gallia and Cantaloupe) on a farm located in south Honduras for export to the USA and the EU. The plants treated with a biological product (rows on the left of field), based on an advanced technology (use of "elite" strains selected on the basis of 'omics generated information and grown directly on-farm by a specifically developed method), performed better or as well as the conventional chemically-based treatment (rows on the right). Note that the vegetation of rows on the left side of field is more abundant of, or at least equivalent to, that of rows on the right. The regular daily application of fungicides, antibiotics, insecticides and nematocides was completely interrupted during the last half of the 60 day crop cycle, and substituted with fresh cultures of the two BCAs plus an occasional spray of botanicals to control mildew, thus permitting a more than 50 % reduction of the total chemical input. An average of approximately 15–20 % yield increase and complete lack of chemical residues on the fruits have now been consistently obtained in the past 5 years. Cultures of the different BCAs are produced in bioreactors built on site (pictures on the right taken in Honduras and Cost Rica), then often directly connected to the irrigation system (Pictures taken by M. Lorito)

variability. All of these properties are already considered now in modern screening programs and could be combined with a few strain mixtures or in new hybrid strains obtained by fusing isolates with different and complementary beneficial traits. The presence and activity of the BCAs will be easily tracked by using newly developed detection kits.

New "exotic" strains will be discovered and selected for their activity against specific stresses, their particular effect on individual crops, their enhanced compatibility with some agrochemicals, their adaptability to global environmental changes, or their ability to target new invasive pathogens.

New strains will be naturally selected or produced by hybridization for dedicated applications to staple or particularly important crops. The ability to positively respond to treatment with *Trichoderma* and other BCAs will be used by plant breeders for the production of new cultivars that will fully benefit from this symbiotic interaction.

A plethora of new formulations, mostly in a liquid form, facilitating the application without the negative impact of drying on propagule viability, will appear. Some of these will be specifically dedicated to hydroponics, forestry and seed treatment.

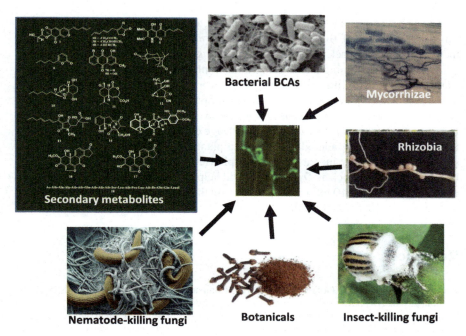

Fig. 36.2 Examples of useful combinations of *Trichoderma* (central picture) with other beneficials. Some of these mixtures are already available as commercial formulations (i.e. *Trichoderma* + Rhizobia or *Trichoderma* + Mycorrhizae), whereas other co-applications are *de facto* obtained by the concurrent use of different products. The increasing understanding of the interaction mechanisms and the evaluation of the effects *in vivo* will provide new opportunities for developing combined products and novel IPM solutions. Pictures taken by the authors or available from Internet were used

New biopesticides and biofertilizers based on *Trichoderma* metabolites, both secondary and proteins, will be used in alternative or in combination with the living fungus, similar to the case of Bt and Bt toxin. The advantages include: both soil and foliar pathogen inhibition, positive effects on the plant obtained by using very low doses, more precise dose-response correlation, low susceptibility to environmental factors, full compatibilty with agrochemicals, excellent shelf life, easy integration with current agricultural practices, endless possibilities of discovering new synergistic mixtures, etc.

Synergistic combinations of *Trichoderma* and other beneficials (including Rhizobia and Mycorrhizae), plus metabolites, botanicals, inorganics, seaweeds, chitosans, animal-derived products, etc. with extended activity against diseases caused by microbes, insects and nematodes, will appear on the market. These new multi-agent formulations will be constructed, differently from the actual products, on a solid scientific basis and on a deep knowledge of the interaction mechanisms involved (Fig. 36.2).

The opening of new registration pipelines, that may accommodate the multi-function properties of BCAs such as *Trichoderma*, may be expected; for instance there is not a single track to register a strain acting both as a biofertilizer and biopesticide, which is becoming a constraint in the implementation of new good products. Many of these possibilities may already become a reality in the near future, while the use of genetically engineered *Trichoderma*, although very promising and appealing, appears more distant ahead. The technology is available and already commonly used in many laboratories worldwide. The knowledge on the role of specific genetic traits that could be usefully altered, either in the plant or in the fungus, has never been so great and is rapidly increasing. To paraphrase a famous quote from N. Bourlag, beneficial microbes such as *Trichoderma* will help us to realize the necessary miracle of feeding a world of 10 billion people in a way that is sustainable for our ecosystem and survival on the planet Earth.

References

Atanasova L, Le Crom S, Gruber S et al (2013) Comparative transcriptomics reveals different strategies of *Trichoderma* mycoparasitism. BMC Genomics 14:121

Battaglia D, Bossi S, Cascone P et al (2013) Tomato below ground–above ground interactions: *Trichoderma longibrachiatum* affects the performance of *Macrosiphum euphorbiae* and its natural antagonists. Mol Plant Microbe Interact 26:1249–1256

Bernard E, Larkin RP, Tavantzis S et al (2012) Compost, rapeseed rotation, and biocontrol agents significantly impact soil microbial communities in organic and conventional potato production systems. Appl Soil Ecol 52:29–41

Brotman Y, Landau U, Cuadros-Inostroza A et al (2013) *Trichoderma*-plant root colonization: escaping early plant defense responses and activationof the antioxidant machinery for saline stress tolerance. PLoS Pathog 9(3)

Chaverri P, Samuels GJ (2013) Evolution of host affiliation and substrate preference in the cosmopolitan fungal genus *Trichoderma* with evidence of interkingdom host jumps. Evolution 67:2823–2837

Contreras-Cornejo HA, Macías-Rodríguez L, Cortés-Penagos C et al (2009) *Trichoderma virens*, a plant beneficial fungus, enhances root biomass production and promotes lateral root growth through an auxin-dependent mechanism in Arabidopsis. Plant Physiol 149:1579–1592

Cumagun CJR (2012) Managing plant diseases and promoting sustainability and productivity with *Trichoderma*: the Philippine experience. J Agric Sci Technol 14:699–714

Dean R, Van Kan JAL, Pretorius ZA et al (2012) The Top 10 fungal pathogens in molecular plant pathology. Mol Plant Pathol 13:414–430

Druzhinina IS, Seidl-Seiboth V, Herrera-Estrella A et al (2011) *Trichoderma*: the genomics of opportunistic success. Nat Rev Microbiol 9:749–759

Ha TN (2010) Using *Trichoderma* species for biological control of plant pathogens in Viet Nam. J Issaas 16:17–21

Harman GE, Howell CR, Viterbo A et al (2004) Trichoderma species—opportunistic, avirulent plant symbionts. Nat Rev Microbiol 2:43–56

Harman GE, Obregón MA, Samuels GJ et al (2010) Changing models for commercialization and implementation of biocontrol in the developing and developed world. Plant Dis 94:928–938

Hermosa R, Belen Rubio M, Cardoza RE et al (2013) The contribution of *Trichoderma* to balancing the costs of plant growth and defense. Int Microbiol 16:69–80

Keswani C, Mishra S, Sarma B et al (2014) Unraveling the efficient applications of secondary metabolites of various *Trichoderma* spp. Appl Microbiol Biotechnol 98:533–544

Lorito M, Woo SL, Garcia FI et al (1998) Genes from mycoparasitic fungi as a source for improving plant resistance to fungal pathogens. PNAS 95:7860–7865

Lorito M, Woo SL, Harman GE et al (2010) Translational research on *Trichoderma*: from 'omics to the field. Annu Rev Phytopathol 48:395–417

Martinez-Medina A, Fernandez I, Sánchez-Guzmán MJ et al (2013) Deciphering the hormonal signalling network behind the systemic resistance induced by *Trichoderma harzianum in* tomato. Front Plant Sci 4:206

Mastouri F, Björkman T, Harman GE (2012) *Trichoderma harzianum* enhances antioxidant defense of tomato seedlings and resistance to water defecit. Mol Plant Microbe Interact 9:1264–1271

Mukherjee PK, Horwitz BA, Herrera-Estrella A et al (2013) *Trichoderma* research in the genome era. Annu Rev Phytopathol 51:105–129

Mukherjee PK, Horwitz BA, Kenerley CM (2012) Secondary metabolism in *Trichoderma*—a genomic perspective. Microbiology 158:35–45

Perazzolli M, Roatti B, Bozza E et al (2011) *Trichoderma harzianum* T39 induces resistanceagainst downy mildew by priming for defence without costs for grapevine. Biol Control 58:74–82

Ramão-Dumaresque AS, de Araújo AS, Talbot NJ et al (2012) RNA interference of endochitinases in sugarcane endophyte *Trichoderma virens* 223 reduces its fitness as a biocontrol agent of pineapple disease. PLoS One 7(10):e47888

Roldán A, Albacete P, Jose A (2011) The interaction with arbuscular mycorrhizal fungi or *Trichoderma harzianum* alters the shoot hormonal profile in melon plants. Phytochemistry 72:223–229

Schirmböck M, Lorito M, Wang Y et al (1994) Parallel formation and synergism of hydrolytic enzymes and peptaibol antibiotics: molecular mechanisms involved in the antagonistic action of *Trichoderma harzianum* against phytopathogenic fungi. Appl Environ Microbiol 60:4364–4370

Seidl V, Marchetti M, Schandl R et al (2006) Epl1, the major secreted proteinof *Hypocrea atroviridis* on glucose, is a member of a strongly conserved protein family comprising plant defence response elicitors. FEBS J 273:4346–4359

Shoresh M, Harman GE (2008) The relationship between increased growth and resistance induced in plants by root colonizing microbes. Plant Signaling Behav 3:737–739

Shoresh M, Harman GE, Mastouri F (2010) Induced systemic resistance and plant responses to fungal biocontrol agents. Annu Rev Phytopathol 48:21–43

Spiegel Y, Sharon E, Bar-Eyal M et al (2007) Evaluation and mode of action of *Trichoderma* isolates as biocontrol agents against plant-parasitic nematodes. Bulletin OILB/SROP 30:25

Studholme D, Harris J, Le Cocq B et al (2013) Investigating the beneficial traits of *Trichoderma hamatum* GD12 for sustainable agriculture—insights from genomics. Front Plant Sci 4:258

Tripathi P, Singh P, Mishra A et al (2013) *Trichoderma*: a potential bioremediator for environmental clean up. Clean Techn Environ Policy 4:1–10

Tucci M, Ruocco M, De Masi L et al (2011) The beneficial effect of *Trichoderma* spp. on tomato is modulated by the plant genotype. Mol Plant Pathol 12:341–354

Vinale F, Sivasithamparam K, Ghisalberti EL et al (2008) A novel role for *Trichoderma* secondary metabolites in the interactions with plants. Physiol Mol Plant Pathol 72:80–86

Vinale F, Sivasithamparam K, Ghisalberti EL et al (2012) Trichoderma secondary metabolites that affect plant metabolism. Natural Product Commun 7:1545–1550

Vos C, De Cremer K, Cammue B et al (2014) The toolbox of *Trichoderma* spp. in biocontrol of *Botrytis cinerea* disease. Mol Plant Pathol (in press)

Woo SL, Donzelli B, Scala F et al (1999) Disruption of the *ech42* (endochitinase-encoding) gene affects biocontrol activity in *Trichoderma harzianum* P1. Mol Plant Microbe Interact 12:419–429

Yedidia I, Srivastva AK, Kapulnik Y et al (2001) Effect of *Trichoderma harzianum* on microelement concentrations and increased growth of cucumber plants. Plant Soil 235:235–242

Chapter 37
Agrobacterium, The Genetic Engineer

Paul J. J. Hooykaas

Abstract Agrobacteria are common soil and rhizosphere bacteria. Most strains are saprophytes, but strains harboring a tumor inducing plasmid (Ti plasmid) are pathogenic and can induce tumors on plants, called crown galls. The disease may lead to growth retardation and eventually the death of the host plant and thus can cause severe damage in horticulture. Crown galls form a favorable niche for *Agrobacterium* as they produce specific chemicals called opines which the bacteria can use for growth. Nowadays, *Agrobacterium tumefaciens* is best known as a natural genetic engineer, which is based on the molecular mechanism which it employs to induce crown gall. This involves the transfer of an oncogenic segment of the Ti plasmid (the T-DNA) to plant cells and its stable maintenance as part of one of the plant chromosomes. Expression of genes on the T-DNA is responsible for the formation of a tumor.

37.1 Crown Gall Disease

The soil bacterium *Agrobacterium tumefaciens* is the causative agent of crown gall. This plant disease is characterized by the formation of tumorous overgrowths at wound sites on plants (Fig. 37.1). In nature tumors are formed often at the root crown, hence the name crown gall. However, the bacterium can induce tumors effectively also on stems, roots or leaves, when these are wounded. The host range is broad: tumors are formed on many dicotyledonous plant species, but not on monocots. The disease causes severe damage in horticulture. Effective biological control has been developed based on non-tumorigenic *Agrobacterium* strains producing bacteriocins (such as agrocin 84) that selectively kill tumor inducing strains.

Crown gall cells are tumorous as they are able to grow in *in vitro* culture in the absence of the plant growth regulators auxin and cytokinin, in contrast to normal plant cells. In the tumors specific chemical compounds are present that are characteristic

---

P. J. J. Hooykaas (✉)
Institute of Biology, Sylvius Laboratory, Leiden University,
PO Box 9505, 2300 RA Leiden, The Netherlands
Tel.: + 31 71 527 4933
e-mail: p.j.j.hooykaas@biology.leidenuniv.nl

© Springer International Publishing Switzerland 2015
B. Lugtenberg (ed.), *Principles of Plant-Microbe Interactions*,
DOI 10.1007/978-3-319-08575-3_37

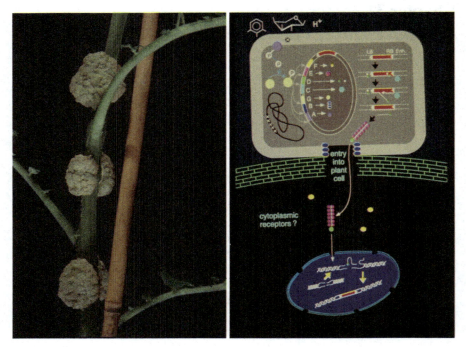

Fig. 37.1 Crown gall tumor formation at wound sites on a potato plant (*left*) and the molecular mechanism underlying T-DNA transfer into plant cells (*right*). Pictures were taken from pages C1 (made by dr HRM Schlaman) and C3 in The Rhizobiaceae. Molecular biology of Model Plant-Associated Bacteria (1998). Reproduced by permission of the publisher

for the tumors and not found in normal plant cells or tissues. These compounds, with the generic name of opines, are condensation products of amino acids and keto acids such as pyruvate and α-keto-glutaric acid or sugars. Tumor inducing agrobacteria, but almost no other bacteria, are able to catabolize the opines. The tumors therefore form a favorable niche for *Agrobacterium*.

The bacterium can only infect plants at wound sites, where it can penetrate the plant and attach to plant cells. Subsequently, the bacterium introduces part of its own DNA into these plant cells. The transferred DNA (T-DNA) is derived from a large plasmid called the Ti (tumor inducing) plasmid, which determines the virulence of the bacterium. The T-DNA is responsible for the tumorous character of the crown gall cells and for its ability to synthesize opines. This is because the T-DNA contains a set of genes that is expressed in plant cells, including genes that mediate the synthesis of the plant growth regulators auxin and cytokinin and genes that mediate the production of the opines.

Agrobacteria belong to the α-Proteobacteria and are most related to the nitrogen fixing bacteria of the genus *Rhizobium*. Initially classification was based on the virulence properties: *A. tumefaciens* for tumor inducing strains, *A. rhizogenes* for root inducing strains and *A. radiobacter* for avirulent strains. After it became clear that the virulence properties were determined by a transmissible plasmid this classification was abandoned.

Below I shall provide more detailed information on *Agrobacterium* and the molecular mechanism of tumor formation. For further information the reader is referred to the books edited by Kahl and Schell (1982); Spaink et al. (1998); Nester et al. (2005); Tzfira and Citovsky (2008). Also the following reviews give a detailed account of several aspects of the system, briefly discussed in this chapter: Bevan and Chilton (1982); Binns and Thomashow (1988); Braun (1978); Dessaux et al. (1993); Gelvin (2000); Hooykaas and Beijersbergen (1994); Morris (1986); Nester et al. (1984).

37.2   The Ti Plasmid

Ti plasmids generally have a size of about 200,000 base pairs. The T-DNA only represents a minor part of it and has a size of about 20,000 base pairs. In some Ti plasmids the T-DNA is divided over two segments, which are independently transferred to plant cells. Only one of such T-DNAs contains the genes provoking autonomous cell proliferation, while the other harbors genes for opine synthesis. Genes located elsewhere on the Ti plasmid confer on *Agrobacterium* the ability to catabolize the opines which are produced in the tumors. Moreover, the Ti plasmids are conjugative plasmids which are mobile especially in plant tumors. This is because not only the opine catabolic genes, but also the transfer genes are activated by the presence of opines. Transfer is also under the control of quorum sensing (see Chap. 7). N-acyl-homoserine lactones (AHLs) are produced by the TraI protein located on the Ti plasmid. These AHLs bind to the transcriptional regulator TraR, which drives expression of the *tra* genes. The *traR* gene itself is under control of a transcriptional regulator which reacts to the presence of its cognate opine and thus conjugation is controlled by opines and occurs *in vivo* specifically in the plant tumors (See for a recent review: Venturi and Fuqua 2013). Besides the T-DNA the Ti plasmid contains another set of genes which are involved in virulence, the virulence (*vir*) genes, which will be discussed below.

37.3   The Virulence Genes

The translocation of T-DNA into plant cells is brought about by a delivery system which is encoded by genes located in the virulence region of the Ti plasmid (Fig. 37.1). The virulence region embraces about 30 genes distributed over a number of operons (Zhu et al. 2000).

Regulation This *vir* regulon is controlled by a 2-component regulatory system consisting of a receptor histidine kinase called VirA and an associated transcriptional regulator called VirG (Winans 1992). The virulence system is only activated in nature when the bacterium encounters plants and more specifically plant wound sites (Stachel et al. 1985). Phenolic compounds which are released at plant wound sites are key inducers of the *vir* genes. They are characterized by the presence of a methoxy group adjacent to the hydroxy group of the phenolic backbone and include well known lignin precursors and breakdown products such as coniferyl alcohol, sinapinic acid, syringaldehyde and acetosyringone (Melchers et al. 1989). The presence of particular sugars such as glucose and galactose, glucuronic acid and galacturonic acid enhances induction. Such sugars bind to a periplasmically located sugar binding protein called ChvE, which in turn interacts with the periplasmic loop of VirA (Cangelosi et al. 1990). Also a low pH (around 5.5) and a temperature below 30 °C are required for *vir*-induction. All the biochemical and biophysical signals are perceived by the VirA receptor, which becomes auto-phosphorylated in the presence of inducing signals. VirA subsequently phosphorylates VirG, which then is able to bind to the *vir*-boxes, specific sequences present in the promoters of the *vir*-genes (Jin et al. 1990). This leads to transcription of the other *vir*-genes.

T-DNA Processing After activation of the *vir*-genes T-DNA processing takes place in the cells. Single strand (ss) DNA breaks are present at the border repeats, which are 25 base pairs sequences that surround the T-DNA, and single stranded copies of the T-DNA (called T-strands) can be isolated from induced bacteria (Stachel et al. 1986). Thus T-DNA transfer resembles the process of bacterial conjugation, by which ssDNA molecules are transferred into recipient cells. In the presence of *vir*-inducers the T-DNA delivery system can indeed mediate the mobilization of *incQ* plasmids between bacteria (Beijersbergen et al. 1992).

The protein that is responsible for the border nicking and the formation of the T-strands is a relaxase called VirD2 (Gelvin 2000). Accessory factors including VirD1, VirC1 and VirC2 are also needed, but their precise role has not been defined as yet. The VirC1 protein binds to a DNA sequence called overdrive, which is located close to the so-called right border repeat from where the production of the T-strands start, and enhances border nicking (Toro et al. 1989). The mobilization of the CloDF13 and *incQ* plasmids by the *vir*-system does not require these *virC* and *virD* functions. These plasmids use their own relaxase system for nicking at the cognate *oriT* sequence and the formation of T-strands (Escudero et al. 2003).

T4SS For translocation of the T-strand into host cells *Agrobacterium* uses a Type4 Secretion System (T4SS) which is encoded by the 11 genes of the *virB* operon (Alvarez-Martinez und Christie 2009). The T4SS forms a gateway for the T-strand from the cytoplasm over the inner and outer membrane and through a large T-pilus structure at the surface of the bacterial cell made up of VirB2 monomers, eventually into the host cell cytoplasm. How the T-pilus contacts host cells and mediates delivery of the T-strand is still unknown. For entrance into the T4SS a coupling protein is required, which is encoded by the *virD4* gene. This large hexameric membrane protein recognizes the substrate and mediates delivery to the first protein of the T4SS (Cascales et al. 2013).

Effector Proteins A number of the remaining virulence proteins such as VirE2 turned out to have a function in the plant host cells. The VirE2 protein is a ssDNA binding protein, which binds to the T-strand in the plant cell to protect it against nucleases and to assist in nuclear delivery (Duckely und Hohn 2003). In the absence of VirE2, transformation is 3–4 logs reduced and the T-DNA is truncated.

The VirF protein was the first bacterial F-box protein discovered. F-box proteins are abundant in eukaryotic cells and involved in targeted proteolysis of specific proteins. Expression of VirF in plant cells could complement *virF* mutants for tumorigenesis. This suggested that VirF was delivered into plant cells by the T4SS and the fact that VirF is not a ssDNA binding protein suggested that delivery of these proteins was independent of that of the T-strand (Regensburg-Tuink and Hooykaas 1993). Direct evidence for independent protein delivery by the *Agrobacterium* T4SS into plant cells was obtained by the CRAfT (Cre Reporter Assay for Translocation) assay, which revealed that some virulence proteins (nowadays also called effector proteins) can be delivered into plant cells by bacterial strains, even by those with a Ti plasmid from which the T-DNA has been removed (Vergunst et al. 2000). The CRAfT assay also allowed the identification of so far unknown virulence effector proteins, including VirE3 and VirD5 (Vergunst et al. 2005). The effector proteins all have a specific 30 amino acids C-terminal, arginine rich transport sequence, which is essential for translocation by the T4SS.

T-DNA transfer Interestingly, the VirD2 protein, which upon nicking remains covalently bound to the T-strand, also has C-terminal sequences outside of its N-terminal relaxase domain which mediate translocation into plant cells. Deletion of these sequences from VirD2 prevents translocation and leads to avirulence. Addition of the C-terminal sequences of one of the effector proteins to such a C-terminally deleted VirD2 protein leads to the restoration of transport and virulence (van Kregten et al. 2009). This suggests that the T4SS mediating T-DNA translocation is in fact evolutionary derived from a protein translocation system, but adapted in such a way that it allows also the translocation of DNA molecules that are covalently attached to proteins which are recognized as substrates of the system.

The T-DNA delivery system is very efficient. When using single cells from a plant (plant protoplasts) in *in vitro* culture up to 40 % of these cells can become transformed during co-cultivation with *Agrobacterium*. This efficiency can be explained by the protection of the delivered DNA molecules by VirE2, and by their immediate targeting to the nucleus. The latter is due to nuclear localization sequences present in VirD2, which mediate binding to α-importins of the host cells (Sheng and Citovsky 1996). The VirE2 protein assists in nuclear targeting as its binding to the T-strand prevents knot formation resulting in a long thin thread that can enter the nuclear pore. Also VirE2 interacts with a host protein called VIP1 (Tzfira et al. 2000), which is a transcription factor involved in the defense response. Upon infection VIP1 is phosphorylated and subsequently moves into the nucleus to activate defense genes. By binding to the VirE2 proteins that coat the T-strand, VIP1 may enhance the nuclear targeting of the long T-complex. *Agrobacterium* apparently has adopted a Trojan horse strategy by abusing defense signaling to bring about transformation (Djamei et al. 2007).

37.4 Applications

The *Agrobacterium* DNA delivery system has become the prime system for the genetic modification of plants. This is due to the low cost involved as there is no expensive apparatus required such as with other delivery methods (particle gun, electroporator), but even more so because of the relative precision of the system. As compared to other systems in general a lower number of DNA copies are integrated in the genome (reducing gene silencing problems) and mostly these copies are intact. The host range for which the system can be used is extremely broad. Not only plants, but also yeasts and fungi turned out to be a host for *Agrobacterium*-mediated transformation (Bundock et al. 1995; de Groot et al. 1998). Nowadays, the *Agrobacterium* system is also the prime transformation system for many fungi (Michielse et al. 2005).

References

Alvarez-Martinez CE, Christie PJ (2009) Biological diversity of prokaryotic type IV secretion systems. Microbiol Mol Biol Rev 73:775–808

Beijersbergen A, den Dulk-Ras A, Schilperoort RA et al (1992) Conjugative transfer by the virulence system of *Agrobacterium tumefaciens*. Science 256:1324–1327

Bevan MW, Chilton MD (1982) T-DNA of the *Agrobacterium* Ti and Ri Plasmids. Annu Rev Genet 16:357–384

Binns AN, Thomashow MF (1988) Cell biology of Agrobacterium infection and transformation of plants. Annu Rev Microbiol 42:575–606

Braun AC (1978) Plant tumors. Biochim Biophys Acta 516:167–191

Bundock P, den Dulk-Ras A, Beijersbergen A et al (1995) Trans-kingdom T-DNA transfer from *Agrobacterium tumefaciens* to *Saccharomyces cerevisiae*. EMBO J 14:3206–3214

Cangelosi GA, Ankenbauer RG, Nester EW (1990) Sugars induce the Agrobacterium virulence genes through a periplasmic binding protein and a transmembrane signal protein.Proc Natl Acad Sci U S A 87:6708–6712

Cascales E, Atmakuri K, Sarkar MK et al (2013) DNA substrate-induced activation of the Agrobacterium VirB/VirD4 type IV secretion system. J Bacteriol 195:2691–2704

De Groot MJA, Bundock P, Hooykaas PJJ et al (1998) *Agrobacterium tumefaciens*-mediated transformation of filamentous fungi. Nat Biotechnol 16:839–842

Dessaux Y, Petit A, Tempé J (1993) Chemistry and biochemistry of opines, chemical mediators of parasitism. Phytochem 34:31–38

Djamei A, Pitzschke A, Nakagami H et al (2007) Trojan horse strategy in Agrobacterium transformation: abusing MAPK defense signaling. Science 318:453–456

Duckely M, Hohn B (2003) The VirE2 protein of *Agrobacterium tumefaciens*: the Yin and Yang of T-DNA transfer. FEMS Microbiol Lett 223:1–6

Escudero J, den Dulk-Ras A, Regensburg-Tuink TJG et al (2003) VirD4-independent transformation by CloDF13 evidences an unknown factor required for the genetic colonization of plants via Agrobacterium. Mol Microbiol 47:891–901

Gelvin SB (2000) Agrobacterium and plant genes involved in T-DNA transfer and integration. Annu Rev Plant Physiol Plant Mol Biol 51:223–256

Hooykaas PJJ, Beijersbergen AGM (1994) The virulence system of *Agrobacterium tumefaciens*. Annu Rev Phytopathol 32:157–179

Jin S, Roitsch T, Christie PJ et al (1990) The regulatory VirG protein specifically binds to a cis-acting regulatory sequence involved in transcriptional activation of *Agrobacterium tumefaciens* virulence genes. J Bacteriol 172:531–537

Kahl G, Schell JS (1982) Molecular biology of plant tumors. Academic, New York

Melchers LS, Regensburg-Tuïnk AJG, Schilperoort RA et al (1989) Specificity of signal molecules on the activation of Agrobacterium virulence gene expression. Mol Microbiol 3:969–977

Michielse CB, Hooykaas PJ, van den Hondel CA et al (2005) Agrobacterium-mediated transformation as a tool for functional genomics in fungi. Curr Genet 48:1–17

Morris RO (1986) Genes specifying auxin and cytokinin biosynthesis in phytopathogens. Annu Rev Plant Physiol 37:509–538

Nester EW, Gordon MP, Amasino RM et al (1984) Crown gall: a molecular and physiological analysis. Annu Rev Plant Physiol 35:387–413

Nester E, Gordon MP, Kerr A (2005) Agrobacterium tumefaciens. From plant pathology to biotechnology. APS, St Paul

Regensburg-Tuïnk AJG, Hooykaas PJJ (1993) Transgenic *N.glauca* plants expressing bacterial virulence gene virF are converted into hosts for nopaline strains of *A.tumefaciens*. Nature 363:69–71

Sheng J, Citovsky V (1996) Agrobacterium-plant cell DNA transport: have virulence proteins, will travel. Plant Cell 8:1699–1710

Spaink HP, Kondorosi A, Hooykaas PJJ (1998) The Rhizobiaceae. Molecular biology of model plant-associated bacteria. Kluwer, Dordrecht

Stachel SE, Messens E, van Montagu M et al (1985) Identification of the signal molecules produced by wounded plant cells that activate T-DNA transfer in *Agrobacterium tumefaciens*. Nature 318:624–629

Stachel SE, Timmerman B, Zambryski P (1986) Generation of single-stranded T-DNA molecules during the initial stages of T-DNA transfer from *Agrobacterium tumefaciens* to plant cells. Nature 322:706–712

Toro N, Datta A, Carmi OA et al (1989) The *Agrobacterium tumefaciens virC1* gene product binds to overdrive, a T-DNA transfer enhancer. J Bacteriol 171:6845–6849

Tzfira T, Citovsky V (2008) Agrobacterium. From biology to biotechnology. Springer, New York

Tzfira T, Rhee Y, Chen MH et al (2000) Nucleic acid transport in plant-microbe interactions: the molecules that walk through the walls. Annu Rev Microbiol 54:187–219

van Kregten M, Lindhout BI, Hooykaas PJJ et al (2009) Agrobacterium-mediated T-DNA transfer and integration by minimal VirD2 consisting of the relaxase domain and a type IV secretion system translocation signal. Mol Plant-Microbe Interact 22:1356–1365

Venturi V, Fuqua C (2013) Chemical signaling between plants and plant-pathogenic bacteria. Annu Rev Phytopathol 51:17–37

Vergunst AC, Schrammeijer B, den Dulk-Ras A et al (2000) VirB/D4-dependent protein translocation from Agrobacterium into plant cells. Science 290:979–982

Vergunst AC, van Lier MCM, den Dulk-Ras A et al (2005) Positive charge is an important feature of the C-terminal transport signal of the VirB/D4-translocated proteins of Agrobacterium. Proc Natl Acad Sci U S A 102:832–837

Winans SC (1992) Two-way chemical signaling in Agrobacterium -plant interactions. Microbiol Rev 56:12–31

Zhu J, Oger PM, Schrammeijer B et al (2000) The bases of crown gall tumorigenesis. J Bacteriol 182:3885–3889

Chapter 38
Take-All Decline and Beneficial Pseudomonads

David M. Weller

Abstract Crops lack resistance to many soil borne pathogens and rely on antagonistic microbes recruited from the soil microbiome to protect their roots. Disease-suppressive soils, the best examples of microbial-based defense, are soils in which a pathogen does not establish or persist, establishes but causes little or no disease, or establishes and causes disease at first but then the disease declines with successive cropping of a susceptible host. Take-all decline (TAD) controls take-all disease of wheat caused by *Gaeumannomyces graminis* var. *tritici*. TAD is a spontaneous reduction in the incidence and severity of take-all occurring with monoculture of wheat or barley following a severe disease outbreak. TAD suppressiveness is transferable, eliminated by soil pasteurization, and reduced by growing non-host crops. It results from the build-up of populations of 2,4-diacetytlphloglucinol (DAPG)-producing *Pseudomonas* spp. to a threshold density of at least 10^5 CFU g^{-1} root. TAD protects wheat against take-all on millions of hectares worldwide.

38.1 Defense of Plant Roots Against Pathogens

The rhizosphere is enriched and supported by rhizodeposition, leading to the loss of as much as 21 % of the plant's photosynthetically fixed carbon from the roots (Chap. 3).These nutrients select for a complex rhizosphere microbiome that contributes to important plant functions like growth, vigor and defense (Pieterse et al. 2014).

Microbial-Based Plant Defense Beneficial rhizosphere microorganisms that directly promote plant growth or defend roots against pathogens are recruited from the bulk soil. Although the soil type is the most important factor determining the composition of the rhizosphere microbial community, the plant also plays a key role in shaping the microbiome and in recruiting and enriching beneficial microorganisms

D. M. Weller (✉)
United States Department of Agriculture-Agricultural Research Service,
Root Disease and Biological Control Research Service, Pullman,
Washington 99164-6430, USA
Tel.: +1-509-335-6210
e-mail: David.Weller@ars.usda.gov

© Springer International Publishing Switzerland 2015
B. Lugtenberg (ed.), *Principles of Plant-Microbe Interactions*,
DOI 10.1007/978-3-319-08575-3_38

because the quantity and quality of root exudates are genetically regulated by the plant. Unlike for many foliar diseases, crop plants lack resistance to some of the most common soil borne fungal pathogens and many species of pathogenic nematodes. As a result, roots rely on microbial-based defenses consisting of microbial "defenders" or "body guards" that antagonize soil borne pathogens and/or initiate induced systemic resistance (ISR), as the first and sometimes the only line of defense against soil borne diseases (Weller et al. 2007).

Disease-Suppressive Soils Suppressive soils provide the best examples of natural microbial-based defense. They are soils in which, because of their microbial makeup and activity, a pathogen does not establish or persist, establishes but causes little or no disease, or establishes and causes disease at first but then the disease declines with successive cropping of a susceptible host crop (Weller et al. 2002; Weller et al. 2007). Suppression may be *general* or *specific*, with the former owing to the collective competitive and antagonistic activity of the total soil microbiome, and the latter owing to a specific group(s) of microorganisms acting against a specific pathogen and often on a specific crop. *General suppression* is not transferable, is reduced by soil steaming; and is enhanced by practices that increase soil microbial activity. Highly effective *specific suppression* is superimposed over general suppression, is transferable, and is eliminated by soil pasteurization. Suppressiveness is a continuum from general to specific, with the former often giving rise to the latter in response to certain cropping practices. In *conducive soils*, disease readily occurs. Suppressive soils with activity against fungi, oomycetes, bacteria or nematodes occur worldwide, and some, like Fusarium wilt suppressive soils are *long standing* because the suppressiveness is a natural characteristic of the soil, its origins are unknown, and it persists in the absence of the plant. Others like *take-all decline* (TAD) soils are *induced* because suppressiveness is initiated and sustained by monoculture of susceptible crops. In general, the microbial basis of most suppressive soils worldwide remains poorly understood (Weller et al. 2007). However, the application of new 'omics' approaches combined with classical methods of characterizing suppressive soils is providing a platform to more rapidly identify the microbial basis of suppression (Mendes et al. 2011; Bakker et al. 2013).

38.2   Take-all

Take-all, caused by the soilborne fungus *Gaeumannomyces* graminis (Sacc.) von Arx & Olivier var. *tritici* Walker (*Ggt*), is one of the most important root diseases of wheat worldwide (Cook 2003). It develops at a soil pH of 5.5–8.5 and is most severe where wheat is grown under high precipitation or irrigation (Cook 2003; Freeman and Ward 2004). However, it also occurs in wheat grown under dryland conditions (Cook 2003). *Ggt* survives saprophytically as mycelium in dead roots and tiller bases, the inoculum source for the next crop (Paulitz et al. 2002; Cook 2003; Freeman and Ward 2004). Primary infection of the roots of seedlings occurs with the growth of dark runner hyphae on the root surface. Hyaline hyphae penetrate into the cortex and colonize the vascular tissue, causing characteristic black lesions. Runner hyphae continue to grow over the root surface, to other roots, and upward to the crown and stem bases. Early infection of the plant ultimately causes yellowing of lower leaves, stunting, and premature death of plants in patches. Crop rotation and tillage are effective approaches to manage take-all, but trends in modern farming systems are toward less tillage (to control erosion) and two or three crops of wheat before a break crop (for economic reasons). Both of these practices greatly exacerbate take-all. Wheat varieties lack resistance to take-all, and methods of chemical control, although available, have had only moderate success in controlling the disease. The take-all pathogen also attacks barley, rye, triticale, and other related grasses, but to a lesser extent than wheat (Cook 2003). *G. graminis* var. *avenae* also attacks wheat and causes typical take-all symptoms but it has a broader host range and more commonly is a problem on oats and bent grass.

38.3 Take-All Decline

TAD, the best characterized example of *induced specific suppression*, is defined as the spontaneous reduction in the incidence and severity of take-all and increase in yield occurring with continuous monoculture of wheat or barley and following a severe attack of the disease (Hornby 1998; Weller et al. 2002). TAD is a worldwide phenomenon (Weller et al. 2002) and it is widely used to manage take-all. For example, in the Pacific Northwest (PNW) of the USA, about 0.8 million ha of wheat suffer little damage from take-all, probably due to TAD (Cook 2003).

Although TAD development follows a consistent pattern everywhere, the previous cropping history, environmental conditions and soil factors in a field impact the robustness of the take-all suppressiveness and the length of time before its onset, which can vary significantly but averages 4–6 years (Weller et al. 2002). TAD suppressiveness is transferable to conducive soils (Raaijmakers and Weller 1998; Weller et al. 2002), eliminated by pasteurizing or fumigating the soil (Raaijmakers and Weller 1998; Weller et al. 2002), reduced or eliminated by growing a non-host crop (Weller et al. 2002), and regained when wheat or barley is again grown. Interestingly, soils in some fields that have transitioned into TAD are *imprinted with a memory* of take-all suppressiveness. Thus, TAD soils cropped for long stretches to non-host crops, resulting in a reduction in suppressiveness, are rapidly reactivated to a state of suppression by only a couple of cycles of wheat (even in the greenhouse), thus bypassing the need for years of continuous wheat monoculture to regain suppression (Weller and Yang, unpublished). In addition, the suppressiveness in TAD soils that have been stored dry in cans for over 25 years, can be readily reactivated by again planting one or two cycles of wheat (Allende-Molar and Weller, *unpublished*).

Microbial Basis of Take-all Suppressiveness Since the discovery of TAD, microbial changes in the bulk soil or rhizosphere resulting in inhibition of *G. graminis* var. *tritici* have most commonly been reported as the mechanism(s) of take-all suppression (Hornby 1998; Weller et al. 2002). The types of antagonists responsible for TAD have been thought to differ throughout the world because microbes from different taxonomic groups have biocontrol activity against take-all and *G. graminis* var. *tritici* is very sensitive to different types of antagonism (destruction of hyphae, cross protection, antibiosis etc.) (Weller et al. 2002). However it is now known that in fields in the U.S and The Netherlands, TAD results from the build-up of populations of 2,4-diacetytlphloglucinol (DAPG)-producing fluorescent *Pseudomonas* spp. to a density above 10^5 CFU g^{-1} root, the threshold required for take-all control (Raaij- makers and Weller 1998; Weller et al. 2002, 2007). DAPG producers are commonly found in the soil microbiome at low densities, but infection of wheat roots by *G. graminis* var. *tritici* leads to a dramatic enrichment in their population size, a process that is repeated over and over during wheat monoculture. The specific suppression in TAD soils is lost when DAPG-producing *Pseudomonas* spp. are eliminated, and conducive soils gain suppressiveness when DAPG producers are introduced via mix- ing in small amounts (1–10 % w/w) of TAD soil (Raaijmakers and Weller 1998). *G. graminis* var. *tritici* is highly sensitive to DAPG (Kwak et al. 2009), which is pro- duced in the rhizosphere in TAD soils (Raaijmakers et al. 1999). Control of take-all does not occur with DAPG-deficient mutants and is positively related to the sen- sitivity of the *G. graminis* var. *tritici* isolate to the antibiotic (Kwak et al. 2009). Although DAPG producers are the primary drivers of TAD, a major gap still exists in our understanding of how other components of the microbiome help to modulate and promote the development of TAD suppressiveness (Sanguin et al. 2009; Bakker et al. 2013).

2,4-Diacetylphloroglucinol and Biocontrol DAPG, a very broad spectrum polyke- tide antibiotic, is a key determinant in the biocontrol of a wide range of root and seedling diseases: for example, root rots of tobacco and tomato, Pythium damping- off of cucumber, and take-all of wheat by *P. protegens* CHA0; damping-off of sugar beet and cyst nematodes and soft rot of potato by *P. fluorescens* F113; and take-all by *P. fluorescens* Q2-87 and SSB17 and *P. brassicacearum* Q8r1-96 (Weller et al. 2007). DAPG is also known to be a strong inducer of resistance to foliar pathogens when produced on the roots (Pieterse et al. 2014). The DAPG biosynthetic locus in- cludes *phlACBDE* and *phlHGF*, which function in synthesis, export and regulation. These genes are conserved among all DAPG-producing fluorescent *Pseudomonas* spp. (Weller et al. 2007).

38.4   A Broad Role for DAPG in the Defense of Roots

Enrichment of DAPG producers by crop monoculture is not limited to cereals. Evidence is rapidly emerging that globally in agroecosystems, strains of DAPG- producing pseudomonads form part of the foundation of elite communities of microbial defenders of roots within the larger rhizosphere microbiome (Weller et al. 2007). For example, DAPG producers contribute to the long-standing suppressive- ness of soils to root rot of tobacco in the Morens region near Payerne, Switzerland (Weller et al. 2002). Population densities of DAPG producers at greater than the threshold needed for disease suppression occurred on both pea and wheat roots grown in a pea monoculture soil (> 30 year) from Mt. Vernon, WA, USA and this soil provides control not only of Fusarium wilt of pea but also take-all (Landa et al. 2002). Threshold densities of DAPG producers also occurred on roots of both flax and wheat grown in a monoculture flax soil (103 years) from Fargo, ND, but were not detected on plants grown in soil from an adjacent field that had been in a long crop rotation for the same amount of time. Cropping systems research needs to focus on identifying rotations and varieties that can sustain populations of DAPG producers and thus retain disease suppressiveness after crop monoculture is broken (McSpadden Gardener et al. 2005). Exciting evidence is also accumulating that DAPG may play a role in defense against diseases in natural ecosystems. For example, *Lysobacter gummosus* appears to protect the red-backed salamander against natural infection by the fungal pathogen causing chytridiomycosis, a serious skin disease, by producing DAPG (Brucker et al. 2008).

38.5 Diversity Among DAPG Producers

DAPG-producing pseudomonads that belong to the *P. fluorescens* complex (Loper et al. 2012) show a considerable amount of genetic diversity (Weller et al. 2007). Currently, 22 different genotypes of DAPG producers (designated A-T, PfY and PfZ) have been defined by whole-cell repetitive sequence-based (rep)-PCR analysis or restriction fragment length polymorphism (RFLP) and phylogenetic analysis of *phlD*, a key gene in the DAPG biosynthesis locus (De La Fuente et al. 2006; Landa et al. 2002; Landa et al. 2006). It appears that there are at least six more genotypes that have not been fully described. More recently, whole genome sequencing has confirmed these groupings (Loper et al. 2012). DAPG-producing pseudomonads are now placed in three species: *P. fluorescens*, *P. brassicacearum* and *P. protegens*.

Host Preference Several genotypes of DAPG producers typically occur in a field, but usually only one or two genotypes dominate on the roots of a crop grown in that soil. For example, although genotypes D, B, E and L occur in PNW TAD fields, D-genotype (*P. brassicacearum*) isolates comprise 60–90 % of the DAPG producers and are primarily responsible for take-all suppression (Weller et al. 2007). The ability of *P. brassicacearum* to be the primary driver of TAD lies in the mutual *affinity* or *preference* that this bacterium and wheat roots have for each other. This affinity allows *P. brassicacearum* to aggressively colonize the wheat rhizosphere far better than other genotypes present in the soil microbiome and to maintain threshold densities throughout the growing season (Raaijmakers and Weller 1998, Weller et al. 2007). Of the 22 described genotypes of DAPG producers, several besides genotype D have shown a crop preference. For example, P- and K-genotype isolates have an affinity for pea and wheat, respectively. The ability of crops to enrich for genotypes of DAPG producers best adapted to colonize their roots is strikingly demonstrated in two adjacent fields at Fargo, North Dakota, USA, each with greater than 100 years of continuous monoculture flax or wheat, respectively. Continuous culture of flax resulted in a dominance of F- (41 %) and J- (39 %) genotype isolates in the rhizosphere, whereas growth of wheat led to a dominance of D-genotype isolates (77 %).

38.6 Robustness of Take-all Suppressiveness

Major gaps in our knowledge of TAD still exist as to the basis for variations in the amount of time required before TAD onset, fluctuations in the robustness of suppressiveness among fields and years, and longer term breakdowns of suppression (Hornby, 1998; Kwak and Weller 2012). Kwak et al. (2009) explored the possibility that *Ggt* isolates develop tolerance to DAPG in TAD fields after many years of monoculture, thus lessening the ability of DAPG producers to suppress the pathogen. However, although *Ggt* isolates within a given field differed in antibiotic sensitivity, isolates from TAD fields and conducive fields did not differ significantly from each other (Kwak et al. 2009). It is not likely that *Ggt* will develop tolerance in the field to DAPG because the antibiotic attacks multiple basic cellular pathways in the pathogen (Kwak and Weller 2012).

Wheat Cultivars Affect Suppressiveness Variations in TAD onset and robustness may be due to the differential ability of wheat cultivars to initiate and sustain take-all suppressiveness. For example, wheat cultivars differ in their supportiveness of population sizes and genotypes of DAPG-producing pseudomonads, the amount of DAPG produced in the rhizosphere, and expression of defense genes when colonized by DAPG producers. Crops and cultivars also affect expression of genes in the *phl* operon of DAPG producers in the rhizosphere (Weller et al. 2007; Kwak and Weller 2012; Maketon et al. 2012). At this time, no wheat breeding program is focused on improving the supportiveness of wheat cultivars to microbes involved in natural disease suppressiveness.

38.7 Conclusions

By the year 2050, there will be 9–11 billion people on Earth to feed using the same amount or less land and water as is available today. Farmers are being challenged to produce more, but to do so using sustainable cropping practices and less fertilizer and pesticides. Soil-borne pathogens that cause root and crown rots, seed and seedling damping-off and wilts continue to be major barriers to the expansion of the production of food, fiber and ornamental crops. Soil-borne pathogens are challenging to control because plants often lack resistance to them and chemical seed treatments are usually only effective during the seedling phase of the disease. The use of broad-spectrum gaseous soil fumigants (methyl bromide, chloropicrin, metham sodium) has become increasingly unacceptable in agriculture. It is now thought that the rhizosphere holds the key to the next *Green Revolution*, whereby innovative new varieties and management practices will allow plants to be far more capable of recruiting and utilizing microbes in the soil microbiome for growth promotion and disease defense. Disease-suppressive soils like TAD are not only sustainable and practical controls for soil-borne pathogens, but also are model systems for elucidating how the roots signals, enrich, and sustain elite microbial defenders from the soil microbiome to answer the plant's *cry for help* when it is attacked.

Acknowledgements Parts of this work were supported by the Orville A. Vogel Wheat Research Fund, Washington State University, Pullman and the Organisation for Economic Cooperation and Development (OECD), Paris. Mention of trade names or commercial products in this publication is solely for the purpose of providing specific information and does not imply recommendation or endorsement by the U.S. Department of Agriculture. USDA is an equal opportunity provider and employer. I thank Dr. Linda S. Thomashow for reviewing this manuscript.

References

Bakker PAHM, Berendsen RL, Doornbos RF et al (2013) The rhizosphere revisited: root microbiomics. Front Plant Sci 4:165

Brucker RM, Baylor CM, Walters RL et al (2008) The identification of 2,4-diacetylphloroglucinol as an antifungal metabolite produced by cutaneous bacteria of the salamander *Plethodon cinereus*. J Chem Ecol 34:39–43

Cook RJ (2003) Take-all of wheat. Physiol Mol Plant Path 62:73–86

Cook RJ (2006) Toward cropping systems that enhance productivity and sustainability. Proc Natl Acad Sci U S A 103:18389–18394

Cook RJ (2007) Take-all decline: a model system in biological control and clue to the success of intensive cropping. In: Vincent C, Goettel M, Lazarovits G (eds) Biological control, a global perspective, CABI Publishing UK, pp 399–414

Freeman J and Ward E (2004) *Gaeumannomyces graminis*, the take-all fungus and its relatives. Mol Plant Pathol 5:235–252

Hornby D (1998) Take-all of cereals: a regional perspective. CAB International, Wallingford

Kwak, Y-S, Weller DM (2012) Take-all of wheat and natural disease suppression. J Plant Pathol 29:125–135

Kwak Y-S, Bakker PAHM, Glandorf DCM et al (2009) Diversity, virulence and 2,4-diacetylphloroglucinol sensitivity of *Gaeumannomyces graminis* var. *tritici* isolates from Washington State. Phytopathology 99:472–479

Landa BB, Mavrodi OV, Raaijmakers JM et al (2002) Differential ability of genotypes of 2,4-diacetylphloroglucinol-producing *Pseudomonas fluorescens* strains to colonize the roots of pea plants. Appl Environ Microbiol. 68:3226–3237

Loper JE, Hassan KA, Mavrodi DV et al (2012) Comparative genomics of plant-associated *Pseudomonas* spp: insights into diversity and inheritance of traits involved in multitrophic interactions. PLoS Genet 8(7):e1002784

Maketon C, Fortuna A-M, Okubara PA (2012) Cultivar-dependent transcript accumulation in wheat roots colonized by *Pseudomonas fluorescens* Q8r1-96 wild type and mutant strains. Biol Control 60:216–224

McSpadden Gardener BB, Gutierrez LJ et al (2005) Distribution and biocontrol potential of *phlD*+ pseudomonads in corn and soybean fields. Phytopathology 95: 715–724

Mendes R, Kruijt M, de Bruijn I et al (2011) Deciphering the rhizosphere microbiome for disease-suppressive bacteria. Science 332:1097–1100

Paulitz TC, Smiley RW, Cook RJ (2002) Insights into the prevalence and management of soilborne cereal pathogens under direct seeding in the Pacific Northwest, U. S. A. Can J Plant Path 24:416–428

Pieterse, CMJ, Zamioudis C, Berendsen RL et al (2014) Induced systemic resistance by beneficial microbes. Annu Rev Microbiol (*in press*)

Raaijmakers JM and Weller DM (1998) Natural plant protection by 2,4-diacetylphloroglucinol-producing *Pseudomonas* spp. in take-all decline soils. Mol Plant-Microbe Interact 11:144–152

Raaijmakers JM, Bonsall RF, Weller DM (1999) Effect of population density of *Pseudomonas fluorescens* on production of 2,4-diacetylphloroglucinol in the rhizosphere of wheat. Phytopathology 89:470–475

Sanguin H, Sarniquet A, Gazengel K et al (2009) Rhizosphere bacterial communities associated with disease suppressive stages of take-all decline in wheat monoculture. New Phytol 184:694–707

Weller DM, Raaijmakers JM, McSpadden Gardener BB et al (2002) Microbial populations responsible for specific suppressiveness to plant pathogens. Annu Rev Phytopathol 40:309–348

Weller DM, Landa, BB, Mavrodi OV et al (2007) Role of 2,4-diacetylphloroglucinol-producing fluorescent *Pseudomonas* spp. in the defense of plant roots. Plant Biol 9:4–20

Chapter 39
The Oomycete *Phytophthora infestans*, the Irish Potato Famine Pathogen

Charikleia Schoina and Francine Govers

Abstract The oomycete *Phytophthora infestans* is a filamentous plant pathogen that causes the late blight disease in potato worldwide. It has been a favorite subject of study since the Great Irish Famine in the 1840s and is considered to be a model species for oomycetes. Its genome of over 240 Mb has a remarkable organization with gene dense regions interspersed with gene poor regions, the latter harboring genes involved in host specificity and virulence. These genes are key players in the arms race with the host. They can easily mutate to avoid recognition by immune receptors and their abundance shows that *P. infestans* possesses an impressive arsenal of weapons for attacking potato and is likely hard to beat.

39.1 Potato and Potato Late Blight

The domestication of potatoes (*Solanum* spp.) probably started at least 10,000 years ago around Lake Titicaca in modern-day Peru and Bolivia. Since 1400 BC, when the earliest farmers settled in the Andes, potato production has been of major importance for the Andean societies. In the sixteenth century the Spanish conquered South-America and one of the treasures that they brought to Europe was the potato. By the late 1700s, potato cultivation was widespread in Europe and today potato is the third most important food crop worldwide. The first reports of potatoes being vulnerable to disease appeared in Belgium in June 1845. With an unprecedented speed a mysterious disease spread over Western Europe and wiped out the entire potato crop. By mid-October of that same year it had reached Ireland, a country where the socioeconomic structure forced the poor peasants to solely rely on potato for their daily food. This led to the Great Irish Famine, a disaster that caused a turning

F. Govers (✉) · C. Schoina
Laboratory of Phytopathology, Wageningen University, Droevendaalsesteeg 1,
6708 PB Wageningen, The Netherlands
Tel.: +31 317 483 138
e-mail: francine.govers@wur.nl

C. Schoina
Tel.: +31 317 483 881
e-mail: charikleia.schoina@wur.nl

© Springer International Publishing Switzerland 2015
B. Lugtenberg (ed.), *Principles of Plant-Microbe Interactions*,
DOI 10.1007/978-3-319-08575-3_39

point in history and gave birth to Irish America. Apart from the 1 million people that died, another one and a half million settled as refugees in North America. At that time the concept that microbes could cause plant diseases was unknown. The sudden appearance of late blight was blamed on abiotic factors like hidden volcanoes, the steam machine, electricity or the wet summer and even on the devil. Reverend M. J. Berkeley however, put forward the hypothesis that the mould flourishing on the potato foliage was the cause and not the consequence of the disease. He succeeded in drawing a rather accurate picture of a fungus-like creature growing as mycelium inside a potato leaf and releasing spore-bearing hyphae through the stomata (Large 1940) (Fig. 39.1). This creature was *Phytophthora infestans,* widely known as the Irish potato famine pathogen and nowadays responsible for yearly economic losses of over 3 billion € worldwide (Fry 2008). It belongs to the oomycetes, a group of diverse organisms that, similar to fungi, grow as mycelium and produce spores to propagate. Oomycetes are best known as plant pathogens –there are over a hundred *Phytophthora* spp. and many downy mildews and *Pythium* spp.- but the group also comprises animal and microbial pathogens as well as saprophytes (Kroon et al. 2012; Kamoun et al. 2014). As such they occupy similar ecological niches as fungi. However, during evolution oomycetes evolved completely independently from fungi and this is reflected in differences in e.g. cell wall composition (see Chap. 6), actin cytoskeleton, biochemical and metabolic pathways, and mating systems (Judelson and Blanco 2005) .

39.2   Tackling the Pathogen

In order to understand how a pathogen attacks and colonizes its hosts one needs to be able to investigate the molecular and cellular machinery of the organism and to gain insight into the type of components that the pathogen produces to cause disease. Most pathogenicity factors are secreted and often produced specifically or at higher levels during interaction with the host. Already in the early 1990s the first attempts were made to identify putative pathogenicity factors using differential gene expression profiling. At around the same time the first successful DNA transformation of *P. infestans* was reported and in the years to follow the molecular toolbox was expanded with protocols suitable for gene function analyses by a targeted 'knock-down' approach based on RNA interference or by over expression (van West et al. 1999). In all cases phenotypic analyses of multiple independent transformants is required to demonstrate that an aberrant phenotype is consistently correlated with reduced or increased transcript levels of the target gene and not the result of disruption of another unrelated gene. In case the transgene carries a fluorescent tag, overexpression transformants can be analyzed by microscopy to visualize the subcellular location of the encoded protein. The transgene can also be modified with targeted deletions or potential 'gain- or loss-of-function' mutations or its anticipated function can be studied in other organisms enabling even more in depth analyses at the biochemical or structural level. Although all these approaches have been and are being used

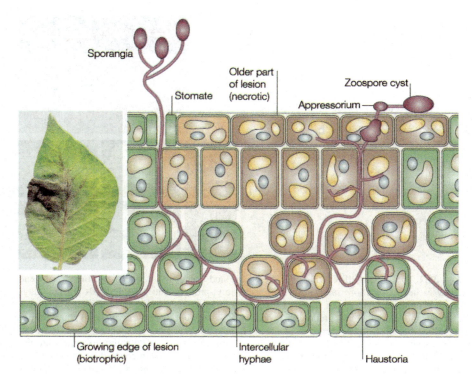

Fig. 39.1 A typical late blight lesion has a necrotic center with heavy sporulation surrounded by a water soaked zone. Outside these zones the pathogen continues to invade healthy cells and the lesion expands further. The schematic figure shows the vegetative life cycle of *P. infestans*. Infection starts when a spore lands on the leaf and germinates. The germ tube forms an appressorium and an emerging penetration peg enters the epidermal cell. From there hyphae colonize the inner cell layers where they grow in between the plant cells and produce finger-like protrusions that penetrate the plant cell. These so-called haustoria are specialised structures that facilitate the delivery of effectors into plant cells and the uptake of nutrients from plant cells. *P. infestans* is a hemi-biotrophic pathogen that needs living plant tissue for the initial phases of colonization, the biotrophic phase. Gradually the plant cells die and the leaf necrotizes. In this phase hyphae escape through the stomata and produce numerous spores named sporangia that easily detach and disperse by wind or water. A sporangium that finds a new host can either germinate directly and initiate a new cycle or, at lower temperatures, undergo cleavage resulting in a zoosporangium from which 6–8 flagellated spores are released. These zoospores can swim for several hours but once they touch a solid surface they encyst and germinate to initiate new infections. Under favourable conditions the pathogen can complete the cycle from infection to sporulation in 4 days. In the field this cycle is repeated multiple times during one growing season resulting in billions of spores and a continuous increase of disease pressure. Besides leaves also stems and tubers get infected and *P. infestans* can continue to flourish on the decaying plant material. If not managed properly, infected seed potatoes or waste on refuse piles are often the sources of inoculum for new infections in spring. An alternative route for surviving the winter is via oospores, resting spores that can survive in soil for many years and are produced upon mating. Most *Phytophthora* spp. are homothallic which means they produce sexual spores without the need for a partner. *P. infestans* is heterothallic; isolates are either A1 or A2 mating type and sex organs only develop when isolates of opposite mating type sense the sex hormone produced by the mate. The schematic figure is reproduced from Judelson and Blanco (2005) by permission of the publisher

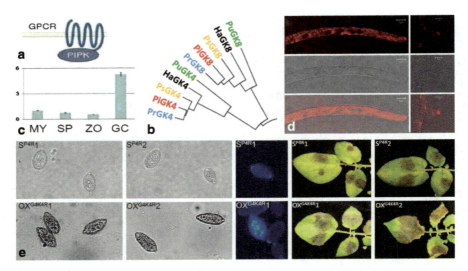

Fig. 39.2 The GPCR-PIPK gene *PiGK4*. A case study of gene discovery and functional gene analyses in *Phytophthora infestans*. Heterotrimeric G-proteins and phospholipids are key players in evolutionary conserved signalling networks in eukaryotes and have been shown to play a role in pathogenicity. A comprehensive inventory of *Phytophthora* genes involved in phospholipid signalling revealed that several indeed encode highly conserved proteins whereas others encode proteins with conserved catalytic domains but combined with other domains. Examples are the GPCR-PIPKs (GKs), novel proteins that are composed of a G-protein coupled receptor (GPCR) domain fused to a phosphatidylinositol phosphate kinase (PIPK) domain (**a**). Based on this domain structure GKs are anticipated to link G-protein and phospholipid signalling. It could be that the GPCR, which normally transmits extracellular signals over the membane to the heterotrimeric G-proteins, bypasses the G-proteins and activates directly the PIPK domain to phosphorylate either lipids or other proteins. Phylogenetic analysis showed that each *Phytophthora* spp. has 12 GKs that are more highly conserved between species than between GKs within one species (**b**). *GK* genes are differently expressed during the life cycle (**c**). *PiGK4* is most highly expressed in germinated cysts whereas other *GKs* are higher expressed in zoospores. Microscopic imaging of *P. infestans* transformants expressing PiGK4 with a fluorescent tag showed the subcellular location of PiGK4 (**d**). In other transformants the *PiGK4* expression was silenced (*S*) or overexpressed (*OX*). Phenotypic characterisation demonstrated a role for PiGK4 in spore development, sporangial cleavage, hyphal elongation and virulence (**e**). In these transformants the transgene is expressed in its own genetic background. Expression in heterologous hosts is another approach to study the gene of interest. For PiGK4 *Escherichia coli* was used as host organism to determine the orientation of the seven transmembrane domains that are characteristic for a canonical GPCR. For more details see Hua et al. (2013). To test if the PIPK domain in PiGK4 has the anticipated enzyme activity, baker's yeast is used as heterologous host. PiGK4 should be able to restore a PIPK deficient yeast mutant in a similar way as the endogenous yeast PIPK gene. Panels **c**, **d** and **e** are reproduced from Hua et al. 2013 by permission of the publisher

successful (Fig. 39.2), manipulating genes in *P. infestans* remains a challenge. Transformation efficiencies are low, primary transformants are often heterokaryons with only a subset of the nuclei carrying the transgene and transgenes are not always stable or expression is lost.

Because of these limitations there was an enormous drive to exploit other means that would lead to insight into the mechanisms underlying pathogenicity. The genomics era created new opportunities and as early as 1999, when DNA sequencing was still costly and filamentous plant pathogens were hardly subjected to large scale sequencing, the first set of a thousand *P. infestans* Expressed Sequence Tags (ESTs) was published (Kamoun et al. 1999). This already uncovered a gene repertoire far more complicated than was envisioned. One example is elicitin, a 10 kD secreted protein highly abundant in culture medium and an elicitor of necrosis in tobacco that was a holy grail in plant-pathogen interaction research for many years. It turned out to be a member of an extensive family with other members also eliciting necrosis (Jiang et al. 2006). Until today, their function in pathogenicity remains elusive. They bind sterols and since *Phytophthora* spp. cannot synthesize sterols themselves, they may have a role in snatching sterols from the environment. Silencing the elicitin *inf1* gene in *P. infestans* did not change virulence on tomato and potato and no detrimental effects were observed. Since elicitins are ubiquitous in *Phytophthora*, they have the characteristics of a pathogen associated molecular pattern (PAMP). Recently a receptor for INF1 was identified in a wild potato species (Du et al. 2014). It is a membrane associated receptor-like protein with extracellular leucine-rich repeats, typical for a pattern recognition receptor, and enhances resistance to late blight when expressed in cultivated potato. This receptor is a new holy grail, this time for plant breeders, to exploit it for boosting late blight resistance in new cultivars.

The great leap forward was the completion of two *Phytophthora* genome projects in 2006 (Govers and Gijzen 2006). For the first time the whole repertoire of potential pathogenicity genes could be mined. The usual suspects were the ones encoding hydrolases such as e.g. proteases, cutinases, lipases or pectinases, as well as protease inhibitors, protein toxins, and ABC transporters. Other genes were suspicious because of their evolutionary trajectory. This could be expansion in number in comparison to close relatives that are not pathogenic, acquisition specifically in *Phytophthora* or oomycetes as is the case for example for the elicitins, or domain shuffling or fusion resulting in proteins that have uncommon domain compositions. Oomycetes have a relatively high proportion of such novel proteins and several of these are truly oomycete-specific (Seidl et al. 2011). They may have evolved along with the metabolic, biochemical and structural features that are characteristic for oomycetes and as such they could be ideal targets for disease control.

Comparison of the genomes of two species attacking different hosts revealed a large family of genes encoding highly divergent secreted proteins that share a common motif in the N-terminus (Jiang et al. 2008). These proteins were coined RXLR effectors based on the amino acid composition of the shared motif. A few years later the *P. infestans* genome was sequenced and included in the comparison (Haas et al. 2009). This genome of 240 Mb, at least twice the size of that of the other two species and much larger than that of most fungal plant pathogens, has a high repeat content of 74 % and a typical bipartite organization with gene-dense regions or 'gene islands' where highly conserved genes are located, and gene-scarce regions or 'gene deserts'. The latter are full of repeats but scattered in these deserts are the RXLR effector genes that, as we know now, play a role in virulence and host specificity of these pathogens.

39.3   Controlling the Disease

To prevent late blight infection most farmers use chemical control. When the disease pressure is high they have to spray their crop once every week to be effective. Because of the adverse effects of chemicals on the environment and the emergence of fungicide resistant *Phytophthora* strains, there is an urgency to find alternatives and late blight resistant potato cultivars are high in demand. Already in the early twentieth century breeders made their first attempts to cross late blight resistance traits into cultivated potato. Although they succeeded, the resistance didn't hold longer than a few years. New *P. infestans* races emerged and some potato lines became susceptible to one race and others to another race (Fry 2008). Genetic analyses confirmed that potato and *P. infestans* interact according to the 'gene-for-gene' model that was postulated in the 1940s to explain differential responses to pathogens within the same plant species. In retrospect, we now understand the reason for the rapid loss of resistance. Nearly all late blight resistance genes encode cytoplasmic 'nucleotide-binding leucine-rich repeat' (NLR) proteins that initiate a resistance response at the moment they encounter the presence of an RXLR effector. As described above, these effectors are encoded by genes located in the most dynamic regions of the genome. The RXLR motif functions as a host cell targeting motif, a kind of ZIP code that tells the protein where to go (Whisson et al. 2007). By an unknown mechanism the numerous RXLR effectors are delivered into the host cell. Their function is to suppress host defense by manipulating the host cell machinery for the wellbeing of the pathogen and as such it is logical that resistance (R) proteins ring the alarm bell when they sense RXLR effectors nearby. The presence inside the plant cell of only one matching pair of R protein and RXLR effector is already sufficient for initiating a hypersensitive response culminating in localized cell death that arrests further growth of the pathogen (Fig. 39.3). The key players in the classical 'gene-for-gene' model are the RXLR effectors and the NLRs. The response of *P. infestans* to a newly introduced resistant cultivar is to rapidly evade recognition by the novel NLR and it does so in many different ways. For example, the matching RXLR gene can be deleted of modified by point mutations or frame shift mutations or its expression can be suppressed by gene silencing (Vleeshouwers et al. 2011; Kasuga and Gijzen 2013) . In other cases the effector is still produced but its activity is suppressed, likely by other RXLR effectors. Knowing how easily an RXLR effector can adapt has predictive value for the durability of the matching NLR gene. For example, already a long time ago breeders experienced that resistance derived from certain wild potato species is less durable than that from other wild potato species and we can now explain this by the flexibility of the RXLR effectors matching the NLRs from those species: deletions, point mutations or frame shift mutations suggesting redundancy in function, versus suppression of effector activity. In the latter case it seems that the pathogen cannot simply get rid of the effector without losing viability and, as a result, the matching NLR confers resistance that is more durable (Vleeshouwers et al. 2011). The current strategy is to stack multiple NLRs in one cultivar and then preferably NLRs that recognize RXLR effectors with diverse activities. So far, only a few host targets

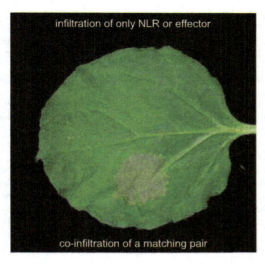

Fig. 39.3 Functional analysis by *in planta* expression. For secreted *Phytophthora* proteins with an anticipated virulence function heterologous expression in plants by *Agrobacterium* mediated transformation is an ideal method for functional analyses. Many of the RXLR effectors promote virulence because they have the ability to suppress defense in the host. This can be monitored by transiently expressing the effector gene in leaves. RXLR effector genes can also be co-expressed with NLR genes. Co-expression of a matching 'gene-for-gene' pair results in a hypersensitive response. Expression in plants is also used for finding host targets of effectors or substrates of proteases, and for analyzing protein-protein interactions between effector and host target

of RXLR effectors are known but it is already clear that these effectors have the capacity to manipulate the host cell machinery at all levels and at different sites. It remains to be seen whether plant resistance conferred by NLRs can fully control late blight. Although very powerful, RXLR effectors are just part of the weaponry that *Phytophthora* utilizes to damage the plant. Besides the cell wall degrading enzymes and proteases there are also numerous secreted proteins with unknown activity, some that during infection stay in the apoplast and others, such as Crinklers that, similar to RXLR effectors, are delivered into the host cell.

For biologists *P. infestans* and its relatives in the oomycete lineage are intriguing; there is still a lot to discover that can broaden our view of life on earth. For the farmers *P. infestans* remains a nuisance that they want to get rid of. The challenge is to bridge the gap: deepen our knowledge on the biology and exploit this knowledge for developing durable and environmental friendly strategies to control late blight.

References

Du J, Verzaux E, Chaparro-Garcia A et al (2014) Elicitin recognition confers resistance to the Irish potato famine pathogen. (To be submitted)

Fry W (2008) *Phytophthora infestans*: the plant (and R gene) destroyer. Mol Plant Pathol 9:385–402

Govers F, Gijzen M (2006) *Phytophthora* genomics: the plant destroyers' genome decoded. Mol Plant-Microbe Interact 19:1295–1301

Haas BJ, Kamoun S, Zody MC et al (2009) Genome sequence and analysis of the Irish potato famine pathogen *Phytophthora infestans*. Nature 461:393–398

Hua C, Meijer HJG, De Keijzer J et al (2013) GK4, a G-protein-coupled receptor with a phosphatidylinositol phosphate kinase domain in *Phytophthora infestans*, is involved in sporangia development and virulence. Mol Microbiol 88:352–370

Jiang RHY, Tyler BM, Whisson SC et al (2006) Ancient origin of elicitin gene clusters in *Phytophthora* genomes. Mol Biol Evol 23:338–351

Jiang RHY, Tripathy S, Govers F et al (2008) RXLR effector reservoir in two *Phytophthora* species is dominated by a single rapidly evolving superfamily with more than 700 members. Proc Natl Acad Sci 105:4874–4879

Judelson HS, Blanco FA (2005) The spores of *Phytophthora*: weapons of the plant destroyer. Nat Rev Microbiol 3:47–58

Kamoun S, Hraber P, Sobral B et al (1999) Initial assessment of gene diversity for the oomycete pathogen *Phytophthora infestans* based on expressed sequences. Fungal Genet Biol 28:94–106

Kamoun S, Furzer O, Jones JDG et al (2014) The top 10 oomycete pathogens in molecular plant pathology. Mol Plant Pathol. doi:10.1111/mpp.12190

Kasuga T, Gijzen M (2013) Epigenetics and the evolution of virulence. Trends Microbiol 21:575–582

Kroon LPNM, Brouwer H, De Cock AWAM et al (2012) The genus *Phytophthora* anno 2012. Phytopathol 102:348–364

Large EC (1940) The advance of the fungi. Alden Press, Oxford

Seidl MF, Van den Ackerveken G, Govers F et al (2011) A domain-centric analysis of oomycete plant pathogen genomes reveals unique protein organization. Plant Physiol 155:628–644

Van West P, Kamoun S, Van't Klooster JW et al (1999) Internuclear gene silencing in *Phytophthora infestans*. Mol Cell 3:339–348

Vleeshouwers VGAA, Raffaele S, Vossen JH et al (2011) Understanding and exploiting late blight resistance in the age of effectors. Ann Rev Phytopathol 49:507–531

Whisson SC, Boevink PC, Moleleki L et al (2007) A translocation signal for delivery of oomycete effector proteins into host plant cells. Nature 450:115–118

Chapter 40
Bacillus, A Plant-Beneficial Bacterium

Rainer Borriss

Abstract Plant growth promotion and biocontrol of plant pathogens are features of *Bacillus* inoculants applied for a more sustainable agriculture. Recent results mainly obtained with *Bacillus amyloliquefaciens* FZB42 and other representatives of the *B. amyloliquefaciens plantarum* subspecies support the hypothesis that stimulation of plant induced systemic resistance (ISR) by bacterial metabolites produced in the vicinity of plant roots is the key mechanism in the biocontrol action of Gram-positive endospore-forming bacteria, whereas a direct effect of the numerous antimicrobial metabolites in suppressing pathogens in the vicinity of plant roots seems to be of minor importance.

40.1 Overview About General Properties and Taxonomy

Several representatives of the Gram-positive *Bacillus* spp. and *Paenibacillus* spp. are able to colonize plants and to develop thereby beneficial actions on plant growth and health. At present, Bacilli are by far the most widely used bacteria on the biopesticide market (Borriss 2011). This is mainly due to their ability to produce durable endospores, which allows the preparation of stable bioformulations with a long shelf-life. Especially members of the *B. subtilis* species complex, such as *B. subtilis*, *B. amyloliquefaciens*, and *B. pumilus*, have been proven to be efficient in plant growth- promotion and biocontrol against plant pathogens. *B. subtilis* and *B. amyloliquefaciens* strains are difficult to distinguish, and several bioagents declared as containing *B. subtilis* spores are in fact representatives of the plant-associated *B. amyloliquefaciens* subsp. *plantarum* (Borriss et al. 2011).

The *Bacillus subtilis* Group *B. subtilis* is the model organism of Gram-positive bacteria. The strictly aerobe *B. amyloliquefaciens plantarum*, represented by its type strain FZB42^T, is distinguished from other representatives of the *B. subtilis* group by its large capacity to synthesize non-ribosomally a high number of polyketides and

R. Borriss (✉)
ABiTEP GmbH, Glienicker Weg 185, Berlin, Germany
e-mail: Rainer.Borriss@rz.hu-berlin.de

© Springer International Publishing Switzerland 2015
B. Lugtenberg (ed.), *Principles of Plant-Microbe Interactions*,
DOI 10.1007/978-3-319-08575-3_40

lipopeptides. Examples of commercial products (biocontrol or biofertilizer) containing *B. amyloliquefaciens plantarum* as their main active ingredient are Kodiak™ (Bayer Crop Science), Companion (Growth Products Ltd.), BioYield™ (Bayer Crop Science), INTEGRAL® (BASF), VAULT® (BASF), SERENADE Max® (Bayer Crop Science), CEASE[(R)] (BioWorks, Inc.), RhizoVital® (ABiTEP GmbH), RhizoPlus® (ABiTEP GmbH), Double Nickel 55™ (Certis USA), and Amylo-X® (Certis USA). See also Table 40.1 for commercial *Bacillus* products for agriculture.

B. licheniformis and **B. pumilus** are other members of the *B. subtilis* group. By contrast to *B. subtilis* and *B. amyloliquefaciens*, they are facultative anaerobes. Biocontrol agents based on *B. licheniformis* SB3086 are Green Releaf and EcoGuard (Novozyme Biologicals Inc.). *B. pumilus* strain GB34 (Yield Shield, Bayer Crop Science) is used as an active ingredient in agricultural fungicides. Other EPA registered biofungicides are SONATA (Bayer Crop Science), and GHA 180 (Premier Horticulture).

Other Bacilli, not Belonging to the B. subtilis Species Complex, also stimulate plant growth and health. *B. firmus* GB126 isolated from cultivated soil is used to control root-knot nematodes in glasshouse and field grown vegetable crops (BioNem AgroGreen, originally from Israel, later acquired by Bayer Crop Science; EPA registered nematicide). Certis USA is developing a product based on B. firmus named BmJ WG. The biofungicide BioArc is prepared from *B. megaterium,* the largest representative of the genus *Bacillus.*

Paenibacillus spp. The PGPR *Paenibacillus polymyxa,* formerly known as *Bacillus polymyxa,* can promote plant growth by producing plant hormones, such as IAA, cytokinins, gibberellins, and ethylene, and volatile compounds. The facultative anaerobe is capable of fixing nitrogen, and of synthesing many antibacterial and antifungal secondary metabolites. NH is a registered fungicide prepared from *Paenibacillus polymyxa* AC-1 by Green Biotech Company Ltd.

The PGPR *P. mucilaginosus* is able to degrade insoluble soil minerals with the release of nutritional ions, such as potassium and phosphorous. Similar to *P. polymyxa, P. mucilaginosus* is also capable of fixing nitrogen.

In the following, I will shortly highlight the different traits of Bacilli involved in their beneficial effect on plants, mainly by using results obtained during the last decade with FZB42[T], which has been successfully commercialized by ABiTEP GmbH (http://www.abitep.de/de/), but is also used as a model strain for scientific research (Borriss 2011).

40.2 Root Colonization

The ability of FZB42 to colonize the rhizoplane is a precondition for plant growth-promotion. Using a GFP-tagged derivative (Fan et al. 2011) the fate of bacterial root colonization was recently studied. The bacterium behaves different in colonizing root surfaces of different plants. FZB42 colonized preferentially root tips when colonizing

Table 40.1 Examples for commercial use of *Bacillus* based bioformulations in agriculture

Trade name	*Bacillus* strain	Known properties	Company
Kodiak™	*Bacillus subtilis* GB03	EPA-registered (71065–2) biological and seed treatment fungicide	Bayer Crop Science, former Gustafsson LLC
Companion	*Bacillus subtilis* GB03	EPA-registered (71065–2) biofungicide, prevent and control plant diseases. It produces a broad-spectrum lturin antibiotic that disrupts the cell-wall formation of pathogens, and it triggers an advantageous Induced Systemic Resistance (ISR) in plants, whereby a plant's natural immune system is activated to fight plant diseases	Growth Products Ltd., White Plains, NY 10603
Yield Shield	*Bacillus pumilus* GB34 (=INR7)	EPA-registered biofungicide (264–985), Suppression of root diseases caused by Rhizoctonia and Fusarium	Bayer Crop Science, previously Gustafsson
BioYield™	*B. amyloliquefaciens* GB99 + Bacillus subtilis GB122	Combination of strong ISR activity (GB99) with phytostimulaton (GB122)	Bayer Crop Science, previously Gustafsson
Subtilex®, INTEGRAL®	*Bacillus subtilis* MBI600	EPA-registered (71840–8.) biofungicide, provides protection against soil-borne pathogens such as Rhizoctonia solani, Pythium spp. and Fusarium spp. to help prevent damping-off and other root diseases	Becker Underwood, Saskatoon, Canada acquired by BASF
VAULT®	*Bacillus subtilis* MBI600	Produced by "BioStacked®" technology, enhancing growth of soy beans and pea nuts	Becker Underwood, Saskatoon, Canada
	Bacillus pumilus BU F-33	EPA-registered (71840-RG, -RE, 2013) plant growth stimulator, induced systemic resistance	Becker Underwood, Saskatoon, Canada
SERENADE Max	*Bacillus subtilis* QST713	EPA-registered (69592–11) biofungicide, Annex 1 listing of the EU agrochemical registration directive (91/414)	Bayer Crop Science, previously AgraQuest
SERENADE SOIL(R)	*Bacillus subtilis* QST713	EPA-registered (69592-EI, 2012) biofungicide for food crops	Bayer Crop Science, previously AgraQuest

Table 40.1 (continued)

Trade name	Bacillus strain	Known properties	Company
SERENADE Optimum®	*Bacillus subtilis* QST713	EPA-registered (2013) biofungicide/bactericide for prevention. It works by stopping spore germination, disrupting cell membrane and inhibiting attachment of the pathogen to leaves. For use in leafy and fruiting vegetables, strawberries and potatoes. Active against fungal (Botrytis, Sclerotinia), and bacterial pathogens (Xanthomonas and Erwinia)	Bayer Crop Science, previously AgraQuest
CEASE[(R)]	*Bacillus subtilis* QST713	Aqueous suspension biofungicide, recommended for leafy and fruiting vegetables, herbs and spices, and ornamentals	BioWorks, Inc., Victor, New York, USA
SONATA®	*Bacillus pumilus* QST2808	EPA-registered (69592–13) biofungicide, powdery mildew control	Bayer Crop Science, previously AgraQuest Inc
RhizoVital®	*Bacillus amyloliquefaciens* FZB42	Biofertilizer, plant growth promoting activity, provides protection against various soil borne diseases, stimulation of ISR	ABiTEP GmbH, Berlin
RhizoPlus®	*Bacillus subtilis*	Plant growth-promoting rhizobacterium and biocontrol agent. It can be used for potatoes, corn, vegetables, fruits and also turf	ABiTEP GmbH, Berlin
Taegro®	*Bacillus subtilis* FZB24	EPA-registered biofungicide. FZB24 has been originally isolated by FZB Berlin, the forerunner of ABiTEP GmbH. Registration as a biofungicide for the US was performed by Taegro Inc. and then sold to Novozymes without agreement with ABiTEP GmbH where the product is still offered	Syngenta, Basel, previously Novozyme, Davis, California and Earth Biosciences
POMEX	*Bacillus subtilis* CMB26	Microbial fungicide, control and inhibition germination effect on powdery mildew, *Cladosporium fulvum* and *Botrytis cinerea*	NIN Co. Ltd.,
	Bacillus subtilis CX9060	EPA-registered 71840-RG,-RE (2012) fungicide, bactericide for food crops, turf and ornamentals	Certis Columbia, MD USA

Table 40.1 (continued)

Trade name	*Bacillus* strain	Known properties	Company
Easy Start® TE-Max	*Bacillus subtilis* E4-CDX	Rhizosphere bacterium that competes with harmful pathogens for space around the roots of the grass plant. Once established this unique strain physically protects the roots and inhibits the advance of soil borne fungi	COMPO Expert GmbH, Münster, Germany
Double Nickel 55™	*B. amyloliquefaciens* D747	EPA-registered (70051-RNI, 2011) a broad spectrum preventive biofungi-cide for control or suppression of fungal and bacterial plant diseases (Powdery mildew, *Sclerotinia, Botrytis, Alternaria*, bacterial leaf spot, bacterial spot and speck, Fire blight, *Xanthomonas, Monilinia*	Certis Columbia, MD USA
Amylo-X®	*B. amyloliquefaciens* D747	Annex 1 listing of the EU agrochemical registration directive. Launched to Italy by Intrachem Bio Italia SpA for control of *Botrytis* and other fungal diseases of grapes, strawberries and vegetables, and bacterial diseases such as fire blight in pome fruit and PSA in kiwi fruit	Certis Columbia, MD USA/Intrachem Bio Italia SpA
BmJ WG	*Bacillus mycoides* BmJ	It works entirely as a microbial SAR activator with no direct effect on the plant pathogen itself. Under development	Certis Columbia, MD USA
	Bacillus pumilus GHA 181	EPA-registered fungicide (2012), Food crops, seeds, ground cover, and ornamentals	Premier Horticulture
BioNem	*Bacillus firmus* GB-126	EPA-registered (2008), suppressing plant pathogenic nematodes, *Bacillus firmus* creates a living barrier that prevents nematodes from reaching the roots	AgroGreen, Israel acquired by Bayer Crop Science

The US governmental EPA registration does not depend on successful field trials; it is only necessary to demonstrate that no negative effects are connected with the use of the biofungicide

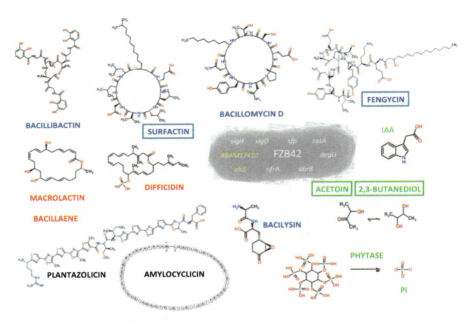

Fig. 40.1 Secondary metabolites with biocontrol or PGP activities produced by *B. amyloliquefaciens* FZB42. Genes involved in plant root colonization (*white*) and plant growth promotion (*yellow*) are listed within the bacterial cell. The cyclic lipopeptides (cLP, *blue*) surfactin, bacillomycin D, and fengycin are nonribosomally synthesized by modularly organized, giant peptide synthetases (NRPSs). Antibacterial polyketides (PK, *red*) are synthesized by membrane-anchored, polyketide megasynthases. Synthesis of PKs and cLPs is dependent on functional phosphor-panthetheinyl-transferase Sfp. NRPSs are also involved in synthesis of the dipeptide bacilysin (*blue*) and the Fe^{2+} siderophore bacillibactin (*blue*). The plant growth-promoting metabolites acetoin, 2,3-butanediol, and indole-3-acetic acid (IAA) are shown in green. Extracellular phytase (*green*) makes phosphate fixed in phytate accessible for plant nutrition. Other extracellular enzymes, which are degrading macromolecules, and supporting the biofertilizer function of FZB42, are ß-glucan and xylan hydrolases, amylases, and proteases, for example. Bacterial metabolites, involved in stimulating plant induced systemic resistance (ISR), are framed

Arabidopsis thaliana. In lettuce, bacterial colonization occurred mainly on primary roots and root hairs, as well as on root tips and adjacent border cells. Essential genes for root colonization are involved in surfactin production, motility, biofilm formation, and stress response (Fig. 40.1). Mutants containing a transposon insertion in the *nfrA* gene, encoding a putative nitro/flavin oxidoreductase, were unable to persist on the surface of lettuce roots, most likely due to their inability to develop an appropriate response against the plant's stress reactions (Budiharjo et al. 2014).

The Rhizosphere Competence of FZB42 was studied by using a combination of field and greenhouse trials. FZB42 is able to effectively colonize the rhizosphere (6.61–7.45 Log_{10} CFU g^{-1} root dry mass) within the growth period of lettuce in the field. However, the cell number (CFU) of FZB42 per gram of soil decreased to 14 % of the initial number of cells after 5 weeks of field cultivation (Chowdhury et al. 2013).

The same samples were analyzed more deeply by mapping of the metagenome sequences corresponding to FZB42. The method called 'fragment recruitments' was used to track persistence of the inoculant FZB42 within the lettuce rhizosphere. Five weeks after inoculation, the DNA fragments corresponding to FZB42 were still traceable, but their number was reduced to around 55 % of the initial number (Kröber et al. 2014). The results obtained with the two different methods indicate that the inoculant strain FZB42 was less competitive than the indigenous community members.

40.3 Plant-Growth Promotion

Although the ability of FZB42 to support growth of potato, maize, cotton, tobacco, leafy and fruiting vegetables, and ornamentals is well documented (Borriss 2011), our knowledge about the molecular basis of the 'biofertilizer' effect of beneficial plant-associated Bacilli are far from complete. Several traits (Fig. 40.1) are involved in the complex interplay between root-colonizing bacteria and plant.

1. **Tryptophan-Dependent Synthesis of Indole-3-Acetic Acid.** Inactivation of genes involved in tryptophan biosynthesis and in a putative tryptophan-dependent IAA biosynthesis pathway led to reduction of both IAA levels and plant growth-promoting activity in the respective mutant strains (Idris et al. 2007). Notably, seed treatment with FZB42 increased root production, an indicator of auxin production, but significantly repressed root Pi uptake at low environmental Pi concentrations (Talboys et al. 2014).

2. **Volatiles, such as 2,3-Butanediol and Acetoin**, released by *B. subtilis* and *B. amyloliquefaciens*, enhance plant growth. To synthesize 2,3-butanediol, pyruvate is converted to acetolactate by acetolactate synthase (AlsS), which is subsequently converted to acetoin by acetolactate decarboxylase (AlsD). FZB42 mutant strains, deficient in the synthesis of volatiles due to mutations in the *alsD* and *alsS* genes, were impaired in plant growth-promotion (Borriss 2011).

3. **Phytase-Producing Bacteria Enhance Phosphorous Availability**. Soil phosphorous is an important macronutrient for plants. Improved phosphorous nutrition is achievable by 'mobilization' of phosphorous fixed as insoluble organic phosphate in phytate (myo-inositol-hexakisphosphate); see also Chap. 24. The extracellular 3-phytase of the PGP *B. amyloliquefaciens* FZB45 hydrolyzed phytate to InsP5 and phosphate *in vitro* (Fig. 40.1). A phytase-negative mutant strain, whose *phyA* gene was disrupted, did not stimulate plant growth under phosphate limitation (Idris et al. 2002). Further experiments under field conditions revealed that FZB45 only stimulates plant growth when phytate is present in soils which are poor in soluble phosphate.

Other mechanisms that are involved in biofertilizer function of Bacilli include nitrogen fixation, mineral solubilization, and secretion of macromolecule degrading enzymes (Borriss 2011).

40.4   Biocontrol by Antimicrobial Compounds

B. amyloliquefaciens FZB42 was successfully applied to suppress the plant pathogen *Rhizoctonia solani* on lettuce (Chowdhury et al. 2013). Genome analysis revealed that nearly 10 % of the FZB42 genome is devoted to synthesizing antimicrobial metabolites and their corresponding immunity genes (Borriss 2013). This antibiotic arsenal (Table 40.2) makes *B. amyloliquefaciens* FZB42 and related *B. amyloliquefaciens plantarum* strains promising microbial biopesticides.

Cyclic Lipopeptides   Five gene clusters involved in non-ribosomal synthesis of c-LPs and of the iron-siderophore bacillibactin were identified in the genome of FZB42 (Table 40.2). Three of the respective gene clusters were assigned to the syntheses of surfactin, fengycin, and bacillomycin D. The iturin bacillomycin D was identified as the most powerful fungicide produced by FZB42. An early surfactin secretion could be of biological relevance since this c-LP, although less fungitoxic than iturins and fengycins, is essential for moving of the bacteria on plant tissues and for matrix formation in biofilms (Chen et al. 2009).

Polyketides   The three gene clusters encoding the modularly organized polyketide synthases (PKS) for syntheses of bacillaene, macrolactin, and difficidin cover nearly 200 kb. Difficidin is the most effective antibacterial compound produced by FZB42^T, but also macrolactin and bacillaene possess antibacterial activity. Difficidin is efficient in suppressing the plant pathogenic bacterium *Erwinia amylovora,* which causes fire blight disease in orchard trees. Macrolactin A (MA) and 7-O-succinyl macrolactin A (SMA), polyene macrolides containing a 24-membered lactone ring, show antibiotic effects superior to those of teicoplanin against vancomycin-resistant enterococci and methicillin-resistant Staphylococcus aureus. MA and SMA are currently being evaluated in preclinical studies in Korea as anti-tumor agents.

Bacilysin   Another product of non-ribosomal synthesis, the dipeptide bacilysin was found as also being involved in the suppression of *Erwinia amylovora*. Recent experiments demonstrated that bacilysin, besides difficidin, is efficient in suppressing *Microcystis aeruginosa*, the main causative agent of cyanobacterial bloom in lakes and rivers (Liming Wu et al. *unpublished*).

Ribosomally Synthesized Antimicrobial Peptides remained unknown in *B. amylolique-faciens plantarum* for a long time with one remarkable exception: mersacidin, a B-type lantibiotic, was detected in strain HIL Y85, later classified as being *B. amyloliquefaciens plantarum* (Herzner et al. 2011). Ribosomally synthesized antibacterial peptides (bacteriocins) were detected in FZB42 by using a mutant strain devoid in non-ribosomal synthesis of polyketides, lipopeptides and bacilysin, which still possessed some remaining antibiotic activity. *Plantazolicin* (PZN) displayed antibacterial activity towards closely related gram-positive bacteria, especially against *B. anthracis*. In addition, PZN displayed a moderate nematicidal activity (Liu et al. 2013). Due to its extensive degree of modification, PZN is well protected from premature

Table 40.2 Genes and gene cluster encoding for secondary metabolites in selected *Bacillus* spp.

Metabolite	Occurrence	Gene cluster	Size	Effect against	Reference
Sfp-dependent non-ribosomal synthesis of lipopeptides					
Surfactin	BAP, BAA, BSU	*srfABCD*	32.0 kb	Virus	Stein 2005
Iturin	BAP, BAA, BSU	*bmyCBAD*	39.7 kb	Fungi	Chen et al. 2007
Fengycin	BAP, BSU	*fenABCDE*	38.2 kb	Fungi	Chen et al. 2007
Polymyxin	PPO	*pmxABCDE*	40.7 kb	Bacteria	Niu et al. 2013
Fusaricidin	PPO	*fus GFEDCBA*	32.4 kb	Fungi	Li and Jensen 2008
Bacillibactin	BAP, BAA, BSU	*dhbABCDEF*	12.8 kb	Bacterial competitors	Chen et al. 2007
Sfp-dependent non-ribosomal synthesis of polyketides					
Macrolactin	BAP	*mlnABCDEFGHI*	53.9 kb	Bacteria	Chen et al. 2007
Bacillaene	BAP, BAA, BSU	*baeBCDE, acpK, baeGHIJLMNRS*	74.3 kb	Bacteria	Chen et al. 2007
Difficidin	BAP	*dfnAYXBCDEFGHIJKLM*	71.1 kb	Bacteria	Chen et al. 2007
Sfp-independent non-ribosomal synthesis					
Bacilysin	BAP, BSU	*bacABCDE, ywfG*	6.9 kb	Bacteria, cyanobacteria	Chen et al. 2007
Ribosomal synthesis of processed and modified peptides (bacteriocins)					
Plantazolicin	BAP FZB42	*pznFKGHIAJC DBEL*	9.96 kb	*B. anthrax*, nematodes	Scholz et al. 2011
Amylocyclicin	BAP FZB42	*acnBACDEF*	4.49 kb	Closely related bacteria	Scholz et al. 2014
Mersacidin	BAP Y2	*mrsK2R2FGEAR1DMT*	12 kb	Gram-positive bacteria	Stein 2005
Amylolysin	BAP GA1	*amlAMTKRIFE*	9.36 kb	Gram-positive bacteria	Arguelles Arias et al. 2014
Subtilin	BSU ATCC 6633	*spaBTCAIFGRK*	12 kb	Closely related bacteria	Stein 2005
Ericin	BAP A1/3	*eriBTCASIFEGRK*	12.5 kb	Closely related bacteria	Stein 2005
Sublancin	BSU	*sunAT bdbA yolJ bdbB*	4.5 kb	Closely related bacteria	Stein 2005
Subtilosin A	BSU	*sboA albABCDEFG*	7.0 kb	Closely related bacteria	Stein 2005

BAP *B. amyloliquefaciens plantarum*, BAA *B. amyloliquefaciens amyloliquefaciens*, BSU *B. subtilis subtilis*, PPO *Paenibacillus polymyxa* degradation by peptidases within the plant rhizosphere (Scholz et al. 2011). A circular bacteriocin, named *amylocyclicin*, was recently identified (Scholz et al. 2014). The peptide suppressed growth of the plant pathogenic actinobacterium *Clavibacter michiganensis* and of several Gram-positive bacteria.

Importance of Secondary Metabolites for Biocontrol For a long time one has thought that the plant protective activity of FZB42 and other PGPR is due to the antibiotic activity of a wide array of antibiotic compounds upon growth under laboratory conditions. However, in recent years, this became doubtful due to pioneering work of Ongena et al. They investigated antibiotic production by MALDI MSI in a gnotobiotic system in which the plantlet and the associated *B. amyloliquefaciens* S499, a close relative of FZB42, were growing on a gelified medium covering the MALDI target plate. Surfactins were detected during early biofilm formation in the rhizosphere in relatively high amounts, representing more than 90 % of the whole c-LP production. In contrast, the synthesis of iturin and fengycin was delayed until the end of the aggressive phase of colonization (Debois et al. 2014).

40.5   Induced Systemic Resistance

Due to the low concentration of antimicrobial compounds detectable in the rhizosphere, it is tempting to speculate that ISR is the main factor for suppressing plant pathogens by PGPR Bacilli. ISR occurs when the plant's defense mechanisms are stimulated and primed to resist infection by pathogens (Doornbos et al. 2012). It has been demonstrated that *Bacillus* derived volatiles and cLPs trigger ISR.

Volatiles Several *Bacillus* PGPR strains emit VOCs that can elicit plant defenses. Exposure to VOCs consisting of 2,3-butanediol and acetoin (3-hydroxy-2-butanone) from PGPR *Bacillus amyloliquefaciens* activates ISR in plants (see Chap. 8). In this context it is worth to mention that expression of AlsS of FZB42, involved in the synthesis of acetoin (Fig. 40.1), was triggered in the presence of maize root exudate (Kierul et al. *unpublished*), suggesting that root exudates play a role in the elicitation of acetoin biosynthesis in FZB42.

   Circular lipopeptides surfactin and fengycin act as elicitors of host plant immunity and contribute to increased resistance toward further pathogenesis ingress in bean and tomato plants (Raaijmakers et al. 2010).

40.6   Effect of *Bacillus* Inoculants on the Environment

The impact of beneficial *Bacillus* inoculants on the root microbiome is important for their plant health effect. Terminal-restriction fragment length polymorphism, T-RFLP, and metagenome analyses of lettuce rhizosphere samples inoculated with *B. amyloliquefaciens* FZB42 vs. non-treated samples revealed that the inoculant strain only had a minor impact on the community structure within this habitat, while inoculation with the pathogen *R. solani* did significantly change the rhizosphere microbial community structure (Chowdhury et al. 2013; Kröber et al. 2014). A significant increase in gamma-proteobacterial diversity was detected in samples inoculated with the pathogen. However, in the presence of FZB42 this increase was less distinct, suggesting a selective compensation of the impact of a pathogen on the indigenous plant-associated microbiome by FZB42 (Erlacher et al. 2014). The results of these metagenome studies suggest that the application of the commercially available inoculant strain FZB42 can be considered as a safe method to promote the health of the economically important lettuce plant and reduce severity of infections by phytopathogens like *R. solani.*

40.7 Conclusions

The beneficial effect of *Bacillus* PGPR on plant health relies on at least three main factors:

1. In previously published studies the set of secondary metabolites described here was suspected to mediate mainly the antibiosis function of *Bacillus* bioinoculants. However, the amounts of the relevant antibiotics found in the vicinity of plant roots were relatively low, making a significant antibiosis function doubtful.
2. These metabolites were also suspected to induce changes within the microbial rhizosphere community, which might affect the health of environment and plant. However, sequence analysis of rhizosphere samples revealed only marginal changes in the root microbiome, suggesting that secondary metabolites are not the key factor in protecting plants from pathogenic microorganisms. On the other hand, adding FZB42 to lettuce plants compensate, at least in part, global changes in the community structure caused by the pathogen, indicating an interesting mechanism of plant protection by beneficial Bacilli.
3. Recent results support hypothesis, that stimulation of plant ISR by bacterial metabolites, such as VOCs and c-LPs, produced in the vicinity of plant roots, is the key mechanism in the biocontrol action of Bacilli.

References

Arguelles Arias A, Ongena M, Devreese B et al (2014) Characterization of amylolysin, a novel lantibiotic from *Bacillus amyloliquefaciens* GA1. PLoS One 8(12): e83037. doi:10.1371/journal.pone.0083037

Borriss R (2011) Use of plant-associated *Bacillus* strains as biofertilizers and biocontrol agents, In: Maheshwari DK (ed). Bacteria in agrobiology: plant growth responses. Springer, Germany, pp 41–76

Borriss R (2013) Comparative analysis of the complete genome sequence of the plant growth-promoting bacterium *Bacillus amyloliquefaciens* FZB42 In: de Brujn FJ (ed) Molecular microbial ecology of the rhizosphere, vol 2. Wiley-Blackwell, Hoboken, pp 883–898

Borriss R, Chen XH, Rueckert C et al (2011) Relationship of *Bacillus amyloliquefaciens* clades associated with strains DSM7^T and FZB42^T: a proposal for *Bacillus amyloliquefaciens* subsp. *amyloliquefaciens* subsp. nov. and *Bacillus amyloliquefaciens* subsp. *plantarum* subsp. nov. based on complete genome sequence comparisons. Int J Syst Evol Microbiol 61:1786–1801

Budiharjo A, Chowdhury SP, Dietel K et al (2014) Transposon mutagenesis of the plant-associated *Bacillus amyloliquefaciens* ssp. *plantarum* FZB42 revealed that the nfrA and the RBAM17410 genes are involved in plant-microbe interactions. PLoS One 9(5): e98267. doi:10.1371/journal.pone.0098267

Chen XH, Koumoutsi A, Scholz R et al (2007) Comparative analysis of the complete genome sequence of the plant growth-promoting bacterium *Bacillus amyloliquefaciens* FZB42. Nat Biotechnol 25:1007–1014

Chen XH, Koumoutsi A, Scholz R et al (2009) Genome analysis of *Bacillus amyloliquefaciens* FZB42 reveals its potential for biocontrol of plant pathogens. J. Biotechnol. 140:27–37

Chowdhury SP, Dietel K, Rändler M et al (2013) Effects of *Bacillus amyloliquefaciens* FZB42 on lettuce growth and health under pathogen pressure and its impact on the rhizosphere bacterial community. PLoS One 8(7):e68818. doi: 10.1371

Debois D, Jourdan E, Smargiasso N et al (2014) Spatiotemporal monitoring of the antibiome secreted by *Bacillus* biofilms on plant roots using MALDI mass spectrometry imaging. Anal Chem 86(9):4431–4438 doi: 10.1021/ac500290s

Doornbos RF, van Loon LC, Bakker PA (2012) Impact of root exudates and plant defense signaling on bacterial communities in the rhizosphere. A review. Agron Sustain Dev 32:227–243

Erlacher A, Cardinale M, Grosch R et al (2014) The impact of the pathogen *Rhizoctonia solani* and its beneficial counterpart *Bacillus amyloliquefaciens* on the indigenous lettuce microbiome. Front Microbiol 5:175. doi: 10.3389/fmicb.2014.00175

Fan B, Chen XH, Budiharjo A et al (2011) Efficient colonization of plant roots by the plant growth promoting bacterium *Bacillus amyloliquefaciens* FZB42, engineered to express green fluorescent protein. J Biotechnol 151: 303–311

Herzner AM, Dischinger J, Szekat C et al (2011) Expression of the lantibiotic mersacidin in *Bacillus amyloliquefaciens* FZB42. PLoS One 6(7): e22389. doi:10.1371/journal.pone.0022389

Idriss, EES, Makarewicz O, Farouk A et al (2002) Extracellular phytase activity of *Bacillus amyloliquefaciens* FZB 45 contributes to its plant growth-promoting effect. Microbiology 148:2097–2109

Idris EES, Iglesias DJ, Talon M et al (2007) Tryptophan dependent production of indole-3-acetic acid (IAA) affects level of plant growth promotion by *Bacillus amyloliquefaciens* FZB42. Mol Plant Microbe Interact 20:619–626

Kröber M, Wibberg D, Grosch R et al (2014) Effect of the strain *Bacillus amyloliquefaciens* FZB42 on the microbial community in the rhizosphere of lettuce under field conditions analyzed by whole metagenome sequencing. Front Microbiol 5:252 doi: 10.3389/fmicb.2014.00252

Li J, Jensen SE (2008) Nonribosomal biosynthesis of fusaricidins by *Paenibacillus polymyxa* PKB1 involves direct activation of a D-amino acid. Chem Biol 15: 118–127

Liu Z, Budiharjo A, Wang Pet al (2013) The highly modified microcin peptide plantazolicin is associated with nematicidal activity of *Bacillus amyloliquefaciens* FZB42. Appl Microbiol Biotechnol 97:10081–90

Niu B, Vater J, Rueckert C (2013) Polymyxin P is the active principle in suppressing phytopathogenic *Erwinia* spp. by the biocontrol rhizobacterium *Paenibacillus polymyxa* M-1. BMC Microbiology 13:137

Raaijmakers J, De Bruin I, Nybroe O et al (2010) Natural functions of cyclic lipopeptides from *Bacillus* and *Pseudomonas*: more than surfactants and antibiotics. FEMS Microbiol Rev 34:1037–1062

Scholz R, Molohon KJ, Nachtigall J et al (2011) Plantazolicin, a novel microcin B17/streptolysin S-like natural product from *Bacillus amyloliquefaciens* FZB42. J Bacteriol 193:215–224.

Scholz R, Vater J, Budiharjo A et al (2014) Amylocyclicin, a novel circular bacteriocin produced by *Bacillus amyloliquefaciens* FZB42. J Bacteriol 196:1842–1852

Stein T (2005) *Bacillus subtilis* antibiotics: structures, syntheses and specific functions. Mol Microbiol 56:845–857

Talboys PJ, Owen DW, Healey JR et al (2014) Auxin secretion by *Bacillus amyloliquefaciens* FZB42 both stimulates root exudation and limits phosphorus uptake in *Triticum aestivum*. BMC Plant Biol 14:51

Chapter 41
Soybean Production in the Americas

Woo-Suk Chang, Hae-In Lee and Mariangela Hungria

Abstract Soybean (*Glycine max* (L.) Merr.) is one of the most important legume crops in the world. Approximately 80 % of the world's soybean is produced by countries in North and South America. Biological nitrogen fixation (BNF) in soybeans, due to the symbiosis with *Bradyrhizobium*, is economically and ecologically beneficial because it reduces the need for synthetic N-fertilizers. This chapter describes history and trends of soybean production, influence of soybean BNF, and development of inoculants to increase the crop yield in North and South America.

41.1 History of Soybean Cultivation in North and South America

Farmers grow soybeans for various purposes such as human food, animal feed, and industrial applications. Since the soybean contains an average of 40 % protein and 20 % oil, it can be a great nutrient source for both humans and animals. A number of soy foods such as tofu, soy sauce, and soymilk are popular in many Asian countries, while in the United States most soybeans for human consumption are used to produce edible oil products such as cooking oils, margarine, mayonnaise, and vegetable shortening. In South America, in addition to the use of oil, soybean-based milk and soups are part of the daily meals of children in public schools. Soybeans are largely added to a variety of processed foods, from meat-derived to crackers. They are

W.-S. Chang (✉) · H.-I. Lee
Department of Biology, University of Texas-Arlington, Arlington, Texas 76019, USA
Tel.: + 1 8172723280
e-mail: wschang@uta.edu

H.-I. Lee
Tel.: + 1 8172723264
e-mail: hilee@uta.edu

M. Hungria
Embrapa Soja, Cx. Postal 231, Londrina, Paraná 86001–970, Brazil
Tel.: + 55 4333716206
e-mail: mariangela.hungria@embrapa.br

© Springer International Publishing Switzerland 2015
B. Lugtenberg (ed.), *Principles of Plant-Microbe Interactions,*
DOI 10.1007/978-3-319-08575-3_41

also considered as an excellent protein source for livestock and are used to produce industrial products such as biodiesel, soy-based lubricants, and soy inks.

Soybean originated in East Asia. It was first cultivated in China around 1100 BC and had spread to Japan and many other countries by the first century AD. In the 1980s, the genus *Glycine* was split into the subgenera *Glycine* (wild perennial species) and *Soja* (including both wild-*Glycine soja* Sieb. and cultivated-*Glycine max* genotypes). *G. max* was introduced into Europe (Paris) only in 1740. It was first introduced to North America from China by Samuel Bowen in 1765 (Hymowitz and Harlan 1983). Soybean became a popular crop in the U.S. between the mid-19th to early 20th centuries. The American Soybean Association was founded in 1920 by soybean farmers and extension workers. World War II kindled the prosperity of the soybean farming in the U.S. as the drastic increase in demand for lubricants and oils by the war increased the soybean demand. The U.S. has been the leading country for soybean production in the world. In Canada, soybeans were first cultivated at the Ontario Agricultural College in 1881, and at present Canada is the world's 7th largest soybean producing country.

Glycine spp. were introduced to South America by the end of the 19th century (Argentina, 1880; Brazil, 1882). Seeds from the U.S., Brazil, Argentina, and Japan were taken to Paraguay in the 1920s, and later to Bolivia, Colombia, Uruguay, and Venezuela. Commercial scale production started in the 1940s in Brazil and Argentina, which led to increased production by the 1960s. One major event for soybean expansion in South America was breeding of soybean cultivars with a long juvenile period, which allowed for the production at very low latitudes.

41.2 Soybean Production in North and South America

World soybean production has been rapidly increasing since 1990, mainly due to increased production in North and South America. The top five countries for soybean production include the U.S., Brazil, Argentina, China, and India, which represent more than 90 % of world production (Fig. 41.1).

The soybean cultivation area in the U.S. has rapidly expanded since the mid-20th century and reached about 31 million ha in 2013. Soybeans are produced in more than one third of the states but mostly in the Midwestern states Iowa, Illinois, Minnesota, and Indiana (order of top producing states). The soybean cultivation area declined in 2007, when many farmers turned to the cultivation of corn to supply the growing bioethanol industry (Salvagiotti et al. 2008). Nevertheless, the total production of soybean and the crop yield has been increasing steadily, mostly due to improved varieties and advances in biotechnology. The national average yield in the U.S. has increased from 1581 kg ha^{-1} in 1960 to 2919 kg ha^{-1} in 2013. In Canada, the average crop yield has reached 3300 kg ha^{-1} in 2012. Bioengineered soybeans are one of the most successful crops commercially in North and South America. They account for 93 % of the soybean produced in the U.S., and for about 90 and 100 % in Brazil and Argentina, respectively. Most of these bioengineered soybeans are improved

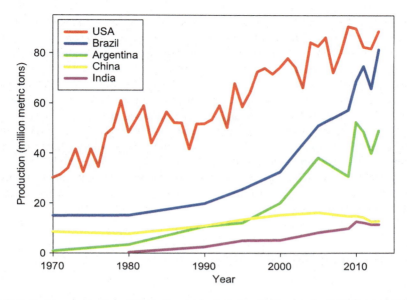

Fig. 41.1 Soybean production of the top 5 countries in the world. The data for the U.S. were retrieved from the USDA statistics for the annual production from 1970 to 2013 (http://www.nass.usda.gov), while the data for the other countries represent occasional years (1970, 1980, 1990, 1995, 2000, 2005, and 2009–2013)

varieties with high resistance against herbicides, and now double resistance to both herbicides and insecticides.

Soybean production continues to grow at impressive rates in the South American countries that account for more than half of the global production (Fig. 41.1); estimates are that soon Brazil may replace the U.S. as the leading producer in the world. National average yield in Brazil increased from 1166 kg ha^{-1} in 1968/1969 to 3115 kg ha^{-1} in 2010/2011. The potential soybean production has been estimated at 8000 kg ha^{-1}, and field trials in North and South America have reported yields of 4000 to 6000 kg ha^{-1}. In addition, while not scientifically proven, there are reports of U.S. farm yields exceeding 10,000 kg ha^{-1} (Hungria and Campo 2004; Hungria et al. 2005, 2006; Hungria and Mendes 2014).

41.3 Nitrogen Fixation Associated with Soybean in North and South America

In 1981, LaRue and Patterson estimated that the average amount of BNF associated with soybean in the U.S. might not exceed 75 kg N ha^{-1} (Larue and Patterson 1981). However, Salvagiotti and colleagues (2008) recently reviewed soybean nitrogen fixation more comprehensively by analyzing 637 data sets from field studies published in international journals between 1966 and 2006. They calculated the average BNF in soybeans to be as high as 111–125 kg N ha^{-1}. Although their analysis was performed with international data sets from many different countries, almost half of the data sets were from studies in the U.S. and Canada.

Interestingly, Herridge et al. (2008) reported that the percentage of N derived from air (%Ndfa) by soybean in the U.S. is on average 60 % of the total Ndfa, which indeed gives around 140 kg N ha^{-1}. Estimates of soybean-associated BNF contribution to the total Ndfa in South America are greater, around 80 % in both Brazil and Argentina (Herridge et al. 2008). This would correspond with 190 kg N ha^{-1}; however, there are reports of contributions higher than 300 kg N ha^{-1} (Hungria et al. 2006). The lower contribution of BNF in the U.S. might be attributed to heavy applications of N-fertilizers in agriculture, with leftovers for the soybean, as well as to low adoption of inoculation by the farmers.

41.4   Inoculants and N-fertilizers

Bradyrhizobium   In the Americas all inoculant strains belong to the genus Bradyrhizobium, including the species *B. japonicum*, *B. diazoefficiens*, and *B. elkanii* (Delamuta et al. 2013). The *Bradyrhizobium*-soybean symbiosis requires specific signal exchange between the two partners. Soybean secretes isoflavonoids, such as genistein and daidzein, into the rhizosphere and these substances trigger nodulation (*nod*) genes in *Bradyrhizobium*. These *nod* genes encode Nod factors, which initiate root hair curling and formation of infection threads. *Bradyrhizobium* cells invade the host plant root cells through the infection threads and subsequently form a new organelle (i.e., the nodule), in which bacteria develop into bacteroids. Within the nodule, the bacteroids can express oxygen-sensitive nitrogenase (*nif*) genes due to the microoxic condition (see Chap. 23).

The Agricultural Research Service (ARS) of the U.S. Department of Agriculture (USDA) has collected *Rhizobium* isolates, including *Bradyrhizobium*, since the early 1900's, and the collection is well known internationally. As of March 2014, the Germplasm Resources Information Network (GRIN) database of the USDA/ARS reports 534 strains isolated from soybeans. Many of the strains were isolated by research programs in the 1930s and 1940s, and the USDA began to produce inoculants for the small research field in the 1950s and 1960s. The need for a *Rhizobium* culture collection was emphasized as the importance of BNF was recognized as a means to supplement hydrocarbon-based fertilizers.

Inoculants   The first U.S. patent for pure cultures of rhizobia to be used in conjunction with artificial inoculation was issued in 1896. Two years later the first inoculant company, Nitragin, was established in the U.S. Inoculants for soybean were commercialized by the company in the early 1900s. The search for an adequate medium to carry rhizobia focused on peat which ultimately became globally established as the "gold standard". By the 1950s, inoculant industries expanded into South America.

Particularly, in the last two decades there has been a shift away from peat towards liquid formulations, as they are easily applied to the seeds and preferred by the farmers. However, concerns about liquid inoculants were raised long ago; Burton and Curley (1965) reported inferior performance of liquid-based inoculants even though 2.5 times as many rhizobia as in peat-based inocula were applied to the seeds. Several decades later, although protective molecules, adhesives and several polymers have been introduced into inoculants, innovation in liquid, gel or other non-solid formulation has been modest. Therefore, a need for a second generation of inoculants has arisen in the soybean industry (more information in section 41.5).

Most farmers in South America are convinced of the benefits of soybean reinoculation with elite strains selected through research programs and registered in governmental agencies. This results in a market size estimated at over 50 million doses of inoculants applied by about 60 % of the farmers (some farmers use multiple doses, especially in first-year planting areas). In contrast, the use of inoculants in the U.S. is a common practice for only about 15 % of the soybean production area, and is perhaps due to low prices of N-fertilizers or a perception of their insignificant performance. An evaluation of the effect of inoculants used in Midwestern states, including Indiana, Iowa, Minnesota, Nebraska, and Wisconsin, between 2000 and 2008 revealed that the use of inoculants was not successful in enhancing crop yield or economic return when used in soils that already had a history of soybean cultivation (De Bruin et al. 2010). Rising prices of N-fertilizers and concerns about the resultant nitrate contamination of the environment may shift this inoculation panorama.

Competitiveness and effectiveness of inoculants have been considered key properties to guarantee the success of nitrogen fixation with the soybean crop. There are long time concerns, however, about highly competitive but low effective indigenous/naturalized *Bradyrhizobium* population in the soybean rhizosphere (Baldwin and Fred 1929). One main example is of the USDA 123 serocluster in the Midwestern U.S., given rise to 60 to 80 % of the nodules, while more effective inoculant strains result in only 10 to 20 % of the nodules formed. This serogroup is also a concern in Canada, Korea, and Brazil (Hungria and Mendes 2014). Several laboratories have tried to identify bacterial properties implicated in competitiveness, such as mobility, chemotaxis, exopolysaccharides, bacteriocins, capacity to respond to several substrates. Interestingly, molecular approaches have also been considered (Hungria et al. 2006); however, application of new strategies or genetically engineered strains to agriculture has been limited.

Reports of the impossibility of displacing competitive strains established in the soils (Thies et al. 1991) have probably discouraged research to select elite strains, and farmers to adopt inoculation in the U.S. Currently, in South America probably more than 90 % of the areas cropped to soybean have been previously inoculated, showing naturalized bradyrhizobial populations ranging from 10^3 to up to 10^6 cells g^{-1} of soil. Dozens of field experiments performed in the last 20 years have consistently shown that reinoculation of soybean results in yield increases. The analysis of sets of experiments indicates that the average increase in yield due to annual reinoculation is 8 % and 14 % in Brazil and Argentina, respectively, compared to the non-inoculated control (Hungria et al. 2006; Hungria and Mendes, 2014). In addition, it is worth noting that massive reinoculation reported in South America may cause the replacement of persistent strains with more efficient strains (Hungria et al. 2006; Hungria and Mendes 2014).

N-fertilizers Less than 40 % of the soybean cultivation area in the U.S. is supplemented with chemically synthesized fertilizers. Nevertheless, as soybean represents a high profit crop, there is also increasing pressure for farmers to use N-fertilizers. Soybeans apparently do not require additional N-fertilizers in normal soil conditions due to BNF with its symbiotic partner *Bradyrhizobium*. Additionally, one of the factors confounding our understanding of both the efficiency of inoculants and the use of N-fertilizers is the effect of crop rotation. High residual soil N as a result of corn cultivation, usually in the Midwest states (i.e., Corn Belt area), may influence interpretation of soybean yield when soybean is planted behind corn (Stewart Smith, *personal communication*). In South America, studies performed over the last two decades demonstrated that soil N availability as low as 20–40 kg of N ha^{-1} may decrease nodulation and N$_2$ fixation in soybeans with no benefits to crop yield. In addition, no benefits have been reported with the addition of 30 to 50 kg of N ha^{-1} at flowering, early or late pod filling stages. Indeed, applications of up to 400 kg N ha^{-1}, split across ten applications did not result in higher yields in comparison to inoculation with elite strains (Hungria et al. 2006; Hungria and Mendes 2014).

41.5   A Second Generation of Inoculants

Promising results have been reported with the use of other microorganisms and molecules as inoculants for soybean. Co-inoculation with plant-growth-promoting rhizobacteria (PGPR) such as *Azospirillum brasilense* may increase yield (Hungria et al. 2013). One of the most exciting new concepts is to take advantage of the molecular dialogue between plant and bacterium. Spaink and colleagues (Spaink et al. 1992) separated Nod metabolites from several *Rhizobium* and *Bradyrhizobium* strains using thin-layer chromatography (TLC) and found that common *nod* genes, such as *nodABC*, play a key role in Nod metabolite production. The Nod metabolite produced by *Bradyrhizobium* strain USDA 110 was subsequently identified as a lipochitooligosaccharide (LCO) signaling molecule, similar to those from *Rhizobium* species (Sanjuan et al. 1992). The LCO from USDA 110 was able to promote the growth of soybean at a low concentration (100 nM) in hydroponic conditions (Souleimanov et al. 2002). The Novozymes' patented LCO molecule product, Optimize®, has been commercially available both in North and South America. Apparently, field responses with LCO have shown slight but consistent increases in soybean yields by an average of ca. 2–3 %, although the increases may depend on specific conditions (Leibovitch et al. 2002). Additionally, it can bring other benefits such as promoting root growth (Souleimanov et al. 2002). The positive effects of Nod factors or bacterial metabolites can be also observed in non-legumes such as corn (Liang et al. 2013; Marks et al. 2013).

41.6   Perspectives on Economical and Ecological Benefits of BNF with Soybeans

Enhancement of soybean BNF has profound economical and ecological benefits by reducing the use of chemical N-fertilizers that are a significant source of greenhouse gas emissions and cause degradation of water sources ranging from groundwater, where derivatives of nitrogen cause the "blue-baby" syndrome (Knobeloch et al. 2000), to oceanic systems where increased nitrogen leads to hypoxia (lack of oxygen) and large scale "dead zones" in the Northern Gulf of Mexico (Burkart and James 1999).

The enormous demand for N by the soybean crop results from the need of about 80 kg of N per 1000 kg of grains; considering that the efficiency of N-fertilizers is rarely more than 50 %, it is calculated that there is a need of 480 kg N-fertilizer ha^{-1}, or about 700 kg of urea, the most broadly used N-source (Hungria et al. 2006). In an exercise to quantify the global contribution of BNF with the soybean crop, Herridge et al. (2008) concluded that 16.4 Tg of N is fixed annually, representing 77 % of the N fixed by all crop legumes. Considering the high price of N-fertilizers in Brazil—70 % of which is imported—in combination with the area cropped, the average BNF rate, and the national average yield, BNF is estimated to save about US$ 15 billion yearly. Not least important, by using a conservative rate of 4.5 kg of e-CO$_2$ kg^{-1} (CO$_2$ equivalent) of N-fertilizer, the replacement of BNF by N-fertilizers in Brazil would result in the emission of about 45 million tons of e-CO$_2$(Hungria and Mendes, 2014). The importance of this approach, specifically in South America, with a strong partnership between plant breeders and microbiologists towards improving the contribution of BNF should be highlighted. Outstanding symbiotic performances as those reported in South America can be lost in a few years if plant breeders and microbiologists do not continue the long-term and successful partnership towards increasing BNF contribution. A continuous pressure from companies to supply N-fertilizers to the soybean crop claiming higher yields can also have profound impacts in BNF contribution. Another important consideration is the increasing use of pesticides and other chemicals in the seed treatment, which can drastically affect *Bradyrhizobium* survival and impair BNF (Mendes and Hungria 2014). On the contrary, disclosure and dissemination of the successful results such as those achieved in South America can stimulate more farmers to use inoculants.

Acknowledgement   Critical review and editing was provided by Dr. Thomas Chrzanowski from the Department of Biology at UT-Arlington

References

Baldwin IL, Fred EB (1929) Strain variation in the root nodule bacteria of clover, *Rhizobium trifolii*. J Bacteriol 17:17–18
Burkart MR, James, DE (1999) Agricultural-nitrogen contributions to hypoxia in the Gulf of Mexico. J Environ Qual 28: 850–859.

Burton JC, Curley RL (1965) Comparative efficiency of liquid and peat-base inoculants on field-grown soybean (*Glycine max*). Agron J 57:379–381

De Bruin JL, Pedersen P, Conley SP et al (2010) Probability of yield response to inoculants in fields with a history of soybean. Crop Sci 50:265–272

Delamuta JR, Ribeiro RA, Ormeno-Orrillo E et al (2013) Polyphasic evidence supporting the reclassification of *Bradyrhizobium japonicum* group Ia strains as *Bradyrhizobium diazoefficiens* sp. nov. Int J Syst Evol Microbiol 63:3342–3351

Herridge DF, Peoples MB, Boddey RM (2008) Global inputs of biological nitrogen fixation in agricultural systems. Plant Soil 311:1–18

Hungria M, Campo RJ (2004) Economical and environmental benefits of inoculation and biological nitrogen fixation with soybean: situation in South America. In: World Soybean Research Conference, 7., International Soybean Processing and Utilization Conference, 4., Brazilian Soybean Congress, 3. Proceedings, Embrapa Soja, Londrina, Brazil, pp 488–498

Hungria M, Mendes IC (2014) Nitrogen fixation with soybean: the perfect symbiosis?. In: de Bruijn F (ed) Biological nitrogen fixation. Wiley-Blackwell, New Jersey. (Wiley Publisher, Hoboken)

Hungria M, Franchini JC, Campo RJ et al (2005) The importance of nitrogen fixation to soybean cropping in South American. In: Werner D, Newton WE (eds) Nitrogen fixation in agriculture, forestry, ecology, and the environment. Springer, Dordrecht, pp 25–42

Hungria M, Campo RJ, Mendes IC et al (2006) Contribution of biological nitrogen fixation to the N nutrition of grain crops in the tropics: the success of soybean (*Glycine max* L. Merr.) in South America. In: Singh RP, Shankar N, Jaiwal PK (eds) Nitrogen nutrition and sustainable plant productivity. Studium Press/LCC, Houston, pp 43–93

Hungria M, Nogueira MA, Araujo RS (2013) Co-inoculation of soybeans and common beans with rhizobia and azospirilla: strategies to improve sustainability. Biol Fertil Soils 49:791–801

Hymowitz T and Harlan JR (1983) Introduction of soybean to North America by Bowen, Samuel in 1765. Econ Bot 37:371–379

Knobeloch L, Salna B, Hogan A et al (2000) Blue babies and nitrate-contaminated well water. Environ Health Perspect. 108:675–678.

Larue TA, Patterson TG (1981) How much nitrogen do legumes fix. Adv Agron 34:15–38

Leibovitch S, Migner P, Zhang F et al (2002) Evaluation of the effect of SoyaSignal technology on soybean yield [*Glycine max* (L.) Merr.] under field conditions over 6 years in eastern Canada and northern United States. J Agron Crop Sci 187(4):281–292

Liang Y, Cao Y, Tanaka K et al (2013) Nonlegumes respond to rhizobial Nod factors by suppressing the innate immune response. Science 341:1384–1387

Marks BB, Megías M, Nogueira MA et al (2013) Biotechnological potential of rhizobial metabolites to enhance the performance of *Bradyrhizobium japonicum* and *Azospirillum brasilense* inoculants with the soybean and maize crops. Appl Microbiol Biotechnol Express 3:21

Salvagiotti F, Cassman KG, Specht JE et al (2008) Nitrogen uptake, fixation and response to fertilizer N in soybeans. A review. Field Crops Res 108:1–13

Sanjuan J, Carlson RW, Spaink HP et al (1992) A 2-O-methylfucose moiety is present in the lipo-oligosaccharide nodulation signal of *Bradyrhizobium japonicum*. Proc Natl Acad Sci U S A 89:8789–8793

Souleimanov A, Prithiviraj B, Smith DL (2002) The major Nod factor of *Bradyrhizobium japonicum* promotes early growth of soybean and corn. J Exp Bot 53:1929–1934

Spaink HP, Aarts A, Stacey G et al (1992) Detection and separation of *Rhizobium* and *Bradyrhizobium* Nod metabolites using thin-layer chromatography. Mol Plant Microbe Interact 5:72–80

Thies JE, Singleton PW, Bohlool BB (1991) Influence of size of indigenous rhizobial populations on establishment and symbiotic performance of introduced rhizobia on field-grown legumes. Appl Environ Microbiol 57:19–28

Part VIII
Future Prospects and Dreams

Chapter 42
Exploring the Feasibility of Transferring Nitrogen Fixation to Cereal Crops

Muthusubramanian Venkateshwaran

Abstract Among the land plants, legumes are unique as they establish symbiotic relationship with soil-borne, nitrogen-fixing bacteria known as rhizobia to meet their nitrogen demand. This symbiotic interaction is considered a promising component of sustainable agriculture due to its economic and ecological benefits. However, the scope of this symbiosis, which is currently limited to legumes, needs to be extended to non-legumes, in particular to economically important cereal crops, to achieve sustainable production of staple crops. This chapter explores the feasibility of transferring the symbiotic nitrogen fixing machinery to non-legume crops.

42.1 Definitions

Pre-penetration Apparatus During infection by arbuscular mycorrhizal fungi, after appressorium formation, a precise succession of processes that are coordinated by the nucleus leads to the formation of the prepenetration apparatus (PPA). This apparatus appears as a cytoplasmic column containing microtubule and microfilament bundles, very dense endoplasmic reticulum cisternae and a central membrane thread. Only after the PPA column has been formed does the fungus enter and grow across the cell (see also Chap. 25).

Pre-infection Thread During rhizobial invasion in legume roots, the cytoplasms in these activated outer cortical cells align with each other, giving rise to columns of cytoplasmic bridges called preinfection threads (PITs) through which the inwardly growing infection thread propagates.

Spontaneous Nodules In certain legumes, such as alfalfa, nodules are formed in the roots in the absence of rhizobia or an infection thread. These white single-to-multi-lobed nodules contain a nodule meristem, cortex, endodermis, central zone and vascular strands. The deregulation of calcium- and calmodulin-dependent protein

M. Venkateshwaran (✉)
School of Agriculture, University of Wisconsin-Platteville,
Platteville, WI 53818, USA
Tel.: + 001-608-3421898
e-mail: venkateshwam@uwplatt.edu

© Springer International Publishing Switzerland 2015   403
B. Lugtenberg (ed.), *Principles of Plant-Microbe Interactions*,
DOI 10.1007/978-3-319-08575-3_42

kinase leads to the formation of spontaneous nodules in the model legume *Lotus japonicus*.

42.2 Commonality in the Molecular Mechanism of the Legume-Rhizobia and Plant-Arbuscular Mycorrhizal Fungi Symbioses

More than 80 % of land plants establish a symbiotic association with arbuscular my-corrhizal (AM) fungi to meet the demand of the plant for phosphorous and other micronutrients. In addition, legumes establish a symbiotic association with rhizo-bia to meet the demand of the plant for nitrogen. These mutualistic associations are established by signal exchanges between plants and microbes. Legumes secrete flavonoids and isoflavonoids, which induce the expression of nodulation (*nod*) genes and the production and secretion of symbiotic signalling molecules, known as Nod factors. Similarly, strigolactones, which are secreted by the majority of land plants, trigger spore germination in AM fungi and the subsequent release of Myc factors. Both Nod and Myc factors are lipochitooligosachcharide (LCO) molecules that are perceived by LysM-type receptor kinases that are expressed in plant root hairs. Nod factor receptors have been characterized in model legumes, such as *Lotus japonicus* (NFR1 and NFR5) and *Medicago truncatula* (NFP and LYK3), along with an LRR receptor-like kinase (LjSYMRK/MtDMI2) that acts as co-receptor. Although Myc factor receptors have not yet been identified, in *M. truncatula*, a LysM-type receptor kinase (LYR1) is being investigated as a Myc factor receptor candidate because its expression is upregulated during AM symbiosis. These LysM-type receptor kinases are conserved across all land plants. Cross-talk in Nod and Myc factor perception has been reported in *M. truncatula* (Rose et al. 2012). The perception of rhizobial Nod factors and AM fungal Myc factors triggers symbiotic signalling events, such as nuclear calcium spiking, i.e., oscillations in the calcium concentration in and around the nucleus. Nuclear calcium spiking is regulated by nuclear cation channels (LjPOL-LUX/MtDMI1, LjCASTOR), a calcium pump (MtMCA8) and yet-to-be-identified calcium channels. Nod and Myc factor-induced calcium spiking is decoded by a calcium- and calmodulin-dependent protein kinase (LjCCaMK/MtDMI3), which is a central regulator of nodulation- and mycorrhization-associated gene expression. This signalling pathway that is shared between legume nodulation and AM is termed the common symbiotic pathway (CSP). This early symbiotic signalling pathway has been thoroughly discussed in many recent reviews (Venkateshwaran et al. 2013; Delaux et al. 2013; Oldroyd et al. 2013). The commonality in the nodulation- and mycorrhization-associated symbiosis pathway (SYM pathway) suggests the possi-bility for expanding the already existing mycorrhization-specific SYM pathway to accommodate and regulate nodulation-specific events in non-legumes (Charpentier and Oldroyd 2010; Venkateshwaran and Ané 2011; Oldroyd and Dixon 2014; Rogers and Oldroyd 2014; Fig. 42.1).

42.3 Specificity Associated with the Legume-Rhizobia Symbiotic Signalling Pathway

Although commonality exists between the legume-rhizobia and AM symbiotic signalling pathway, specificity exists in the signalling pathway that is crucial for the establishment of root nodule symbiosis (RNS; Venkateshwaran et al. 2013 and Delaux et al. 2013; Fig. 42.1). There is a logical correlation between the existence of genes that belong to the symbiosis tool kit and the prevalence of either or both nodulation and mycorrhization (Venkateshwaran et al. 2013; Delaux et al. 2013; Delaux et al. 2014). Plants belonging to the Brassicaceae, such as *Arabidopsis thaliana*, cabbage and cauliflower, lack many key genes of this symbiosis tool kit and are, thus, unable to establish even AM association, one of the most prevalent symbioses among land plants (Fig. 42.1) Comparative phylogenomic studies performed among legumes, non-legume AM hosts (corn and tomato), and non-legume non-AM hosts (Brassicas) have uncovered the impact of symbiotic associations on the host genome evolution and have helped identify key genes that are required for nodulation (Delaux et al. 2014). Therefore, engineering nitrogen fixation in cereals largely relies on the molecular dissection of the symbiotic signalling pathway, which in turn relies on genetic, genomic and comparative phylogenomic studies.

42.4 Transferring Nitrogen Fixation to Non-Legumes

Often considered as a search for the Holy Grail, engineering nitrogen fixation in cereals has been the dream of scientists for many decades, with the ultimate goal of dissecting the molecular mechanism of RNS to reconstitute this symbiotic signalling machinery in non-legume plants. Although this goal is long-term and extremely challenging, such a transfer of the legume-rhizobia symbiotic machinery to non-legume crops could significantly minimize the exogenous application of energy-expensive nitrogen fertilizers. In addition to reducing the cost of cultivation for food, feed and bioenergy, this strategy would also provide protection for the environment against pollution from chemical nitrogen fertilizers. This task will be extremely challenging but the availability of advanced molecular tools in rhizobia and host plants make this goal more realistic now than ever before. Several strategies are currently underway by international research groups and can be broadly grouped into the following two major approaches.

Engineering Cereal Crops to Mimic Legumes to Accommodate Endosymbiotic Nitrogen-Fixing Rhizobia The high specificity of the legume-rhizobia interaction makes it difficult to nodulate one group of legumes with the symbiotic partner of another group. Most rhizobia have a very narrow host range, making studies across species difficult. However, the rhizobial strain NGR234, with a wide host range of 353 legume species representing 122 genera, seems an interesting candidate for the transfer of root nodule symbiosis (RNS) to non-legume crops. To reconstitute the

Fig. 42.1 The elaborate symbiotic signalling pathway (*SYM pathway*) that is conserved in legumes allows these plants to associate with both rhizobia and AM fungi, whereas the SYM pathway that is present in non-legume cereal crops enables these plants to associate with AM fungi alone but not with nitrogen-fixing rhizobia. Plant species belonging to Brassicaceae lack a sophisticated SYM pathway and are, therefore, unable to establish associations with either rhizobia or AM fungi

RNS machinery in non-legume plants, the genetic make-up of the host needs to be fine-tuned to allow major nodulation events, such as infection thread formation, intracellular uptake and the formation of root nodules (Markmann and Parniske 2009).

The prevalence of CSP in the majority of land plants suggests that this pathway could be adjusted to accommodate RNS. The homologue of MtDMI2/LjSYMRK is necessary for another nitrogen-fixing root endosymbiosis, actinorhizal symbiosis (ARS), in *Casuarina glauca* and in the cucurbit *Datisca glomerata* (Gherbiet al. 2008; Markmann et al. 2008). The events leading to the intracellular accommodation of the symbiont share remarkable similarities in these three mutualistic interactions (RNS, AM and ARS). Prior to invasion by the respective endosymbiont, the host cell prepares to accommodate the invading partner by cytoskeletal and organellar rearrangements and nuclear movements. Similarities exist in the formation of the 'pre-penetration apparatus' (PPA) in AM symbiosis and the 'pre-infection threads' (PIT) in RNS and ARS. Because AM symbiosis is prevalent in the majority of land plants, the components that are necessary for the intracellular accommodation of symbionts may already exist in non-legume crops. Therefore, in my opinion, engineering a receptor-mediated perception of nitrogen-fixing symbionts in non-legume plants is not just a dream and will be possible in the near future (Venkateshwaran and Ané 2011).

The recognition of symbiotic partners by the host is a key factor for the establishment of mutualistic interactions. Such microsymbiont recognition is mediated by receptor kinases in plants. The LysM receptor kinases NFR1 and NFR5 in *L. japonicus* and NFP and LYK3 in *M. truncatula* are Nod factor receptors. The introduction of *L. japonicus* Nod factor receptors into *M. truncatula* enables the latter to enter into a symbiotic association with the *L. japonicus* symbiont, *Mesorhizobium loti*. Such an extreme specificity of receptors towards Nod factors needs to be relaxed by genetic manipulations, which would enable the host plant to recognize a wide array of rhizobial species. The possibilities for the presence of additional Nod factor receptor complexes and crosstalk between Nod and Myc factor perception have been reported in *M. truncatula* (Rose et al. 2012). Homologues of LysM receptor kinases are present in non-legume plants, such as rice and *A. thaliana*, where they trigger the defence reaction upon the perception of chitin oligomers, common elicitors of fungal pathogens. Even a non-AM host, such as Arabidopsis, can perceive rhizobial Nod factors that suppress microbe-associated molecular pattern (MAMP)-triggered immunity in the plant (Liang et al. 2013). Utilizing such pre-existing recognition machinery without compromising the defence-related role of this machinery presents additional challenges. Alternatively, Nod factor-independent infection strategies, such as those observed in certain species of the photosynthetic *Bradyrhizobium*, have also been considered as avenues for non-legume plants. Although it is not clear whether LysM receptor kinases play a role in the recognition of the yet unknown actinorhizal factors, the dual infection in *Parasponia* (Ulmaceae) by both Frankia- and Nod factor-dependent rhizobia suggests this possibility (Venkateshwaran and Ané, 2011).

Rhizobial infection and nodule organogenesis are two separable events in RNS. The genes that are involved in the spatio-temporal synchronization of these two distinct events are mandatory for successful RNS. The further characterization of mutants, such as *IPD3/CYCLOPS* or *NIN*, might identify the coordinators of epidermal (infection) and cortical (infection and nodule organogenesis) events. As previously discussed, the commonalities between legume root nodule symbiosis and actinorhizal symbiosis should be considered to identify the key components and missing links in non-nodulating plants (Markmann and Parniske 2009). Likewise, the molecular dissection of the dissimilarities between nodulating and non-nodulating legumes seems a promising approach to identify the master regulators of RNS that are missing in non-legume plants (Sprent and James 2007). The molecular dissection of the RNS machinery in the non-legume plant *Parasponia* could also provide clues as to the master players in symbiotic signalling that are lacking in non-nodulating plants.

The chemical reaction of biological nitrogen fixation (BNF; see also Chap. 23) is catalyzed by the molybdenum-dependent nitrogenase enzyme, which comprises two component proteins, the 'Fe protein' and the 'MoFe' protein (Burgess and Lowe 1996). Several genes are involved in the synthesis and assembly of functional nitrogenase in rhizobia, albeit with a high level of variations among different species. Photosynthetic *Bradyrhizobium* and *Azorhizobium caulinodans* carry approximatly 15 *nif* genes that are required for nitrogenase synthesis, while *Rhizobium* and *Sinorhizobium* possess 8 and 9 genes, respectively. Some of the core *nif* genes that are conserved in the majority of the rhizobia are *nifH* (dinitrogenase reductase), *nifDK* (α and β subunits of dinitrogenase), *nifEN* (scaffold for iron-molybdenum cofactor), *nifB* (P-cluster) and the regulatory gene *nifA*. The high level of intricacy and sophistication in the functionality of nitrogenase in BNF enables legumes to occupy a unique niche among land plants. The possibility of introducing these genes encoding functional nitrogenase assembly in non-legume host plants, such as cereals, and expressing the functional protein complexes in a tissue-specific manner (roots or spontaneous nodules) for *de novo* BNF would be another promising strategy.

In addition to having the genes that are necessary for rhizobial infection, nodule organogenesis and their coordination, legumes possess symbiosis-specific leghaemoglobin genes for the protection and function of nitrogenase activity. Homologues of leghaemoglobin genes are present in the majority of land plants, including the evolutionarily older Bryophytes and Pteridophytes. In non-legume plants, including rice and maize, homologues of leghaemoglobins are present as non-symbiotic haemoglobins (nsHbs). Such nsHbs from *Parasponia* are expressed in the nodules that are formed during rhizobial interaction and facilitate the diffusion of oxygen to bacteroids. The role of nsHbs in symbiosis in actinorhizal plants, such as *Causarina glauca, Alnus glutinosa* and *Myrica gale*, is still unknown. Similarly, leghaemoglobin genes are induced during AM, but their role in this widespread ancestral symbiosis is unknown. These observations provide encouraging evidence that nsHbs can be utilized in the assembly of the nodulation pathway in non-legume plants.

Engineering Natural Endophytes of Cereal Crops to Mimic the Nitrogen Fixing Ability of Rhizobia Another approach in enabling endosymbiotic BNF in non-legume crops is to manipulate the microsymbiont/natural endophytes. The endophytic colonization of plant roots by rhizobia has been demonstrated several times in the past decade (Gutierrez-Zamora and Martinez-Romero 2001; Lundberg et al. 2012), suggesting that rhizobia can utilize the pre-existing infection machinery in non-legume plants to colonize these plants as endophytes. Many plants, including the non-AM host Arabidopsis, naturally accommodate endophytes in their roots (see also Chap. 5). These endophytes belong to a wide range of genera, including Rhizobiaceae.

Several endophytic nitrogen-fixing bacteria have been isolated from sugarcane (*Saccharum* spp.) tissues, including *Gluconacetobacter diazotrophicus*, *Herbaspirillum seropedica*, and *H. rubrisubalbicans*. These intracellular associations seem to be beneficial for the plant, as bacteria promote plant growth when inoculated in sugarcane plantlets by nitrogen fixation and/or the production of hormones, such as auxin and gibberellin (Sevilla et al. 2001; see also Chap. 26). The plant mechanisms that permit bacterial colonization in an endophytic (intracellular spaces and vascular tissues) and non-pathogenic manner, establishing a beneficial association, are not yet clear. Different sugarcane genotypes have different rates of BNF, suggesting that plant genetic factors might control the process of bacterial recognition, colonization, and/or nitrogen fixation (Urquiaga et al. 1992).

Rhizobium sp. strain IRBG74 is a nitrogen-fixing symbiont in the Agrobacterium/Rhizobium clade that nodulates the aquatic legume *Sesbania* sp. and can infect rice endophytically, improving plant growth, health, and yield, making it a good model system for determining the mechanisms of Rhizobium-cereal interactions. Recently, the sequence of the IRBG74 genome, which is composed of a circular chromosome, a linear chromosome, and a symbiotic plasmid (pIRBG74a), has been released (Crook et al. 2013). This strain carries all of the necessary *nod* genes for the production and secretion of Nod factors that mimic Myc factor LCOs (*unpublished*). Preliminary results suggest that the Nod factors of *Rhizobium* sp. IRBG74 can trigger nuclear calcium spiking in *M. truncatula* and that rice is able to perceive the signalling molecules of this strain (*unpublished*), further confirming that the search for natural endophytes of cereals, particularly those that belong to the Rhizobium clade with an ability to fix atmospheric nitrogen, is the best alternate avenue to achieving the goal of transferring nitrogen fixation to cereal crops. The release of the IRBG74 genome sequence enables the manipulation of this strain at the molecular level to enhance its nitrogen-fixing ability in rice.

42.5   Conclusion

Enabling endosymbiotic nitrogen fixation in cereal crops is a promising component of sustainable agriculture. Although this goal is challenging and long-term, it could be achieved by manipulating host and microbial partners. The advent of molecular tools in both host plants and microsymbionts and our ever-expanding knowledge of the symbiotic signalling pathways provide hope that the re-constitution of the root nodule symbiotic machinery in non-legume plants is a feasible yet distant goal.

Acknowledgements I thank Dr. Jean-Michel Ané, University of Wisconsin-Madison, USA for his guidance in the preparation of this chapter and the figure. I thank Kari Parthasarathy for the critical reading of the chapter. I thank Katherine Baldwin for technical help in the preparation of the Fig. 42.1

References

Burgess BK, Lowe DJ (1996) Mechanism of molybdenum nitrogenase. Chem Rev 96:2983–3012

Charpentier M, Oldroyd G (2010) How close are we to nitrogen-fixing cereals? Curr Opin Plant Biol 13:556–564

Crook MB, Mitra S, Ané JM et al (2013) Complete genome sequence of the *Sesbania* symbiont and rice growth-promoting endophyte *Rhizobium* sp. Strain IRBG74. Genome Announc 6: e00934–13

Delaux PM, Séjalon-Delmas N et al (2013) Evolution of the plant–microbe symbiotic "toolkit". Trends Plant Sci 18:298–304

Delaux PM, Varala K, Edger PP et al (2014) Comparative phylogenomics uncovers the impact of symbiotic associations on host genome evolution. PLoS Gen 10(7):e1004487

Gherbi H, Markmann K, Svistoonoff S et al (2008) SymRK defines a common genetic basis for plant root endosymbioses with arbuscular mycorrhiza fungi, rhizobia, and Frankia bacteria. Proc Natl Acad Sci U S A 105:4928–4932

Gutierrez-Zamora ML, Martinez-Romero E (2001) Natural endophytic association between *Rhizobium etli* and maize (*Zea mays* L.). J Biotech 91:117–126

Liang Y, Cao Y, Tanaka K et al (2013) Non-legumes respond to rhizobial Nod factors by suppressing the innate immune response. Science 341:1384–1387

Lundberg DS, Lebeis SL, Paredes SH et al (2012) Defining the core *Arabidopsis thaliana* root microbiome. Nature 488:86–90

Markmann K, Parniske M (2009) Evolution of root endosymbiosis with bacteria: how novel are nodules? Trends Plant Sci 14:77–86

Markmann K, Giczey G, Parniske M (2008) Functional adaptation of a plant receptor-kinase paved the way for the evolution of intracellular root symbioses with bacteria. PLoS Biol 6:e68

Oldroyd GE (2013) Speak, friend, and enter: signalling systems that promote beneficial symbiotic associations in plants. Nat Rev Microbiol 11:252–263

Oldroyd GE, Dixon R (2014) Biotechnological solutions to the nitrogen problem. Curr Opin Biotech 26:19–24

Rogers C, Oldroyd GE (2014) Synthetic biology approaches to engineering the nitrogen symbiosis in cereals. J Exp Bot 65:1939–1946

Rose CM, Venkateshwaran M, Volkening JD et al (2012) Rapid phosphoproteomic and transcriptomic changes in the rhizobia-legume symbiosis. Mol Cell Proteomics 11:724–744

Sevilla M, Burris RH, Gunapala N et al (2001) Comparison of benefit to sugarcane plant growth and $^{15}N_2$ incorporation following inoculation of sterile plants with *Acetobacter diazotrophicus* wild-type and *Nif*-mutant strains. Mol Plant Microbe Interact 14:358–366

Sprent JI, James EK (2007) Legume evolution: where do nodules and mycorrhizas fit in? Plant Physiol 144:575–581

Urquiaga S, Cruz HS, Boddey RM (1992) Contribution of nitrogen fixation to sugarcane: nitrogen-15 and nitrogen balance estimates. Soil Sci Soc AM J 56:105–114

Venkateshwaran M, Ané JM (2011) Legumes and nitrogen fixation: Physiological, molecular, evolutionary perspective and applications. In: Hawkesford MJ & Barraclough PB (eds) The molecular basis of nutrient use efficiency in crops. Wiley, USA, pp 457–489

Venkateshwaran M, Volkening JD, Sussman MR et al (2013) Symbiosis and the social network of higher plants. Curr Opin Plant Biol 16:118–127

Chapter 43
The Minimal Rhizosphere Microbiome

Jos M. Raaijmakers

Abstract The rhizosphere provides a home to numerous (micro) organisms that in turn may affect plant growth, development, and tolerance to abiotic and biotic stresses. How plants shape the rhizosphere microbiome has been subject of many past and present studies with the ultimate goal to identify plant genetic traits that select and support beneficial microorganisms. Novel 'omics technologies have provided more in-depth knowledge of the diversity and functioning of the rhizosphere microbiome and significant advances are being made to uncover mechanisms, genes and metabolites involved in the multitrophic interactions in the rhizosphere. To better understand this intriguing complexity, both reductionists' and systems ecology approaches are needed to identify the biotic and abiotic factors involved in microbiome assembly. Here, different strategies are discussed to re-shape the rhizosphere microbiome in favour of microbial consortia that promote root development and plant growth, and that prevent the proliferation of pests and diseases.

43.1 Introduction

Currently more than one third of the crop yields worldwide are lost due to abiotic and biotic stress factors, such as drought, salinity, pests and diseases. Future increases in crop yields will have to be achieved on sub-optimal soils with reduced input of fertilizers and pesticides (*'more with less'*). These challenges have increased the awareness of the importance of the plant microbiome (i.e. the collective communities of microorganisms on and in plants, their genomes and interactions) for improved and sustainable agricultural practices. Plants are colonized by an astounding number of microorganisms that can have profound effects on seed germination, plant growth and development, nutrition, diseases and productivity. In this context, plants can be viewed as superorganisms that rely in part on their microbiome for specific functions

J. M. Raaijmakers (✉)
Department of Microbial Ecology, Netherlands Institute of Ecology,
Droevendaalsesteeg 10, 6708 PB Wageningen, The Netherlands
Tel.: + 31317473497
e-mail: j.raaijmakers@nioo.knaw.nl

© Springer International Publishing Switzerland 2015
B. Lugtenberg (ed.), *Principles of Plant-Microbe Interactions,*
DOI 10.1007/978-3-319-08575-3_43

and traits. In return, plants deposit a substantial part of their photosynthetically fixed carbon into their direct surroundings (spermosphere, rhizosphere, phyllosphere), thereby feeding the microbial community and influencing their activities and composition (Mendes et al. 2013). For many plant-associated microorganisms, however, there is still little knowledge of their impact on plant growth and health. Hence, deciphering the plant microbiome is critical to identify beneficial microorganisms that can be used as an integral component of future agriculture and horticulture.

43.2   The Rhizosphere

The rhizosphere is the narrow zone surrounding and influenced by plant roots via the release of so-called rhizodeposits (i.e. exudates, border cells, mucilage) (Lynch 1990). The nutrients, trace elements, volatile organic compounds and other metabolites deposited by plant roots attract many (micro)organisms such as bacteria, archaea, fungi, nematodes and protozoa, making the rhizosphere a hot spot of microbial activity and interactions (Raaijmakers et al. 2009; Buée et al. 2009; Mendes et al. 2013). Following the terminology used for microorganisms colonizing the human body, the collective communities of microorganisms on and also inside plant root tissue, their genomes and interactions are now referred to as the rhizosphere microbiome. Over the past five decades, numerous studies have shown that specific members of the rhizosphere microbiome can affect plant growth and development, plant nutrition and stress tolerance (Berendsen et al. 2012; Mendes et al. 2013; Philippot et al. 2013). In this context, Cook et al. (1995) postulated that plants may 'cry for help' by selectively stimulating microorganisms that protect them from invading pathogens. Rhizosphere microorganisms that have been well studied for their beneficial effects on plant growth and health are the nitrogen-fixing bacteria, mycorrhizal fungi, plant growth-promoting rhizobacteria (PGPR), and saprophytic and mycoparasitic fungi (Mendes et al. 2013). For the vast majority of rhizosphere (micro)organisms, however, there is still little to no understanding of their metabolic potential and functions. This lack of knowledge has led to numerous studies to catalogue microbial communities in the rhizosphere of different plant species, to elucidate which microbes are active during plant development and to unravel which functions and biosynthetic pathways are displayed in time and space (Mendes et al. 2013; Philippot et al. 2013).

To go beyond 'collecting stamps', several meta-'omics' approaches (transcriptomics, proteomics, metabolomics) have been and are still being developed to identify gene transcripts, proteins and metabolites in the rhizosphere. For example, Wang et al. (2011) adopted a metaproteomics approach to unravel interactions between plants and rhizosphere microorganisms in different cropping systems. They found, among others, that approximately half of the bacterial groups classified by proteomic analysis were not found in the DNA-based metagenomic analyses of the rhizosphere bacterial community and vice versa (Wang et al. 2011), emphasizing the need to improve the resolution and sensitivity of these approaches. Also, stable isotope probing (Prosser et al. 2006) has provided new opportunities to identify microorganisms that are metabolically active in the rhizosphere. These and other technologies revealed that also fungi make up a significant part of the rhizosphere microbial biomass, especially during flowering and senescence (Hannula et al. 2010). Hence, top-down approaches such as metagenomics and bottom-up approaches targeting individual microbial species or strains should be integrated to provide a comprehensive coverage and understanding of the microbial community and their activities as a whole (see also Zengler and Palsson 2012; Mendes et al. 2013).

43.3 Shaping the Rhizosphere Microbiome

Several species and strains of rhizobacterial and fungal genera, including *Bacillus*, *Pseudomonas*, *Collimonas*, *Trichoderma*, *Piriformospora* and nonpathogenic *Fusarium oxysporum*, have been shown to promote plant growth and to protect plants from stress by different mechanisms (Lugtenberg and Kamilova 2009; Raaijmakers et al. 2009; Raaijmakers and Mazzola 2012; Chap. 3) . These include biofertilization (Chaps 23, 24 and 25), stimulation of root growth (Chap. 26), antibiosis (Chap. 18), induced systemic resistance (Chap. 14), parasitism and rhizoremediation (Chap. 29). These mechanisms are well documented for rhizobacteria belonging to the Proteobacteria and Firmicutes, i.e. *Pseudomonas* (Chap. 18) and *Bacillus* (Chap. 40), as well as for the mycoparasitic fungi *Trichoderma* (Chap. 36) and *Gliocladium*. Hence, there is a major interest to develop strategies that re-shape the rhizosphere microbiome in favour of microorganisms that promote root development and plant growth, and that prevent the proliferation of pests and diseases.

The first and most obvious strategy to re-direct the microbial composition and activities in the rhizosphere is changing the quality and/or quantity of root exudates via plant breeding or via genetic modification. This form of '*rhizosphere engineering*' requires detailed knowledge of the exudate composition (spatial, temporal) and their effects on microbial growth and activity (Bakker et al. 2012). Although our understanding of exudate chemistry and microbial interactions in the rhizosphere has improved considerably, there are, to my knowledge, no specific breeding programs yet that evaluate plant lines for their broad interaction with the rhizosphere microbiome. More than a decade ago, Smith et al. (1999) investigated the genetic basis in plants for interactions with beneficial rhizobacteria. They discovered substantial variation among recombinant inbred lines of tomato and identified loci that were associated with growth of and disease suppression by a beneficial *Bacillus cereus*. Rudrappa et al. (2008) further showed that plants can stimulate, via malic acid, the protective effects of a beneficial *Bacillus subtilus* strain in the rhizosphere. Similarly, Neal et al. (2012) showed that a beneficial *Pseudomonas putida* strain was attracted to 2,4-dihydroxy-7-methoxy-1,4-benzoxazine-3-one (DIMBOA), the allelopathic compound that is exuded in relatively high quantities from roots of young maize seedlings. These and other studies exemplify that specific phenotypic traits and genetic variation in host plant species can be exploited to enhance beneficial associations of plants with rhizosphere microorganisms. To date, however, our knowledge of root exudation *in situ* is still too limited to provide specific targets that can be used in plant breeding programs.

The second strategy to re-direct the rhizosphere microbiome is to introduce selected beneficial microorganisms at high densities in soil, onto seeds or other planting materials (Mendes et al. 2013). Over the past decades, many bacterial and fungal strains with different beneficial traits have been studied for their ability to boost plant performance and to control pests and diseases. Although there are several successful cases (e.g. *Agrobacterium radiobacter, Bacillus subtilis*), many of the promising microbes tested to date were less effective in disease control than their chemical counterpart and therefore not commercially attractive enough for product development and implementation in practice. The observed inconsistency in performance of various promising microbial agents has been attributed to various reasons, including poor establishment on/in seed or plant tissue, poor survival or lack of expression of the desired microbial trait/activity at the right time and place.

43.4   Reconstructing a 'Minimal Rhizosphere Microbiome'

To date, there has been a strong emphasis on '*one-microbe-at-a-time*' applications, whereas many ecosystem functions, including nutrient cycling and disease suppression, are generally driven by the (sequential) activity of microbial consortia. Furthermore, several microorganisms only exhibit a specific activity when they are part of a consortium (Garbeva and de Boer 2009; Garbeva et al. 2011). Hence, the use of assemblages of different rhizosphere microorganisms with complementary or synergistic traits may provide a much more effective and consistent effect. This concept of so-called 'reconstructed microbiomes' or 'synthetic communities' (De Roy et al. 2013; Grosskopf and Soyer 2014) is gaining momentum not only in plant-microbe interactions but also in the fields of probiotics and natural product discovery. However, to find and select the right players and microbial composition of a rhizosphere consortium for a specific function (e.g. disease suppression) is still a puzzle and requires more fundamental understanding of the temporal and spatial dynamics of the rhizosphere microbiome, the chemistry, the underlying communication and beneficial activities.

Natural disease suppressive soils (Chap. 38) provide a very good 'model system' to unravel and design the optimal microbial consortium to protect plants from infection by soil-borne pathogens. Studies by Kyselkova et al. (2009); Mendes et al. (2011) and Rosenzweig et al. (2012) on soils suppressive to different fungal and bacterial plant pathogens pinpointed multiple bacterial genera that were more abundant in the suppressive than in the corresponding disease conducive soils. Although the potential role of the identified bacterial communities in disease suppressiveness was addressed for only a few genera, these studies do provide a framework to reconstruct microbial consortia for disease control. Clearly, there is a need for a community systems approach to resolve the interplay between individual community members, the host plant and the soil environment (Zengler and Palsson 2012). In this context, Kinkel et al. (2011) proposed a co-evolutionary framework for inducing or managing natural disease suppressiveness of soils. They also argued that control of different plant pathogens on different crops most likely requires a different subset of microorganisms (Kinkel et al. 2011). Ideally, the ultimate goal is to design a so-called '*minimal rhizosphere microbiome*' that is effective against multiple soil-borne pathogens in different agro-ecosystems. Based on the concept of the minimal genome (Moya et al. 2009; Juhas et al. 2011), the minimal rhizosphere microbiome is defined here as the minimal set of microorganisms, microbial traits and genomes that are needed to effectively and consistently execute a specific function in the rhizosphere, e.g. protection of plant roots against fungal infections or rhizoremediation of toxic compounds (Mendes et al. 2013).

For controlling plant diseases, designing a separate minimal rhizosphere micro-biome for each of the major pathogen groups (bacteria, fungi, nematodes, oomycetes) may be feasible. This assumption is based on the fact that various studies have pointed to common players and mechanisms in different soils that are naturally suppressive to specific fungal pathogens. For example, *Pseudomonas* species have been shown to contribute to suppressiveness of soils to either Fusarium wilt disease or to take-all disease of wheat (Weller et al. 2002). Furthermore, the onset of natural disease suppressiveness of soils follows a similar pattern for different fungal pathogens (Weller et al. 2002), suggesting that similar cues, mechanisms and microbes play a role in the transition of a soil from a conducive to a suppressive state. An in-depth understanding of the shifts in community composition and microbial activities during this transition will be required to select the right microbiome members. Selection and assembly of minimal rhizosphere microbiomes should be based on functional traits and genes rather than on taxonomic classification only (Burke et al. 2011; Boon et al. 2013). Using a modelling approach, Scheuring and Yu (2012) suggested three easy steps to assemble a beneficial microbiome. In their models, the first step is that the new host's microbiome starts with a higher proportion of beneficials either by vertical transmission or by a higher immigration rate. The second step involves a high resource supply from the host to the beneficials, which in turn (third step) fuels intense interference competition via antibiotic production leading to competitive dominance of the beneficial microbes. Although Scheuring and Yu (2012) focused primarily on antibiosis as a key function of a beneficial microbiome, their models are highly instrumental to identify major processes that drive assembly of a beneficial microbiome. Whether these models could also be used for other important traits of beneficial rhizosphere microbiomes, such as parasitism, induced resistance and resource competition, remains to be determined.

In conclusion, the rhizosphere is a diverse and dynamic habitat with multiple microorganisms that affect plant growth, development and tolerance to abiotic and biotic stresses. To better understand the multitrophic interactions in the rhizosphere, both reductionists' and systems biology/ecology approaches are needed to resolve the underlying mechanisms involved in microbiome assembly and activity.

References

Bakker MG, Manter DK, Sheflin AM et al (2012) Harnessing the rhizosphere microbiome through plant breeding and agricultural management. Plant Soil 360:1–13

Berendsen RL, Pieterse CMJ, Bakker P (2012) The rhizosphere microbiome and plant health. Trends Plant Sci 17:478–486

Boon E, Meehan CJ, Whidden C et al (2013) Interactions in the microbiome: communities or organisms and communities of genes. FEMS Microbiol Rev 38:90–118.

Buee M, De Boer W, Martin F et al (2009) The rhizosphere zoo: an overview of plant-associated communities of microorganisms, including phages, bacteria, archaea, and fungi, and of some of their structuring factors. Plant Soil 321:189–212

Burke C, Steinberg P, Rusch D, Kjelleberg S, Thomas T (2011) Bacterial community assembly based on functional genes rather than species. Proc Natl Acad Sci U S A 108:14288–14293

Cook RJ, Thomashow LS, Weller DM et al. (1995) Molecular mechanisms of defense by rhizobacteria against root disease. Proc Natl Acad Sci U S A 92:4197

De Roy K, Marzorati M, Van den Abbeele P et al. (2013) Synthetic microbial ecosystems: an exciting tool to understand and apply microbial communities. Environ Microbiol doi:10.1111/1462–2920.12343

Garbeva P, de Boer W (2009) Inter-specific interactions between carbon-limited soil bacteria affect behavior and gene expression. Microb Ecol 58:36–46

Garbeva P, Silby MW, Raaijmakers JM et al (2011) Transcriptional and antagonistic responses of *Pseudomonas fluorescens* Pf0–1 to phylogenetically different bacterial competitors. ISME J 5:973–985

Grosskopf T, Soyer OS (2014) Synthetic microbial communities. Curr Opin Microbiol 18:72–77

Hannula SE, de Boer W, van Veen JA (2010) In situ dynamics of soil fungal communities under different genotypes of potato, including a genetically modified cultivar. Soil Biol Biochem 42:2211–2223.

Juhas M, Eberl L, Glass JI (2011) Essence of life: essential genes of minimal genomes. Trends Cell Biol 21:562–568

Kinkel LL, Bakker MG, Schlatter DC (2011) A coevolutionary framework for managing disease-suppressive soils. Annu Rev Phytopathol 49:47–67

Kyselková M, Kopecký J, Frapolli M et al (2009) Comparison of rhizobacterial community composition in soil suppressive or conducive to tobacco black root rot disease. ISME J 3:1127–1138

Lugtenberg B, Kamilova F (2009) Plant-growth-promoting rhizobacteria. Annu Rev Microbiol 63: 541–556

Lynch JM (1990) The rhizosphere. Wiley, New York

Mendes R, Kruijt M, de Bruijn I et al (2011) Deciphering the rhizosphere microbiome for disease-suppressive bacteria. Science 332:1097–1100

Mendes R, Garbeva P, Raaijmakers JM (2013) The rhizosphere microbiome: significance of plant beneficial, plant pathogenic, and human pathogenic microorganisms. FEMS Microbiol Rev 37:634–663

Moya A, Gil R, Latorre A et al (2009) Toward minimal bacterial cells: evolution vs. design. FEMS Microbiol Rev 33:225–235

Neal AL, Ahmad S, Gordon-Weeks R, Ton J (2012) Benzoxazinoids in root exudates of maize attract *Pseudomonas putida* to the rhizosphere. PLOS One 7:e35498

Philippot L, Raaijmakers JM, Lemanceau P, Van der Putten WH (2013) Going back to the roots: the microbial ecology of the rhizosphere. Nat Rev Microbiol 11:789–799

Prosser JI, Rangel-Castro JI, Killham K (2006) Studying plant-microbe interactions using stable isotope technologies. Curr Opin Biotechnol 17:98–102

Raaijmakers JM, Paulitz TC, Steinberg C et al (2009) The rhizosphere: a playground and battlefield for soilborne pathogens and beneficial microorganisms. Plant Soil 321:341–361

Raaijmakers J, Mazzola M (2012) Diversity and natural functions of antibiotics produced by beneficial and pathogenic soil bacteria. Annu Rev Phytopathol 50:403–424

Rosenzweig N, Tiedje JM, Quensen JF et al (2012) Microbial communities associated with potato common scab-suppressive soil determined by pyrosequencing analyses. Plant Dis 96:718–725

Rudrappa T, Czymmek KJ, Pare PW, Bais HP (2008) Root-secreted malic acid recruits beneficial soil bacteria. Plant Physiol 148:1547–1556

Scheuring I, Yu DW (2012) How to assemble a beneficial microbiome in three easy steps. Ecol Lett 15:1300–1307

Smith KP, Handelsman J, Goodman RM (1999) Genetic basis in plants for interactions with disease-suppressive bacteria. Proc Natl Acad Sci U S A 96:4786–4790

Wang HB, Zhang ZX, Li H et al (2011) Characterization of metaproteomics in crop rhizospheric soil. J Proteome Res 10:932–940

Weller DM, Raaijmakers JM, Gardener BBMS, Thomashow LS (2002) Microbial populations responsible for specific soil suppressiveness to plant pathogens. Annu Rev Phytopathol 40: 309–348

Zengler K, Palsson BO (2012) A road map for the development of community systems (CoSy) biology. Nat Rev Microbiol 10:366–372

Chapter 44
The Edible Plant Microbiome: Importance and Health Issues

Gabriele Berg, Armin Erlacher and Martin Grube

Abstract Plants live together with microbial communities to form tight interactions that are essential for the performance and survival of the host. In recent decades, many studies have discovered a vast plant-associated microbial diversity. However, even though plants are a substantial part of a balanced diet including raw-eaten vegetables, fruits and herbs, the plant-associated microbial diversity has been largely ignored in this context. We hypothesize that the edible plant microbiome and its diversity can be important for humans as (i) an additional contributor to the diversity of our gut microbiome, and (ii) as a stimulus for the human immune system. Two specific examples for plant microbiomes, of lettuce and banana, are discussed in comparison with other relevant studies to explore these hypotheses. Moreover, the biotechnological potential of the edible plant microbiome is evaluated.

44.1 Plant-Associated Microbial Diversity

All Food Plants are Associated with a High Diversity of Microorganisms This diversity is still currently only partly characterized and is, to a certain degree, specific for the host species or even cultivars of food plants (Berg and Smalla 2009). This diversity is also specific for each microhabitat of plants which are usually distinguished as: the rhizosphere (roots), the phyllosphere (leaves), the caulosphere (stem), the anthosphere (flowers), the carposphere (fruits), and the endosphere

G. Berg (✉) · A. Erlacher
Institute of Environmental Biotechnology, Graz University of Technology, 8010 Graz, Austria
Tel.: + 43664608738310
e-mail: Gabriele.berg@tugraz.at

A. Erlacher · M. Grube
Institute of Plant Sciences, University of Graz, 8010 Graz, Austria
Tel.: + 43 (316) 873-4312-8423
e-mail: armin.erlacher@tugraz.at

M. Grube
Tel.: + 43 (0)316 380-5655
e-mail: martin.grube@uni-graz.at

© Springer International Publishing Switzerland 2015
B. Lugtenberg (ed.), *Principles of Plant-Microbe Interactions*,
DOI 10.1007/978-3-319-08575-3_44

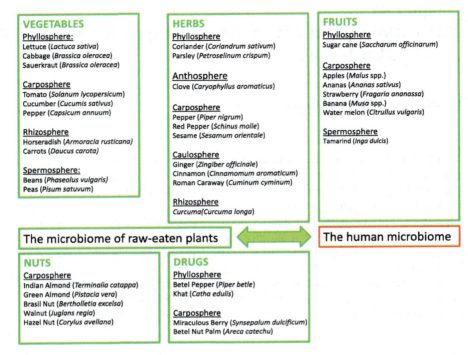

VEGETABLES

Phyllosphere:
Lettuce (*Lactuca sativa*)
Cabbage (*Brassica oleracea*)
Sauerkraut (*Brassica oleracea*)

Carposphere
Tomato (*Solanum lycopersicum*)
Cucumber (*Cucumis sativus*)
Pepper (*Capsicum annuum*)

Rhizosphere
Horseradish (*Armoracia rusticana*)
Carrots (*Daucus carota*)

Spermosphere:
Beans (*Phaseolus vulgaris*)
Peas (*Pisum satuvum*)

HERBS

Phyllosphere
Coriander (*Coriandrum sativum*)
Parsley (*Petroselinum crispum*)

Anthosphere
Clove (*Caryophyllus aromaticus*)

Carposphere
Pepper (*Piper nigrum*)
Red Pepper (*Schinus molle*)
Sesame (*Sesamum orientale*)

Caulosphere
Ginger (*Zingiber officinale*)
Cinnamon (*Cinnamomum aromaticum*)
Roman Caraway (*Cuminum cyminum*)

Rhizosphere
Curcuma(Curcuma longa)

FRUITS

Phyllosphere
Sugar cane (*Saccharum officinarum*)

Carposphere
Apples (*Malus spp.*)
Ananas (*Ananas sativus*)
Strawberry (*Fragaria ananassa*)
Banana (*Musa spp.*)
Water melon (*Citrullus vulgaris*)

Spermosphere
Tamarind (*Inga dulcis*)

The microbiome of raw-eaten plants

The human microbiome

NUTS

Carposphere
Indian Almond (*Terminalia catappa*)
Green Almond (*Pistacia vera*)
Brasil Nut (*Bertholletia excelsa*)
Walnut (*Juglans regia*)
Hazel Nut (*Corylus avellana*)

DRUGS

Phyllosphere
Betel Pepper (*Piper betle*)
Khat (*Catha edulis*)

Carposphere
Miraculous Berry (*Synsepalum dulcificum*)
Betel Nut Palm (*Areca catechu*)

Fig. 44.1 Interactive microbiomes. Examples of the edible plant microbiome of fruits, herbs, nuts and drugs. We eat all parts of the plants including the phyllosphere (*lettuce, cabbage*), the rhizosphere (*carrots, turnip*), the carposphere (*tomato, banana*) as well as seeds (*beans, peas*) including all endospheres

(all inner parts). Although the discovery of specific microbiomes is primarily associated with the rhizosphere, there are currently only a few other compartments where species-specific diversity was detected, e.g. the carposphere (Leff and Fierer 2013). Given the estimated number of 370 000 species of higher plants, a great deal of work is still required before the details of global plant microbiome diversity will be fully understood.

Plants are a basic and substantial part of our daily diet. Vegetables, fruits, herbs, nuts, and medicinal herbs belong to the raw-eaten plants; several examples of each group are shown in Fig. 44.1. Our food thus comprises all parts of plants that include their microbial habitats and microhabitats that can be colonized by up to 10^4–10^{10} microorganisms per gram of plant. These microbial habitats include the phyllosphere (lettuce, cabbage), the rhizosphere (carrots, turnip), the carposphere (tomato, banana), as well as the seeds (beans, peas) and corresponding endospheres. An initial study published by Leff and Fierer (2013) demonstrated that fruits and vegetables harbored distinct and diverse bacterial communities, and interestingly showed that vegetables or fruits grown primarily close to the soil surface (i.e., sprouts, spinach, lettuce, tomatoes, peppers, and strawberries) appear to share communities characterized by high relative abundance of *Enterobacteriaceae*. The authors concluded that humans are exposed to substantially different bacteria depending on the types of fresh produce they consume.

What could the effect of this exposure be on humans? Plant-associated microorganisms could have both a direct and indirect influence on human health. Indirect positive effects are linked to organisms that enhance the quality factors (including the content of active principles). Only a few examples are known for such effects, and most are related to medicinal plants and their bioactive substances (Köberl et al. 2013). For example, microorganisms are involved in the production of antimicrobial substances, e.g. taxol in endophytic fungi of *Taxus baccata* (Garyali et al. 2013), apigenin in *Chamomille matricaria* (Schmidt et al. 2014), or maytansine in *Putterlickia verrucosa* (Wings et al. 2013). Moreover, fruit-associated bacteria seem to influence the aroma expression in strawberries, where *Methylobacterium* treatment has been shown to enhance the production of aromatic furaneol substances (Verginer et al. 2010a). Evidence was also provided by Verginer et al. (2010b) for an influence of grape-associated microorganisms on the aroma of wine, indicating that the "terroir" effect can to some extent be attributed to bacteria. The indirect negative effects caused by plant-associated microorganisms are well-studied. The outbreak of plant pathogens is often associated with a microbiome shift and accompanied with minor pathogens. They do not only contribute to bad odor and taste, but also to the expression of mycotoxins which are among the world's most toxic and carcinogenic compounds (Wu et al. 2014). They have been responsible for numerous foodborne diseases and epidemics throughout history including *Claviceps purpurea*, the causative agent for the infamous Saint Anthony's Fire in Medieval times that occurred after eating contaminated bread (Belser-Ehrlich et al. 2013). Although such problems could be primarily solved by food hygiene, *Fusarium* mycotoxins still play an important role for our health (Wu et al. 2014). There is still very little knowledge concerning the long term effects of bioactive compounds at low concentration, and only recently has evidence been introduced for endophytes that produce novel and still poorly understood compounds. New technologies will contribute to increase the detection rate of specific beneficial plant-microbe interactions that are also relevant for human health.

What do we know about the direct effects of plant microbiomes that we consume along with our food? Most of our existing knowledge concerns fermented food, such as yoghurt as the foremost example for sources of probiotic strains. However, a substantial part of our plant diet is consumed fresh and may possibly include trillions of microorganisms during each meal. Even after washing or rinsing food surfaces, a substantial number of bacteria is expected to enter the body with our food. Our primary hypothesis is that the edible plant microbiome and its microbial diversity is important for humans as: (i) a contributor to the diversity within our gut microbiome, and (ii) as a stimulus for our immune response. We will present two examples for crop-associated microbiomes which are eaten raw by humans: of lettuce and banana. Furthermore, we will discuss our hypotheses as well as the impact of microbial diversity in general.

The Specific Structure of the Lettuce Microbiome Lettuce species such as *Lactuca sativa* L., *Eruca sativa* Mill., and their varieties belong to the most important raw-eaten vegetables world-wide and are a substantial part of a balanced, healthy diet. Several beneficial effects on health and lifestyle are attributed to the consumption of lettuce as it contains several vitamins, and is also a source of manganese and high amounts of dietary fibers. The relatively low amount of carbohydrates and fats correlates with its low calorie value. Lettuce provides habitats for a diverse range of microbes (Rastogi et al. 2012; Rastogi et al. 2013). Lettuce-associated microorganisms have currently only made it into the headlines in the context of scattered pathogen outbreaks. There are two crucial features that may be responsible for lettuce's vulnerability to pathogens: the variability and specificity of the associations within the microbial communities. Overall, a proportion of 12.5 % cultivar-specific bacteria were identified for the rhizosphere of eight different *Lactuca sativa* cultivars as well as the wild relative *L. serriola*. In addition, a large core microbiome was identified that includes 68 operational taxonomic units from nine major phyla (*Proteobacteria* the most abundant), and represents 48.8 % of the microbiome. A correlation analysis showed that within the lettuce microbiome co-occurrence prevailed over co-exclusion. Although predominant taxa (e.g. *Pseudomonas, Flavobacterium*, and *Sphingomonadaceae*) showed positive interactions, they were not necessarily involved in highly correlated modules of species. This loose bacterial network observed for lettuce allowed allochthonous organisms to colonize lettuce to interactive niching in microbial communities.

Little is known about the impact of biotic factors on the lettuce microbiota. Our hypothesis was that any disturbance of the native microbiomes (i) can induce drastic shifts in the community and that each pathogen outbreak (ii) could be accompanied by "minor", less virulent pathogens. In mesocosm and field experiments by using a combined approach including network analyses of 16S rRNA gene amplicon libraries and FISH microscopy (see Chap. 31), we found substantial impacts detectable as microbiome shifts by a plant pathogenic fungus, herbivorous gastropoda, or visiting pets. Although the genera *Enterobacter, Stenotrophomonas, Pseudomonas*, and *Acinetobacter* form a core microbiome, all three disturbing factors induced significant shifts in the community and increased species richness. In *Lactuca*, this was strongly correlated with an increase of *Enterobacter* and in *Eruca* with *Escherichia/Shigella* and *Pantoea*—all genera contain potential pathogens. A bacterial diversity associated with leaves is detectable by cultivation and bacterial DNA analysis, but very few bacteria are detected on the surface as only a few colonies occupy cavities along the external surface and in the vicinity of stomata (Fig. 44.2a). Through colonization experiments, we revealed unexpected colonization patterns of enteric species in lettuce leaves and found that bacterial populations do not colonize the surface, but rather intrude into the endosphere (Fig. 44.2b).

The Specific Structure of the Banana Microbiome Bananas and plantains are among the most important crops in the tropics and sub-tropical regions world-wide. Microhabitat-specific microbial communities for the rhizosphere, phyllosphere, and endosphere of bananas grown in three different traditional farms in Uganda were

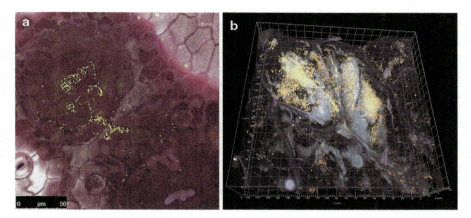

Fig. 44.2 Visualization of the lettuce microbiome. **a** Naturally occurring Gammaproteobacterial micro-colonies on the lettuce surface and in the vicinity of stomata visualized by FISH and CLSM. **b** Colonization patterns on lettuce leaves treated with *E. coli* cells. Both experiments are explained in detail in Erlacher et al. (2014)

detected (Rossmann et al. 2010). Interestingly, the banana stem endosphere showed the highest bacterial counts (up to 10^9 gene copy numbers g^{-1}), and *Enterobacteriacaea* provided 1/3 of the total bacteria. They comprise 14 genera including potential human pathogens, (*Escherichia, Klebsiella, Salmonella, Yersinia*) plant pathogens (*Pectobacterium*), but also disease-suppressive bacteria (*Serratia*). This dominant role of enterics can be explained by their permanent nature and the vegetative propagation of banana plants, as well as the addition of human and animal manure in traditional cultivations.

44.2 The Edible Plant Microbiome: Diversity and Human Health

Concerning our first hypothesis of a link between the plant and human gut microbiome, there is an interesting overlap between the plant and human gut microbiome with respect to species composition and function (Ramírez-Puebla et al. 2013). Recent studies showed that the stomach is colonized by a higher diversity of microbial species than has long been expected, and explained by the hostile conditions of low pH values. The stomach milieu thus does not pose a strict barrier for microbial passage as was previously thought (von Rosenvinge et al. 2013). Even though the effects of probiotics are often controversially discussed, it has now been shown that strains, including probiotics, survived the stomach passage to establish successfully in the gut (Iqbal et al. 2014). David et al. (2014) also recently provided additional evidence for the survival of foodborne microbes (both animal- and plant-based diet) after transit through the digestive system, and that foodborne strains may have been metabolically active in the gut.

Our second hypothesis is that bacteria, associated with our diet, such as *Enterobacteriaceae*, act as stimuli for our immune system. Recently, Hanski et al. (2012) showed a correlation between bacterial diversity and atopy as shown through significant interactions with *Enterobacteriaceae*. Furthermore, they showed a positive association between the abundance of *Acinetobacter* and Interleukin-10 expression in peripheral blood mononuclear cells in healthy human individuals. Interleukin-10 is an anti-inflammatory cytokine and plays a central role in maintaining immunologic tolerance to harmless substances (Lloyd and Hawrylowicz 2009). Endotoxin derived from Gram-negative bacteria, such as *Enterobacteriaceae*, is known to have allergy-protective and immunomodulatory potential (Doreswamy and Peden 2011).

Microhabitats of plants are a reservoir for *Enterobacteriaceae* (Leff and Fierer 2013, Rastogi et al. 2012), which also include potentially human pathogenic bacteria such as human enteric pathogens (Brandl 2006). Particularly after intermediate disturbances, these human enteric pathogens are enhanced (Erlacher et al. 2014). Although outbreaks of enteric pathogens associated with fresh produce in the form of raw or minimally processed vegetables and fruits have recently increased (Holden 2010), the ecology of enteric pathogens outside of their human and animal hosts is less understood (Teplitski et al. 2011). If plants are a natural reservoir of *Enterobacteriaceae*, then these bacteria must have been a "natural" part of our diet for a long time. Taking into account how many vegetables and fruits are eaten by people worldwide, these outbreaks seem to be more of an accident than the norm, particularly considering that traditionally, food was not processed and sterilized before eating. A function of the plant-associated microbiome as an immunostimulant or "natural vaccination" is more likely than their pathogenic role.

44.3 Conclusions

Members of the prokaryotic and eukaryotic domains of life are often tied together by intricate interactions. While past research has paid much more attention to the pathogenic interactions, the results obtained over the last decade have taught us much more about a beneficial balance between microorganisms and their hosts (Blaser et al. 2013). It seems that in developing these interactions, diversity plays an incredibly important role. Diversity is intrinsically correlated with a low incidence of pathogen outbreaks in both plants and humans. Where does microbial diversity come from? The plants themselves as well as their secondary metabolites and microbiomes co-evolved together; microbes contribute to the diversification of plants and *vice versa* and continue to add to the high plant-associated microbial diversity. Interesting examples are medical as well as endemic plants which harbor a unique microbiome (Zachow et al. 2009; Köberl et al. 2013). Conversely, crops cultivated in intensive agriculture are often characterized by a reduced diversity in comparison with organic agriculture or natural ecosystems. In the past, breeding strategies induced a specific microbiome as cultivar-specificity was very often reported (Berg and Smalla 2009). Our lettuce example revealed a higher diversity in comparison to its wild ancestor as well as a loose bacterial co-occurrence network in the modern cultivars. This could explain its susceptibility for pathogens as well as for biocontrol agents. Efficient biocontrol approaches were already shown for lettuce (Scherwinski et al. 2008; Erlacher et al. 2014). The enhancement of plant-associated microbial diversity is important for the sustainability of future agriculture. In addition, for human food and health, microbial diversity is an important issue, and we should take care of plant-associated diversity and produce our food in a way that is optimal for this purpose. Biotechnological strategies can be developed to contribute to this purpose. For example, "microbiome therapies" are a promising method to maintain or enhance plant-associated microbial diversity in combination with quality control (Gopal et al. 2013). Another interesting example is the biocontrol agent *Bacillus amyloliquefaciens* FZB42, which was able to enhance the overall plant-associated diversity (Erlacher et al. 2014). Next generation microbial inoculants should take both the diversity as well as human health issues into consideration (Berg et al. 2013), and someday in the future should have the potential to control plant diseases, generally enhance microbial diversity, and stimulate our immune system.

References

Belser-Ehrlich S, Harper A, Hussey J, Hallock R (2013) Human and cattle ergotism since 1900: symptoms, outbreaks, and regulations. Toxicol Ind Health 29:307–316

Berg G, Smalla K (2009) Plant species and soil type cooperatively shape the structure and function of microbial communities in the rhizosphere. FEMS Microbiol Ecol 68:1–13

Berg G, Zachow C, Müller H, Philipps J et al (2013) Next-generation bio-products sowing the seeds of success for sustainable agriculture. Agronomy 3:648–656

Blaser M, Bork P, Fraser C et al (2013) The microbiome explored: recent insights and future challenges. Nat Rev Microbiol 11:213–217

Brandl MT (2006) Fitness of human enteric pathogens on plants and implications for food safety. Annu Rev Phytopathol 44:367–392

David LA, Maurice CF, Carmody RN et al (2014) Diet rapidly and reproducibly alters the human gut microbiome. Nature 505:559–563

Doreswamy V, Peden DB (2011) Modulation of asthma by endotoxin. Clin Exp Allergy 41:9–19

Erlacher A, Cardinale M, Grosch R et al (2014) The impact of the pathogen *Rhizoctonia solani* and its beneficial counterpart *Bacillus amyloliquefaciens* on the indigenous lettuce microbiome. Front Microbiol 5:175. doi:10.3389/fmicb.2014.00175

Garyali S, Kumar A, Reddy MS (2013) Taxol production by an endophytic fungus, *Fusarium redolens*, isolated from Himalayan yew. J Microbiol Biotechnol 23:1372–1380

Gopal M, Gupta A, Thomas GV (2013) Bespoke microbiome therapy to manage plant diseases. Front Microbiol 4:355

Hanski I, von Hertzen L, Fyhrquist N et al (2012) Environmental biodiversity, human microbiota, and allergy are interrelated. Proc Natl Acad Sci U S A 109:8334–8339

Holden NJ (2010). Plants as reservoirs for human enteric pathogens. CAB Rev Perspect Agric Vet Sci Nutr Nat Resour 47:1–11

Iqbal MZ, Qadir MI, Hussain T et al (2014) Review: probiotics and their beneficial effects against various diseases. Pak J Pharm Sci 27:405–415

Köberl M, Schmidt R, Ramadan EM et al (2013) The microbiome of medicinal plants: diversity and importance for plant growth, quality and health. Front Microbiol 4:400

Leff JW, Fierer N (2013) Bacterial communities associated with the surfaces of fresh fruits and vegetables. PLoS One 8:e59310

Lloyd CM, Hawrylowicz CM (2009) Regulatory T cells in asthma. Immunity 31:438–449

Ramírez-Puebla ST, Servín-Garcidueas LE, Jiménez-Marín B et al (2013) Gut and root microbiota commonalities. Appl Environ Microbiol 79:2–9

Rastogi G, Sbodio A, Tech JJ et al (2012) Leaf microbiota in an agroecosystem: spatiotemporal variation in bacterial community composition on field-grown lettuce. ISME J 6:1812–1822

Rastogi G, Coaker GL, Leveau JH (2013) New insights into the structure and function of phyllosphere microbiota through high-throughput molecular approaches. FEMS Microbiol Lett 348:1–10

Rossmann B, Müller H, Smalla K et al (2010) Banana-associated microbial communities in Uganda are highly diverse but dominated by Enterobacteriaceae. Appl Environ Microbiol 78:4933–4941

Scherwinski K, Grosch R, Berg G (2008) Effect of bacterial antagonists on lettuce: active biocontrol of *Rhizoctonia solani* and negligible, short-term effects on nontarget microorganisms. FEMS Microbiol Ecol 64:106–116

Schmidt R, Köberl M, Mostafa A et al (2014) Effects of bacterial inoculants on the indigenous microbiome and secondary metabolites of chamomile plants. Front Microbiol 5:64

Teplitski M, Warriner K, Bartz J et al (2011) Untangling metabolic and communication networks: interactions of enterics with phytobacteria and their implications in produce safety. Trends Microbiol 19:121–127

Verginer M, Siegmund B, Cardinale M et al (2010a) Monitoring the plant epiphyte *Methylobacterium extorquens* DSM 21961 by real-time PCR and its influence on the strawberry flavor. FEMS Microbiol Ecol 74:136–145

Verginer M, Leitner E, Berg G (2010b) Production of volatile metabolites by grape-associated microorganisms. J Agric Food Chem 58:8344–8350

Von Rosenvinge EC, Song Y, White JR et al (2013) Immune status, antibiotic medication and pH are associated with changes in the stomach fluid microbiota. ISME J 7:1354–1366

Wings S, Müller H, Berg G et al (2013) A study of the bacterial community in the root system of the maytansine containing plant *Putterlickia verrucosa*. Phytochemistry 91:158–164

Wu F, Groopman JD, Pestka JJ (2014) Public health impacts of foodborne mycotoxins. Annu Rev Food Sci Technol 5:351–372

Zachow C, Berg C, Müller H et al (2009) Fungal diversity in the rhizosphere of endemic plant species of Tenerife (Canary Islands): relationship to vegetation zones and environmental factors. ISME J 3:79–92

Chapter 45
From Nodulation to Antibiotics

Eva Kondorosi

Abstract The importance and future potential of biological nitrogen-fixation is widely recognized. A further, yet unexplored and less known value of nitrogen-fixing root nodules is the presence of hundreds of plant peptides with antimicrobial activities and novel modes of action. These nodule-specific plant peptides are only produced in the Rhizobium-infected symbiotic cells where they abolish the endosymbiont's cell division ability, transforming them to non-cultivable polyploid nitrogen-fixing cells. The symbiotic cationic peptides are able to kill a wide range of microbes, including important human pathogens. The peptides are highly stable and their interactions with multiple bacterial targets reduce the probability to develop resistance. Worldwide spreading of antibiotic resistant microbes became a major cause of mortality. Therefore there is an urgent need for novel antibiotics. The characteristics and multifaceted action of symbiotic peptides make them excellent antibiotic candidates.

45.1 Introduction

This chapter focuses on host-governed terminal differentiation of the endosymbionts and the roles of host-symbiotic peptides. The morphology and physiology of nitrogen-fixing Rhizobium bacteria, called bacteroids are not uniform (Kondorosi et al. 2013). In certain legumes the bacteroids are similar to the cultured bacteria which maintain their cell proliferation potential and can return to the free-living life-style. In other legumes, the endosymbionts undergo drastic irreversible morphological and physiological changes, transforming them to non-cultivable enlarged polyploid nitrogen-fixing cells. Terminal bacteroid differentiation is host-controlled and has multiple origins in the *Leguminosae* family indicating that it might be advantageous for the host plants, for example because of more efficient nitrogen-fixation or consumption of the bacteroid's cell content during senescence (Oono et al. 2010). These bacteroids are significantly larger than the free-living bacteria and their shape

E. Kondorosi (✉)
Institute of Biochemistry, Biological Research Centre of the Hungarian Academy of Sciences, Temesvári krt. 62, Szeged 6726, Hungary
Tel.: +36 62 599 673
e-mail: kondorosi.eva@brc.mta.hu

© Springer International Publishing Switzerland 2015
B. Lugtenberg (ed.), *Principles of Plant-Microbe Interactions,*
DOI 10.1007/978-3-319-08575-3_45

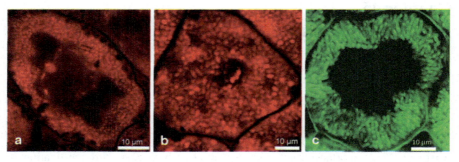

Fig. 45.1 Different forms of nitrogen-fixing bacteroids. **a** similar to free-living cells, **b** swollen, and **c** elongated morphology. Bacteroids were stained with PI (*red color*) in **a** and **b**, and with Syto9 (*green color*) in **c** (Kondorosi et al. *unpublished*)

can be elongated and even branched or spherical (Fig. 45.1). Cell growth is associated with the amplification of the bacterial genome (Mergaert et al. 2006). The degree of genome amplification varies in the different legume hosts and its significance has not been elucidated yet. The features of terminal bacteroid differentiation and the plant effectors have been discovered in the *Sinorhizobium meliloti-Medicago truncatula* symbiosis. The *M. truncatula* nodules are of the indeterminate type in which cells below the nodule meristem become infected with *S. meliloti* which first multiplies in the young symbiotic cells but, once in the older cells of the infection zone, the endosymbiont's cell division becomes arrested and definitively lost while in parallel their elongation begins. The fully differentiated nitrogen fixing bacteroids are 5- to 10-fold longer than the cultured bacteria and their tripartite genome (the chromosome and two megaplasmids) is uniformly amplified at least 20-fold (Mergaert et al. 2006). Comparison of nodule transcriptomes with reversible and terminal bacteroid differentiation revealed the presence of several hundreds of small genes encoding nodule-specific secreted peptides in the genome of those legumes in which terminal differentiation of bacteroids occurred (Alunni et al. 2007). During their translation, these peptides enter the endoplasmic reticulum where the signal peptidase cleaves off the signal peptide and the processed mature peptide is delivered via the secretory pathway to the endosymbionts. The essential role of nodule-specific peptides has been demonstrated in a signal peptidase mutant of *M. truncatula* in which the unprocessed peptides remains in the endoplasmic reticulum and the lack of their interaction with the endosymbionts abolished bacteroid differentiation (Van de Velde et al. 2010).

45.2 *Medicago truncatula* Symbiotic Peptides: The NCR and the GRP Families

In the *M. truncatula* genome, 600 small genes code for secreted nodule specific peptides. The *NCR* family, encoding nodule-specific cysteine rich (NCR) peptides, consists of more than 500 members (Mergaert et al. 2003; Nallu et al. 2014). The peptides are composed of a relatively conserved signal peptide and a 30–50 amino acid long mature peptide with four or six cysteine residues at conserved positions (Mergaert et al.2003). The highly diverse amino acid composition and sequences, and thereby charge and hydrophobicity of the NCR peptides, result in a range of anionic, neutral and cationic peptides with pI values from 3 to more than 11. The NCR peptides are unique but show structural resemblance to defensins, the largest class of antimicrobial peptides (AMPs) in plants that protect the plants from bacterial and fungal pathogens (Silverstein et al. 2005).

The GRP family is less numerous and comprises nodule-specific glycine rich peptides (GRPs) (Kevei et al. 2002; Alunni et al. 2007). GRPs also contain a signal peptide while the mature peptides have more than 100 amino acid residues and, similarly to NCRs, could be cationic, neutral or anionic. GRPs represent a unique class of glycine-rich proteins which share certain similarity with glycine-rich antimicrobial peptides but evolved for symbiosis in galegoid legumes.

Both the NCR and the GRP genes are exclusively expressed in the *S. meliloti*-infected symbiotic nodule cells. However, different sets of genes are activated during the early and late stages of symbiotic cell development. A single symbiotic cell produces all the 600 peptides during its maturation but at a given stage probably only several dozens. Combined and consecutive actions of these peptides drive bacteroid differentiation.

45.3   NCR Peptides Operate With Numerous Bacterial Targets Affecting Multiple Pathways

Due to the extremely high numbers and recent discovery of the nodule-specific peptides, the knowledge on the peptide actions is rudimentary. It is unclear how the peptides enter the bacteria. Cationic peptides can interact with the bacterial membranes and penetrate the cells without pore formation but neutral and anionic peptides, which did not show interaction with the membranes, are also present in the bacteroids (Van de Velde et al. 2010; Farkas et al. 2014). The presence of early NCR peptides in the fully differentiated nitrogen-fixing bacteroids indicates their high stability and resistance against bacterial proteases. Therefore, it is possible that the functions of the peptides may be required beyond the site of their production. The present studies are focused on those whose absence provokes symbiotic phenotypes or which have well-defined *in vitro* activities. While the peptides are produced effectively in the polyploid symbiotic cells, experiments with heterologous expression systems were so far unsuccessful. Therefore chemically synthetized peptides (mostly NCRs), labeled with various tags allowing their detection and affinity purification, have been used for functional analysis. The NCR-bound protein partners have been identified with LC-MS/MS and Western analysis. In the following the bacterial targets of the cationic NCR247 peptide will be presented as an example for NCR actions (Farkas et al. 2014). NCR247 is one of the smallest members of the family which exhibits antimicrobial activities *in vitro* and its bioinformatics analysis predicted exceptional protein-binding properties. Identifying its bacterial targets revealed unexpectedly complex interactions and various ways to modulate the bacterium's physiology at multiple sites.

NCR247-FtsZ Interaction: Inhibition of Bacterial Cell Division NCR247 is expressed in those symbiotic cells where bacterial cell division becomes arrested and the endosymbionts undergo remarkable elongation and genome amplification (Farkas et al. 2014). One of the NCR247's interactors was the FtsZ, the primary protein required for septum formation and cell division in bacteria. FtsZ is present in the cytosol but preceding cell division it polymerizes and forms a Z-ring at the future site of bacterial cell division which serves as a platform for septum assembly. When the Z-ring is absent, there is no cell division while filamentous growth of microbes can occur. Treatment of *S. meliloti* with NCR247 abolished Z-ring formation demonstrating that one of the NCR247 peptide's functions is to inhibit bacterial cell division probably by preventing FtsZ polymerization. In agreement with this finding, Penterman et al. (2014) showed that treatment of *S. meliloti* with sublethal levels of NCR247 provoked cell division block, attenuated expression of critical cell cycle regulators and antagonized Z-ring function.

NCR247-Ribosomal Protein Interactions: Negative Effect on Bacterial Protein Synthesis The most numerous interactors of NCR247 are ribosomal proteins. From the NCR247-treated bacteria, 14 small subunit and 12 large subunit proteins were identified in the NCR247 complex. As NCR247 has exceptional protein-binding capacity, it might interact directly with all these ribosomal proteins but more likely a few—yet unidentified—ribosomal proteins are the primary targets that bind additional ribosomal proteins. Ribosomal proteins are known targets of many antibiotics (like streptomycin, neomycin, spectinomycin, tetracyclin, hygromycin B, chloramphenicol, erythromycin, etc...) and these interactions interfere with different phases of translation and inhibit protein synthesis (Wilson 2014). Similarly, NCR247 slows down or fully inhibits protein synthesis in a concentration-dependent manner. The protein synthesis is ongoing in the nitrogen-fixing bacteroids, but the complexity of the proteome is reduced and might be influenced by the NCR peptides. Transcriptome analysis of bacteria and bacteroids revealed in the bacteroids 10 to 20-fold lower expression of ribosomal protein genes as well as altered relative abundance of transcripts that could provoke ribosome diversification in the bacteroids. Accordingly, NCR247 attracted fewer ribosomal proteins from the bacteroid extracts. Ribosomal heterogeneity represents another level of translational regulation in bacteria which in the polyploid bacteroid genome could be a major determinant in selective translation of symbiosis-specific proteins (Byrgazov et al. 2013). Treatment of bacteria with NCR247 resulted also in down-regulation of ribosomal protein genes contributing to ribosome diversification in the bacteroids. Thus NCR247, by controlling the expression of ribosomal protein genes and interfering with translation, plays multiple roles in the reduced and specific proteome of the bacteroids.

NCR247-GroEL Interaction: Multiplication of Bacterial Targets   Another binding partner of NCR247 is the bacterial chaperone GroEL. GroEL interacts with hundreds of proteins and is required for their proper folding (Kerner et al. 2005). GroEL has multiple roles in various symbioses; it is required for maintenance of endosymbionts in the host cells and in the rhizobia. It is required for full induction of nodulation genes and for the assembly of the nitrogenase complex. Moreover, its interaction with NCR247 shed light on its role in efficient bacterial infection and terminal bacteroid differentiation (Farkas et al. 2014). While the significance of the GroEL-NCR247 interaction needs to be discovered, NCR247—and perhaps other NCRs as well—may modify the affinity of GroEL toward various substrate proteins and thereby may influence the repertoire of active proteins in the endosymbiont in a host-governed manner.

45.4   Antimicrobial Properties of NCR Peptides

The definitive loss of the endosymbiont's cell division ability and structural resemblance of the nodule-specific symbiotic peptides to antimicrobial peptides raised the possibility that at least some of them may have antimicrobial activities outside their natural nodule environment. Using a set of chemically synthesized mature NCR peptides it has been proven that cationic NCRs have indeed antimicrobial activities, killing not only the symbiotic bacterium partner but also other microbes. The strength and spectrum of cell killing activities are dependent on both the charge and the amino acid sequence of the peptides. This is illustrated on the example of two cationic peptides NCR335 (pI: 11.22) and NCR247 (pI: 10.15), both of which have antimicrobial activities at 50 μg/ml concentration but differing in their antibacterial spectrum (Tiricz et al. 2013). These peptides eliminated the bacteria more rapidly than the classical antibiotics kanamycin and tetracycline. Rapid killing of bacteria was associated with increased permeability and disintegration of the bacterial membranes, allowing penetration of the membrane impermeable dye, propidium iodide (PI) into the cells. Membrane damage is a characteristic killing action of the positively charged cationic AMPs whose interaction with the negatively charged bacterial membranes leads to pore formation and subsequently to cell lysis. However, in the symbiotic cells the bacteroids remain alive and the bacteroid membrane is intact, indicating that the peptide concentrations during symbiosis might be significantly lower than those used in the *in vitro* assays where the actions of peptides are different from the membrane damaging AMPs.

Besides the bactericidal effect, several NCR peptides proved to be efficient killers of various fungi, including important human and plant pathogens (unpublished data from the laboratory). The bactericidal and fungicidal activities were usually associated in the tested NCRs but some peptides exhibited primarily anti-fungal or anti-bacterial activities.

45.5   Urgent Need for New Antibiotics

Since their introduction $\sim$70 years ago, antibiotics have been the most powerful weapons against microbial invaders. The use of antibiotics such as penicillin, streptomycin, tetracycline and chloramphenicol correlates with a sudden increase of the human population. These classical antibiotics cured and even led to the eradication of many infectious diseases, saving the lives of millions. However the uncontrolled use of antibiotics in humans and in livestock led to the appearance of a large number of antibiotic-resistant bacteria which became a major cause of mortality and morbidity worldwide. The original antibiotics and their successors are largely ineffective against the antibiotic resistant bacteria. Therefore new antimicrobial compounds are needed with novel modes of action, good efficacy and low resistance profile. "The 10 × 20 initiative" aims to develop ten new systemic antibacterial drugs by 2020, with particular focus on the ESKAPE pathogens (*Enterococcus faecalis, Staphylococcus aureus, Klebsiella pneumoniae, Acinetobacter baumannii, Pseudomonas aeruginosa* and *Enterobacteriaceae*) (Bassetti et al. 2013).

45.6   Antimicrobial Peptides as Potential Antibiotics Candidates

AMPs are produced by all living organisms and are effectors of the innate immune system. AMPs are composed of 10 to 50 amino acids and can be cationic or anionic. The cationic AMPs show broad-spectrum antimicrobial activities against bacteria and fungi but anionic AMPs can have also antimicrobial activities or can improve the activities of cationic AMPs (Guilhelmelli et al. 2013). The classical mode of their antimicrobial activities is membrane damage but several AMPs have intracellular targets and attack microbes simultaneously at multiple sites which provokes the microbe's death. These parallel alternative mechanisms for killing and reduce the tendency for resistance development. However, microbes co-evolved with AMPs developed strategies to resist AMPs by surface modifications, biofilm formation, or the AMPs' increased efflux, proteolytic degradation, trapping or neutralization. Nevertheless, resistance to AMPs is less common than to conventional antibiotics. These characteristics of AMPs draw increasing attention on them as promising therapeutic drugs for the treatment of infectious diseases.

45.7   Benefits of NCRs: Novel Therapeutic Drugs and Versatile Applications

The symbiotic plant peptides in the nitrogen-fixing nodules represent a gold mine of bioactive molecules. At least one third of them are cationic and potentially antimicrobial peptides. Microbes—except the endosymbionts—have never been exposed to these symbiotic peptides, thus microbial resistance could not have evolved against them. Broad spectrum activities of NCRs include killing of antibiotic resistant and ESKAPE pathogens and their long persistence in the endosymbionts indicates remarkable stability of these plant peptides against proteolytic degradation. Moreover, treatment of human cell cultures or injection of mice with NCRs so far did not provoke symptoms of cytotoxicity (unpublished data from the laboratory) while many antibiotics used in therapy—severe side effects and toxicity. These beneficial properties of the symbiotic plant peptides open multiple possibilities for their application.

Therapeutic Potential Antimicrobial NCRs are new antibiotic drug candidates. The most active and preferably the smallest peptides that are non-toxic for human cells can function as lead molecules for drug design. Antibacterial and antifungal NCRs could be used without modifications for topical treatments of skin, nail and epithelial infections. The oral or intravenous administration of NCRs requires, however, further exploratory studies on whether the peptides remain stable and active in the body. It is known that the activity of cationic AMPs is attenuated in the serum and inhibited by Mg^{2+} and Ca^{2+} ions and by high salt concentrations. Similarly, NCRs are also sensitive to high salt, Mg^{2+} and Ca^{2+} levels. Thus their internal use may require their stabilization and/or specific delivery to the site of infection.

Non-Medical Applications Pests and pathogens reduce the crop yields with 30–80 % (Becker-Ritt and Carlini 2012). Therefore, hundreds of synthetic organic compounds are used in agriculture for protection of plants against microbes and pests. The NCRs, based on their broad range activity against Gram-positive and Gram-negative bacteria and fungi, could be alternatives for the organic chemical products. Constitutive or organ/tissue-specific production of selected NCR peptides could greatly increase the plant's resistance against pathogens; but this would require production of transgenic plants at a large scale which is not feasible for several reasons. A realistic alternative solution is the external treatments of plants with the most effective broad-range antimicrobial NCRs. Spraying the plants with peptides, obtained by chemical synthesis or heterologous expression, could serve for prevention and elimination of pathogens.

Among many others, a further possible application could be in the food industry where microbial contamination, unavoidably present on the raw materials, should be decreased and optimally eliminated. To date there are no standardized, widely accepted methods for perfect elimination of microbial contamination. The fast acting, non-toxic antimicrobial NCR peptides introduced at various phases of food processing could contribute to solving this problem.

Acknowledgements   I am grateful to Pal Venetianer for critical reading of the manuscript. Our work is supported by the "SYM-BIOTICS" Advanced Grant of the European Research Council to EK (grant number 269067).

References

Alunni B, Kevei Z, Redondo-Nieto M et al (2007) Genomic organization and evolutionary insights on GRP and NCR genes, two large nodule-specific gene families in *Medicago truncatula*. Mol Plant Microbe Interact 20:1138–1148

Bassetti M, Merelli M, Temperoni C et al (2013) New antibiotics for bad bugs: where are we? Ann Clin Microbiol Antimicrob. 12:22. doi:10.1186/1476-0711-12-22

Becker-Ritt AB, Carlini CR (2012) Fungitoxic and insecticidal plant polypeptides. Biopolymers 98:367–384

Byrgazov K, Vesper O, Moll I (2013) Ribosome heterogeneity: another level of complexity in bacterial translation regulation. Curr Opin Microbiol 16:133–139

Farkas A, Maróti G, Durgö H et al (2014) *Medicago truncatula* symbiotic peptide NCR247 contributes to bacteroid differentiation through multiple mechanisms. Proc Natl Acad Sci U S A 111:5183–5188. doi:10.1073/pnas.1404169111

Guilhelmelli F, Vilela N, Albuquerque P et al (2013) Antibiotic development challenges: the various mechanisms of action of antimicrobial peptides and of bacterial resistance. Front Microbiol 4:353

Kerner MJ, Naylor DJ, Ishihama Y et al (2005) Proteome-wide analysis of chaperonin-dependent protein folding in *Escherichia coli*. Cell 122:209–220

Kevei Z, Vinardell JM, Kiss GB et al (2002) Glycine-rich proteins encoded by a nodule-specific gene family are implicated in different stages of symbiotic nodule development in *Medicago* spp. Mol Plant Microbe Interact 15:922–931

Kondorosi E, Mergaert P, Kereszt A (2013) A paradigm for endosymbiotic life: cell differentiation of *Rhizobium* bacteria by host plant factors. Annu Rev Microbiol 67:611–628

Mergaert P, Nikovics K, Kelemen Z et al (2003) A novel family in *Medicago truncatula* consisting of more than 300 nodule-specific genes coding for small, secreted polypeptides with conserved cysteine motifs. Plant Physiol 132:161–173

Mergaert P, Uchiumi T, Alunni B et al (2006) Eukaryotic control on bacterial cell cycle and differentiation in the Rhizobium-legume symbiosis. Proc Natl Acad Sci U S A 103:5230–5235

Nallu S, Silverstein KAT, Zhou P et al (2014) Patterns of divergence of a large family of nodule cysteine-rich peptides in accessions of *Medicago truncatula*. Plant J 78:697–705

Oono R, Schmitt I, Sprent JI et al (2010) Multiple evolutionary origins of legume traits leading to extreme rhizobial differentiation. New Phytol 187:508–520

Penterman J, Abo RP, De Nisco NJ et al (2014). Host plant peptides elicit a transcriptional response to control the *Sinorhizobium meliloti* cell cycle during symbiosis. Proc Natl Acad Sci U S A 111:3561–3566. doi:10.1073/pnas.1400450111

Silverstein KA, Graham MA, Paape TD et al (2005) Genome organization of more than 300 defensin-like genes in Arabidopsis. Plant Physiol 138:600–610

Tiricz H, Szücs A, Farkas A et al (2013) Antimicrobial nodule-specific cysteine-rich peptides induce membrane depolarization associated changes in *Sinorhizobium meliloti*. Appl Environ Microbiol. 79:6737–6746. doi:10.1128/AEM.01791-13

Van de Velde W, Zehirov G, Szatmari A et al (2010) Plant peptides govern terminal differentiation of bacteria in symbiosis. Science 327:1122–1126

Wilson DN (2014) Ribosome-targeting antibiotics and mechanisms of bacterial resistance. Nat Rev Microbiol 12:35–48. doi:10.1038/nrmicro3155

Index

© Springer International Publishing Switzerland 2015
B. Lugtenberg (ed.), *Principles of Plant-Microbe Interactions,*
DOI 10.1007/978-3-319-08575-3

CPSIA information can be obtained
at www.ICGtesting.com
Printed in the USA
LVOW05*2046181117
556811LV00002B/24/P

9 783319 085746